Remote Sensing and Geosciences for Archaeology

Special Issue Editor
Deodato Tapete

MDPI • Basel • Beijing • Wuhan • Barcelona • Belgrade

MDPI

Special Issue Editor
Deodato Tapete
Italian Space Agency (ASI)
Italy

Editorial Office
MDPI AG
St. Alban-Anlage 66
Basel, Switzerland

This edition is a reprint of the Special Issue published online in the open access journal *Geosciences* (ISSN 2076-3263) from 2017–2018 (available at: http://www.mdpi.com/journal/geosciences/special_issues/archaeology).

For citation purposes, cite each article independently as indicated on the article page online and as indicated below:

Lastname, F.M.; Lastname, F.M. Article title. *Journal Name* **Year**, *Article number*, page range.

First Edition 2018

ISBN 978-3-03842-763-6 (Pbk)
ISBN 978-3-03842-764-4 (PDF)

Table of Contents

About the Special Issue Editor

Deodato Tapete is a Researcher in Earth Observation in the Scientific Research Division of the Italian Space Agency (ASI). With more than 10 years of research experience in Earth Sciences and Remote Sensing applied to the study and conservation of cultural heritage, Dr Tapete is specialized in Synthetic Aperture Radar imaging and interferometry for deformation monitoring, hazard assessment and archaeological prospection. His publications in top-ranked journals (e.g., Remote Sensing of Environment and Journal of Archaeological Science) are highly cited across the community. He has led projects focused on advanced technologies to study renowned UNESCO World Heritage Sites, including the historic centers of Rome, Florence and Naples in Italy, Valletta in Malta, the Complex of Koguryo Tombs in Democratic People's Republic of Korea, the Lines and Geoglyphs of Nasca and Palpa and the Historic Sanctuary of Machu Picchu in Peru. He is a Fellow of the Higher Education Academy (FHEA).

Preface to "Remote Sensing and Geosciences for Archaeology"

The use of remote sensing techniques, by means of either ground-based instrumentations or airborne or space-borne sensors, for archaeological studies, namely, archaeological remote sensing, already has a long history of research, scientific publications, and implementation in the field. Therefore, archaeological remote sensing is not a novel discipline.

However, the following three triggers are currently stimulating advances in methodologies for data acquisition, signal processing, and the integration and fusion of extracted information:

(i) the technological development of sensors for data capture;
(ii) the accessibility of new remote sensing and Earth Observation data;
(iii) the awareness that a combination of different techniques can lead to the retrieval of diverse and complementary information to characterize landscapes and objects of archaeological value and significance.

This book is the first volume of, hopefully, a series of MDPI books dedicated to Remote Sensing and Geosciences for Archaeology.

I had the pleasure of assessing and recommending the publication of the twenty-one papers that are collected in this book, written by renowned experts and scientists from across the globe, to showcase the state-of-the-art and forefront research in archaeological remote sensing and the use of geoscientific techniques to investigate archaeological records and cultural heritage.

In the hope that readers will use these contributions to learn new methodologies and take inspiration for new research and applications, I express my sincere gratitude to all the authors, editors and reviewers for their help during this editorial project and Ms Alma Wu, Geosciences™ Assistant Editor, for her dedication and continued support.

Deodato Tapete
Special Issue Editor

geosciences

MDPI

Editorial

Remote Sensing and Geosciences for Archaeology

Deodato Tapete

Italian Space Agency (ASI), Via del Politecnico snc, 00133 Rome, Italy; deodato.tapete@asi.it

Received: 20 January 2018; Accepted: 22 January 2018; Published: 25 January 2018

Abstract: Archaeological remote sensing is not a novel discipline. Indeed, there is already a suite of geoscientific techniques that are regularly used by practitioners in the field, according to standards and best practice guidelines. However, (i) the technological development of sensors for data capture; (ii) the accessibility of new remote sensing and Earth Observation data; and (iii) the awareness that a combination of different techniques can lead to retrieval of diverse and complementary information to characterize landscapes and objects of archaeological value and significance, are currently three triggers stimulating advances in methodologies for data acquisition, signal processing, and the integration and fusion of extracted information. The Special Issue *"Remote Sensing and Geosciences for Archaeology"* therefore presents a collection of scientific contributions that provides a sample of the state-of-the-art and forefront research in this field. Site discovery, understanding of cultural landscapes, augmented knowledge of heritage, condition assessment, and conservation are the main research and practice targets that the papers published in this Special Issue aim to address.

Keywords: remote sensing; optical; SAR; geophysics; terrestrial laser scanning; point cloud; GIS; archaeological prospection; pattern recognition; condition assessment

1. Introduction

The use of remote sensing techniques, by means of either ground-based instrumentations or airborne or space-borne sensors, for archaeological studies—namely, "archaeological remote sensing"—has already a long history of research, scientific publications, and implementation in the field.

Renowned advantages include, but are not limited to:

- The estimation of parameters and surface/subsurface properties without direct contact with the object of study (i.e., non-invasiveness);
- The capability of making remote observations, thereby preventing risks for the operator and reducing costs of in situ investigations;
- The possibility to revisit in time and carry out iterative workflows of data analysis for the purposes of monitoring and condition assessment (e.g., multi-temporal change detection).

The abundant literature (also recently re-examined in review papers, e.g., [1,2]), the proliferation of special issues in international journals (e.g., [3,4]), and the publication of specialist books and manuals (e.g., [5–7]) are three clear indicators suggesting that remote sensing for archaeology is an established discipline, which attracts great interest across different scientific communities (e.g., image analysts, Earth observers, Geographic Information System (GIS) experts, archaeologists, heritage conservators) and has reached a level of maturity by which we are now in the position to assess its achievements and perspectives for future advances.

Therefore, while there is no doubt about the value and existing capability of remote sensing to allow the discovery of new sites, investigation of cultural landscapes, condition assessment of heritage assets, and monitoring and modeling of impacts due to natural hazards and human threats, there is still the need for translating this expertise, spread across the globe, into capacity, best practices, and tools made available widely. In a recent book review [8], for example, I highlighted that this field still lacks of shared standardized methods of data processing tailored for the specific requirements of users.

On the other side, remote sensing is commonly used in archaeology jointly with other geoscientific methods, ranging from geophysical survey methods to GIS, not to forget traditional methodologies of ground-truth and historical data collection. There is a variety of ways researchers and practitioners combine remote sensing and geosciences. Frequently, this depends on local expertise and available instrumentation. Workflows from data capture to data analysis are specifically designed by each research team to suit the questions they aim to address in their case studies. However, lessons learnt from implementation in specific geographic and research contexts can contribute to form a shared methodological basis for wider application.

In the above scenario, the call for papers for publication in the Special Issue *Remote Sensing and Geosciences for Archaeology* that I launched in March 2017 aimed to collect articles on recent work, experimental research, or case studies outlining the current state-of-the-art in at least one of the following topics of remote sensing and geosciences for archaeology:

- archaeological prospection
- digital archaeological fieldwork
- GIS analysis of spatial settlement patterns in modern landscapes
- assessment of natural or human-induced threats to conservation
- education and capacity building in RS for archaeology.

2. Facts and Figures of the Special Issue

A total of 27 submissions were received for consideration of publication in the Special Issue from late April to early December 2017. After rigorous editorial checks and peer-review processes involving external and independent experts in the field, the acceptance rate was 78%.

The published Special Issue contains a collection of 21 articles; one is a review paper on piezoelectric/seismoelectric methods to identify near-surface targets [9], and six are feature papers that were solicited for submission to provide either an overview of specific domains of archaeological remote sensing (e.g., passive airborne optical imaging, [10]) or demonstrations of state-of-the-art remote sensing data [11], research methodologies [12,13], and processing tools and workflows [14,15] for landscape archaeology and archaeological prospection.

Figure 1 compares the geographic distribution of the authors and research teams publishing in the Special Issue (Figure 1a) and of the case studies and demonstration sites (Figure 1b). This is, of course, a sample of the whole scientific community working on remote sensing and geosciences for archaeology and therefore not an exhaustive representation. However, it already provides a glimpse of the widespread expertise of experimental research and field practice, and proves how widely remote sensing is applied to investigate and preserve archaeological heritage.

Figure 2 is a word cloud of the disciplines and scientific domains the authors of the papers published in the Special Issue belong to, as inferred from their affiliations. "Archaeology" is predominant and "geosciences" is the second most populated discipline, as expected due to the specific target of the Special Issue. Nevertheless, it appears that other disciplines are well represented, including "remote sensing", "engineering", "environmental sciences", "cultural heritage", "history", and "geography". This is a further proof of the multi-disciplinary nature of archaeological remote sensing.

In general, two situations are most frequently observed:

1. Groups of different professionals join their efforts to combine skills of image processing and computing science with arts and humanities expertise;
2. Remote sensing specialists or environmental scientists dedicate their research to archaeological topics.

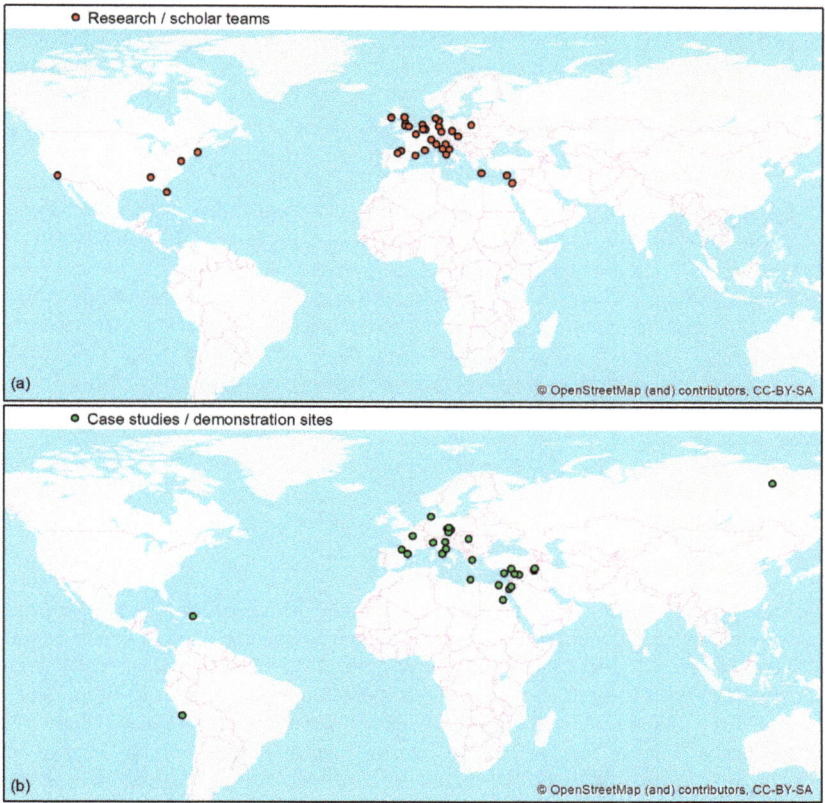

Figure 1. Geographic distribution of: (**a**) authors and research teams publishing in the Special Issue; (**b**) case studies and demonstration sites that are discussed in the papers.

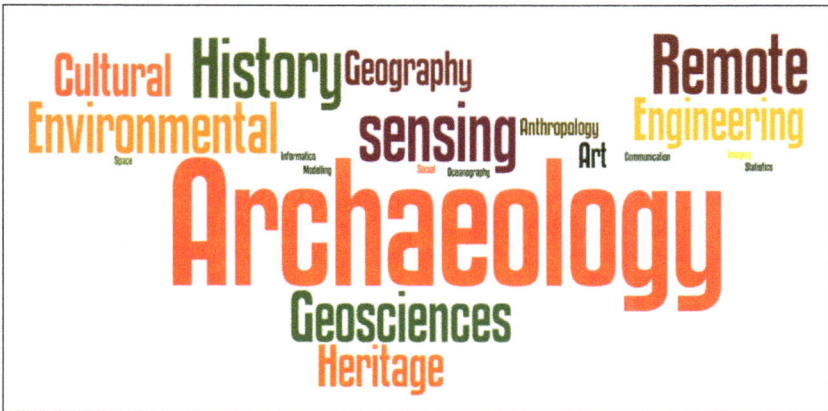

Figure 2. Word cloud of the disciplines and scientific domains the authors publishing in this Special Issue belong to [9–29], as inferred from their affiliations.

Similar evidence has been found in review exercises of specific domains of archaeological remote sensing based on decades of published literature [1]. In this regard, it is worth mentioning that the Special Issue includes examples of international, collaborative efforts between teams of scholars, academics, and/or cultural institutions for remote sensing-based studies on archaeology and heritage of developing countries or crisis zones [12–14].

3. Overview of the Special Issue Contributions

The papers published in the Special Issue cover a wide spectrum of techniques: satellite Synthetic Aperture Radar (SAR) imaging [11,16,17]; optical remote sensing from high to very-high resolution space-borne sensors [12,13,18–21]; geophysics including magnetometer surveys, Ground Penetrating Radar (GPR), geoelectric resistivity measurements, multi frequency Electromagnetic Induction (EMI), piezoelectric/seismoelectric methods [9,22,23]; photogrammetry [24]; terrestrial laser scanning and three-dimensional (3D) point clouds [25,26]. Some of the papers also explore the benefits of combining different approaches, e.g., optical remote sensing, aerial imagery and geophysical prospection [27,28], 3D surveys, and reconstruction in GIS environment [29]. In other cases, the focus is the demonstration of automatic detection tools and workflows [14,15].

Each paper explores successes and challenges that are consequential from the use of the above remote sensing and geoscientific techniques. In the following paragraphs, the main features of the Special Issue papers are highlighted.

3.1. Satellite Optical Remote Sensing

Building upon the renowned capability of satellite optical imagery (mostly sourced from commercial providers) to be an objective source of information to assess the condition of heritage at risk, the contributions by Danti et al. [12] and Rayne et al. [13] demonstrate how archaeologists and heritage professionals—with skills of remote sensing or supported by image analysts—can exploit large volumes of very high-resolution images to undertake systematic regional-scale efforts of damage mapping and assessment, leading to site-scale investigations and multi-temporal change detection. Both the teams of scholars have addressed the challenge of building site databases that could enable spatial and temporal querying of the results. These papers discuss how their experiences contribute to the standardization of methodologies for digital recording and metadata compilation, the use of shared terminology, and accurate classification with a logged level of uncertainty.

In particular, the implementation in the iconic sites of Nimrud in northern Iraq, Palmyra, and the Old City of Mosul in Syria offers a selection of incidents that highlight the results that Danti et al. [12] achieved with a methodology integrating satellite-based assessments with ground-based observations and open-source information. The methodology is flexible enough for addressing aspects of cultural heritage crises in other conflict zones, as it offers various alternatives for providing or publicly attributing sources of reliable information in cases where direct ground-based observations are unfeasible.

Rayne et al. [13], instead, focus on the value of combining multi-temporal satellite imagery with published data to create a detailed set of database records for a single site. They also discuss how the mapping of site disturbances at an extensive scale across cultural landscapes should integrate multi-temporal image interpretation, land-use mapping, and field surveys.

Anthropogenic impacts on the preservation of archaeological heritage are also the subject of the other three papers, two of which present results achieved with very high-resolution (open-source) optical satellite and aerial imagery [18,20], and the third paper freely accessible Landsat time series [19].

Interestingly, the geographic focuses of the studies by Chyla [19] and Parcak et al. [20] are located in Egypt. Despite the different data and methodologies used, both papers show examples of impacts on Egyptian archaeological heritage due to the expansion of agriculture and urbanization, which represent sources of threat for conservation in addition to natural processes characterizing the local fluvial environment. The authors of these papers demonstrate how satellite-based assessments can be used to support decision-making before sites disappear or are irreversibly damaged.

For early warning and prevention, Agapiou et al. [18] propose a novel methodology for the detection of archaeological looting incidents based on a set of indices and processing operations (e.g., vegetation indices, fusion, automatic extraction after object-oriented classification) of high-resolution WorldView-2 multispectral satellite imagery and RGB high-resolution aerial orthorectified images.

3.2. Airborne Optical Remote Sensing

From the papers introduced in the previous section, it is clear how archaeologists rely on satellite imagery. However, airborne optical remote sensing still stands as another important source of information, especially for archaeological reconnaissance, prospection, and landscape archaeology. Despite operational costs and permission to fly, airborne facilities are highly competitive solutions for archaeologists who want to investigate cultural landscapes, owing to the spatial resolution of the onboard sensors and the fact that targeted surveys can be undertaken instead of imagery acquisition according to pre-defined observational scenarios and fixed acquisition parameters.

Nevertheless, inherent biases and limitations need to be accounted for. After a clear statement about the concept of 'landscape archaeology' and an agile introduction to 'passive' remote imaging, Verhoeven [10] provides an insightful discussion of these biases, in particular those caused by sub-par sampling strategies, cost, instrument availability, and post-processing issues. Technological and methodological aspects are clearly explained within the overall framework of the archaeological theory relying on the capture, documentation, and interpretation of archaeological records, in conjunction with field data, in order to corroborate an archaeological hypothesis. The paper therefore sets out where we are in this domain of archaeological remote sensing. As such, it is worth reading not only by mature scholars, but also by beginners who can learn research methodology and how to go beyond the mere use of these technologies as black boxes.

3.3. Satellite and Airborne Synthetic Aperture Radar (SAR) Imaging

SAR imaging has been used for archaeology since the 1980s. A quick search through the specialist literature proves that, already a decade ago, the remote sensing community had achieved a good understanding of the capabilities of SAR for archaeological prospection, either via implementation on real case studies or in laboratory experiments (see specific SAR papers in [6]). Furthermore, a recent review exercise clearly highlighted a ramp-up in indexed peer-review publications using SAR data in archaeology in the last 30 years [1]. Therefore, it cannot be stated that SAR is novel in archaeological remote sensing. Instead, the novelty nowadays lies in the use of new imaging modes (mostly from new or current space missions) and higher spatial resolution and more accurate derived products, as well as the development of methodologies integrating multi-source SAR data. It is with regard to these aspects that the papers published in the Special Issue [11,16,17] offer interesting evidence of scientific research achievements.

Gade et al. [17] presents probably the first published example of the use of high- to very high-resolution SAR images acquired by the German Aerospace Center (DLR) constellation TerraSAR-X, including the Staring Spotlight mode, to detect and monitor the condition of lost coastal heritage in intertidal flats. The application on a constantly changing environment—where conditions of visibility of the archaeological remains and traces are discontinuous—proves how the tasking of satellite acquisitions is as important as the capability to extract features from the images after signal processing.

The TerraSAR-X constellation is also an example of a SAR mission equipped with twin satellites (TerraSAR-X and TanDEM-X in this case) providing data for the generation of high-resolution and accurate Digital Elevation Models (DEMs) with different spatial resolutions (e.g., processed at 2 m from High-Resolution Spotlight mode). In Cilicia, Rutishauser et al. [11] were able to extract cross-sections of height values (profiles) for the detailed analysis of the topography across riverbeds and paleo-channels from the TanDEM-X DEM mosaic covering the study region. The DEM-based analysis of silted up riverbeds and historic CORONA imagery allowed the first indications for the reconstruction of former river channels, when there was only limited coring available.

In the current international scenario of SAR missions, the European Space Agency (ESA) constellation Sentinel-1 cannot be forgotten. This is a space mission providing free and consistently acquired time series of SAR images with a large swath, high revisiting time of up to six days and a spatial resolution of 5 m by 20 m in Interferometric Wide (IW) swath mode. Sentinel-1 data therefore can be valuable to feed into studies of landscape archaeology and wide-area monitoring activities. It is with this purpose that Comer et al. [16] experimented with Sentinel-1 images to detect and measure landscape disturbance threatening the world-renowned archaeological features and ecosystems of the Lines and Geoglyphs of the Nasca and Palpa World Heritage Site in Peru. Multi-scale and multi-band analysis was also achieved thanks to targeted airborne surveys that were undertaken with the National Aeronautics and Space Administration (NASA)/Jet Propulsion Laboratory (JPL) Uninhabited Aerial Vehicle Synthetic Aperture Radar (UAVSAR) platform. In particular, these data allowed the authors to assess the impact on local heritage due to a known event that occurred in December 2014.

3.4. Automated Methods of Data Processing

As recalled in Section 1, a step forward in the use of remote sensing data is the development of robust automated methods for data processing to extract features of archaeological significance or to estimate parameters, and ease their uptake by the user community. The contributions by Sonnemann et al. [14] and Traviglia and Torsello [15] provide an interesting glimpse of the research currently being carried out on this topic.

In particular, Sonnemann et al. [14] applied a chain of image registration with a set of pre-processed, very diverse datasets, to combine multispectral bands to feed two different semi-automatic direct detection algorithms: a posterior probability and a frequentist approach. The method was tailored to generate maps that, with statistical significance, could address the question about the probability of finding archaeological evidence across the landscape and identifying sites in the Dominican Republic. Although not validated, the trials presented in the paper demonstrate the experimental attempt made by the authors towards automation in the context of multi-source data (e.g., satellite imagery, airborne surveys) which is one of the current challenges in remote sensing and geosciences.

Traviglia and Torsello [15] instead applied a workflow of feature extraction and detection implemented in ©Matlab to estimate the location and periodicity of dominant linear features on georeferenced images depicting the cultural landscape in northern Italy that had been engineered with land cadastration since Roman times. The trials suggest that this approach provides the accurate location of target linear objects and alignments signaled by a wide range of physical entities with very different characteristics, which also can be later interpreted against historical documentations, old maps, and in situ archaeological evidence.

3.5. Geophysical Techniques of Archaeological Prospection

The level of penetration of geophysical techniques in archaeological practice is such that several manuals and textbooks are published and disseminate best practices on how to carry out geophysical surveys in archaeological field evaluation according to established and shared standards [30,31]. The type of information that these techniques collect to investigate sites of archaeological potential is broadly classifiable as "geoscientific", since the outputs from the field surveys are measurements of parameters and properties of soil, subsurface materials, and buried objects.

In this regard, the contributions published in this Special Issue offer an interesting sample of case studies scattered across Europe.

Garcia-Garcia et al. [22] analyze the validity of a geoarchaeological core survey to check the archaeological interpretations based on geophysical results in the Roman site in Auritz/Burguete and Aurizberri/Espinal, Navarre, Spain.

Křivánek [23] showcases examples from the Czech Republic where different geophysical techniques were combined (i.e., surface magnetometer, resistivity survey) and integrated with other remote sensing methods (e.g., aerial photo-documentation, LiDAR) to address archaeological questions in different archaeological and environmental contexts. One of the sites is interestingly located close to a modern infrastructure. Achievements and limitations are discussed.

In addition, Agapiou et al. [27] focus on integration. In particular, in the demonstration site of Vésztő-Mágor Tell in the eastern part of Hungary, the authors run a workflow of data integration and fusion consisting of nine steps, including data capture, regression models to examine more than 70 different vegetation indices, cross-check and validation with GPR, and in situ magnetic gradiometry measurements.

Kalayci et al. [28] integrated geomagnetic, electromagnetic induction and GPR, historic aerial imagery, and Remotely Piloted Aerial Systems, as well as very high-resolution space-borne sensors, to discover differences in layouts of Early and Neolithic settlements in Thessaly, central Greece, and to investigate commonalities as a way to understand Neolithic use of space. In this way, the authors detected and characterized different aspects of the hindered prehistoric settlements that could have been overlooked by using only one geophysical approach.

The last contribution of the Special Issue focused on geophysics is a review by Eppelbaum [9] of piezoelectric/seismoelectric methods to capture piezoelectric and seismo-electrokinetic phenomena manifested by electrical and electromagnetic processes that occur in rocks under the influence of elastic oscillations triggered by shots or mechanical impacts. Reporting some examples from mining geophysics in Russia and an ancient metallurgical site in Israel, the author demonstrates that piezoelectric/seismoelectric anomalies may be analyzed quantitatively via advanced and reliable methodologies developed in magnetic prospection.

3.6. Laser Scanning, 3D Reconstruction, and GIS

Significant technological advancement is currently being achieved in the remote sensing and geoscience domains of laser scanning, manipulation of cloud points, and modeling. As rightly recalled by Poux et al. [26], point clouds and derivatives are changing the way curators, cultural heritage researchers, and archaeologists investigate heritage and collaborate on its understanding.

Guidi et al. [29], in this regard, provide a demonstrative example. Based on a proper mix of quantitative data originated by current 3D surveys and historical sources, such as ancient maps, drawings, archaeological reports, restrictions decrees, and old photographs, the authors were able to achieve a diachronic reconstruction of the ancient Roman Circus of Milan that, at present, is completely covered by the urban fabric of the modern city. The hypothesis of temporal evolution and transformation of the monument is presented via an easy-to-understand visual output.

The accessibility of information stored in point clouds and derived models is crucial for dissemination among end-users. Poux et al. [26] propose a classification method that relies on hybrid point clouds from both terrestrial laser scanning and dense image matching, and feeds into a WebGL prototype enabling different heritage actors to interact in a collaborative way. The geometric reconstruction, therefore, serves as a digital platform for storing and sharing relevant information and for easing communication with and between end-users.

Multi-disciplinarity is also at the base of the paper published by Drap et al. [24]. Medieval archaeologists and computer science researchers collaborated towards a connection between 3D spatial representation and archaeological knowledge, by integrating observable (material) and non-graphic (interpretive) data.

In addition to promoting understanding, digital 3D recording enables timely, iterative, and repeatable documentation of heritage to inform decision-making on its conservation. Corso et al. [25] show how terrestrial laser scanning can be used by practitioners and heritage bodies to document and monitor the heritage for which they are responsible; to map present conditions and assess visible impacts of weathering and deterioration processes; and to extract information to design mitigation and preservation measures.

4. Key Messages for Future Research

The wide portfolio of methodologies, data, and techniques presented in the contributions published in this Special Issue proves that remote sensing and geosciences for archaeology are currently vibrant research and practice domains, with expertise spread across the globe and teams fully exploiting the capability of remote sensing to investigate sites and landscapes in different geographic, social, and environmental contexts.

It is clear that there is no barrier for techniques—based on either space-borne, airborne, or ground-based instrumentation and sensors—to be employed in the field, to carry out experimental test of new functions and facilities, or to exploit known capabilities to accomplish professional, research, and institutional tasks (e.g., prospection, recording, condition assessment).

In this regard, satellite imagery (particularly from optical sensors and open-source platforms) is a clear example, since it has become a facility fully embedded in standardized routines for mapping and monitoring heritage. SAR, on the other side, is a stimulating research arena. While there are still difficulties for this technology to be utilized by non-expert users, the technological developments offering novel imaging modes and processing methods are triggering forefront research paving the way for a wider spectrum of applications.

Landscapes remotely sensed using different bands of the electromagnetic spectrum are ideal test sites for researchers and archaeologists to focus on signal processing and analysis to extract information not otherwise achievable, except via ground investigations. When archaeologists have access to different geophysical techniques, they tend to combine them to improve the level of data acquisition and the amount of information. However, this abundance of information leads, on one side, to the challenge of developing strategies and methods to fuse information and, on the other, to the need to isolate and extract the relevant information from the unnecessary and redundant.

Automation plays an important role, since it allows the operator subjectivity to be counterbalanced, and the time spent for data processing to be decreased. As a consequence, archaeologists and image analysts can concentrate more on archaeological interpretation and product generation.

In this regard, more efforts should be made towards the sharing of processing routines tailored to archaeological applications and their embedding within software or open platforms. Moreover, there is a need for more publications showing successful stories of the conversion of experimental methodologies into best practices, as well as the discussion of 'bad practice' examples where current methodologies are not adequate enough and improvements are needed.

Interestingly, the majority of the Special Issue papers prove that different professionals tend to team up to share expertise and technical skills to better address archaeological questions and/or facilitate the dissemination and sharing of information.

Acknowledgments: The Guest Editor thanks all the authors, *Geosciences*' editors, and reviewers for their great contributions and commitment to this Special Issue. A special thank goes to Alma Wu, *Geosciences*' Assistant Editor, for her dedication to this project and her valuable collaboration in the design and setup of the Special Issue.

Conflicts of Interest: The author declares no conflict of interest.

References

1. Tapete, D.; Cigna, F. Trends and perspectives of space-borne SAR remote sensing for archaeological landscape and cultural heritage applications. *J. Archaeol. Sci. Rep.* **2016**. [CrossRef]
2. Agapiou, A.; Lysandrou, V. Remote sensing archaeology: Tracking and mapping evolution in European scientific literature from 1999 to 2015. *J. Archaeol. Sci. Rep.* **2015**, *4*, 192–200. [CrossRef]
3. Satellite remote sensing in archaeology: Past, present and future perspectives. *J. Archaeol. Sci.* **2011**, *38*, 1995–2002.
4. Lasaponara, R.; Masini, N. Special Issue: Satellite Radar in Archaeology and Cultural Landscape. *Archaeol. Prospect.* **2013**, *20*, 71–162. [CrossRef]
5. Lasaponara, R.; Masini, N. *Satellite Remote Sensing: A New Tool for Archaeology*; Springer: Dordrecht, The Netherlands, 2012; ISBN 9789048188017.

6. Wiseman, J.; El-Baz, F. (Eds.) *Remote Sensing in Archaeology*; Interdisciplinary Contributions to Archaeology; Springer: New York, NY, USA, 2007; ISBN 978-0-387-44453-6.

7. Parcak, S.H. *Satellite Remote Sensing for Archaeology*; Routledge: London, UK; New York, NY, USA, 2009; ISBN 9780415448772.

8. Tapete, D.; Donoghue, D. Satellite Remote Sensing: A New Tool for Archaeology By RosaLasaponara and NicolaMasini (eds). Springer-Verlag, Heidelberg, 2012. ISBN 978-90-481-8801-7. Price: £117.00 (hardback). Pages: 364. *Archaeol. Prospect.* **2014**, *21*, 155–156. [CrossRef]

9. Eppelbaum, L. Quantitative examination of piezoelectric/seismoelectric anomalies from near-surface targets. *Geosciences* **2017**, *7*, 90. [CrossRef]

10. Verhoeven, G. Are We There Yet? A Review and Assessment of Archaeological Passive Airborne Optical Imaging Approaches in the Light of Landscape Archaeology. *Geosciences* **2017**, *7*, 86. [CrossRef]

11. Rutishauser, S.; Erasmi, S.; Rosenbauer, R.; Buchbach, R. SARchaeology—Detecting Palaeochannels Based on High Resolution Radar Data and Their Impact of Changes in the Settlement Pattern in Cilicia (Turkey). *Geosciences* **2017**, *7*, 109. [CrossRef]

12. Danti, M.; Branting, S.; Penacho, S. The American Schools of Oriental Research Cultural Heritage Initiatives: Monitoring Cultural Heritage in Syria and Northern Iraq by Geospatial Imagery. *Geosciences* **2017**, *7*, 95. [CrossRef]

13. Rayne, L.; Bradbury, J.; Mattingly, D.; Philip, G.; Bewley, R.; Wilson, A. From Above and on the Ground: Geospatial Methods for Recording Endangered Archaeology in the Middle East and North Africa. *Geosciences* **2017**, *7*, 100. [CrossRef]

14. Sonnemann, T.; Comer, D.; Patsolic, J.; Megarry, W.; Herrera Malatesta, E.; Hofman, C. Semi-Automatic Detection of Indigenous Settlement Features on Hispaniola through Remote Sensing Data. *Geosciences* **2017**, *7*, 127. [CrossRef]

15. Traviglia, A.; Torsello, A. Landscape Pattern Detection in Archaeological Remote Sensing. *Geosciences* **2017**, *7*, 128. [CrossRef]

16. Comer, D.; Chapman, B.; Comer, J. Detecting Landscape Disturbance at the Nasca Lines Using SAR Data Collected from Airborne and Satellite Platforms. *Geosciences* **2017**, *7*, 106. [CrossRef]

17. Gade, M.; Kohlus, J.; Kost, C. SAR Imaging of Archaeological Sites on Intertidal Flats in the German Wadden Sea. *Geosciences* **2017**, *7*, 105. [CrossRef]

18. Agapiou, A.; Lysandrou, V.; Hadjimitsis, D. Optical Remote Sensing Potentials for Looting Detection. *Geosciences* **2017**, *7*, 98. [CrossRef]

19. Chyla, J. How Can Remote Sensing Help in Detecting the Threats to Archaeological Sites in Upper Egypt? *Geosciences* **2017**, *7*, 97. [CrossRef]

20. Parcak, S.; Mumford, G.; Childs, C. Using open access satellite data alongside ground based remote sensing: An assessment, with case studies from Egypt's delta. *Geosciences* **2017**, *7*, 94. [CrossRef]

21. Malinverni, E.; Pierdicca, R.; Bozzi, C.; Colosi, F.; Orazi, R. Analysis and Processing of Nadir and Stereo VHR Pleiadés Images for 3D Mapping and Planning the Land of Nineveh, Iraqi Kurdistan. *Geosciences* **2017**, *7*, 80. [CrossRef]

22. Garcia-Garcia, E.; Andrews, J.; Iriarte, E.; Sala, R.; Aranburu, A.; Hill, J.; Agirre-Mauleon, J. Geoarchaeological Core Prospection as a Tool to Validate Archaeological Interpretation Based on Geophysical Data at the Roman Settlement of Auritz/Burguete and Aurizberri/Espinal (Navarre) †. *Geosciences* **2017**, *7*, 104. [CrossRef]

23. Křivánek, R. Roman Comparison Study to the Use of Geophysical Methods at Archaeological Sites Observed by Various Remote Sensing Techniques in the Czech Republic. *Geosciences* **2017**, *7*, 81. [CrossRef]

24. Drap, P.; Papini, O.; Pruno, E.; Nucciotti, M.; Vannini, G. Ontology-Based Photogrammetry Survey for Medieval Archaeology: Toward a 3D Geographic Information System (GIS). *Geosciences* **2017**, *7*, 93. [CrossRef]

25. Corso, J.; Roca, J.; Buill, F. Geometric Analysis on Stone Façades with Terrestrial Laser Scanner Technology. *Geosciences* **2017**, *7*, 103. [CrossRef]

26. Poux, F.; Neuville, R.; Van Wersch, L.; Nys, G.-A.; Billen, R. 3D Point Clouds in Archaeology: Advances in Acquisition, Processing and Knowledge Integration Applied to Quasi-Planar Objects. *Geosciences* **2017**, *7*, 96. [CrossRef]

27. Agapiou, A.; Lysandrou, V.; Sarris, A.; Papadopoulos, N.; Hadjimitsis, D. Fusion of Satellite Multispectral Images Based on Ground-Penetrating Radar (GPR) Data for the Investigation of Buried Concealed Archaeological Remains. *Geosciences* **2017**, *7*, 40. [CrossRef]

28. Kalayci, T.; Simon, F.-X.; Sarris, A. A Manifold Approach for the Investigation of Early and Middle Neolithic Settlements in Thessaly, Greece. *Geosciences* **2017**, *7*, 79. [CrossRef]
29. Guidi, G.; Gonizzi Barsanti, S.; Micoli, L.; Malik, U. Accurate Reconstruction of the Roman Circus in Milan by Georeferencing Heterogeneous Data Sources with GIS. *Geosciences* **2017**, *7*, 91. [CrossRef]
30. David, A. Geophysical Survey in Archaeological Field Evaluation. Available online: https://historicengland.org.uk/images-books/publications/geophysical-survey-in-archaeological-field-evaluation/ (accessed on 20 January 2018).
31. Oswin, J. *A Field Guide to Geophysics in Archaeology*; Springer: Berlin/Heidelberg, Germany, 2009; ISBN 3540766928.

geosciences

MDPI

Article

The American Schools of Oriental Research Cultural Heritage Initiatives: Monitoring Cultural Heritage in Syria and Northern Iraq by Geospatial Imagery

Michael Danti [1,*], Scott Branting [2] and Susan Penacho [1]

[1] ASOR Cultural Heritage Initiatives, 650 Beacon Street 2nd Floor, Boston, MA 02215, USA; asormaps@bu.edu
[2] Department of Anthropology at the University of Central Florida, 4000 Central Florida Blvd, Orlando, FL 32816, USA; scott.branting@ucf.edu
* Correspondence: michaeldanti@asor.org; Tel.: +1-857-990-3139

Received: 1 August 2017; Accepted: 23 September 2017; Published: 28 September 2017

Abstract: The American Schools of Oriental Research Cultural Heritage Initiatives (ASOR CHI) continues to address the cultural heritage crisis in Syria and Northern Iraq by: (1) monitoring, reporting, and fact-finding; (2) promoting global awareness; and (3) conducting emergency response projects and developing post-conflict rehabilitation plans. As part of this mission, ASOR CHI, through a public–government collaboration with the United States of America (US) Department of State, has been provided with access to hundreds of thousands of satellite images, some within 24 h of the image being taken, in order to assess reports of damage to cultural heritage sites, to discover unreported damage, and to evaluate the impacts of such incidents. This work is being done across an inventory of over 13,000 cultural heritage sites in the affected regions. The available dataset of satellite imagery is significantly larger than the scales that geospatial specialists within archaeology have dealt with in the past. This has necessitated a rethinking of how the project uses satellite imagery and how ASOR CHI and future projects can more effectively undertake the important work of cultural heritage monitoring and damage assessment.

Keywords: endangered cultural heritage; remote sensing; large dataset; crowd-sourcing information; condition assessment; real-time processing; Syria; Iraq; conflict; Nimrud; Palmyra; Mosul

1. Introduction

The American Schools of Oriental Research (ASOR) established the Cultural Heritage Initiatives (CHI) in 2014 to assist in addressing the current cultural heritage crises in the conflict zones of Syria and Northern Iraq, the worst such catastrophe since the Second World War. Since that time, on a daily basis, CHI has documented new incidents of looting, theft, damage, and destruction. Sustained ground and aerial combat, intensified by long-standing ethno-sectarian tensions and international intervention, have resulted in widespread damage and destruction to individual heritage sites and whole urbanscapes. Extremists such as the so-called Islamic State (ISIS) have deliberately destroyed hundreds of ancient monuments, mosques, churches, shrines, cemeteries, and other sites, as part of a systematized campaign of cultural cleansing, enacted to advance radical ideologies and to achieve more worldly military, political, and economic objectives. Years of warfare and instability have subjected local populations to unspeakable suffering, abysmal living conditions, and abject poverty. Millions of Syrians and Iraqis are internally displaced, living in makeshift camps or even archaeological ruins, or have undertaken the perilous journey to live abroad. Criminal activity inevitably peaks when such appalling conditions co-occur with rampant regional corruption, transnational organized crime, and predatory terrorist networks. One seemingly inevitable tragedy has been the systematic pillaging of the region's renowned cultural repositories, private collections, and archaeological sites, as locals struggle to support their families by trading away irreplaceable cultural assets for a pittance to exploitive mobsters, warlords, and terrorists, seeking easy profits from the global illicit art and antiquities markets. Hundreds of archaeological sites have been mined for antiquities, resulting in untold losses of archaeological data in the ancient Near East, home to the world's earliest known agricultural communities and literate state-level societies and the wellspring of several of the world's major religions and powerful empires. The loss to our global cultural patrimony is staggering and highlights the importance of rethinking the current international response within the modalities of cultural security and cultural property protection during conflicts (for recent overviews see [1–4]).

The CHI project began in August of 2014, with a focus on cultural heritage within Syria, but has since expanded to include Northern Iraq and Libya. The core mission of CHI entails monitoring and fact-finding activities, disseminating results to the United States of America (US) Department of State (DOS) and the public, implementing emergency response projects, developing post-conflict rehabilitation plans, and producing public outreach and education initiatives. In order to undertake these activities, the project has synthesized expansive data collected by its wide-ranging international network of heritage experts and analysts, including activists and institutions in the conflict zone of the Middle East and North Africa, from three principal sources: news outlets and social media, in-country contacts, and satellite imagery [5–8]. While the intersection of all three sources of information has proven critical to CHI's success, this article will primarily focus on the analysis enabled by the third of these sources—satellite imagery—within the modality of monitoring and assessing cultural heritage damage to the built environment, in Syria and Northern Iraq. CHI's ongoing data acquisition and analysis of the impacts of the Syrian and Iraqi conflicts on cultural heritage in real time, represents the first such comprehensive effort borne out of a public–government collaboration.

Results of CHI have been made available since August 2014 in weekly, bi-weekly, and monthly reports, submitted to the US Department of State and subsequently appearing on the CHI website in redacted form [9]. Additional special reports have been compiled in response to particular events or tactics of significance to cultural heritage, and some of these also appear on the CHI website. In addition to reporting, public and private presentations in various venues form an important component of promoting the awareness of impacts to cultural heritage. Our overriding vision entails empowering local communities to preserve and protect cultural resources through the establishment of broad and diverse coalitions. Such nimble, adaptive, and cost-effective responses appear to be the future of the field and form an integral part of broader international humanitarian conflict resolution, and post-conflict peacebuilding and recovery efforts [10].

The use of satellite imagery is not a new method in either cultural heritage monitoring or archaeology. Since the 1970s, archaeologists have made use of satellite imagery—which has grown out of extensive prior use of aerial photography—for discovering new cultural heritage sites and contextualizing both newly discovered sites and known sites within their broader cultural, political, and environmental landscapes ([11] (p. 33); [12] (pp. 18–28); [13] (p. 27)). However, fundamental changes in access to geospatial data granted to the CHI team in 2014 have greatly facilitated this work and have also offered a glimpse into future monitoring and research trajectories. Furthermore, the methods implemented by CHI to analyze and present reliable and verifiable deliverables under time-sensitive conditions represent a fundamental departure from traditional archaeological research projects. This article will discuss these changes and how they enable the ongoing work of CHI. Cultural heritage case studies, monitored by the project, will illustrate the importance of these developments and the challenges faced in monitoring and assessing cultural heritage during times of instability and conflict.

2. The ASOR CHI Methodology and Geospatial Data

The ASOR CHI methodology was initially developed in 2014, in response to a fundamental need to integrate both the geospatial and non-geospatial portions of the project. The geospatial team was tasked with analyzing satellite imagery to discover and document previously unreported damage to cultural heritage sites, or to confirm and, when possible, document the extent of reported damage uncovered by the non-geospatial reporting teams, through their use of networks of individuals located in-country, or by monitoring social and traditional media reports. This integrated method has proven powerful in thousands of cases, including those discussed in greater depth in the results section. The geospatial team has discovered unreported cases of cultural heritage destruction. They have also supported the reporting team by confirming, detailing, and correcting information on heritage incidents reported in the media or from private sources. Meanwhile, the reporting team has been able to make use of in-country contacts to, when possible, visit the location and assess, on the ground, the events first discovered by the geospatial team. Together this integrated pairing has provided detailed and verifiable reports from within the conflict zones in a timely fashion.

It should be stressed that this methodology was developed within the context of an existing conflict situation that encompassed a broad geographic area spanning all of Syria and large portions of Northern Iraq. This is a major difference between ASOR CHI and the Endangered Archaeology in the Middle East and North Africa (EAMENA) project, which started a year later, in 2015, and which strives to monitor an even larger area, most of which falls in countries that have been relatively peaceful during the past two years [14,15]. Both the open conflict, and the broad and expanding geographic scope of the CHI project's coverage, presented major challenges to successful implementation. Conflict areas create limitations in communication with in-country contacts as well as risks to their lives. Data from satellite platforms, high above the reach of the conflict, are necessary to provide information in areas where communication is impossible or where there are significant physical risks to individuals on the ground. The broad territorial expanse, encompassing many subregions lacking authoritative or standardized heritage site inventories, necessitated an immense amount of constantly refreshing geospatial data. From CHI's inception, it was apparent that access to resources well beyond the capabilities of freely provided satellite imagery, through platforms such as Google Earth or Bing Maps, would be needed to keep up with the fast-moving conflict situation. At the request of ASOR, in 2014, access to hundreds of thousands of satellite images, purchased by the US government, was granted through the mechanism of the collaborative agreement between CHI and the DOS. This benefit, gained from the public–government collaboration, went well beyond the access that had previously been granted to other archaeological projects.

The core geospatial dataset provided to the project is comprised of a subset of all DigitalGlobe orthorectified imagery available through the EV WebHosting service [16]. The imagery is available for download or for direct linkage through an ArcMap add-on that provides web map and tile services. The accuracy of the rectification of the imagery provided by the service has proven to be more than sufficient in most cases for immediate overlay and analysis, without the need for subsequent rectification by CHI. The geographic scope of the dataset has grown with CHI's increasing scope of work to encompass Syria and Northern Iraq, as well as more recently, Libya. The available image sets include WorldView-1 to WorldView-4 and GeoEye-1 satellite collections and are available to CHI as single band panchromatic and true-color pan-sharpened images at less than 50 cm resolution. The subset of imagery frequently changes, both with older images being removed and newly-collected imagery being added. Collection dates for the available imagery range from just prior to the start of the Syrian conflict in 2011, to the present. The dataset also includes ongoing new data collections and the regular tasking of satellites incorporating CHI's requests and lists of endangered cultural heritage sites—some requested sites have been collected and processed within 24 h of a major cultural heritage event. This rate of incorporation of new imagery is essential to timely analysis and reporting within the context of a rapidly and dynamically evolving conflict, and is not available through publicly available platforms, such as Google Earth.

The amount of geospatial data, while essential, presents a significant change to prevailing imagery analysis routines within archaeology. Typically, in terms of high resolution images, single images or sets of up to a few dozen images have been used within archaeological research for detecting, investigating, or monitoring cultural heritage locations [17,18]. The requirements for monitoring the rampant looting of southern Iraq following the second Gulf War raised the upper threshold of image datasets to 1000–2000 images [19]. This functional cap can largely be attributed to the cost of the imagery. Since the 1970s, freely available lower resolution satellite images, like Landsat, or higher resolution satellite images from earlier decades, like the declassified US spy satellites, have seen much more widespread use in regional archaeological applications involving higher numbers of images ([11] (p. 33); [13,20–22]). Likewise, large amounts of periodically updated imagery, available freely through online platforms, like Google Earth, have inspired archaeological projects on countrywide or larger scales [15,23,24]. For CHI, access to hundreds of thousands of images through the public–government collaboration necessitated some adaptation of prevailing methodologies, particularly in time-sensitive analysis situations with a quick turnaround for reporting, and have been the catalyst for even further methodological developments.

Initially CHI was tasked with simultaneously assessing large quantities of geospatial data and assembling an inventory of cultural heritage sites within the area of work. To achieve the latter objective and proceed with site monitoring and assessment activities, CHI required the locational and descriptive data for thousands of cultural heritage sites. No available comprehensive inventory had been undertaken for these areas, and so CHI compiled one from existing, overlapping inventories and by sorting through centuries of published material. This task would have been impossible without the support of collaborators within CHI, public inventories such as the ANE Placemarks for Google Earth [25], and networks of contacts willing to share sizable inventories of subsets of the area such as the Computational Research on the Ancient Near East (CRANE) Project [26], Ross Burns [27], The Fragile Crescent Project [28], and the Deutsches Archäologisches Institut (DAI) [29]. Merging these datasets posed some challenges. In many cases, these different inventories and other published sources did not agree on precise site locations or site names, which were inconsistently recorded going back centuries. This necessitated significant cross-checking between the inventories and manual reconfirmation of most locations using the imagery dataset at CHI's disposal and the assistance of personnel from the CRANE Project. It should be stressed that all of this inventory creation took place in the midst of open conflict, which severely curtailed access on the ground to cultural heritage site locations, unlike the impressive inventories that have been assembled by EAMENA in other portions of the Middle East [14]. As of 30 June 2017, the inventory consists of 13,186 unique cultural heritage sites across Syria, Northern Iraq, and Libya. While the core of the inventory consists of archaeological sites and monuments, it also includes other important heritage sites, such as mosques and churches, historic houses, and museums and libraries located within these geographic areas. This inventory continues to be a work in progress, with site locations revised in light of new information and additional research. It also remains of vital importance that access to this inventory is restricted, though the CHI project has shared this expanding inventory with other groups undertaking cultural heritage monitoring such as the United Nations Institute for Training and Research's Operational Satellite Applications Program (UNITAR-UNOSAT) [30], the CRANE Project, the Fragile Crescent Project, Shirín [31], Ross Burns, the DAI, and EAMENA. An inventory like this would be of use to not only those monitoring cultural heritage, but also by those seeking to intentionally loot or destroy these same sites. Balancing security risks with the need for access to the inventory by CHI members, collaborators, and others is an issue that will continue into the future.

With the inventory and access to geospatial data, CHI developed a cultural property protection and preservation methodology, building on existing methodologies employed for archaeological research projects, and workflows to analyze and present information gained from geospatial and non-geospatial datasets, for use by local stakeholders, cultural heritage professionals, activists, law enforcement, decision makers, and policy makers. Typical workflows depend on the source of the report of damage, either originating with the satellite imagery that is being monitored by the geospatial team, or with on-the-ground or media sources being monitored by the reporting team. For damage reports originating from the geospatial team, typical workflows include using trained analysts to visually assess new satellite imagery daily, over inventoried cultural heritage sites, as the imagery becomes available within the EV WebHosting service. Changes impacting the cultural heritage site detected between consecutive temporal images are recorded as spatial and attribute data, and are categorized according to a CHI schema of threats and damage, which expanded on the MEGA Jordan Guidelines' Threats and Disturbances Schema developed by the Getty Conservation Institute and World Monuments Fund [32,33]. The schema expansions by CHI included new threats and damages categories to cover military conflict, looting, and intentional destructions of cultural heritage. CHI chose the Middle Eastern Geodatabase for Antiquities (MEGA) - Jordan schema because of its standardization, which facilitates data sharing with partnering projects, and because it underpins the Arches open-source heritage inventory and management system, chosen by CHI and other projects as a cross-collaborative data sharing platform [34]. In 2014–2015, an initial rapid assessment protocol was used to quickly assess tens of thousands of images for the backlog of incidents that occurred between 2011 and 2014. These initial assessments identified the presence of damage, bracketed the dates of the damage incidents, identified damage type(s) according to the CHI schema, and categorized the severity of damage, along with defining the extent of the cultural heritage location. This initial rapid assessment protocol was altered as the backlog became more manageable. The current protocol includes adding background information concerning the particular cultural heritage site and cross-checking the incident against data being generated by the non-geospatial reporting team. This cross-checking can also include engaging local in-country sources to visit the site and further assess the damage if the dangers of the conflict allow such access.

Workflows are different in cases where the report of damage originates from in-country individuals or the social and traditional media sources being monitored by the reporting team. In these cases, the incorporation of these non-spatial sources of information are critical to the overall CHI methodology and its ultimate reporting activities [6]. Not only must CHI rapidly and regularly produce a diverse range of reports that cover a broad geographic area, subjected to intense damage and destruction, but these reports must address the needs of a diverse range of cultural property protection and preservation modalities and, most importantly, prove to be reliable and verifiable under time-sensitive conditions. Given the irregularities in data dissemination on heritage incidents in the conflict zone, new data may prove or disprove our published analyses in a matter of a few days—a frequent potentiality for famous heritage sites such as Palmyra or urban environments such as Aleppo and Mosul—or it may take months or even years for new information to surface, as is often the case for remote rural sites under the control of radical extremist groups, such as the so-called Islamic State.

To address these needs, information from conflict-zone sources or from media reports must intersect the geospatial data assessment workflow at several points depending on circumstances of data availability and reliability. CHI often develops initial reports of heritage incidents through these other channels, and we then assess available satellite imagery using the standard workflow described above to confirm, refine, or refute these reports. This is especially important since open-source streams, such as social media and online news sites, often provide near-instantaneous coverage but frequently contain inaccurate, propagandistic, and deliberately falsified information. Furthermore, ground-based sources provide highly reliable information, such as reports from CHI in-country site assessment teams, but they usually can only provide localized information, given the difficulty of travel and access to sites in active war zones. The integration of information from each of these data streams is key to establishing a higher confidence level in the reporting of a given incident during the assessment process. Over a three-year period, we have steadily refined our methods and have continued to achieve a high degree of verification and reliability, based on the constant re-evaluation of previous CHI incident reporting. This does not mean that CHI has been able to identify every instance of cultural heritage damage within the conflict, but it has expanded and reshaped knowledge of this important element in the conflict and it has often been the source for information subsequently disseminated by traditional news outlets.

The integration between in-country sources and the satellite imagery also carries with it further benefits and risks. Cultural heritage has regularly been at the center of conflicts in the Middle East and North Africa, vastly elevating danger levels for in-country experts and activists attempting to address looting and cultural heritage damage and destruction. Beyond the daily risks associated with conducting cultural property protection and preservation initiatives in active war zones, multiple radical extremist groups have enacted brutal campaigns of cultural cleansing, involving the deliberate targeting of both cultural assets and personnel. Ethno-sectarian tensions fuel violence against noncombatants and have created complex and constantly shifting zones of political and military control—even short-distance travel can be perilous. Complex, entangled networks of terrorists, criminal organizations, and highly corrupt state and non-state actors in the conflict zones of Syria and Northern Iraq have engaged in the looting of archaeological sites, thefts of cultural property from private and public collections, and smuggling—investigating these crimes is fraught with hazards. At times the satellite imagery assessments have been used to keep people out of danger by assessing areas that would have been too dangerous for a person to access, such as active combat zones, illegal border crossings used by smugglers and terrorists, and archaeological sites occupied by military forces or controlled by criminal gangs. At other times, satellite assessments have provided an alternative and publishable source of information about an incident in which it would otherwise have been too dangerous to reveal that an in-country contact had provided the original information. At the same time, when the situation on the ground becomes less dangerous, even if months or years after the incident, ground truthings of assessments have been an essential and powerful component in further verifying information in the CHI satellite imagery assessments. This power of the integration of geospatial and non-geospatial components within the methodology of CHI will, it is hoped, be a model which can be replicated, customized, and enhanced for monitoring cultural heritage within future conflict situations.

In terms of refining methods, CHI has been working to address key issues encountered since 2014. With a distributed core team and numerous collaborators throughout the world, the sharing of data, and especially geospatial data, is one key issue. A collaboration between the Getty Conservation Institute and CHI allowed our team to assist in specifying key elements of Version 4 of the Arches open-source heritage management software that was developed by the Getty Conservation Initiative (GCI) and the World Monuments Fund [34]. This includes a tile server component that will allow internal sharing of geospatial information among members of the team and their collaborators. Spatial queries, as well as non-spatial queries, of the assembled information will also contribute to long-term archival aspects of CHI.

A second new direction utilized by CHI in collaboration with University of California San Diego's Center for Cyber-Archaeology and Sustainability (CCAS) has been a pilot project, which has crowdsourced preliminary assessments of cultural heritage through the TerraWatchers portal [35]. The project utilizes the inventory developed by CHI, and therefore the dozens of participants in the crowdsourcing effort are vetted and trained prior to their participation. Participants do not directly access the DigitalGlobe imagery available to CHI through EV Webhosting, given access limitations, but rather work with Google Earth/Maps data and with a subset of DigitalGlobe data available through Qualcomm Institute's Big Pixel Initiative [36]. To date, the TerraWatchers collaboration has trained 131 students from the Universities of California at San Diego, Merced, and Berkeley, in assessing damage via satellite imagery. While the project is still assessing the overall results, during the project's first phase from 7 April 2016 to 5 May 2016, the participants made 4500 observations on over 3500 individual sites. They correctly identified damage based on looting, modern development, agricultural encroachment, and military-based earthworks and trenching. Students were most accurate in identifying modern settlements and burials on sites, while they had more difficulty identifying roadworks and mining or quarrying at archaeological sites. Through the process of training, students learned to identify specific forms of damage; the crowdsourcing project went from 7% accuracy during its first trial run to 39% accuracy [37].

Another new direction that has been developed by CHI in collaboration with the University of Central Florida's Center for Research in Computer Vision (CRCV), is automation of change detection analysis within the hundreds of thousands of satellite images. The project has undertaken initial change identification for incidents such as looting, bomb damage, and collapsed heritage, with the primary goal of prioritizing new images for analysts' attention based on the likelihood that a new image contains evidence of such events. This would help to mitigate the bottleneck created by the quantity of imagery in relation to the numbers of analysts (limited by cost factors) and is in line with similar work being undertaken by archaeologists elsewhere in the world [38–40].

3. Results

While all of the redacted incident reports compiled by CHI are available online [41], a selection of incidents that highlight the methodologies and results of this work are presented here by way of illustration. In addition to the reporting series published online, our team of geospatial analysts has also visually assessed 6662 heritage sites in Syria, Iraq, and Libya—assessments which are continually being updated and refined. Utilizing the most recently available satellite imagery, each heritage site was assessed for damage occurring since the start of the conflict and assigned a percentage of total visible damage. Figure 1 displays these assessments, broken down by site type. Although our reporting series focuses on damaged sites, the majority of assessed heritage sites display no visible damage—a total of 63%. The second highest percentage of damage falls under some damage, between 10% and 60%, at 26% of total assessed sites.

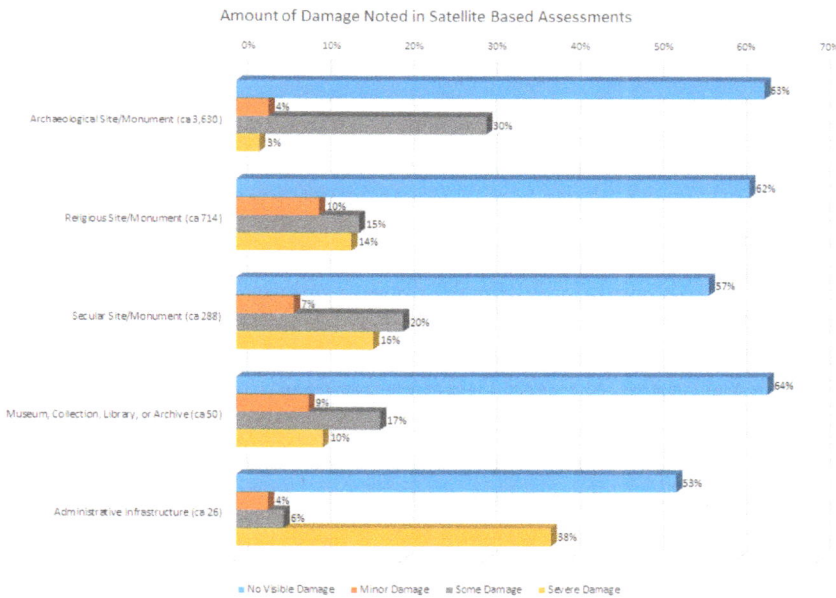

Figure 1. Satellite based assessments of heritage sites within the Cultural Heritage Initiatives (CHI) Inventory (mainly Syria and northern Iraq) according to site types and levels of assessed damage (ASOR CHI; 26 July 2017).

As of the 30 June 2017, CHI has produced 870 reports of cultural heritage damage in Syria and Northern Iraq over the past three years. This covers 1100 unique cultural heritage sites that have been affected by the ongoing conflicts in these areas. Over the last three years, CHI has recorded damage to heritage sites due to a variety of disturbances, including military activity and human activity, such as illegal excavations, agriculture, and urban encroachment. Each damage incident is assigned a pattern of damage, based on the primary cause of the destruction or damage. The full list of damage patterns can be seen in Figure 2. Military activity ranks as the most frequent damage source, primarily incidents caused by explosives—mainly artillery strikes and airstrikes—as well as from gunfire.

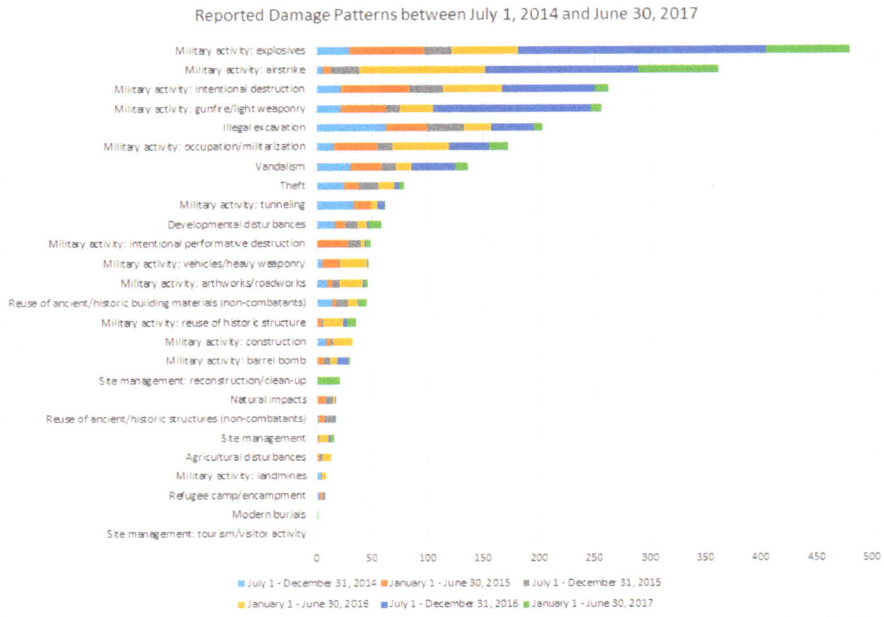

Figure 2. Damage patterns reported by CHI in the ongoing weekly and monthly report series since the project's inception in 2014. The horizontal bars representing incident totals (horizontal axis) by causes of damage are subdivided (stacked) according to report dates rather than the date(s) that the incidents occurred, to account for those incidents currently without known dates (ASOR CHI; 26 July 2017).

During the first six months of the project, CHI documented twice as many reported illegal excavations of archaeological sites as compared to explosive damage to heritage sites. During this period, the conflict kinetics were less intense, but also, at project startup, the overall number of reported conflict-related looting incidents was artificially elevated, given that such activity had been ongoing since late 2011–2012—although looters were especially active in 2014 and early 2015—and the evidence of illegal open-pit excavation in this largely arid and unvegetated region is highly visible and readily identifiable. Conversely, the spread of the practice of tunnel looting later in the conflict, presented some challenges for satellite based site assessment and has likely resulted in a slight underreporting of the incidence of looting activity assessed through satellite imagery. In the subsequent 32 months of the project, damage patterns flipped, with increasing damage to heritage sites due to explosives, primarily artillery shelling. In addition, airstrikes, the second- ranked cause of damage, dramatically increased starting from 1 January 2015, as aircraft from the Syrian Arab Republican Guard, Russian Military, US-led Coalition Forces, and Iraqi Government Forces carried out major offensives within Syria and Iraq. Intentional destruction of heritage sites, primarily carried out by ISIS, represents the third most common cause of damage to heritage sites, impacting most site categories (i.e., archaeological, religious, and secular). Such acts were prominent in reporting from 1 January to 30 June 2015, as ISIS and other groups targeted shrines and other religious sites in Iraq and Syria as well as archaeological sites such as Palmyra, Nimrud, and Hatra.

The analysis of satellite imagery has helped to redress biases in our understanding of the cultural impacts of the conflict in Syria and Northern Iraq, stemming from open source and traditional media coverage. A comparison of causes of damage to the most common site types in the CHI inventory, archaeological sites and monuments versus religious sites and monuments, illustrates two complex and contrasting patterns of heritage damage (Figure 3). Archaeological sites and monuments are impacted by a wide range of factors, including urban development and encroachment, military earthworks and construction, as well as the reuse of ancient structures and buildings by both combatants and noncombatants. The leading causes of damage are illegal excavations and military occupation. Yet traditional media outlets and social media sites have largely focused on covering intentional destruction and looting, rather than the much more complex situation on the ground, in which instability, lack of rule of law and regulation, population displacements and deterioration and neglect play major roles in the loss of cultural assets. Religious sites and monuments, including mosques, churches, and shrines, have been devastated by military explosives and airstrikes, intentional destructions, and gunfire. Although ubiquitous, such incidents have received less media coverage, particularly outside the Middle East, relative to the less frequent spectacles of intentional destructions at famous archaeological sites, despite the deleterious long-term impacts that such attacks exert on the region's communities, conflict resolution, and regional stability.

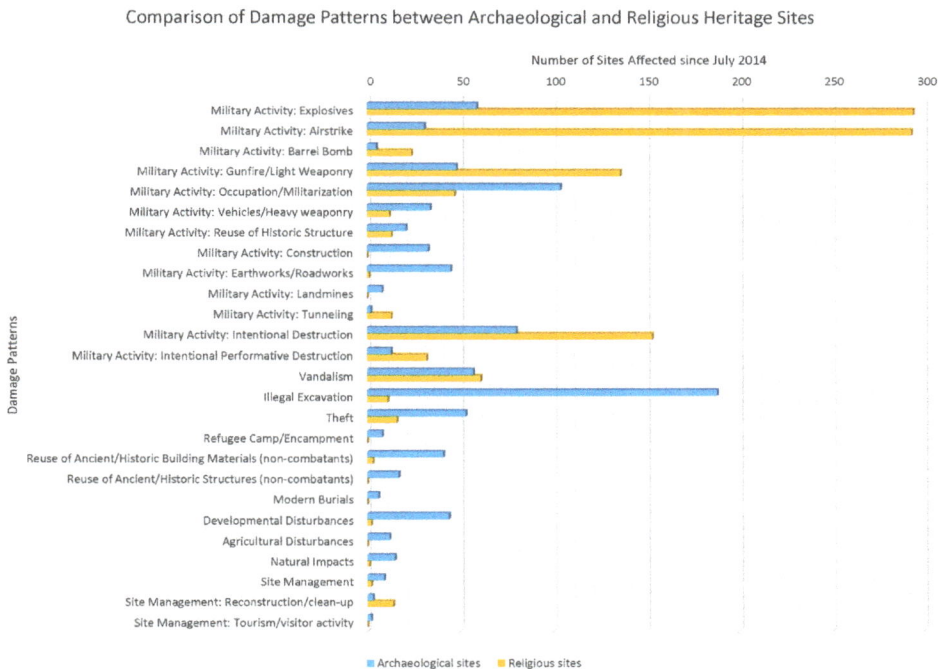

Figure 3. Comparison of the causes of damage to archaeological and religious heritage sites in Syria and Northern Iraq between 1 July 2014 and 30 June 2017 (ASOR CHI; 26 July 2017).

In the first three years of the project, unredacted CHI reporting has been regularly consulted by various government and non-government organizations to assess the overall situation and to develop and implement policies and actions, including but not limited to, presentations and written submissions for multiple US government agencies and congressional committees, the European Union, United Nations Educational, Scientific, and Cultural Organization (UNESCO), the United Nations, the United Nations Security Council and, it is our understanding, for the US National Counterterrorism Center and multiple Presidential Daily Briefs. In terms of raising public awareness, redacted CHI analyses have been featured in hundreds of news reports and public presentations detailing the impacts of the conflict on cultural heritage and cultural memory, identity, and diversity. CHI has supported investigations by activists, journalists, and law enforcement agencies worldwide to recover stolen cultural property, combat looting, and counter violent extremism. Most importantly, the program strives to support Syrian, Iraqi, and Libyan cultural heritage experts and local stakeholders in their struggle to save their beleaguered cultural patrimony. In this regard, CHI monitoring and reporting has supported multiple in-country initiatives to preserve and protect cultural heritage, by reducing risks, mitigating threats, repairing damage and, ultimately, maintaining local access to cultural heritage.

To illustrate the results of the reporting methodology, three case studies are presented below. They were selected to reflect the development of the methodology and how it utilizes workflows, and originated with reports from both the geospatial and non-geospatial portions of the project. These three examples also represent instances in which media organizations utilized ASOR CHI reporting to develop stories to inform the public of damage to cultural heritage as the incidents were transpiring or shortly thereafter. The monitoring of Nimrud between January and April of 2015, and the reports published during that time, are one of the earliest examples from a single cultural heritage site where the CHI-integrated methodology was implemented. The monitoring of Palmyra from 2014 to 2017 provides a series of events across a wider expanse of associated cultural heritage sites that were monitored using this integrated methodology, including incidents that were first made known at the time by CHI in its reports. Finally, the ongoing monitoring of damage to the old city of Mosul shows the power of this integrated approach when it is expanded to an urban scale, encompassing numerous cultural heritage sites.

3.1. Early Implementation of the ASOR CHI Methodology at Nimrud

Nimrud, a multi-period site, best known for its 9th and 8th century BCE occupation as an early capital of the Neo-Assyrian Empire, was one of the first cultural heritage sites where the CHI methodology was fully implemented over an extended period of time. The archaeological site had come under threat prior to the start of the project in June 2014, but between January and April of 2015 the walled Acropolis, and particularly the Northwest Palace of Ashurnasirpal II and its famous sculpted bas reliefs, were the subject of filmed acts of destruction by ISIS (i.e., performative deliberate destructions).

CHI received the earliest reports of intended ISIS attacks from in-country contacts in January 2015, by which time Nimrud had already been added to a short list of priority sites requiring ongoing tasking of satellites for imagery. No evidence of destruction from assessments of satellite imagery were noted until a 7 March image, following increasing reports from 5 March to 7 March, that ISIS had deployed personnel and equipment to destroy standing architecture at Nimrud (Figure 4a,b). This incident was part of a larger pattern of performative deliberate destructions targeting cultural heritage in the Mosul area, including ancient sculptures and replicas in the city's museum and monumental architecture and sculptures at Nineveh, another Neo-Assyrian capital, which attracted widespread attention and worldwide condemnation in late February.

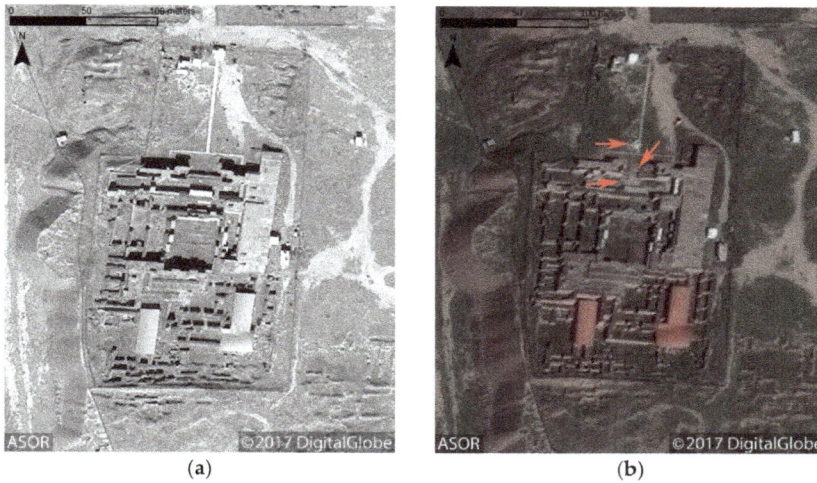

Figure 4. The Northwest Palace at Nimrud in DigitalGlobe satellite imagery (**a**) prior to damage from ISIS with protective roofing over the stone reliefs in situ (DigitalGlobe NextView License; 26 February 2015); (**b**) with arrows indicating a pile of rubble and vehicle tracks within the palace walls (DigitalGlobe NextView License; 7 March 2015).

In the 7 March imagery, CHI noted evidence for cuts in walls at key access points in the Northwest Palace, as well as piles of rubble in areas adjacent to the Throne Room, that exceeded the volumes of missing portions of walls. These rubble piles appeared to consist of freshly broken stone matching the color of stone used in Nimrud's bas reliefs. CHI analysts posited that ISIS had targeted the reliefs lining the entranceway and walls within the Throne Room for destruction. CHI shared this information with in-country and international experts, including the State Board of Antiquities and Heritage of Iraq (SBAH) and UNOSAT, shortly after the release of the imagery to CHI on 8 March, and updates continued as new imagery was released into early April, showing further evidence of ongoing damage. In an image from 1 April, further damage is visible in the Throne Room, evidenced by the removal of a protective cover over the reliefs and heavy vehicle tracks within the palace (Figure 5a). Around 2 April, ISIS demolished the Northwest Palace through the detonation of a series of barrel bombs set along the face of the relief-covered walls. ISIS released a video of this criminal act on 11 April. The destruction appears in satellite imagery taken on 17 April (Figure 5b). Analysis of both the released footage and the satellite image allowed preliminary assessments of the damage, setting up subsequent on-the-ground damage assessments as ISIS was pushed back from the area. The entire event was reported publicly, within CHI Weekly Reports 31, 34, and 36, as well as in a special summary report published on 5 May 2015 [42–45]. This case demonstrates the utility of the integrated CHI methodology and of the public–government collaboration for following cultural heritage threats over an extended period of time. The case highlights the capability of cultural property protection programs to monitor and alert the international community to impending and ongoing attacks on cultural assets. Such situational awareness allows the international community to seize the initiative and to conduct public outreach and awareness activities prior to the online release of extremist propaganda featuring performative deliberate destructions.

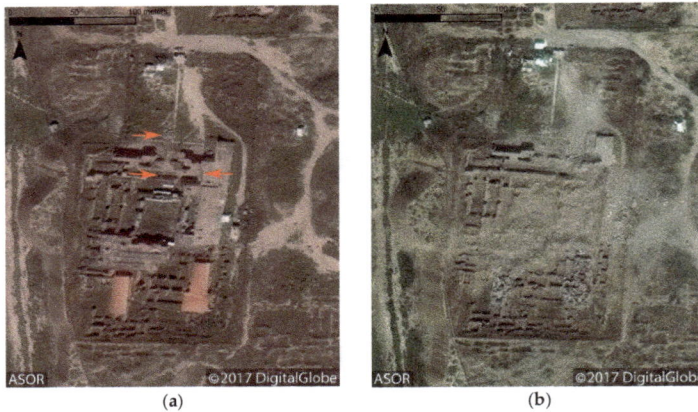

(a)   (b)

Figure 5. Further destruction at the Northwest Palace of Nimrud (**a**) with more vehicle tracks and removal of stone reliefs noted with the red arrows (DigitalGlobe NextView License; 1 April 2015); (**b**) after the detonation of bombs by ISIS within the palace walls (DigitalGlobe NextView License; 17 April 2015).

3.2. Intentional Destructions at Palmyra in 2015

The UNESCO World Heritage Site of Palmyra (the ancient site and adjacent modern town are known as Tadmor in Arabic) has been under heightened threat and damaged during multiple periods of the Syrian conflict due to the area's strategic significance. This desert oasis was controlled by ISIS from May 2015 through March 2016, and again from December 2016 until 2 March 2017 until it was recaptured by Syrian Arab Republic Government (SARG) forces. During the time in which ISIS held Palmyra, the archaeological site was looted and standing architecture was repeatedly targeted for intentional destruction—some performative acts were released in videos and photos as part of ISIS's propaganda campaign.

ISIS committed large numbers of atrocities in the Tadmor area, targeting the town's inhabitants and more modern religious heritage. Most telling of all, prior to targeting Palmyra's ancient monuments for intentional destruction, in fact almost immediately upon capturing the area, the group carried out destructions and vandalisms of Sunni, Sufi, Shia, and Christian heritage. Such acts reveal the organization's prioritization of cultural cleansing and the intimidation and subjugation of modern populations. Many of these sites were located in the remote desert areas surrounding Tadmor, increasing the importance of geospatial analysis for investigating alleged incidents, given the paucity of other information.

The destruction of Palmyra's famous Temple of Baalshamin and Temple of Bel formed the middle stages in this campaign of performative destruction [46,47]. Soon after ISIS took over the site in May 2015, reports began to emerge that ISIS militants had planted explosive devices within the archaeological site, which was confirmed by 23 June 2015 through CHI sources, who reported that locals had seen members of ISIS place "large mines/bombs in the ruins of many buildings in Palmyra", and told Tadmor's residents of their intent to destroy the ruins—using loudspeakers in Tadmor—to gain media coverage and possibly as a deterrent to counterattacks. During this time, ISIS leaders also allegedly lived at the site, to protect themselves from airstrikes, and munitions were stored on-site. ISIS has regularly repurposed heritage sites and cultural and educational buildings for military and political use, throughout the conflict zone. On 23 August 2015, reports began to emerge that ISIS had destroyed the Baalshamin Temple (largely of the 2nd century CE). Soon after, ISIS released photographs showing the temple walls lined with barrels of explosives and the subsequent explosion. DigitalGlobe satellite imagery taken on 27 August 2015 acquired by CHI confirmed this destruction. A few days later, on 30 August 2015, the pattern repeated at the Temple of Bel, which dated to the 1st century CE, with reports of its destruction using explosives. This was confirmed via satellite imagery a few days later (Figure 6a,b). ISIS later published images of the destruction in its online magazine.

(a)   (b)

Figure 6. DigitalGlobe satellite imagery of the Temple of Bel: (**a**) prior to intentional destruction (DigitalGlobe NextView License; 26 June 2015); (**b**) post-destruction with the cella of the temple destroyed (DigitalGlobe NextView License; 2 September 2015).

At the same time as these performative destructions at the Temples of Bel and Baalshamin, ISIS destroyed large parts of the Valley of the Tombs, though less publicly. During three phases, the tallest and most well-known tower tombs, all located on the northern slopes of the Umm al-Belqis, were destroyed by ISIS, using explosives. This act and others suggest ISIS leadership optimizes heritage targeting, in targeted rich environments, based on its perceived significance. The Valley of the Tombs is an area of the Palmyra necropolis located west of the city's ancient walls, containing around 100 tower tombs, hypogea (underground tombs), and funerary temples (tombs built to look like small temples or houses). The most eye-catching monuments in this area are the tower tombs. Often several stories high, each floor of a tower had multiple chambers, containing loculi, or small spaces for individual interments. No reports of this destruction were known prior to their assessment by CHI in DigitalGlobe satellite imagery. Between 26 June 2015 and 27 August 2015 the Tomb of Iamliku was destroyed and the Banai Family Tomb directly to its east was badly damaged (Figure 7a,b).

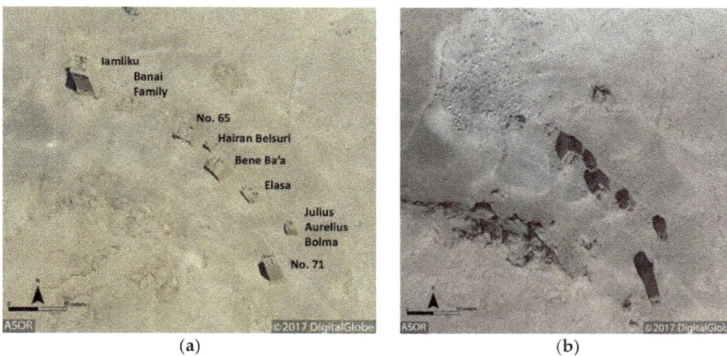

(a)   (b)

Figure 7. DigitalGlobe satellite imagery of the Valley of the Tombs: (**a**) with no visible damage (DigitalGlobe NextView License; 26 June 2015); (**b**) with visible damage to Tomb of Iamliku and Tomb of the Banai Family (DigitalGlobe NextView License; 27 August 2015).

During a second phase of destruction, between 27 August 2015 and 2 September 2015, ISIS destroyed more tower tombs, including the Tomb of Elahbel, the Tomb of Kithoth in the Northern necropolis, and two additional unnamed tombs near the Tomb of Iamliku (Figure 8a). Then, between 2 September 2015 and 30 March 2016, the Tower Tombs of Elasa, Bene Ba'a, Hairan Belsuri, and No. 65 were severely damaged, as seen in DigitalGlobe imagery (Figure 8b). The tomb of Elasa appears to still be standing without damage, while the three structures to the east have all sustained various degrees of damage. Some walls are still standing, but the large rubble piles at their bases indicate some destruction with explosives. ISIS never published photos or videos of the damage done to these monuments. In addition to these destructions, tombs in Western, Southeastern, and Southwestern Necropoli were also intentionally destroyed with explosives, which was only revealed in satellite imagery [48,49]. Lastly, in January 2017, ISIS intentionally destroyed the Tetrapylon and part of the Roman Theater's stage backdrop, which CHI identified, as part of our monitoring of the archaeological site using DigitalGlobe satellite imagery [50,51].

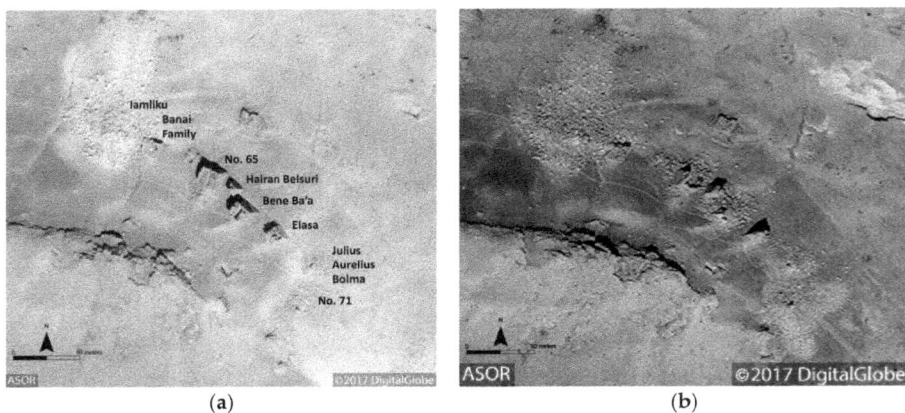

Figure 8. DigitalGlobe satellite imagery of the second and third phases of destruction in the Valley of the Tombs: (**a**) with visible damage to Tomb of Julius Aurelius Bolma and Tomb No. 71 (DigitalGlobe NextView License; 2 September 2015); (**b**) with new damage to the Tombs of Bene Ba'a, Hairan Belsuri, and No. 65 although some of the structures are still intact (DigitalGlobe NextView License; 30 March 2016).

Since the site has been recaptured by SARG forces, cultural heritage professionals have been able to access the site and assess the damage, using small drones as well as on the ground photography. CHI has incorporated these images into its monitoring to better understand the extent of the damage to cultural heritage and evaluate our earlier interpretation of satellite imagery.

The site of Palmyra and other heritage sites in the city of Tadmor and surrounding area have suffered severe damage since 2011. While ISIS has caused the majority of damage, all belligerents in the conflict have committed or been complicit in cultural property crimes in Tadmor or logistical blunders, such as the construction of a military base on the site by Russian forces [52]. Our understanding of these events has heavily depended on rapid-paced geospatial analysis. As in the case of Nimrud, situational awareness did not prevent these incidents, but it is hoped that the documentary efforts of CHI and other organizations will facilitate the rehabilitation of Tadmor and help to bring the perpetrators of these crimes to justice. The careful documentation of Palmyra has provided one of our most comprehensive examples of ISIS cultural cleansing and pillage, which reveals the intentions and priorities of ISIS leadership to persecute and uproot ethnic and religious minorities first, and later, loot and destroy ancient cultural resources, to impose and propagate ideology and finance global terrorism.

3.3. Assessment of Damage to the Old City of Mosul

On 11 June 2017, Iraqi Prime Minister, Haider al-Abadi, announced Iraq's victory over ISIS in the militant group's former Northern Iraqi stronghold of Mosul [53–57]. His announcement followed months of fighting, extensive aerial bombardment, massive infrastructure damage, human displacement, and thousands of civilian casualties [58]. The struggle for Mosul spanned nine months, with the most intense fighting occurring in the labyrinthine confines of the Old City. The final month of military operations brought the worst damage and high numbers of civilian casualties as ISIS militants unleashed waves of car bombings, suicide bombers, and snipers to target both Iraqi forces and Mosul residents [59,60]. The Old City of Mosul, located on the Western Bank of the Tigris, dates back to the Zengid Dynasty (1127–1250 CE) with historic souqs, mosques, churches, and government buildings dating from 1200 to 1800 CE. In June 2014, a force of about 1000 ISIS militants invaded the city [61].

Between June 2014 and late December 2015, ISIS carried out multiple intentional destructions of major cultural heritage sites across Mosul. Using a combination of satellite imagery, media, and in-country sources, CHI has confirmed the total destruction of 32 religious sites across the city of Mosul, including shrines, mosques, churches, and cemeteries, during ISIS's three-year occupation [42,62–66]. In the Old City alone, ISIS conducted 13 intentional destructions during that period, with mosques, churches, and shrines demolished with explosives or heavy machinery. Many of those sites were then cleared of all debris, paved, and used as parking lots by ISIS. After that initial burst of destruction, ISIS militants turned their attention elsewhere until the recent destructions of al-Nuri al-Kabir Mosque and al-Hadba Minaret— ISIS's final cruel acts of retributory violence.

CHI has monitored recapturing operations in Mosul and has documented damage to dozens of cultural heritage sites. Much of the heavy combat during the recapture of the city was focused on the Old City of Mosul (Figure 9). As of 30 June 2017, CHI has reported 23 individual incidents of damage to religious heritage in the Old City of Mosul, including mosques (14 incidents), churches (six incidents), and shrines (three incidents). Of these reported incidents, 16 were ISIS intentional destructions of Muslim and Christian sites during the occupation of the city, and the other seven were due to military explosives, possibly from shelling, heavy artillery, or airstrikes. Information on heritage incidents inside Mosul was not easy to acquire, and CHI regularly relied on available satellite imagery to monitor and confirm destructions of heritage sites.

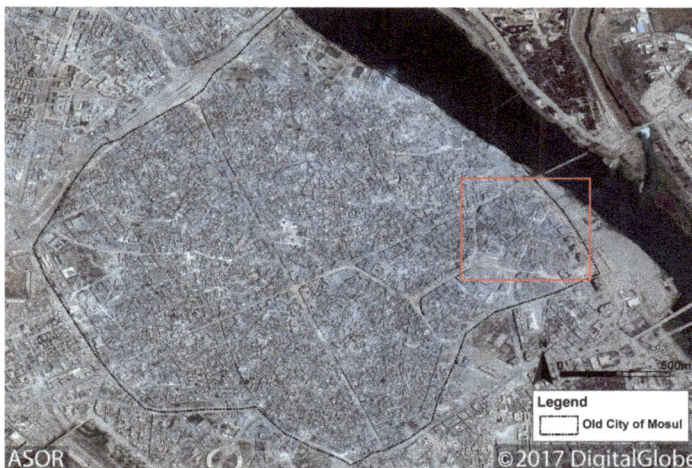

Figure 9. Old City of Mosul with the area of the souqs outlined in red. See below for insets of damage before and after the offensive (ASOR CHI/DigitalGlobe NextView License; 12 July 2017).

During CHI monitoring of the occupation of Mosul, our assessments were aided by the continuing advancements in satellite technology. Urban areas posed a problem when the project first started, since much of the damage was too small to see, even on 46 cm resolution imagery, such as the WorldView-2 satellite. With newer imagery available from WorldView-3, at 34 cm resolution, we are able to focus on damage at the level of single, small buildings and identify and assess areas of damage and destruction with increased spatial precision and interpretive accuracy and detail.

The ongoing bombardment and street warfare continues to leave its mark on the historically and culturally significant locations within the city, none more so than the area of the souqs. Much of this damage has occurred within the last year, beginning in March 2017, as visible in DigitalGlobe satellite imagery. Since May 2016, the streets and courtyards of this area have been systematically covered with metal roofs, including in the final phase over Nineveh Street and Ghazi Street, as a way to provide cover from airstrikes. In March, severe aerial bombardment damaged much of the area, including Bab al-Tub Police Headquarters, Souq al-Alwah Mosque, and Bab al-Tub Mosque. The damage was so severe that pieces of the metal roofing were visible in satellite imagery floating in the Tigris River. As of 22 May 2017, the ongoing conflict has further damaged these sites as well as the al-Aghawat Mosque, al-Pasha Mosque, and the former site of the Madrasa of the Abdal Mosque, which had been previously razed to the ground in an intentional destruction by ISIS and a new construction built on top of it. By the end of the offensive in the Old City, of the eight heritage sites identified in this area, three were destroyed, four were severely damaged, and one showed some damage (Figure 10).

Figure 10. DigitalGlobe satellite imagery of the souq area within the Old City of Mosul: (**a**) prior to the offensive to retake the Old City (DigitalGlobe NextView License; 9 May 2016); (**b**) after recapture by Iraqi forces (DigitalGlobe NextView License; 12 July 2017).

As of 30 June 2017, CHI has assessed the damage to 64 heritage sites within the Old City of Mosul. We have noted 37 heritage sites that exhibit severe damage (60–100% damaged), 12 of which have some damage (10–60%), nine with minor damage (1–10%), and six with no visible damage (Figure 11). These assessments will continue to be updated as more photographs are taken and heritage professionals on the ground complete assessments. In comparison, United Nations Institute for Training and Research's Operational Satellite Applications Program (UNITAR-UNOSAT) identified 5536 affected structures of all types within the Old City from imagery dated 30 June 2017, with almost 500 of those destroyed and 3310 severely damaged [67]. This was an increase of 37% from their previous report, just 14 days prior. Geospatial analysis shows this area of the city has sustained intense damage to all buildings, and the effort to rebuild the lives of those living in Mosul will be a long and difficult process.

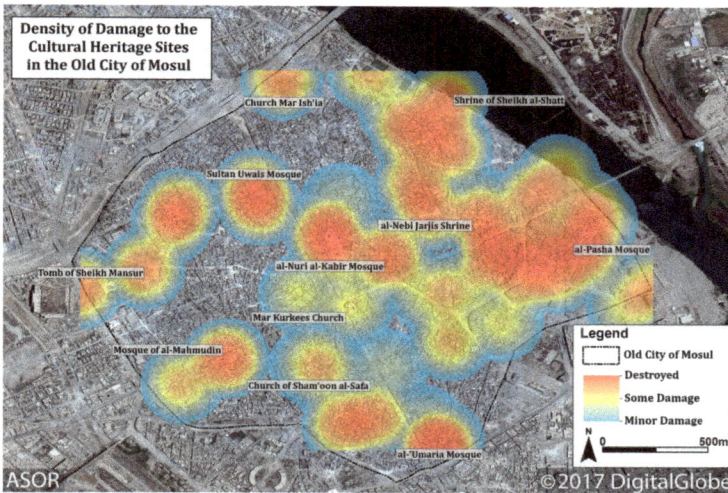

Figure 11. Density of damage to cultural heritage sites within the Old City of Mosul (ASOR CHI/DigitalGlobe NextView License; 12 July 2017).

4. Discussion

Conflict situations present a number of unique difficulties for monitoring and protecting cultural heritage. Foremost among these, is the risk to stakeholders living in the vicinity. They are the essential key to the long-term protection and monitoring of cultural heritage. Particularly in a conflict where cultural heritage has been purposefully targeted for propagandistic, psychological, and strategic goals, the risk of loss and even the death of local stakeholders is a very real and ever present danger. Methodologies that prioritize their safety are essential. While gaining information during the conflict is important, it should never endanger the lives of local stakeholders.

The methodology developed by CHI for integrating assessments of satellite data with ground-based observations and open-source information has wide applicability for addressing aspects of cultural heritage crises in conflict zones. In situations such as those in Syria and Northern Iraq, where direct ground-based observations have often been impossible or carried considerable risk for individuals in the vicinity, the methodology offers various alternatives for providing or publicly attributing sources of reliable information. When the on-the-ground situation allows, the methodology also incorporates ground verification to further strengthen conclusions.

The work of CHI benefits enormously from the unique public–government partnership, made possible by the collaboration with the DOS. While access to enormous numbers of satellite images creates bottlenecks and issues with workflows, it is essential for providing data necessary for producing the reliable and verifiable reports, for which CHI is now known. It furthermore necessitated that CHI find effective ways to address the needs for an authoritative inventory of the locations of cultural heritage and for a way to cross-compare reports of damage being collected by different analysts and organizations. Both of these goals have been attained through a strong network of project collaborators and through access to this large body of satellite imagery. Furthermore, the extended standardized schema of threats and disturbances developed by CHI to address this second need will have widespread applicability in future conflict zones that see military damage, looting, or performative destruction of cultural heritage.

The impact of CHI extends beyond the important role it is playing in Syria, Northern Iraq, and Libya. Projects monitoring cultural heritage crises in future conflicts will have a model in place for how to effectively pair big data from remote sensing with a broad network of collaborators, including area specialists and local stakeholders. New technologies will no doubt change the specifics of the data being used, but the essential methodology employed is flexible enough to incorporate any datasets. Ongoing work by CHI to refine the methodology, through initiatives such as crowd-sourcing and automation, or to press for the further development of cross-collaborative data sharing platforms, like Arches, will likewise impact how such projects are undertaken in future conflict situations and in the post-conflict periods that follow. However, in the end, it is the people involved, and particularly the local stakeholders that will enable the power of cultural heritage to impact our present and future.

Acknowledgments: ASOR CHI has received financial support through two consecutive cooperative agreements awarded by the U.S. Department of State (NEA-PSHSS-14-001 and S-IZ-100-17-CA021). Additional support for complementary aspects of the project has been provided by the J.M. Kaplan Fund, National Endowment for the Humanities, Getty Conservation Institute, and the Whiting Foundation. The University of Central Florida supported work on the automated detection of cultural heritage in satellite imagery through a COS/ORC Seed Grant.

Author Contributions: All authors conceived and designed the methodology, performed the methodology, analyzed the data, and wrote the paper.

Conflicts of Interest: The authors declare no conflict of interest.

References

1. Kila, J.D. *Heritage under Siege: Military Implementation of Cultural Property Protection Following the 1954 Hague Convention*; Brill: Leiden, The Netherlands, 2012.

2. Kila, J.D.; Zeidler, J.A. *Cultural Heritage in the Crosshairs: Protecting Cultural Property during Conflict*; Brill: Leiden, The Netherlands, 2013.

3. Kila, J.D.; Balcells, M. *Cultural Property Crime: An Overview and Analysis of Contemporary Perspectives and Trends*; Brill: Leiden, The Netherlands, 2014.

4. Nemeth, E. *Cultural Security: Evaluating the Power of Culture in International Affairs*; Imperial College Press: London, UK, 2015.

5. Casana, J. Satellite imagery-based analysis of archaeological looting in Syria. *Near East. Archaeol.* **2015**, *78*, 142–152. [CrossRef]

6. Danti, M.D. Ground-based observations of cultural heritage incidents in Syria and Iraq. *Near East. Archaeol.* **2015**, *78*, 132–141. [CrossRef]

7. Danti, M.D. ISIS, war and the threat to cultural heritage in Iraq and Syria: Recent cultural heritage developments: Recent Cultural Heritage Developments. *IFAR J.* **2016**, *16*, 33–46.

8. Danti, M.D. Protecting endangered cultural heritage in Syria and Iraq: The role of international organizations and governments. In Proceedings of the One Hundred Tenth Annual Meeting of the American Society of International Law, Washington, DC, USA, 31 March 2016; Cambridge University Press: Cambridge, UK, 2017; pp. 95–112.

9. ASOR Cultural Heritage Initiatives. Available online: http://www.asor-syrianheritage.org (accessed on 31 August 2017).

10. Danti, M.D. Near Eastern archaeology and the Arab Spring: Avoiding the ostrich effect. *Antiquity* **2014**, *88*, 639–643. [CrossRef]

11. Adams, R.M. *The Heartland of Cities*; University of Chicago Press: Chicago, IL, USA, 1981.

12. Parcak, S.H. *Satellite Remote Sensing for Archaeology*; Routledge: London, UK, 2009.

13. Masini, N.; Lasaponara, R. Sensing the past from space: Approaches to site detection. In *Sensing the Past: From Artifact to Historical Site*; Masini, N., Soldovieri, F., Eds.; Springer: Cham, Switzerland, 2017; pp. 23–60.

14. EAMENA Endangered Archaeology in the Middle East and North Africa. Available online: http://eamena.arch.ox.ac.uk/ (accessed on 31 August 2017).

15. Rayne, L.; Bewley, R. Using Satellite Imagery To Record Endangered Archaeology. In *Remote Sensing and Photogrammetry Society Archaeology Special Interest Group (RSPSoc Archaeology SIG) Newsletter*; 2016; pp. 15–20. Available online: http://eamena.arch.ox.ac.uk/wp-content/uploads/Bewley_and_Rayne_2016.pdf (accessed on 24 September 2017).

16. EV WebHosting Login. Available online: https://evwhs.digitalglobe.com/myDigitalGlobe/login (accessed on 31 August 2017).

17. Beck, A. Google Earth and World Wind: Remote sensing for the masses. *Antiquity* **2006**, *80*. Available online: http://www.antiquity.ac.uk/projgall/beck308 (accessed on 31 August 2017).
18. Contreras, D.A.; Brodie, N. The utility of publicly-available satellite imagery for investigating looting of archaeological sites in Jordan. *J. Field Archaeol.* **2010**, *35*, 101–114. [CrossRef]
19. Stone, E.C. Patterns of looting in southern Iraq. *Antiquity* **2008**, *82*, 125–138. [CrossRef]
20. Branting, S.; Trampier, J. Geospatial data and theory in archaeology: A view from CAMEL. In *Space—Archaeology's Final Frontier? An Intercontinental Approach*; Salisbury, R.B., Keeler, D., Eds.; Cambridge Scholars Publishing: Cambridge, UK, 2007; pp. 272–289.
21. Fowler, M.J. Declassified intelligence satellite photographs. In *Archaeology from Historical Aerial and Satellite Archives*; Hanson, W.S., Oltean, I.A., Eds.; Springer: New York, NY, USA, 2013; pp. 47–66.
22. Nebbia, M.; Leone, A.; Bockmann, R.; Hddad, M.; Abdouli, H.; Masoud, A.M.; Elkendi, N.M.; Hamoud, H.M.; Adam, S.S.; Khatab, M.N. Developing a collaborative strategy to manage and preserve cultural heritage during the Libyan conflict. The case of the Gebel Nāfusa. *J. Archaeol. Method Theory* **2016**, *23*, 971–988. [CrossRef]
23. Parcak, S.; Gathings, D.; Childs, C.; Mumford, G.; Cline, E. Satellite evidence of archaeological site looting in Egypt: 2002–2013. *Antiquity* **2016**, *90*, 188–205. [CrossRef]
24. Casana, J.; Panahipour, M. Satellite-based monitoring of looting and damage to archaeological sites in Syria. *J. East. Mediterr. Archaeol. Heritage Stud.* **2014**, *2*, 128–151. [CrossRef]
25. ANE Placemarks for Google Earth. Available online: http://www.lingfil.uu.se/research/assyriology/earth/ (accessed on 31 August 2017).
26. The CRANE Project: Computational Research on the Ancient near East. Available online: https://www.crane.utoronto.ca/ (accessed on 31 August 2017).
27. Monuments of Syria. Available online: http://monumentsofsyria.com (accessed on 31 August 2017).
28. The Fragile Crescent Project: Settlement Change during the Urban Transition. Available online: https://www.dur.ac.uk/fragile_crescent_project/ (accessed on 31 August 2017).
29. Syrian Heritage Archive Project. Available online: https://arachne.dainst.org/project/syrher (accessed on 31 August 2017).
30. UNITAR's Operational Satellite Applications Programme—UNOSAT. Available online: https://unitar.org/unosat/ (accessed on 31 August 2017).
31. Shirín. Available online: http://shirin-international.org/ (accessed on 31 August 2017).
32. Getty Conservation Institute; World Monuments Fund. *Middle Eastern Geodatabase for Antiquities (MEGA)—Jordan: Guidelines for Completing Site Cards*; Getty Conservation Institute: Los Angeles, CA, USA, 2010.
33. Barnes Gordon, L.; Rouhani, B.; Cuneo, A.; Penacho, S. A methodology for documenting preservation issues affecting cultural heritage in Syria and Iraq. In Proceedings of the Objects, Joint Architecture + Objects, and Joint Objects + Wooden Artifacts Specialty Group Sessions 44th Annual Meeting, Montreal, QC, Canada, 13–17 May 2016; Hamilton, E., Dodson, K., Eds.; Available online: http://resources.conservation-us.org/osg-postprints/postprints/v23/barnesgordon/ (accessed on 24 September 2017).
34. Arches: An Open Source Data Management Platform for the Heritage Field. Available online: https://www.archesproject.org (accessed on 31 August 2017).
35. TerraWatchers: Crowd Sourced Satellite Image Analysis. Available online: http://terrawatchers.org/ (accessed on 31 August 2017).
36. Big Pixel Initiative at UC San Diego. Available online: http://bigpixel.ucsd.edu (accessed on 31 August 2017).
37. Savage, S.; Johnson, A. Terrawatchers, crowdsourcing, and at-risk world heritage in the Middle East. In *Acquisition, Curation, and Dissemination of Spatial Cultural Heritage Data*; Vincent, M., Bendicho, V.M.L.-M., Ioannides, M., Levy, T.E., Eds.; Springer: New York, NY, USA, 2017; in press.
38. Lauricella, A.; Cannon, J.; Branting, S.; Hammer, E. Semi-automated detection of looting in Afghanistan using multispectral imagery and principal component analysis. *Antiquity* **2017**, *91*, 1344–1355, in press. [CrossRef]
39. Bowen, E.F.W.; Tofel, B.B.; Parcak, S.; Granger, R. Algorithmic identification of looted archaeological sites from space. *Front. ICT* **2017**, *4*, 1–11. [CrossRef]
40. Lasaponara, R.; Danese, M.; Masini, N. Satellite-based monitoring of archaeological looting in Peru. In *Satellite Remote Sensing: A New Tool for Archaeology*; Lasaponara, R., Masini, N., Eds.; Springer: Heidelberg, Germany, 2012; pp. 177–193.

41. ASOR Cultural Heritage Initiatives Reports. Available online: http://www.asor-syrianheritage.org/reports (accessed on 31 August 2017).
42. ASOR Cultural Heritage Initiatives Weekly Report 31 (9 March 2015). Available online: http://www.asor-syrianheritage.org/syrian-heritage-initiative-weekly-report-31-march-9-2015 (accessed on 31 August 2017).
43. ASOR Cultural Heritage Initiatives Weekly Report 34 (30 March 2015). Available online: http://www.asor-syrianheritage.org/syrian-heritage-initiative-weekly-report-34-march-30-2015 (accessed on 31 August 2017).
44. ASOR Cultural Heritage Initiatives Weekly Report 36 (13 April 2015). Available online: http://www.asor-syrianheritage.org/syrian-heritage-initiative-weekly-report-36-april-13-2015 (accessed on 31 August 2017).
45. ASOR Cultural Heritage Initiatives Report on the Destruction of the Northwest Palace at Nimrud. Available online: http://www.asor-syrianheritage.org/report-on-the-destruction-of-the-northwest-palace-at-nimrud (accessed on 31 August 2017).
46. ASOR Cultural Heritage Initiatives Weekly Report 55–56 (18 August 2015–1 September 2015). Available online: http://www.asor-syrianheritage.org/asor-cultural-heritage-initiatives-weekly-report-55-56-august-18-2015-september-1-2015 (accessed on 31 August 2017).
47. ASOR Cultural Heritage Initiatives Weekly Report 57–58 (2 September 2015–15 September 2015). Available online: http://www.asor-syrianheritage.org/asor-cultural-heritage-initiatives-weekly-report-57-58-september-2-2015-september-15-2015 (accessed on 31 August 2017).
48. ASOR Cultural Heritage Initiatives Special Report: Update on the Situation in Palmyra. Available online: http://www.asor-syrianheritage.org/special-report-update-on-the-situation-in-palmyra (accessed on 31 August 2017).
49. ASOR Cultural Heritage Initiatives Special Report: The Recapture of Palmyra. Available online: http://www.asor-syrianheritage.org/4290-2 (accessed on 31 August 2017).
50. ASOR Cultural Heritage Initiatives New Damage in Palmyra Uncovered by ASOR CHI. Available online: http://www.asor-syrianheritage.org/new-damage-in-palmyra-uncovered-by-asor-chi (accessed on 31 August 2017).
51. ASOR Cultural Heritage Initiatives Update Palmyra: New Photographs Detail Damage to the UNESCO World Heritage Site of Palmyra. Available online: http://www.asor-syrianheritage.org/update-palmyra-new-photographs-detail-damage-to-the-unesco-world-heritage-site-of-palmyra (accessed on 31 August 2017).
52. ASOR Cultural Heritage Initiatives Weekly Report 93–94 (11 May 2015–24 May 2016). Available online: http://www.asor-syrianheritage.org/asor-cultural-heritage-initiatives-weekly-report-93-94-may-11-2016-may-24-2016/ (accessed on 31 August 2017).
53. BBC News: Battle for Mosul: Iraq PM Abadi Formally Declares Victory. Available online: http://www.bbc.com/news/world-middle-east-40558836 (accessed on 31 August 2017).
54. Haider Al-Abadi: From the Old City We Announce the Liberation of Mosul and Remember the Heroic Sacrifices of Our Armed Forces and Their Families. Available online: https://twitter.com/HaiderAlAbadi/status/884464192023138304 (accessed on 31 August 2017).
55. U.S. Department of Defense. Iraqi Forces Liberate Mosul from ISIS. Available online: https://www.defense.gov/News/Article/Article/1242101/iraqi-forces-liberate-mosul-from-isis/source/GovDelivery (accessed on 31 August 2017).
56. The New York Times: Iraqi Prime Minister Arrives in Mosul to Declare Victory over ISIS. Available online: https://www.nytimes.com/2017/07/09/world/middleeast/mosul-isis-liberated.html (accessed on 31 August 2017).
57. Reuters: Iraqi PM Declares Victory over Islamic State in Mosul. Available online: http://www.reuters.com/article/us-mideast-crisis-iraq-mosul-idUSKBN19V105 (accessed on 31 August 2017).
58. Reuters: Iraq and Allies Violated International Law in Mosul Battle: Amnesty. Available online: https://www.reuters.com/article/us-mideast-crisis-iraq-civilians-idUSKBN19W0CR (accessed on 31 August 2017).
59. Reuters: Islamic State Makes Desperate Stand in Mosul, Commanders Say. Available online: http://www.reuters.com/article/us-mideast-crisis-iraq-mosul-idUSKBN19R2K1 (accessed on 31 August 2017).
60. Reuters: Old City Bears the Brunt of Islamic State's Last Stand in Mosul. Available online: https://www.reuters.com/article/us-mideast-crisis-iraq-destruction-idUSKBN19V20I (accessed on 31 August 2017).
61. ASOR: A Reflection on Three Years of Occupation by ISIL. Available online: http://www.asor.org/news/2017/07/3rd-anniversary-isil (accessed on 31 August 2017).

62. ASOR Cultural Heritage Initiatives Weekly Report 32 (16 March 2015). Available online: http://www.asor-syrianheritage.org/syrian-heritage-initiative-weekly-report-32-march-16-2015 (accessed on 31 August 2017).

63. ASOR Cultural Heritage Initiatives Weekly Report 39 (5 May 2015). Available online: http://www.asor-syrianheritage.org/syrian-heritage-initiative-weekly-report-39-may-5-2015 (accessed on 31 August 2017).

64. ASOR Cultural Heritage Initiatives Weekly Report 41 (19 May 2015). Available online: http://www.asor-syrianheritage.org/syrian-heritage-initiative-weekly-report-41-may-19-2015 (accessed on 31 August 2017).

65. ASOR Cultural Heritage Initiatives Weekly Report 42–43 (2 June 2015). Available online: http://www.asor-syrianheritage.org/syrian-heritage-initiative-weekly-report-42-43-june-2-2015 (accessed on 31 August 2017).

66. ASOR Cultural Heritage Initiatives Weekly Report 47–48 (7 July 2015). Available online: http://www.asor-syrianheritage.org/syrian-heritage-initiative-weekly-report-47-48-july-7-2015 (accessed on 31 August 2017).

67. UNITAR: Damage Assessment of Old City, Mosul, Ninawa Governorate, Iraq (30 June 2017). Available online: http://www.unitar.org/unosat/node/44/2615?utm_source=unosat-unitar&utm_medium=rss&utm_campaign=maps (accessed on 31 August 2017).

geosciences

MDPI

Article

From Above and on the Ground: Geospatial Methods for Recording Endangered Archaeology in the Middle East and North Africa

Louise Rayne [1,*], Jennie Bradbury [2], David Mattingly [1], Graham Philip [3], Robert Bewley [2] and Andrew Wilson [2]

[1]   School of Archaeology and Ancient History, University of Leicester, University Road, Leicester LE1 7RH, UK; djm7@leicester.ac.uk
[2]   School of Archaeology, University of Oxford, 1-2 South Parks Road, Oxford OX1 3TG, UK; jennie.bradbury@arch.ox.ac.uk (J.B.); robert.bewley@arch.ox.ac.uk (R.B.); andrew.wilson@arch.ox.ac.uk (A.W.)
[3]   Department of Archaeology, University of Durham, South Road, Durham DH1 3LE, UK; graham.philip@durham.ac.uk
*   Correspondence: ler14@leicester.ac.uk; Tel.: +44-011-637-36248

Received: 31 July 2017; Accepted: 28 September 2017; Published: 5 October 2017

Abstract: The EAMENA (Endangered Archaeology of the Middle East and North Africa) project is a collaboration between the Universities of Leicester, Oxford and Durham; it is funded by the Arcadia Fund and the Cultural Protection Fund. This paper explores the development of the EAMENA methodology, and discusses some of the problems of working across such a broad region. We discuss two main case studies: the World Heritage site of Cyrene illustrates how the project can use satellite imagery (dating from the 1960s to 2017), in conjunction with published data to create a detailed set of database records for a single site and, in particular, highlights the impact of modern urban expansion across the region. Conversely, the Homs Cairns case study demonstrates how the EAMENA methodology also works at an extensive scale, and integrates image interpretation (using imagery dating from the 1960s to 2016), landuse mapping and field survey (2007–2010) to record and analyse the condition of hundreds of features across a small study region. This study emphasises the impact of modern agricultural and land clearing activities. Ultimately, this paper assesses the effectiveness of the EAMENA approach, evaluating its potential success against projects using crowd-sourcing and automation for recording archaeological sites, and seeks to determine the most appropriate methods to use to document sites and assess disturbances and threats across such a vast and diverse area.

Keywords: archaeology; cultural heritage; Middle East; North Africa; remote sensing

1. Introduction

As a result of innovations in open source geospatial and database technologies and software, archaeologists can now collect and analyse data at unprecedented scales e.g. [1,2]. As Hritz [3] (p. 229) recently pointed out, these developments have also enabled us to develop strategies to ensure better documentation and management of landscapes that are under threat or rapidly disappearing. Despite these advances, access to the data, technology and software required to query, analyse and manage threats is very uneven across the Middle East and North Africa (MENA) region.

The Endangered Archaeology of the Middle East and North Africa (EAMENA) Project [4,5] is documenting archaeological sites and the threats posed to them in an online database that spans 20 countries (an area of roughly 10,000,000 km², see Figure 1). The project uses two main methodological approaches, both designed to promote the recording, protection and understanding of cultural heritage at risk across the MENA region as a whole. First, we focus on accessible, user-friendly and open-source remote sensing technologies and tools. Second, we seek to enhance our data and understanding of risk/damage to sites with more specific analyses using high-resolution data where possible.

EAMENA is a collaborative project between the Universities of Oxford, Leicester and Durham, directed by a group of archaeologists with significant experience of remote sensing and archaeological survey in the MENA region, and supported by a team of post-doctoral researchers who undertake data entry, remote sensing analysis and prepare fieldwork based studies, and who will deliver training. To date, the project has focused on:

- the construction of our database, using the open-source Arches software, and the creation of over 150,000 records
- the detailed analysis of specific causes of damage to archaeological sites in the MENA region
- the initial stages of our training programme [6].

In a second phase of the project, we will develop a series of intensive training courses in the EAMENA methodologies to be attended by heritage professionals from eight MENA region countries.

This article explores the underlying methodological approaches adopted by the EAMENA project. We discuss how EAMENA focuses on the production of accurate and accessible data by applying well established techniques to promote standardisation and replicability; ensuring openness, ease of training, and adoption across the MENA region as a whole. We evaluate this methodology alongside other geospatial methods for heritage recording such as crowd-mapping and automation. The challenges of measuring and dealing with uncertainty are also addressed.

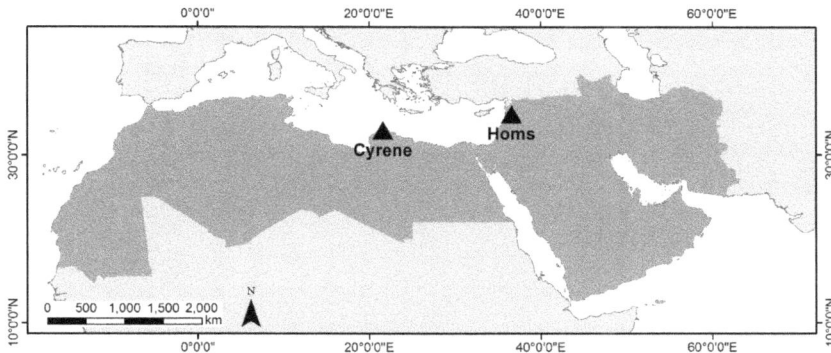

Figure 1. Endangered Archaeology of the Middle East and North Africa (EAMENA) study area (highlighted in grey) and the location of the case studies discussed in this paper.

Remote Sensing and Heritage Recording in the MENA Region

The use of historical aerial photography and satellite imagery has a considerable legacy in the MENA region, developing from the work of Poidebard [7], Stein [8], and others in the early 20th century. Recent projects have revisited historical aerial images [9–11], and conducted new programmes of image collection (for example, the APAAME (Aerial Photographic Archive for Archaeology in the Middle East) project [12–14]). The use of these resources, alongside drone photography, photogrammetry and satellite imagery analysis, is now fairly commonplace. For Middle Eastern landscapes in particular, the declassification in the 1990s of Cold War satellite photography collected in the 1960s–1970s revolutionised this sub-field. This facilitates the mapping of features, especially as many sites have been damaged or destroyed during phases of agricultural and urban expansion in the last 40 years [15–19]. In North Africa, projects focusing on Libya initially made use of the Landsat sensors which have been collecting data since the 1970s (e.g., the Libyan Valleys Project, [20]).

The greater availability of high-resolution modern satellite data since the early 2000s, such as Spot 5/6 and Ikonos, and more recently from sensors with spatial resolutions as high as 0.30 m, such as WorldView 3 and 4, has also allowed projects working in the MENA region to undertake detailed recording of archaeological sites across discrete sub-regions [5,11,15,18]. While these data enable the mapping of complex features to be undertaken, their high cost is prohibitive for most archaeological projects. Free data, such as Google Earth, have allowed the mapping of more extensive areas and have been widely used by archaeologists to identify sites (for example by the Fragile Crescent Project in the Middle East (Durham), and the Trans-Sahara Project in North Africa (Leicester)).

There has also been a growing awareness of the potential of remote sensing to detect and monitor damage and disturbances to archaeological sites and thus a growing emphasis on its use for these purposes [21–23]. Archaeologists in this region increasingly rely on space-borne data to give a wide-scale view of heritage. For example, projects have made use of imagery offering a wide spectral range, for example mapping causes of damage using the multispectral properties of datasets such as Landsat, Sentinel, and higher resolution images (at a cost). SAR (Synthetic Aperture Radar) is also now being used by archaeologists to map problems such as looting (for example [23]).

Heritage projects are currently using several different methods to populate their databases quickly and efficiently in the face of the huge geographical areas involved. For example, photogrammetry and crowd-sourced images are being used to reconstruct the proportions of specific sites (e.g., Curious Travellers [24]), whilst other projects utilise a combined approach, including GIS and remote sensing, to study specific sites or regions (e.g., ASOR (American Schools of Oriental Research) [25]). Crowd mapping projects have developed exciting and innovative training packages to go along with their calls for help (see [26,27]). Crowd-source mapping takes advantage of the easy availability of appropriate technology to turn non-specialists into 'citizen sensors' of geospatial data [28], including information about archaeological sites. It also allows the rapid production of huge amounts of data, though local knowledge and repeated error checking/correction are necessary quality assessment measures [29,30]. There are some problems with using these methods, including errors caused by limited training of mappers, lack of authentication and standardisation, and unequal access to the necessary technology [28]. Before any interpretations of the data can be made, some kind of validation of its quality is necessary. Indeed, from our experience, techniques of image interpretation applied to archaeology have to be learnt over a considerable period of time, practised, refined, applied and re-applied to different areas and environments. Identification of sites based on automated and machine-learning methods to locate the spectral signatures of archaeological sites have also been explored in recent years [31–34]. These methods rely on detecting particular materials identified by a detailed understanding of the nature of archaeological deposits (such as their spectral properties and the shapes of features) in any given location. Menze and Ur [34], for example, used multispectral image-based classification of soils they interpreted as archaeological, and classification of mounded features using DEMs (Digital Elevation Models), to identify tell sites in Syria. While effective at recognising this specific type of site, other archaeological features that exist in the same landscapes are not so easy to classify using algorithms. Moreover, whilst their multispectral image classification was correct 73–97% of the time, when compared with field data, it also produced false positives, identifying features which turned out not to be sites [34] (pp. 781–782). Bennett et al. [33] suggest that automatic feature recognition performed by a computer is arguably more objective than human interpretation. The rapidity of recording that machine learning allows for is also significant, and given the accelerating loss of archaeology that EAMENA is identifying, is a factor to be taken into consideration. There are, however, also some good reasons for skepticism [17].

As with crowd-mapping methods, validation using field data and/or manual checking of the results is required. It can also be difficult to obtain affordable high-resolution satellite imagery covering large areas, while to be really effective high-resolution imagery and elevation data are needed. Another significant issue is the assumption that archaeological sites have standardised, homogenous spectral signatures: Beck et al. [15] dispute this idea. Instead, they argue that spectral properties are a representation of a wide variety of structural forms, building materials, soil and geological conditions, especially across a region as large as the MENA area, and that contrast, i.e., difference from background, is an important determinant of ease of detection [35]. This can lead to the omission of important features, especially when relying solely on automation. Automated detection algorithms also need building and checking by remote-sensing experts, with adaptations for each different region required.

The scale of heritage recording initiatives again varies, from highly detailed assessments of single sites (e.g., ASOR [25], Tapete et al. 2016 [23]), to initiatives similar to our own, which are trying to assess and collate data on disturbances and threats to archaeological sites across the entire MENA region, from Mauritania to Iran [5]. EAMENA is the only project, however, specifically taking an open-source approach using trained interpreters to record the whole region systematically. Whilst there are a number of different projects undertaking similar work we seek, where possible, to minimise any duplication or repetition of data collection. In an effort to avoid duplication, the dataset that was created through the French-British collaborative programme "Historic Environment Record for Syria" is currently being prepared for inclusion within the EAMENA database, while the project is working in collaboration with groups such as SHIRIN (Syrian Heritage in Danger) [36] and the ASOR Syrian Heritage Initiative to encourage the exchange and sharing of data when appropriate. However, the need to work closely with in-country heritage organisations, who may view archaeological inventories as a national resource, requires caution in a situation where digital data can be transferred onwards at the click of a mouse. Even in cases where specific sites or locations are repeatedly analysed by different projects, this work does not devalue the overall goals or success of our project; rather, it opens up opportunities for collaboration and enhancement of data in order to maximise the protection of these sites.

2. Materials and Methods

EAMENA's interdisciplinary, remote-sensing driven methodology has been developed from techniques employed by previous archaeological projects in the MENA region; the Trans-Sahara Project e.g., [11], the Fragile Crescent Project e.g., [1,2], and APAAME e.g., [13,14] amongst others [37]. Our image interpretation methodology, which primarily relies on Google Earth and Bing maps, feeds directly into user-friendly and standardised data entry, ultimately facilitating on-going and future recording of archaeology across the whole MENA region. In addition, the EAMENA project also undertakes detailed assessment and analysis of damage using high-resolution satellite and aerial data for selected areas [5,38]. By doing this, we are able to attain a greater understanding of the main types of damage affecting archaeology and identify the kinds of modern activities that most threaten archaeological sites.

Recording across such an extensive region presents several challenges. A key issue is the need to develop an approach which is consistent. As we will discuss in more detail below, the ways in which different researchers and specialists interpret and record the archaeological record, and in particular interpret aerial/satellite imagery, vary. These difficulties are compounded by the fact that image-interpreters also have to take into account massive geological/environmental variations across the region, which can limit or enhance the visibility of archaeological features and disturbances. The terminologies used to describe archaeological sites also need to be standardised and to account for local variations (for example, the multiple uses of the term *Qasr/Qsur* for a range of mostly fortified sites) across the whole MENA region.

2.1. Datasets Used by EAMENA for Identifying Sites and Mapping Change

One key factor which helps to make our methodology replicable across different parts of the MENA region is our use of open-source data and software. We principally use imagery that is freely available via Google Earth and Bing maps. Importantly, these platforms are accessible in the majority of the MENA region countries, and offer a range of images representing different dates of acquisition. Recording features from a satellite image is subjective, with visibility dependent on both ground and atmospheric conditions at that particular moment in time [15]. EAMENA's use of sources of data which offer a range of images increases the possibility of recording a site, even where ground survey data is lacking. This also allows for a process of validation to be undertaken where the initial data needs checking e.g., see [39]. The successful identification of many sites that were previously unknown (e.g., many cairn fields across the MENA region), was aided by the use of multiple images. For example, the use of imagery from different years or seasons meant that ground observation could take place under multiple types of crop-cover, or varied levels of soil moisture.

Given the inherent subjectivity of image-interpretation, EAMENA has developed methodologies to guide analysts through the decision-making process. Through comparison with existing digitised datasets, analysts are able to make interpretations about what the features observed via imagery might represent. Figures 2 and 3 illustrate the typical decision-making process that an analyst will go through, and how they assess and interpret visible feature types.

What features can be identified at this location?

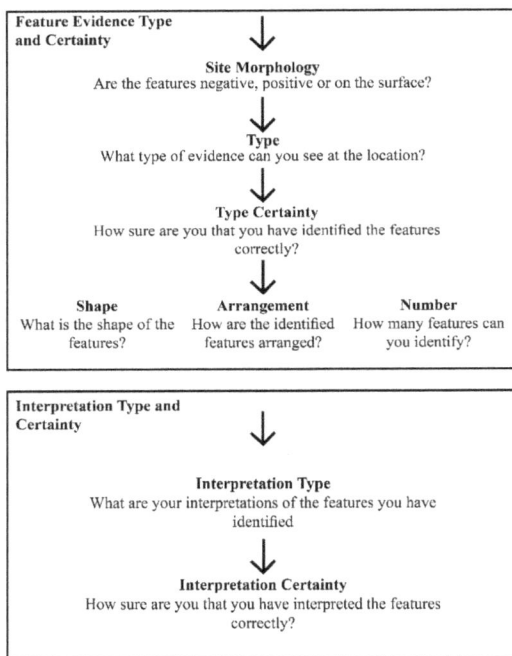

Figure 2. Flow chart of decision making process-draft described above.

Imagery is first examined systematically by trained analysts and recorded using a system based on geographic longitude and latitude and quarter-degree grid squares (each covering roughly 20 km × 30 km). Recording including drawing lines, points and polygons is done within Google Earth and using data which can be imported into GIS packages such as ArcMap and QGIS (e.g., orthorectified satellite images, data available as basemaps etc). Users record any potential features of archaeological interest recognised from these datasets by marking their location before setting up a database record. The parameters which describe the feature, including shape, arrangement, and morphology are then entered, as well as more complex interpretations dealing with form, function and date [38] (Figure 2).

Figure 3. The EAMENA Interpretation of sites and features, using example terminologies from our Evidence Type and Interpretation glossaries. (**a**) In this case the team have identified a tell site (Syria) and (**b**) a qanat/foggara system (Morocco).

One obvious weakness with the methodology described above is the limited time depth allowed by this approach; high resolution imagery available via Google Earth rarely pre-dates 2004, impacting our ability to interpret or identify sites disturbed or destroyed prior to this date. Coverage and availability of high resolution imagery via Google Earth and Bing can also vary across the region. Moreover, in some cases, sites are not visible in any kind of imagery, but have been recorded by published surveys. In other instances, sites have been completely destroyed and historical satellite imagery or field reports are the only remaining sources of information. As demonstrated by Cunliffe [22], damage to archaeology in the MENA region has been taking place over a long timescale. In order to understand when changes might have occurred (or may take place in the future), EAMENA, where possible and cost effective, also uses a range of other freely available and purchased satellite and aerial imagery (Table 1). This allows our mapping in some areas to extend back to the 1940s–1950s. For example, we use historical aerial images held in archives, such as that of the Society for Libyan Studies in Leicester and those freely available on the APAAME website [12]. Hard copies are scanned and georeferenced using appropriate camera models, so that they can be integrated into a GIS and directly compared with more recent imagery.

Table 1. Image datasets used by EAMENA.

Dataset	Spatial Resolution	Examples	Available Dates of Image Acquisition
Aerial photographs	Multiple/unknown	Hunting Aerosurvey images	1930s–2017
Declassified Satellite imagery	2 m–8 m	KH4B, KH4A, KH7	1960s–1970s
Modern low-resolution satellite imagery	10 m–60 m	Landsat 4–8, Sentinel-2	1972–2017
Modern high-resolution satellite imagery	0.3 m–2 m	Pléiades, WorldView, GeoEye	1999–2017

We also make use of declassified satellite imagery (KH7, KH4B) collected for surveillance purposes during the Cold War, which is a useful resource for mapping archaeological features in the condition they were in during the 1960s–1970s [40–43]. With accessibility and open access again in mind, the online Corona Atlas and Referencing System is an important resource [17]. Additional images can also be purchased at a relatively low cost from United States Geological Survey (USGS).

Higher resolution recent digital satellite imagery is particularly useful for more arid areas, where the outlines of many archaeological sites are still visible as standing structural features. EAMENA is using higher resolution imagery from sensors such as the WorldView satellites, the Pléiades constellation, and GeoEye to sample locations in North Africa and the Middle East. Georeferencing is straightforward using sensor models bundled with the imagery, and in some cases high resolution multispectral imagery can be obtained. The main limitation of these data is their high cost which prohibits extensive use of the original images for archaeological purposes (for example, as of 2017, WorldView-3 imagery cost $18 per km^2). However, there is a growing body of material that can be viewed free of charge via Google Earth and used for simple image interpretation. Lower-resolution multi-spectral imagery such as Landsat can be used to map land-cover and land-use since the 1970s. For many locations, comparing all these different images allows changes to a site over a period of at least 50–70 years to be recorded and analysed (see below); the opportunities this offers for future heritage management and conservation should not be underestimated. When deciding whether to purchase new satellite imagery, an assessment of the cost effectiveness and also the potential added value has to be considered. The project does have a small budget to cover the purchase of new areas of satellite imagery, but where possible we aim to use as many freely available or low costs sources as possible. This means that our methods can be reproduced even where there are funding constraints.

Although automated detection of sites is not appropriate to our aims, at a coarser scale EAMENA is utilising semi-automated methods to explore the main threats posed by modern landuse [38]. Importantly these approaches need to be capable of mapping threats across large regions. A standard way of mapping landuse is by the semi-automated classification of multispectral satellite images.

As a starting point, the EAMENA project uses the multispectral properties of imagery including Landsat and Sentinel-2 to map land-use. To identify irrigated cultivation automatically, for example, we have used vegetation index algorithms applied to Landsat images covering a period from the 1970s to the present day. The images were obtained via Google Earth Engine or directly from USGS and GLCF (Global Land Cover Facility) and processed to represent Top-Of-Atmosphere Reflectance. The algorithms used were NDVI (Normalised Difference Vegetation Index) and SAVI (Soil Adjusted Vegetation Index). Based on the properties of vegetation in different spectral bands of the imagery these algorithms identify pixels in the Landsat images most likely to contain vegetation e.g., see [44]. In arid areas (much of our study area) these represent irrigated crops. By performing the SAVI algorithm for multiple images of the same location of different dates we can quantify and measure how the agricultural area has increased over time and identify when any areas of archaeological sites could have been damaged. We are also examining the impact of urban expansion on archaeological sites and have applied algorithms such as NDBI (Normalised Difference Built-up Index) and change analyses to measure modern settlement growth (see Section 3.1 below).

We applied these methods to several areas, including the oasis of al-Jufra in central Libya [38], collaborating with the Trans-Sahara project of Leicester University. The analysis of the cultivated area using the vegetation indices revealed that by 2017 it had expanded by *c*.9500 ha from an initial *c*.600 ha in 1975. Of around 90 archaeological sites recorded in al-Jufra, 47 had been damaged by modern agricultural activity. EAMENA is now developing a methodology using Google Earth Engine see [45] to apply these methods more widely across the MENA region in order to identify the most significant land-use impacts affecting each area.

2.2. Interpretation and Enhancement of Data and Record Creation

For its database [4] the project uses Arches, a freely available open source platform created by the Getty Conservation Institute and World Monuments Heritage Fund. Arches is a customisable platform and we have modified it for our specific project requirements. Data entry can be carried out either manually, or via bulk upload. Our analysts prepare data for batch-upload, or enter them directly into the EAMENA database. With either approach there are important control mechanisms (e.g., standard terminologies and glossaries using drop-down menus) that encourage analysts to check through their data in terms of consistency and accuracy. Once loaded into the database, records can then be further enhanced. Most of the fields in the database use standardised terminologies derived from drop-down menus, and free text fields are used only when absolutely necessary (e.g., for toponyms etc.). This ensures that data are consistent and comparable and thus searchable, even when the database is translated into other languages: it is possible to identify 'equivalent' terms in different languages on a one-to-one basis. Doing so has facilitated the translation of the database into Arabic, and the production of Arabic-language training and support materials.

All EAMENA staff and volunteers are trained image interpreters and there is continual discussion about the identification of both archaeological sites. As Casana [17] (pp. 226, 228, 230–231) highlights, large training samples, weeks of training and, where possible, a first-hand understanding of local settlement histories, archaeologies and environments are all important tools that an image interpreter will be required to use within their work.

In addition to identifying and interpreting archaeological sites, we have developed an approach to assessing damage and threats, both remotely and on the ground. Our analysts first make an assessment of the condition of the site, and the percentage area that has been affected by anything classed as a disturbance. It is important to differentiate between these two variables. A site submerged under the centre of a lake may be 91–100% disturbed (that is, totally covered by the lake), but it may still be classed as being in "Good" condition. Conversely, partially submerged sites located at the edge of a lake, may be recorded as 31–60% disturbed, yet due to their location, at the active edge of the lake, may be classified as in "Poor" overall condition.

Specific disturbance events are also identified via imagery and recorded in the database, including not only the cause of disturbance, but also any identifiable effects. For example, we may identify the cause of the disturbance as inundation, with the possible effects of this including erosion, compaction, waterlogging, as well as structural collapse. Using imagery of multiple dates we can record temporal information about when different disturbance events took place (Figure 4).

Figure 4. Al-Hasakah, Syria. (**a**) The site of Tell Abu Hufur in 1990; (**b**) The site has since been inundated by the West Hasakah dam. However, due to droughts in the area in 2013, the waters receded and the site was again exposed, causing erosion to the deposits.

We also record any identifiable threats and indicate the likelihood of these threats being realised. For example, construction would be recorded as a probable threat for an archaeological site currently located on the edges of a modern town, as with several ancient cemeteries surrounding the Libyan site of Cyrene (record EAMENA-0116807). In contrast, if the nearest town or settlement is several kilometres away from a site, construction may not be considered an imminent threat. Identifying the causes and effects of specific types of disturbance based on remote sensing can be done rapidly, but it does have potential challenges and drawbacks. Causes and/or effects can be wrongly identified or attributed, and in some cases, depending on the resolution of the imagery, certain causes might not be identifiable at all (see case studies for further discussion).

Geospatial interpretations are not without uncertainties, and can be error-checked [1,46–50]. Our project has therefore integrated the concept of 'certainty', into its data-recording models. Using set terminologies (definite to negligible), analysts can indicate how confident they are that something is, for example, archaeological (this is most obviously an issue with potential sites that are identified from imagery alone, i.e., for which there is no confirmation through ground observation) (Figure 5), rather than natural or modern; or how precisely it is located in terms of geographical space and/or correctly interpreted in terms of archaeological categorisation. Certainties can be assigned to locations and extent. This is especially useful in the case of information recorded during field survey where paper maps used were imprecise, the data was collected before accurate GPS data could be gathered (i.e., before selective availability was turned off in 2000), or where the locations recorded are simply incorrect. Building the concept of 'certainty' into our data recording methodology provides an important tool for both researchers and heritage specialists alike, and in particular those who may work with, and seek to refine this information some way into the future. For example, for researchers certainty can also be a tool through which to test hypothetical data extrapolations [1] (pp. 1008–1009), while for heritage specialists certainties can be used as a way to prioritise management and intervention strategies. The various issues outlined above are important if the EAMENA database is to provide an initial basis for Historic Environment Records (HER), which should help national heritage agencies to record, manage and protect cultural heritage in the future.

Figure 5. Different examples of site types and archaeological certainty. All sites have been identified and classified from imagery as (**a**) a site where there is a high likelihood of the feature being archaeological; (**b**) a site where there is a medium likelihood of the feature being archaeological; (**c**) a site where there is a low likelihood of the feature being archaeological; (**d**) a site where there is a negligible likelihood of the feature being archaeological.

3. Results

As the number of trained and authorised users throughout the MENA region increases, the EAMENA database has the potential to be mined for the analysis of large-scale patterns by researchers and policy makers. The concept of "big data" is currently a fashionable topic in many sectors, facilitated by technological advances, and has already been applied to archaeological research [51]. The term "big data", the origins of which are unclear, and its exact parameters hard to define, describes the huge collections of digital data created and stored by any particular organisation, which often transcend normal software and analysis methods, and which offer immense potential for research [52,53].

To date, our project has created detailed records for over 20,000 sites from a total of *c.*150,000 identified sites with partial records. Of the detailed records, over *c.*20% are previously known sites, documented from published surveys or excavations. A further *c.*65% are sites identified from satellite imagery and classed as having a medium or high certainty of being an archaeological 'site' or "feature". The remaining *c.*15% are those with a low or negligible certainty of being archaeological. The database also contains over 50,000 records providing details about the sources (e.g., satellite imagery, aerial photographs or bibliographic sources) consulted by the project. This work is constantly developing, and the team is currently in the process of evaluating many thousands of potential new archaeological sites.

Table 2 presents initial results for a sample of the site interpretations in our glossary, showing that burial features, enclosures and settlements represent a significant proportion of sites. Unsurprisingly, in view of its primary reliance upon remote sensing methods, the project has recorded far fewer rock art sites or temporary camp sites, as these are predominantly not visible from satellite imagery. There are potential implications for the relative interpretability of different site types; for example, settlement sites represented by a collection of buildings are easy to identify and interpret from a satellite image; temporary camps and rock art sites, on the other hand, are less easy to distinguish, both on the ground and via imagery, and as a result are likely to be substantially under-represented in our data. Over time, as the number of records grows and we integrate more field and published data, our database should facilitate an improvement in the documentation, interpretation, and monitoring of site classes that were previously poorly understood.

Table 2. Examples of resource interpretation types (including site types and features).

Interpretation Type	Database Records
Settlement/Habitation Site	3745
Building	1286
Tomb/Grave/Burial	9844
Enclosure	3978
Temporary Camp	25
Inscription/Rock Art/Relief	206
Temple/Sanctuary/Shrine	170

It is already possible to make interpretations about the impacts of particular disturbance types on cultural heritage. For example, our recording process shows that agricultural activity is one of the main causes of damage to archaeological sites across the region (Table 3). While no great surprise perhaps, it is important that decisions around heritage protection are made on the basis of hard evidence, rather than assumptions, or the publicity around infrequent, but high-profile, events. This includes ploughing and damage from levelling in newly-irrigated areas. Figure 6 shows sites damaged by agriculture superimposed over land-cover classes across the MENA region mapped from MODIS NDVI and VIIRS [54,55] derived using Google Earth Engine. Surviving sites in areas conducive to modern agriculture are at significant risk.

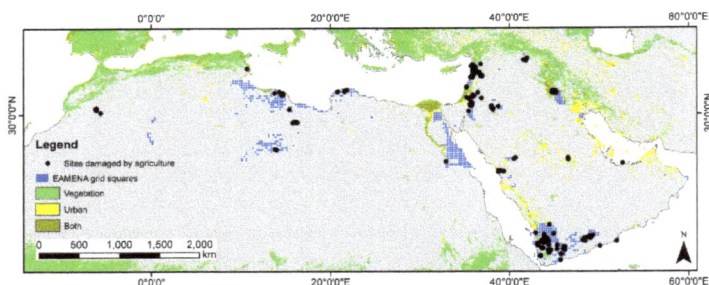

Figure 6. Distribution of EAMENA records with damage caused by agriculture recorded and grid squares with records entered by EAMENA superimposed on land-cover derived from data obtained via Google Earth engine: vegetation (MODIS NDVI, [54]) and areas of human activity based on night-time radiance data (VIRS, [55]).

Table 3. Examples of disturbance types.

Disturbance Type	Database Records
Agricultural/Pastoral	4367
Development	378
Infrastructure/Transport	1228
Industrial/Productive	627
Military	36
Archaeological Excavation	133
Looting	893
Unknown (includes sites where it is not possible to identify the disturbance type, either due to poor imagery resolution, cloud cover, or lack of data)	5886

As our work progresses, we will need to remain critically aware of the implications for researchers of utilising such a large, standardised dataset [53]. Care must be taken to avoid misleading interpolations (e.g., see Fradley and Sheldrick's [56] commentary on Parcak [57]). For example, statistical variations in site density will need to take into account differential preservation levels which depend on landscape contexts. Although recording every possible archaeological site is impossible, as the project continues to develop, we will have to ensure our data are representative of trends across different regions, landscapes and periods. Subjectivity, inherent in many processes of archaeological interpretation, also needs to be taken into account and mitigated against using standardised terminologies.

3.1. Case Study 1: Cyrene: The Impacts of Modern Development on a World Heritage Site and Its Immediate Hinterland

The site of Cyrene in Eastern Libya is a UNESCO (United Nations Educational, Scientific and Cultural Organization) World Heritage site (designation number 190). It faces significant problems arising from present-day activities including expansion of the adjacent town of Shahat and limited enforcement of planning regulations. While the gradual degradation of the archaeology has been an issue for many years, damage to archaeological features has accelerated because of the civil war of 2011 and the subsequent instability.

Cyrene's monuments were first recorded by travellers in the 18th and 19th centuries [58], with more extensive archaeological investigations over the course of the 20th and 21st centuries e.g., [59–65]. Cyrene developed from a Greek colony in the 7th–4th centuries BC, with occupation continuing through the Roman and Byzantine periods [66]. Located beside the modern town of Shahat, Cyrene has a walled circuit although much of the ancient city is outside this area, including large suburban cemeteries (Table 4) and sanctuaries (Figure 7). The site lies on the edge of an escarpment, 8 km from the coast, and is surrounded by arable fields and modern farms. The urban core is fenced and protected, but the suburban zones are vulnerable to a variety of threats [67].

Table 4. Cemeteries of Cyrene. Details from Cassels' notebooks [68].

Location	Approx. Number of Tombs
Northern necropolis	422
Southern necropolis	423
Western necropolis	158
Eastern necropolis	267

Despite current difficulties of access for foreign archaeologists, approaches that combine remote sensing and GIS survey undertaken by Libyan archaeologists have highlighted the severity of the threat (for example, [69]). Several recent projects concerned with heritage protection have examined the risks faced by Cyrene and worked to document it; their published reports have been cited in our database where applicable and instances of damage they describe logged. The Cyrenaica Archaeological Project has undertaken a holistic approach including recording and training [70]. They have noted specific instances of damage, for example caused by weathering and vegetation, also recorded in the EAMENA database. They worked collaboratively with the Department of Antiquities in Shahat to develop a sites and museums database. The Curious Travellers Project [24] is also gathering data to make 3D models of Cyrene using photogrammetric methods.

3.1.1. EAMENA's Methodology for Recording Cyrene

EAMENA has created a detailed set of records describing the site of Cyrene, the nature of the damage and the risks that are affecting it. The site consists of one "parent" record and over 30 sub-records which represent individual structures/features which are part of the overall complex and surrounding features. To build each record, we have used multiple sets of data including aerial photographs and satellite imagery and published and unpublished reports from archaeologists. Details were recorded from these data including the form, morphology, location, interpretation and condition of these features; for example, whether they were of good or poor condition according to the latest information and to what extent they had been impacted on by modern landuse.

A historical analysis of images (Table 5) of Cyrene and its immediate hinterland highlights the impacts that development related expansion has had on archaeology, and the value of examining this over a long period of time (1949–2016). We can map the site and its immediate hinterland in detail using aerial photographs collected by Hunting Surveys dating to 1949 [65]. A KH7 satellite image from 1967 allows further mapping. 39 images on Google Earth dating from 2006–2017 allowed detailed identification of the archaeological features and modern changes. A GeoEye-1 image from 2016 has allowed mapping and spectral analysis. Changes in the size of Shahat over time were made using Landsat images (1986 and 2000) and a Sentinel-2 image (Figure 8). This broad dataset highlights EAMENA's use of a range of sources to populate our database. Some features were detectable in data such as the satellite imagery, but others could only be recorded using the published data. Using both these types of information in conjunction allowed details of instances of damage to be established.

Table 5. Datasets used for recording Cyrene.

Source	Acquisition Date
Aerial photographs, Hunting Aerial Surveys, in the archive of the Society for Libyan Studies	1949
KH7 image, from USGS	1967
39 Google Earth images, variety of unknown sensors	2006–2017
GeoEye-1 image, © DigitalGlobe	2016
Landsat 5 TM image	1986
Landsat 7 ETM+ image	2000
Sentinel-2 image	2017

While many features are visible in the high-resolution satellite images and aerial photographs, there are features which cannot be easily recorded in this way. These include tombs of several different types and morphologies, including rock-cut structures and sarcophagi. Some are located on the slopes of the escarpment and side of wadis, making them particularly invisible to remote sensing methods. There are also specific instances of damage that cannot be identified remotely. This highlights the necessity for EAMENA to use a variety of datasets, where possible backed up by field work. In this case, several sources of published information deriving from surveys, excavations, guides and archival research have been consulted [65] and we have worked closely with a Libyan PhD student at the University of Leicester, Mohamed Omar, who is studying Cyrene's suburbs.

3.1.2. Antiquity to Mid-20th Century AD

Evidence for events which damaged Cyrene prior to the 20th century comes from excavations and historical texts rather than from satellite imagery. Damage to structures in the Mediterranean region were caused in antiquity by earthquakes in the mid-third century AD and in AD 365 [71,72]. These are mentioned in historical sources and confirmed by archaeological and geological evidence [62]. Archaeological excavations, ongoing since the 19th/early 20th century, have disturbed components of the site and have been logged in the EAMENA database as events which may have affected the site's preservation. Restoration and landscaping efforts undertaken during this era have also had a deleterious impact on Cyrene, including tree planting to the north and east of the acropolis during the Italian colonial period [66] (p. 148). The potential effects of vegetation on archaeological features have been noted in our database.

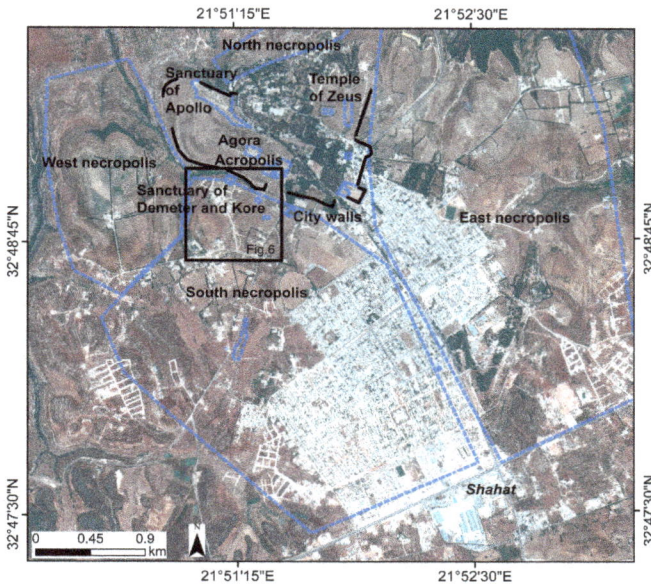

Figure 7. Map of Cyrene showing areas with specific EAMENA records so far. GeoEye-1 image 5 July 2016 © DigitalGlobe, Inc. All Rights Reserved.

3.1.3. Mid-20th–21st Century AD

In addition to other sources, aerial and satellite images can be used to record changes from the first half of the 20th century. The 1949 aerial photographs show that the ancient city and its suburbs were relatively undisturbed by construction and development work at that time, and that Shahat was a small village. It had originally been located on the northern part of the ancient town, but on the advice of the archaeologist Goodchild, its focus was shifted to the south-east, outside the walls [66]. The KH7 image (1967) shows that Shahat had started to expand in its new location by the late 1960s, but the cemeteries and other suburban features still appear to have been relatively unaffected by construction-related work. Since then, however, this area has been particularly at risk, and tomb robbing and vandalism has been recorded by archaeologists from the 1960s onwards [65]. By the 1980s the expansion of Shahat had destroyed most of the Southern Necropolis [66] (p. 151), [68]. That this process is continuing is clearly documented on satellite imagery and confirmed by Libyan archaeologists [69,73]. The recent developments include construction of houses, farms and infrastructure, with evident impacts on structures outside the ancient city walls, especially the cemeteries and the sanctuary of Demeter. Some ancient structures have been bulldozed or otherwise damaged to make way for new constructions, whilst others have been exploited for building materials [69].

The expansion of the present-day town is very apparent on imagery dating from 2006 onwards. Modern roads and farms have encroached on the area of the southern and eastern cemeteries in particular. Structural robbing for building materials is part of this unregulated expansion and is recorded in our database as a source of damage and future risk. Published and unpublished records and reports have also highlighted other issues which affected the site during this period, such as pollution by sewage and rubbish dumping, which are less visible on satellite imagery [69].

3.1.4. Recent Changes

Activities affecting the preservation of Cyrene and its immediate hinterland have accelerated even further over the past five years following the recent conflict, which has seen much illegal and unregulated construction work in Libya [38,67]. The satellite images indicate a further massive increase in the extent of Shahat, for example, demonstrated by a Landsat 5 image (1986) and a Sentinel 2 image (2017) (Figure 8).

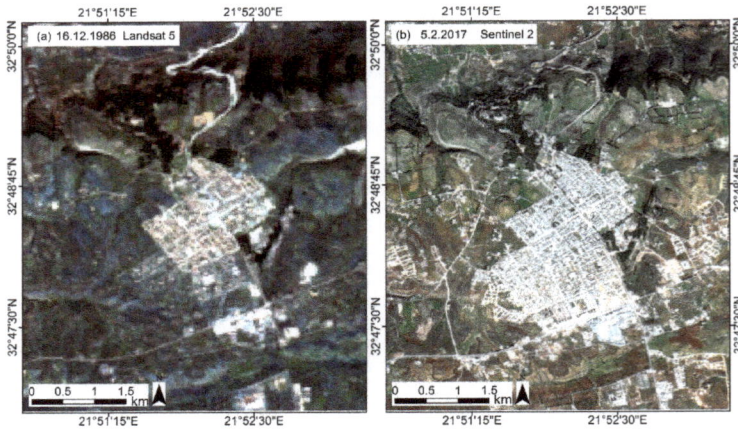

Figure 8. Shahat has expanded significantly in size since 1986. (**a**) A landsat 5 TM image (USGS) 16 December 1986; (**b**) A Sentinel-2 image (ESA), 5 February 2017.

Unsupervised classifications of Landsat and Sentinel-2 images were calculated using ERDAS (Figure 9). Other than a detection of part of the archaeological area on the acropolis, they have picked out pixels representing modern urban activity. These show how the urban area has grown between 1986–2017. The impact of this development was recorded for the sites in our database affected by it. The core area of the town has expanded slightly, especially towards the south-west; however, more dispersed structures and associated infrastructure has spread in all directions, directly threatening the archaeological features in these areas. In the southern necropolis, in particular, there are new farms, buildings and roads. Al-Raeid et al. [69] (pp. 8–9) reported robbing of ancient structures for building materials in this area and the looting and vandalism of tombs. They also noted the effects of continued lack of conservation on these structures. The pattern is similar in the other suburban areas. The impact of processes less identifiable from satellite imagery, including damage caused by vandalism, and water pollution, have also been noted by other sources and are listed by the World Heritage Committee in its most recent documentation [74].

3.1.5. Damage Statistics

The systematic recording by EAMENA of the causes of damage and potential threats allows these problems to be measured. Table 6 presents the results from 38 site records, which were created from interpretation of aerial and satellite images, information from published reports and guides, and from discussions with our Libyan colleagues. It is worth noting that although counted only once here, some of these sites represent large areas containing multiple archaeological features. Table 7 records the proportion of sites recorded as being destroyed, damaged, or of unknown condition.

Archaeological excavations since the 19th–20th centuries have affected at least 24 sites. However, one of the most significant causes of damage to archaeology at Cyrene is modern development which comprises construction of buildings and related infrastructure/transport and utilities (24 sites affected by these categories so far—63% of the records). While the area inside the walls, including the acropolis, the main urban area and the sanctuary of Apollo, is protected from this type of damage, the features outside this zone including the cemeteries are being encroached upon by modern constructions including roads, tracks, farms and houses (Figure 10). This problem has been mapped and recorded across the wider area of Cyrene and Shahat using satellite images showing expansion since the 1960s (e.g., see Figures 8 and 9).

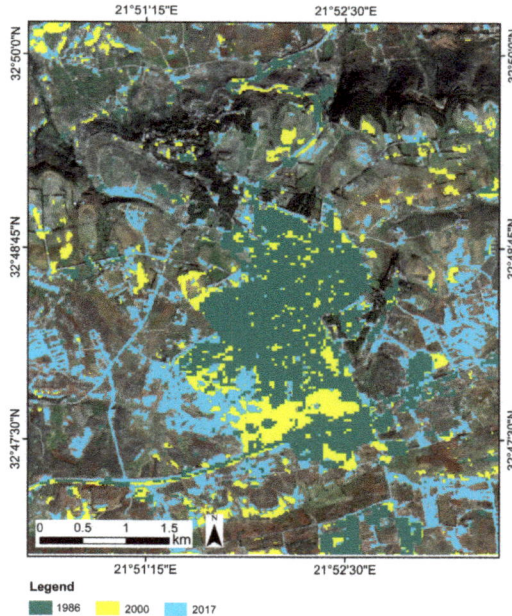

Figure 9. Unsupervised classifications (represented in 3 colours) highlighting the urban areas around Cyrene produced from a Landsat 5 image (1986), a Landsat 7 image (2000), and a Sentinel-2 image (2017). The background is the Sentinel-2 image (2017).

Table 6. Disturbance types logged at Cyrene.

Disturbance Type	Numbers of Sites Affected
Natural	13
Agricultural/Pastoral	7
Development	12
Infrastructure/Transport	8
Utilities	4
Looting	7
Archaeological (e.g., excavations and reconstructions)	24
Unknown	11

Table 7. Condition state of sites logged at Cyrene.

Condition State	Numbers of Sites Affected
Destroyed	0
Poor	8
Fair	22
Good	1
Unknown	7

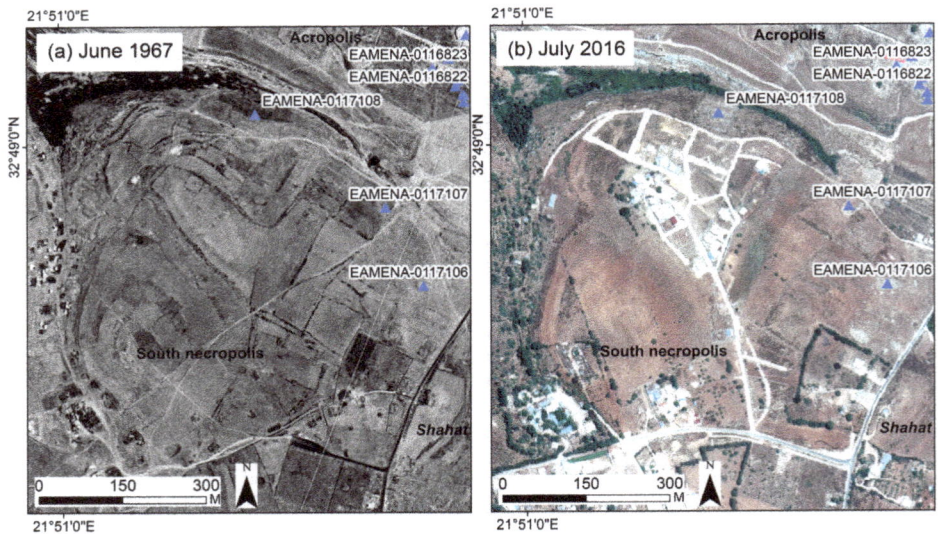

Figure 10. Comparison of (**a**) KH7 (7 June 1967) and (**b**) GeoEye-1 (5 July 2016) images showing damage to the area to the west of Shahat and immediately to the south of Cyrene. GeoEye-1 image 5 July 2016 © DigitalGlobe, Inc. All Rights Reserved.

Although difficult to identify using imagery alone, structural robbing of tombs, as well as deliberate vandalism has affected many sites. Several have been recorded as having been looted (at least seven); a figure that may rise when individual tombs can be logged.

Agricultural activity has also affected the areas surrounding the site. The hippodrome (EAMENA-0116827) has been damaged by long-term agricultural activity including planting and ploughing. Since the recent conflict, the inability to enforce regulations has led to clearing of remains to make way for new fields [67] (p. 156). The category 'natural' is also a significant cause of damage (13 sites so far). This comprises recent issues such as tree growth but also known instances of damage caused by earthquakes in antiquity.

The condition of the sites was recorded using EAMENA terminologies and was assessed using the analysis of the satellite imagery and classifications and the reports of recent visitors. 23 (65%) of the sites could be described as "Good" to "Fair" (Table 7), especially sites nominally protected by their location on the acropolis ridge. This means that they can be regarded as being reasonably stable. However sites surrounding the acropolis were less well preserved with signs of severe structural instability/missing and deteriorating features and were suffering from the consequences of ongoing activity such as structural robbing. In some cases it was not possible to identify the current condition of sites other than noting that they were likely to have been impacted by disturbances.

As described above, EAMENA also records potential threats and risks (Table 8) which could affect archaeological sites in the future. These are recorded based on problems currently affecting sites and analysis of continuing issues in the vicinity. The urban growth identified using the multispectral satellite images (Figures 8 and 9) is an urgent issue. For example, the westward expansion of Shahat is likely to cause further damage to archaeological features in that area, including the tombs of the southern and western cemeteries. Larger features in that zone, including the Sanctuary of Demeter (EAMENA-0117108), are at high risk, for example from structural robbing or even demolition.

Table 8. Potential threats which may affect Cyrene in the future.

Threat Type	Numbers of Sites Affected
Natural	3
Agricultural/Pastoral	10
Development	18
Infrastructure/Transport	7
Utilities	6
Looting	31
Archaeological	0
Unknown	18

Cyrene achieved World Heritage Site status in 1982. The World Heritage Committee has recognised the ongoing threats to the site and have proposed satellite monitoring, field recording, additional security measures and identification of the boundaries of the designated site [74]. However, World Heritage status has not provided tangible protection to Cyrene. Since 2011, often at considerable personal risk, Libyan archaeologists and local people have worked to protect archaeological sites and museums at Cyrene, but so far it has not been possible to enact a solution to the problems [67] (pp. 155–156). Overall, our analysis of multiple datasets shows that while development in the vicinity of Cyrene has been taking place since the 1960s at least, it is now occurring at an especially rapid rate, one that directly threatens surviving features in the hinterland of the site including rock-cut tombs. Archaeological sites close to urban areas should therefore be monitored and recorded as a priority and regular classifications of multispectral imagery performed to track Shahat's growth.

3.2. Case Study 2: Homs Cairns: The Benefits and Challenges of Monitoring Stone Monuments via Remote Sensing

From 2007–2010 a fieldwork project undertaken by one of the current authors mapped and analysed 525 potential burial cairns to the north-west of the modern city of Homs (Syria). This project was undertaken within the framework of the Syrian-British landscape project Settlement and Landscape Development in the Homs Region, and the field data was recorded within its GIS framework. Published overviews of the archaeology of the Homs basalt region in Graeco-Roman [75] and earlier periods [76] contextualise the various monuments in relation to settlement activity and the wider landscape; readers should consult these for further information. Cairns are visible on the ground as piles of stone (Figure 11), and vary considerably in terms of size, structure and form. A well-documented form of monument found throughout the Levant and North Africa, they are also visible via satellite imagery, and can be distinguished as small circular or oval features, in many cases associated with enclosures and other archaeological traces. Additional research as part of a PhD thesis [77,78] identified a further 169,000 potential cairns from an area of *c*.21,000 km^2 (Figure 12) using remotely sensed data spanning the late 1960s to the early 2000s (Corona KH4-B, KH7, historic aerial photographs, Ikonos (panchromatic and multi-spectral). The majority of these (over 90%) were found in association with the local basalt flows to the north and south-west of the modern city of Homs [77,78], whilst a much smaller percentage was found in association with lacustrine marls, limestones, clays and sands.

Figure 11. Image of cairn surveyed in the field in spring 2007. The cairn has a modern shelter constructed on top of it, and areas of structural collapse are visible in the image.

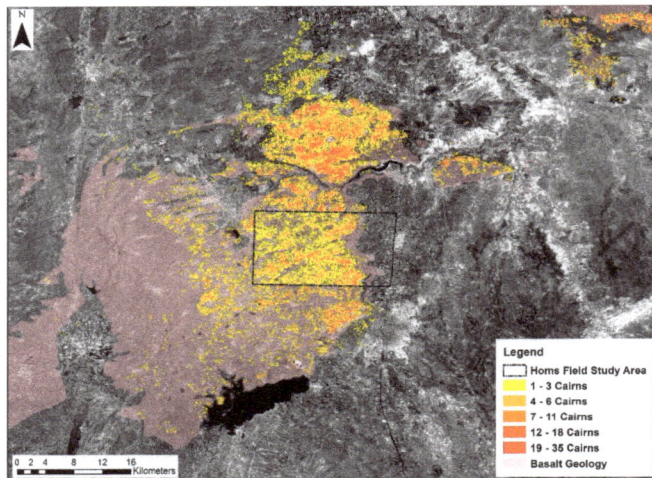

Figure 12. Distribution of cairns identified from remote sensing. Most are found in areas of basalt geology. The area surveyed in the field is indicated by a black rectangle. Cairns are plotted against panchromatic Landsat 7 mosaic (10 April 2005 and 26 June 2007).

Details recorded during the fieldwork included the form, morphology, location and interpretation of these features. These were all collated in a project database, alongside basic information about levels of preservation. For example, a rough measure of "percentage intactness" was recorded for each cairn surveyed in the field (less than 50% intact; more than 50% intact; 100% intact), and notes were made about the potential causes and effects of any identifiable disturbances. A preliminary assessment of recent, pre-conflict, land-use practices (Figure 13), carried out in 2010 using Ikonos panchromatic imagery (from 2002), indicated that over 60% of the archaeological features, including cairns, enclosures and other features, identified from the Corona satellite imagery have been either partly or totally destroyed by clearance or 'de-rocking' operations using heavy machinery, often bulldozers, noted in the field by the authors during fieldwork, with the intention of increasing the cultivable area [79]. The irony is that a practice that was originally supported by development organisations to increase agricultural productivity has been widely adopted at a local level, often on a 'freelance' basis and with little technical or administrative oversight, and now poses a serious risk to the preservation of cultural heritage. Assessments carried out using this imagery, however, also indicated that areas of the study region were still being used for grazing activities and had, as yet, not been cleared or bulldozed.

In 2016 these field records (525 in total) were loaded into the EAMENA database, and updated using the EAMENA methodology. Preliminary disturbance and threat assessments were also recorded for a sample (6975) of the 169,000 potential cairns, identified from satellite imagery, bringing the total recorded from this area in the EAMENA database up to 7000.

In the case of the surveyed cairns, disturbance assessments were generated from the field survey records, which recorded landuse and landcover at the time of data collection. This information was then double checked against the most up to date imagery in Google Earth (2014–2016). For those cairns not visited in the field, a remote characterisation assessment was made, by assessing groups of cairns in relation to their association with different types of landuse.

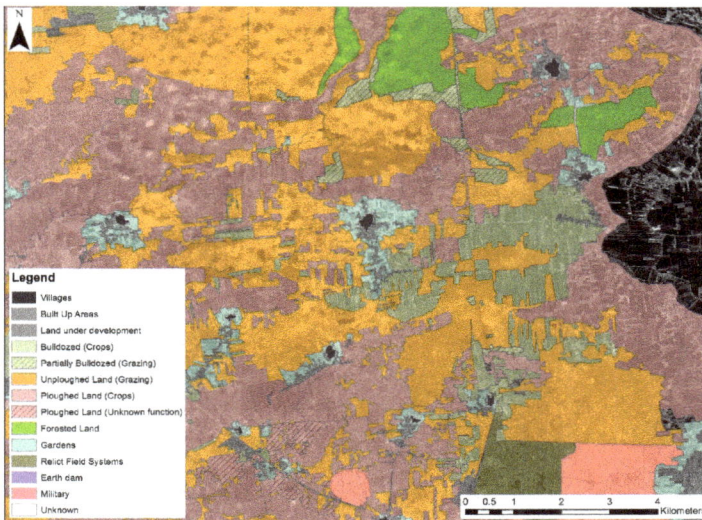

Figure 13. Modern landuse practices based on an assessment of Ikonos panchromatic imagery, acquired 2 March 2002.

As the original 2010 study had indicated, this work demonstrated that whilst 59% of potential cairns showed "No Visible/Known" disturbance causes (Figure 14), nearly 40% were affected by bulldozing or clearance activities (Figure 15). Clearance destroys even substantial surface and sub-surface archaeological features, and creates a "cleared" field, bordered by newly constructed field walls composed of huge basalt boulders, which can easily be identified from satellite imagery. In total 2683 (38%) of digitised cairns were recorded as "Destroyed", while 4159 (59%) were recorded as being in 'Good' condition.

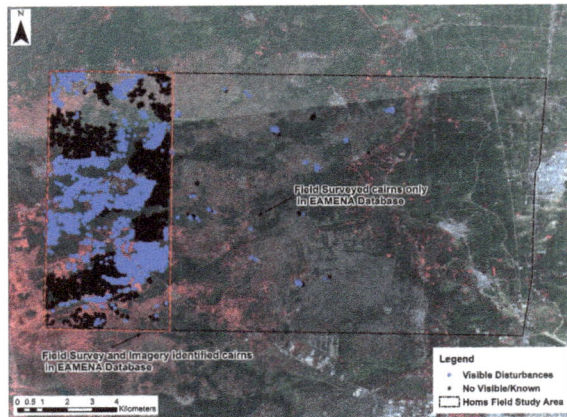

Figure 14. Distribution of disturbed cairns across the study area. Cairns are plotted against a multispectral Landsat 7 image (14 January 2000) processed in Erdas to show landcover. The area marked in red represents the area of cairns identified from satellite imagery, and not visited in the field, which have currently been assessed and recorded in the EAMENA database. Data entry for the cairns identified to the east of this sample area is currently on-going.

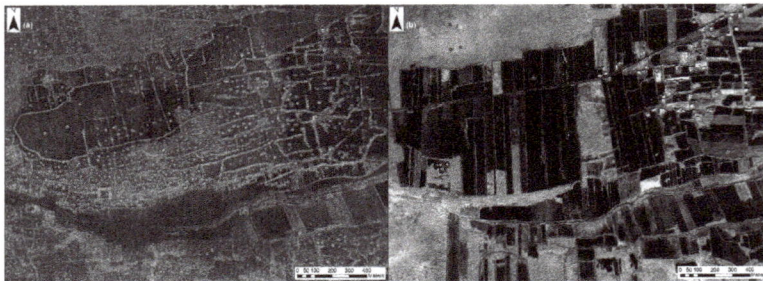

Figure 15. Comparison between Corona KH4-B (17 December 1969) and Panchromatic Ikonos (2 March 2002) showing the changes, and areas of de-rocking and areas that remain 'un-cleared'.

This preliminary and basic assessment has a number of limitations. For example, whilst some cairns identified during ground survey were recorded as being in either a "Fair" or "Poor" condition, the resolution of the imagery means that, more often only two basic condition states can be identified in remote sensing analysis: "Destroyed" or "Good". Using this "broad brush" approach also limits the range of disturbance causes and effects that can be identified. In particular, clearance activities appear as a major disturbance factor. Moreover, the size of the features (generally between 2 m–20 m in diameter), means that the different types of disturbance causes which can be identified from satellite imagery alone are limited. As a result, the number of features affected by other disturbance causes, such as looting, construction and dumping is probably a significant under-estimate.

For example, out of the 104 cairns which were recorded in field survey as having identified disturbance causes, over 70% were affected by recent construction activities, such as the erection of small hides or shelters or modern dumping of cleared material or rubbish. A much lower percentage (*c*.17%) were recorded as having been disturbed by illicit excavations, whether recently or in antiquity, although dumping might have concealed earlier looting activity. Based on the field notes, just under 8% of recorded cairns were associated with a disturbance cause of clearance/bulldozing activities. This low figure is due to a number of factors; firstly, as one of the field survey's main aims was to understand the morphology, chronology and location of these features, the research specifically targeted cairns in areas where they were better preserved, based on 2002 imagery. Thus, while *c.*8% of cairns were categorised as damaged by clearance or bulldozing activities in the intervening five-eight years, based on satellite imagery analysis, this number probably significantly underestimates the overall impact of this disturbance cause at the regional level. It is also apparent that a number of disturbance causes and effects identified by the field survey cannot be identified from satellite imagery alone. By way of example, EAMENA-0059581 was recorded in 2006 as a fully intact, large cairn. The survey returned to the same location in 2007 to find that the feature in question had been illicitly excavated and was most likely not a cairn, but instead a mausoleum dating to the Roman period. The excavation exposed the internal structure of the monument and material, mostly consisting of pottery sherds, was strewn across the area, but no evidence of this disturbance is visible via Google Earth (Figure 16).

Despite these limitations, recent remote sensing analysis allows us to identify broad-scale changes and while the ongoing conflict has rendered these features inaccessible on the ground, we have taken our analysis further. Using imagery from Google Earth, we have been able to revise overall condition assessments for the original surveyed cairns. Out of a total of 525 field-recorded cairns, 127 (24%), required an updated re-assessment, whilst the remaining 398 (76%) showed no significant changes over the seven years since the original study was completed. Unfortunately, the majority of cases involved updates to the disturbance extent and overall condition state. Most required a re-classification of the overall state from "Good", "Fair" or "Poor" to "Destroyed".

Figure 16. EAMENA-0059581 ((**a**) Panchromatic Ikonos 2 March 2002; (**b**) 2007; (**c**) 2008).

Overall, based on these re-analyses, the percentage of cairns listed as '91–100% disturbed' increased to 66% of the sample, with the number of cairns with an 'Unknown' or '1–10% disturbance' extent also increasing (Table 9). This reveals fairly significant changes, with the total number of cairns recorded as showing "91–100% disturbance" increasing from 0 to 85 (16%). Due to the poor resolution of some of the latest available imagery in Google Earth, the number of cairns for which assessment was not possible (e.g., disturbance extent or condition recorded as 'Unknown') also increased.

Table 9. Disturbance Extent (%) based on field survey and remote assessment. The category 'No Visible/Known' includes sites where it was possible to make a disturbance assessment, but no disturbances were visible. 'Unknown' indicates sites were it was **not possible** to make a disturbance assessment due to lack of data, cloud cover and/or poor imagery resolution.

No Visible/ Known	1–10%	11–30%	31–60%	61–90%	91–100%	Unknown	TOTAL
Original Field Assessments for 127 Cairns (2007–2010)							
97	0	18	0	3	0	9	127 cairns
76.4	0.0	4.2	0.0	2.4	0.0	7.1	100%
Updated Remote Assessments for 127 Cairns (2017)							
0	4	1	0	0	85	37	127 cairns
0.0	3.1	0.8	0.0	0.0	66.9	29.1	100%
Unchanged and Updated Assessments for 525 Cairns (2007–2017)							
No Visible/Known	1–10%	11–30%	31–60%	61–90%	91–100%	Unknown	TOTAL
374	10	1	0	0	0	13	Unchanged Assessments (2017)—398 cairns
0	4	1	0	0	85	37	Updated Assessments (2017)—127 cairns
374	14	2	0	0	85	50	TOTAL (2017)—525 cairns
71.2	2.7	0.4	0.0	0.0	16.2	9.5	% of Cairns

The pattern for the overall condition state is very similar (Table 10), with the total number of cairns identified as being in 'Good' condition decreasing from 368 (70%) to 272 (52%) of surveyed cairns. Conversely, the number of 'Destroyed' cairns has increased from 0 (0%) to 84 (16%) during this period.

Table 10. Overall Condition State based on field survey and remote assessment.

Totals	Destroyed	Poor	Fair	Good	Unknown	TOTAL
Unchanged Assessments (2007–2010)—398 cairns	0	46	90	262	0	398
Original Field Assessments (2007–2010)—127 cairns	0	3	18	106	0	127
Updated Remote Assessments (2017)—127 cairns	84	0	1	10	32	127
Old Totals (2007–2010)—525 cairns	0	49	108	368	0	525
Revised Totals (2017)—525 cairns	84	46	91	272	32	525
% from Old Totals (2007–2010) —525 cairns	0	9	21	70	0	100
% from Revised Totals (2017) —525 cairns	16	9	17	52	6	100

This updated analysis also allowed us to confirm, and quantify, a number of disturbance causes which were not originally recorded in the field, although were noted as possibilities. These included evidence for flooding, a disturbance cause that was identifiable through the use of multi-temporal imagery which showed that a number of cairn clusters found in the vicinity of seasonal lakes were likely to have been affected by flooding.

Despite the limitations of using satellite imagery to record and monitor disturbances such as the illicit excavation of stone monuments, this case study illustrates the benefits of using EAMENA's simple remote sensing techniques to continue to monitor monuments and update records in currently inaccessible areas. As the most recent imagery available for this area in Google Earth dates to 2015/2016, it is likely that the disturbance patterns identified here have continued since then.

3.3. Remote Sensing and Field Survey

Field-based validation for many archaeological features in the database may be possible in the long term: the EAMENA database is being made available to individuals and institutions with responsibilities for cultural heritage throughout the MENA region. Its uptake is being facilitated by dedicated training courses and collaborative working. As with any monuments record, database entries can be revisited and updated in the future as necessary.

Over a more immediate timescale, however, we need to ensure that our image interpretation methodology is producing viable data which will help, rather than hinder, the protection efforts of archaeologists in the MENA countries. Ultimately, each filled-in record needs to be a starting point for future detailed recording of site location, ideas and interpretation, and the identification of potential threats. As a cross-check on our methodology we are systematically comparing field-based and remote-based interpretation for select samples of sites. EAMENA is actively collaborating with several projects conducting field survey, for example the Middle Draa Project [80] and Koubba Coastal Survey [81]. Ground survey allows further details about many sites to be added to the database. However, most significantly, it allows us to assess the accuracy of EAMENA's remote-sensing methodology of standardised interpretations and terminologies.

Validation of remote sensing methods by comparing results to interpretations made on the ground is a well-accepted process in the wider field, and there are established statistical and descriptive methods in remote sensing for assessing the accuracy of data such as image classifications [82,83]. Accuracy assessments need clear plans, an unbiased and consistent sampling procedure, and a process of analysing the data. "Classes" assigned to the site from both ground collected data (fieldwork) and image interpretation can be compared, for example by using an error matrix [82] (p. 3) [83]. Although adopted by some projects [34], the process of quantifying the accuracy of image interpretation and remote-sensing has not been widely used by archaeologists, and can be challenging when dealing with multiple levels of image interpretation. In many cases, some archaeological information simply cannot be known without field-based investigation.

While it is beyond the scope of this article to discuss this at length, we outline our field-based validation strategy here. The data comparison below was made by getting an analyst not familiar with the areas concerned, but trained in the EAMENA methods, to identify sites and damage threats in two sample areas for which we have ground data in Morocco and Lebanon. We compared the site records made separately using Google Earth images with interpretations made on the ground, using a simple table to reflect key terminology from our database. We counted the number of sites which matched, had a full or partial match, or did not match. The concordance between the numbers identified using each method is then established (Table 11).

Table 11. Concordance of interpretations table N = 50 sites surveyed with EAMENA methodology.

	Full Match between Image and Ground (%)	Partial or Full Match (Combined) between Image and Ground (%)	No Match (%)
Morphology	78	96	4
Form	50	88	12
Interpretation	32	86	14
Damage	26	68	32
Threats	26	72	28

The results of this exercise are presented in Table 11. This shows the number of exact matches between image and ground interpretations and, given the difficulties of making detailed interpretations from imagery alone, we also counted correlated matches, including instances where the image interpreter simply made a broader interpretation (e.g., "building") than the field-based interpreter (e.g., "house"). The morphology, form and functional interpretation of a site were often easy to identify using imagery. While we were often correct in recognising that a site had been damaged in some way, it was much more challenging to identify the type of damage which had affected it.

A key factor to note is that the EAMENA site terminology extends beyond what may be visible on satellite imagery, as it incorporates categories that derive from ground survey, but are meaningful in an archaeological sense. By the same token, analysts are trained not only in the EAMENA methodologies, but also in the regional archaeological typologies and dating frameworks, which are generally derived from a long history of ground based investigation of sites. In our blind tests, we required an analyst with expertise in the field archaeology of Lebanon but unfamiliar with Morocco to look at that area and one with experience in Morocco to look at the Lebanese data. This probably accounts, in-part, for the lack of exact matches and emphasises the importance of local knowledge, and so highlights some of the significant challenges faced by crowd mapping and automated methods.

As we increase the number of samples used for this validation process, we will be able to refine our methodology based on these results and so identify error thresholds appropriate for application to assessments of archaeological remote recording. Remote classification of modern land-use, for example, can be relatively straightforward and its accuracy easily assessed. Given that many archaeological sites cannot be fully interpreted without ground-based work, especially excavation, EAMENA will seek to establish a more nuanced methodology that is attuned to assessing accuracy of archaeological interpretations. Whilst the details of this are beyond the scope of the current paper, the project is developing its field and imagery validation methods via further blind tests. We will explore the different factors affecting our ability to accurately identify and categorise site types, disturbances and threats, and determine whether our methodology needs to be adapted or refined as a result of this. As archaeological work is likely to rely increasingly upon remote sensing for making interpretations, robust assessment of accuracy is necessary, especially for the large-scale data collection undertaken by our project.

4. Discussion

We have outlined and evaluated three main methods of large-scale heritage recording projects: crowd-mapping; automated detection; and our own methodology which relies on trained image interpreters and the incorporation of a variety of data. These represent different approaches for mapping large regions. There are elements of uncertainty deriving from any of these approaches, because of the need to make decisions about the nature of archaeological features, often remotely. Our comparison of a sample of ground- and image-based interpretations shows that even for trained analysts recording archaeological sites and their condition using imagery can be difficult. Knowledge of local archaeological specifics is clearly as important as technical skills. Dealing with uncertainty is a significant issue for our project because it is inherent in geospatial recording as well as in making archaeological interpretations. Uncertainty arises as a result of missing information, user mistakes, and incorrect information and interpretation. We not only need to recognise features in imagery, taking into account image properties, the sensor's characteristics [84] and seasonal conditions [15] (p. 167), but we need to make decisions about their function in the past. Ultimately, the EAMENA project has developed a standardised and user-friendly way of quantifying levels of thematic certainty in order to avoid presenting misleading information and to allow users of the data to decide how to interpret it. It is necessary to differentiate between interpretations that have been made using a variety of reliable sources, including data gathered in the field, and our recording of features with an unknown function that have been logged only from a single satellite image. In some cases, information simply cannot be known [48].

It is also important to recognise potential limitations of EAMENA's remote-sensing based recording methods. Remote sensing is not always the most appropriate method for identifying archaeological sites. Even the application of labour intensive remote-sensing visual analysis or 'brute 'force' methods as Casana [17] (p. 231) has termed them is in some regions simply not the best option for large-scale site detection and monitoring. For example, trials carried out by the EAMENA team using their methodology in Kuwait revealed that large numbers of 'known' archaeological sites could not be identified via imagery, even when given precise locations for these sites. This is, in part, a result of the limited resolution of imagery available in Google Earth across this area; however, other factors also play a role.

As discussed above, certain types of archaeological sites are simply less readily visible in imagery than others, which could lead to under-recording. Such sites include lithic and pottery scatters and shell middens; the latter being a characteristic site type from Kuwait's coastal environs. As scholars have also clearly demonstrated, the application of remote-sensing techniques in regions with, for example, active sand dunes is also particularly challenging [34,85]. In these cases, remote sensing can aid in pointing towards likely locations for archaeological sites, but cannot necessarily be directly utilised in their detection, mapping and interpretation, without field visits having taken place.

Freely available imagery such as Google Earth has limited potential for spectral analysis, with images being displayed in "true colour" [86]. However, it is clear that, for now at least, the benefits of using freely available satellite imagery in Google Earth and Bing Maps outweigh the negatives. By and large the use of open source software and data allows us to meet our aim of conducting at least a preliminary analysis for most countries in the MENA region. It also ensures that training in our methodology can be carried out at a regional and pan-regional scale. Training and use of the methodology can also be sustained without the continual need for investment by each country or heritage agency in expensive software.

Among the alternatives to our methodology, we believe the scale of the MENA region and diversity of its heritage make automation and crowd mapping approaches quite problematic. We emphasise the importance of analysts working on regional Historic Environment Records to have both a high level of technical skills and relevant knowledge of the regional archaeological record. This requires in-depth and sustainable training initiatives. We are not implying that there is no place for crowd-mapping (for example see [26]) and automated recording within projects using remote sensing. However, we do offer a note of caution, pointing towards the necessity of the rigorous evaluation of any data collected and the need for interpreters familiar with the archaeology of the region they are working in. As the use of remote sensing for site monitoring moves into the mainstream of heritage management, there is also a clear need to develop methods and models for the consistent and comparable recording of damage and disturbances to archaeological sites within specific countries, but also at a wider regional scale.

5. Conclusions

The case studies and results discussed above show that it is possible to apply a standardised recording methodology to the archaeology across the MENA region. Our analyses have revealed that the future shape of the heritage base (i.e., what will remain in existence two decades hence) will be determined less by the spectacular incidents perpetrated by terror groups, than by the widespread and continuing attrition that results from poorly controlled development-related activities, in particular agricultural intensification and urban expansion. These problems have been exacerbated by the reduction in both the monitoring of heritage sites and regulatory enforcement by governmental organisations, and the increased opportunities for land grabbing and unauthorised development that result from recent conflicts. Our database and methodologies, if adopted in MENA countries, will provide heritage agencies there with tools that will enhance both their monitoring and management of heritage assets.

It will not be possible to map everything, but by the end of the current phase of the project in 2020 we hope to have fully enhanced database records completed for a sample of grid squares in each of the countries we are working in. We are also actively seeking and making contact with researchers who are willing to contribute data from their own work and research into the database. Importantly, this work is not just going to be carried out by researchers in the UK and Europe. Data entry will be supported and enhanced by a programme of training courses, with each trainee contributing to and adding new sites to the database (e.g., see [6]). Training is fundamental, not only in terms of image interpretation, but also in terms of data recording and terminologies. In addition, using fieldwork to validate remote recording methods is crucial to increasing the accuracy of the process of assigning interpretation.

Supplementary Materials: The EAMENA database of archaeological sites is available online at http://eamenadatabase.arch.ox.ac.uk/.

Acknowledgments: The EAMENA project is a collaboration between the Universities of Oxford, Leicester, and Durham, under the leadership of Robert Bewley (Project Director, Oxford), Professor Andrew Wilson (Principal Investigator, Oxford), David Mattingly (Co-Investigator, Leicester), and Graham Philip (Co-Investigator, Durham). EAMENA is funded by the Arcadia Fund for documentation, database development and fieldwork, as well as the DCMS/British Council's Cultural Protection Fund (2017–2020) to train heritage professionals from seven countries in the EAMENA methodology. We would like to thank all the members of the EAMENA team for their work and contributions, especially Martin Sterry for his advice and Michael Fradley for the inspiration provided by his blog post (http://eamena.arch.ox.ac.uk/the-difficulty-of-verifying-heritage-damage-reports/). We are grateful to all those who commented on drafts of this paper; any errors remain our own.

Author Contributions: The data for the Cyrene case study was analysed and described by Louise Rayne. The Homs case study data was analysed and described by Jennie Bradbury. The fieldwork validation experiment was designed, analysed and described by Louise Rayne and Jennie Bradbury. The paper was written collaboratively between all six authors. The authors declare no conflict of interest. The founding sponsors had no role in the design of the study; in the collection, analyses, or interpretation of data; in the writing of the manuscript, and in the decision to publish the results.

Conflicts of Interest: The authors declare no conflict of interest.

References

1. Lawrence, D.; Philip, G.; Wilkinson, K.; Buylasert, J.P.; Murray, A.S.; Thompson, W.; Wilkinson, T.J. Regional Power and Local Ecologies: Accumulated Population Trends and Human Impacts in the Northern Fertile Crescent. *Quat. Int.* **2017**, *437 Pt B*, 60–81. [CrossRef]
2. Wilkinson, T.J.; Philip, G.; Bradbury, J.; Dunford, R.; Donoghue, D.; Galiatsatos, N.; Lawrence, D.; Ricci, A.; Smith, S. Contextualizing Early Urbanization: Settlement Cores, Early States and Agro-Pastoral Strategies in the Fertile Crecent during the Fourth and Third Millennia BC. *J. World Prehist.* **2014**, *27*, 43–109. [CrossRef]
3. Hritz, C. Contributions of GIS and Satellite-based Remote Sensing to Landscape Archaeology in the Middle East. *J. Archaeol. Res.* **2014**. [CrossRef]
4. EAMENA. Available online: www.eamena.org (accessed on 22 July 2017).
5. Bewley, R.; Wilson, A.; Kennedy, D.; Mattingly, D.; Banks, R.; Bishop, M.; Bradbury, J.; Cunliffe, E.; Fradley, M.; Jennings, R.; et al. Endangered archaeology in the Middle East and North Africa: Introducing the EAMENA project. In Proceedings of the 43rd Annual Conference on Computer Applications and Quantitative Methods in Archaeology, Siena, Italy, 30 March–3 April 2015; Campana, S., Scopigno, R., Carpentiero, G., Cirillo, M., Eds.; Archaeopress Publishing Ltd.: Oxford, UK, 2016; pp. 919–933.
6. Nikolaus, J.; Mugnai, N.; Rayne, L.; Zerbini, A.; Mattingly, D.; Walker, S.; Abdrbba, M.; Alhaddad, M.; Buzaian, A.; Emrage, A. Training, partnerships, and new methodologies for protecting Libya's cultural heritage. In *Atti del "Red Castle and Museum of Tripolitania" Workshop, Tenutosi a Zarzis (Tunisia), 8–11 Giugno 2015 (Quaderni di Archeologia Della Libya, 21)*; La Rocca, E., Ed.; "L'Erma" di Bretschneider: Rome, Italy, 2017, in press.
7. Poidebard, R.P.A. *La Trace de Rome Dans le Désert de Syrie: Le Limes de Trajan a la Conquête Arabe, Recherches Aériennes, 1925–1932*; Haut-commissariat de la République Francaise en Syrie et au Liban, Service Des Antiquités et des Beaux-arts. Bibliothèque Archéologique et Historique, Tome 18; Geuthner: Paris, France, 1934.
8. Gregory, S.; Kennedy, D. *Sir Aurel Stein's 'Limes Report*'; BAR International Series 272; BAR Publishing: Oxford, UK, 1985.
9. Edwards, D.N. The archaeology of the southern Fazzan and prospects for future research. *Libyan Stud.* **2001**, *32*, 49–66. [CrossRef]
10. Kennedy, D. Aerial Archaeology in the Middle East. *AARG News* **1996**, *12*, 11–15.
11. Mattingly, D.J.; Sterry, M. The first towns in the central Sahara. *Antiquity* **2013**, *87*, 503–518. [CrossRef]

12. Apaame. Available online: www.apaame.org (accessed on 22 July 2017).

13. Kennedy, D.; Bewley, R. *Ancient Jordan from the Air*; CBRL: London, UK, 2004.

14. Kennedy, D.; Bewley, R. Aerial archaeology in Jordan. *Antiquity* **2009**, *83*, 69–81. [CrossRef]

15. Beck, A.; Philip, G.; Abdulkarim, M.; Donoghue, D. Evaluation of Corona and Ikonos high resolution satellite imagery for archaeological prospection in western Syria. *Antiquity* **2007**, *81*, 161–175. [CrossRef]

16. Casana, J.; Cothren, J.; Kalayci, T. Swords into Ploughshares: Archaeological Applications of CORONA Satellite Imagery in the Near East. *Internet Archaeol.* **2012**, *32*. [CrossRef]

17. Casana, J. Regional-scale archaeological remote sensing in the age of big data. Automated site discovery vs. brute force methods. *Adv. Archaeol. Pract. J. Soc. Am. Archaeol.* **2014**, 222–233. [CrossRef]

18. Galiatsatos, N.; Donoghue, D.N.M.; Philip, G. High resolution elevation data derived from stereoscopic CORONA imagery with minimal ground control: An approach using IKONOS and SRTM data. *Photogramm. Eng. Remote Sens.* **2008**, *74*, 1093–1106. [CrossRef]

19. Ur, J.A. *Urbanism and Cultural Landscapes in Northeastern Syria: The Tell Hamoukar Survey, 1999–2001*; Oriental Institute Publications No, 137; Oriental Institute of the University of Chicago: Chicago, IL, USA, 2010.

20. Barker, G.; Gilbertson, D.; Jones, B.; Mattingly, D. *Farming the Desert: The UNESCO Libyan Valleys Archaeological Survey. Volume 1: Synthesis*; Society for Libyan Studies: London, UK, 1996.

21. Bjørgo, E.; Boccardi, G.; Cunliffe, E.; Fiol, M.; Jellison, T.; Pederson, W.; Saslow, C. Satellite-Based Damage Assessment to Cultural Heritage Sites in Syria. UNITAR/UNOSAT. Available online: http://www.unitar.org/unosat/chs-syria (accessed on 1 October 2017).

22. Cunliffe, E. Remote assessments of site damage: A new ontology. *Int. J. Herit. Digit. Era* **2014**, *3*, 453–473. [CrossRef]

23. Tapete, D.; Cigna, F.; Donoghue, D.N. 'Looting marks' in space-borne SAR imagery: Measuring rates of archaeological looting in Apamea (Syria) with TerraSAR-X Staring Spotlight. *Remote Sens. Environ.* **2016**, *178*, 42–58. [CrossRef]

24. Curious Traveller's. Available online: http://www.visualisingheritage.org/locateheritage.php (accessed on 8 September 2017).

25. ASOR. Available online: www.asor-syrianheritage.org/ (accessed on 22 July 2017).

26. Terrawatchers. Available online: http://terrawatchers.org/ (accessed on 22 July 2017).

27. Global Explorer. Available online: www.globalxplorer.org/about (accessed on 22 July 2017).

28. Harris, T.M. Interfacing archaeology and the world of citizen sensors: Exploring the impact of neogeography and volunteered geographic information on an authenticated archaeology. *World Archaeol.* **2012**, *44*, 580–591. [CrossRef]

29. Goodchild, M.F.; Li, L. Assuring the quality of volunteered geographic information. *Spat. Stat.* **2012**, *1*, 110–120. [CrossRef]

30. Heipke, C. Crowdsourcing geospatial data. *ISPRS J. Photogramm. Remote Sens.* **2010**, *65*, 550–557. [CrossRef]

31. Kvamme, K. An examination of automated archaeological feature recognition in remotely sensed imagery. In *Computational Approaches to Archaeological Spaces*; Bevan, A., Lake, M., Eds.; Left Coast Press: Walnut Creek, CA, USA, 2013; pp. 53–68.

32. Harrower, M.J. Survey, Automated Detection, and Spatial Distribution Analysis of Cairn Tombs in Ancient Southern Arabia. In *Mapping Archaeological Landscapes from Space*; Comer, D., Harrower, M., Eds.; Springer: New York, NY, USA, 2013; pp. 259–268.

33. Bennett, R.; Cowley, D.; De Laet, V. The data explosion: Tackling the taboo of automatic feature recognition in airborne survey data. *Antiquity* **2014**, *88*, 896–905. [CrossRef]

34. Menze, B.H.; Ur, J.A. Mapping patterns of Long-Term Settlement in Northern Mesopotamia at a Large Scale. *Proc. Natl. Acad. Sci. USA* **2012**, *109*, E778–E787. [CrossRef] [PubMed]

35. Beck, A. Archaeological site detection: The importance of contrast. In Proceedings of the Annual Conference of the Remote Sensing and Photogrammetry Society, Newcastle upon Tyne, UK, 11–14 September 2007; pp. 307–312.

36. Syrian Heritage in Danger (SHRIN). Available online: http://shirin-international.org (accessed on 11 September 2017).

37. Kennedy, D.; Bishop, M. Google Earth and the archaeology of Saudi Arabia. A case study from the Jeddah area. *J. Archaeol. Sci.* **2011**, *38*, 1284–1293. [CrossRef]

38. Rayne, L.; Sheldrick, N.; Nikolaus, J. Endangered Archaeology in Libya: Recording Damage and Destruction. *Libyan Stud.* **2017**, 1–27. Available online: https://www.cambridge.org/core/journals/libyan-studies/article/endangered-archaeology-in-libya-recording-damage-and-destruction/59C240F04F16EE257ED76D27E02A482C (accessed on 1 October 2017).

39. The Difficulty of Verifying Heritage Damage Reports, EAMENA Blog. Available online: http://eamena.arch.ox.ac.uk/the-difficulty-of-verifying-heritage-damage-reports/ (accessed on 22 July 2017).

40. Casana, J.; Cothren, J. Stereo analysis, DEM extraction and orthorectification of CORONA satellite imagery: Archaeological applications from the Near East. *Antiquity* **2008**, *82*, 732–749. [CrossRef]

41. Conesa, F.C.; Madella, M.; Galiatsatos, N.; Balbo, A.L.; Rajesh, S.V.; Ajithprasad, P. CORONA photographs in monsoonal semi-arid environments: Addressing archaeological surveys and historic landscape dynamics over North Gujarat, India. *Archaeol. Prospect.* **2015**, *22*, 75–90. [CrossRef]

42. Kennedy, D. Declassified satellite photographs and archaeology in the Middle East: Case studies from Turkey. *Antiquity* **1998**, *72*, 553–561. [CrossRef]

43. Rayne, L. Water and Territorial Empires: The Application of Remote Sensing Techniques to Ancient Imperial Water Management in Northern Mesopotamia. Ph.D. Thesis, Durham University, Durham, UK, 2014.

44. Huete, A.R. A Soil Adjusted Vegetation Index. *Remote Sens. Environ.* **1988**, *25*, 295–309. [CrossRef]

45. Google Earth Engine Team, Google Earth Engine: A Planetary-Scale Geospatial Analysis Platform. Available online: https://earthengine.google.com (accessed on 1 October 2017).

46. Mowrer, H.T.; Congalton, R.G. (Eds.) *Quantifying Spatial Uncertainty in Natural Resources: Theory and Applications for GIS and Remote Sensing*; CRC Press: Boca Raton, FL, USA, 2003.

47. Crosetto, M.; Tarantola, S. Uncertainty and sensitivity analysis: Tools for GIS-based model implementation. *Int. J. Geogr. Inf. Sci.* **2001**, *15*, 415–437. [CrossRef]

48. Couclelis, H. The certainty of uncertainty: GIS and the limits of geographic knowledge. *Trans. GIS* **2003**, *7*, 165–175. [CrossRef]

49. Wang, G.; Gertner, G.Z.; Fang, S.; Anderson, A.B. A methodology for spatial uncertainty analysis of remote sensing and GIS products. *Photogramm. Eng. Remote Sens.* **2005**, *71*, 1423–1432. [CrossRef]

50. De Runz, C.; Desjardin, E.; Piantoni, F.; Herbin, M. Using fuzzy logic to manage uncertain multi-modal data in an archaeological GIS. In Proceedings of the International Symposium on Spatial Data Quality-ISSDQ, Enschede, The Netherlands, 13–15 June 2007.

51. Cooper, A.; Green, C. Embracing the complexities of 'big data' in archaeology: The case of the English Landscape and Identities Project. *J. Archaeol. Method Theory* **2016**, *23*, 271–304. [CrossRef]

52. Manyika, J.; Chui, M.; Brown, B.; Bughin, J.; Dobbs, R.; Roxburgh, C.; Byers, A.H. Big Data: The Next Frontier for Innovation, Competition, and Productivity, 2011. Available online: http://www.mckinsey.com/business-functions/digital-mckinsey/our-insights/big-data-the-next-frontier-for-innovation (accessed on 23 June 2017).

53. Boyd, D.; Crawford, K. Critical questions for big data: Provocations for a cultural, technological, and scholarly phenomenon. *Inf. Commun. Soc.* **2012**, *15*, 662–679. [CrossRef]

54. Didan, K. MOD13A1 MODIS/Terra Vegetation Indices 16-Day L3 Global 500m SIN Grid V006. *NASA EOSDIS Land Processes DAAC*. 2015. Available online: https://doi.org/10.5067/modis/mod13a1.006 (accessed on 1 October 2017).

55. Mills, S.; Weiss, S.; Liang, C. VIIRS day/night band (DNB) stray light characterization and correction. In Proceedings of the SPIE Optical Engineering+ Applications, International Society for Optics and Photonics, Prague, Czech Republic, 23 September 2013; p. 88661.

56. Fradley, M.; Sheldrick, N. Satellite imagery and heritage damage in Egypt: A response to Parcak et al. *Antiquity* **2017**, *91*, 784–792. [CrossRef]

57. Parcak, S.; Gathings, D.; Childs, C.; Mumford, G.; Cline, E. Satellite evidence of archaeological site looting in Egypt: 2002–2013. *Antiquity* **2016**, *90*, 188–205. [CrossRef]

58. Beechey, F.W.; Beechey, H.W. *Proceedings of the Expedition to Explore the Northern Coast of Africa: From Tripoly Eastward in MDCCCXXI. and MDCCCXXII., Comprehending an Account of the Greater Syrtis and Cyrenaica; and of the Ancient Cities Composing the Pentapolis*; John Murray: London, UK, 1828.

59. Luni, M. *La Scoperta Di Cirene: Un Secolo di Scavi (1913–2013)*; "L'Erma" di Bretscnheider: Rome, Italy, 2014.

60. Goodchild, R.; Reynolds, J.; Herington, C. The Temple of Zeus at Cyrene. *Pap. Br. Sch. Rome* **1958**, *26*, 30–62. [CrossRef]

61. White, D. Cyrene's Sanctuary of Demeter and Persephone: A Summary of a Decade of Excavation. *Am. J. Archaeol.* **1981**, *85*, 13–30. [CrossRef]

62. White, D.; Reynolds, J. *The Extramural Sanctuary of Demeter and Persephone at Cyrene, Libya: Final Reports: The Sanctuary's Imperial Architectural Development, Conflict with Christianity, and final days*; University of Pennsylvania Museum of Archaeology and Anthropology Philadelphia, for The Libyan Department of Antiquities: Philadelphia, PA, USA, 2012.

63. White, D. The Pennsylvania Museum's Demeter and Persephone Sanctuary Project at Cyrene: A Final Progress Report? *Libyan Stud.* **1989**, *20*, 71–77. [CrossRef]

64. Cassels, J. The cemeteries of Cyrene. *Pap. Br. Sch. Rome* **1955**, *23*, 1–43. [CrossRef]

65. Thorn, J.C. *The Necropolis of Cyrene: Two Hundred Years of Explorations*; L'Erma di Bretschneider: Rome, Italy, 2005.

66. Kenrick, P. *Cyrenaica. Libya Archaeological Guides*; Society for Libyan Studies, Silphium Press: London, UK, 2013.

67. Abdulkariem, A.; Bennett, P. Libyan heritage under threat: The case of Cyrene. *Libyan Stud.* **2014**, *45*, 155–161. [CrossRef]

68. Thorn, D.; Thorn, J.C. *A Gazeteer of the Cyrene Necropolis: From the Original Notebooks of John Cassels, Richard Tomlinson and James and Dorothy Thorn*; L'Erma di Bretschneider: Rome, Italy, 2009.

69. Al-Raeid, F.; Di Valerio, E.; Di Antonio, M.G.; Menozzi, O.; Abdalgadar El Mziene, M.A.S.; Tamburrino, C. The main issues of the Cyrene necropolis and the use of remote sensing for monitoring in the case of the eastern necropolis. *Libyan Stud.* **2016**, *47*, 7–30. [CrossRef]

70. Cyrenaica Archaeological Project. Available online: http://www.cyrenaica.org/ (accessed on 8 September 2017).

71. Goodchild, R. Earthquakes in ancient Cyrenaica. In *Geology and Archaeology of Northern Cyrenaica, Libya*; Barr, F.T., Ed.; Petroleum Exploration Society of Libya: Tripoli, Libya, 1968; pp. 41–44.

72. Di Vita, A. Archaeologists and earthquakes: The case of 365 AD. *Annali Geofisica* **1995**, *506*, 971–976.

73. Mohamed, O.A.; University of Leicester. Personal communication, 2017.

74. Archaeological Site of Cyrene. Available online: http://whc.unesco.org/en/soc/3399 (accessed on 22 July 2017).

75. Newson, P.; Abdulkarim, M.; McPhillips, S.; Mills, P.; Reynolds, P.; Philip, G. Landscape study of dar es-salaam and the wa'ar basalt region north-west of homs, syria. Report on work undertaken during 2005–2007. *Berytus* **2009**, *10*, 1–35.

76. Philip, G.; Bradbury, J. Pre-classical activity in the basalt landscape of the Homs region, Syria: The development of "sub-optimal" zones in the Levant during the Chalcolithic and Early Bronze Age. *Levant* **2010**, *42*, 136–169. [CrossRef]

77. Bradbury, J. Landscapes of Burial? The Homs Basalt, Syria in the 4th–3rd Millennia BC. Ph.D. Thesis, University of Durham, Durham, UK. Unpublished work. 2011.

78. Bradbury, J.; Philip, G. The world beyond the tells: Pre-classical activity in the basalt landscape of the Homs region, Syria. In *Pierres Levées, Stèles Anthropomorphes et Dolmens. Standing Stones, Anthropomorphic Stelae and Dolmens*; BAR International Series 2317; Steimer-Herbert, T., Ed.; Maison de l'Orient et de la Méditerranée Jean Pouilloux: Lyon, France; Archaeopress: Oxford, UK, 2011; pp. 169–180.

79. International Fund for Agricultural Development (IFAD). Syrian Arab Republic: Thematic Study on Land Reclamation through De-Rocking. International Fund for Agricultural Development, Near East and North Africa Division, Programme Management Department. 2010. Available online: www.ifad.org/pub/pn/syria.pdf (accessed on 1 April 2010).

80. Mattingly, D.J.; Bokbot, Y.; Sterry, M.; Cuénod, A.; Fenwick, C.; Gatto, M.; Ray, N.; Rayne, L.; Janin, K.; Lamb, A.; et al. Long-term History in a Moroccan Oasis Zone: The Middle Draa Project 2015. *J. Afr. Archaeol.* **2017**, in press.

81. Bradbury, J.; Sader, H.; McPhillips, S.; Kennedy, M.; Banks, R.; Hoffman, D.; Mardini, M.; Vafadari, A.; Wannessian, B. The Kubba Coastal Survey: First Season Report. *Bull. d'archéol. d'archit. Liban.* **2016**, submitted.

82. Congalton, R.G.; Green, K. *Assessing the Accuracy of Remotely Sense Data Principles and Practices*, 2nd ed.; Taylor and Francis: Hoboken, NJ, USA, 2008.

83. Congalton, R.G. A review of assessing the accuracy of classifications of remotely sensed data. *Remote Sens. Environ.* **1991**, *37*, 35–45. [CrossRef]

84. Fisher, P. The pixel: A snare and a delusion. *Int. J. Remote Sens.* **1997**, *18*, 679–685. [CrossRef]

85. Casana, J.; Herrmann, J.T.; Qandil, H.S. Settlement history in the eastern Rub al-Hali: Preliminary Report of the Dubai Desert Survey (2006–2007). *Arab. Archaeol. Epigr.* **2009**, *20*, 30–45. [CrossRef]

86. Corrie, R.K. Detection of ancient Egyptian archaeological sites using satellite remote sensing and digital image processing. *Proc. SPIE* **2011**, *8181*, 81811B. [CrossRef]

geosciences

MDPI

Article

How Can Remote Sensing Help in Detecting the Threats to Archaeological Sites in Upper Egypt?

Julia M. Chyla [1,2]

[1] Antiquity of Southeastern Europe Research Centre, University of Warsaw, Warsaw 00-927, Poland;
    Julia.chyla@gmail.com; Tel.: +48-604-835-339
[2] Institute of Archaeology, University of Warsaw, Warsaw 00-927, Poland

Received: 27 July 2017; Accepted: 28 September 2017; Published: 2 October 2017

Abstract: The analysis of contemporary and archival satellite images and archaeological documentations presents the possibility of monitoring the state of archaeological sites in the Near East (for example, Palmyra in Syria). As it will be demonstrated in the case of Upper Egyptian sites, the rapid growth of agricultural lands and settlements can pose a great threat to sites localized on the border of fields and the desert. As a case study, the Qena district was chosen, a region of significance for the history of ancient Egypt. To trace the expansion of agriculture and the development of modern settlements, a synthesis of archival maps (from the last 200 years), and archival and contemporary satellite images was created. By applying map algebra to these documents, it was possible to determine areas which may be marked as "Archaeological Hazard Zones". The analysis helped to trace the expansion of agricultural areas during the last 200 years and the influence of both—ancient Egyptians and the Nile—on the local landscape.

Keywords: remote sensing; change detection; archaeology; LANDSAT; digital cultural resources management

1. Introduction

During the Predynastic Period (c. 3150 BC–c. 2686 BC), Upper Egypt was one of the main areas for the development of early civilizations. Archaeological sites important for this period are located in the Qena Bend of the Nile valley, but also to the south and to the north of it, in the Qena district. It includes sites like: Abydos, Huw, Dendera, Naqada, Armant, and Gebelein, the most important places of the Predynastic Period. One of the models of the predynastic settlement pattern [1] suggests that necropoli and settlements were located on the border of the floodplain and the lower desert.

Nowadays, the development of urbanism and agriculture in Upper Egypt poses a great threat to fragile archeological areas located on that former border and can lead to the complete destruction of these unique historical landscapes [2,3]. The spread of agriculture can be traced back to the building of the Aswan High Dam, but also to the period of introducing diesel pumps for the improvement of irrigation systems [4], and to the unstable period after the countrywide unrest in 2011.

A comparison of agricultural lands visible on archival maps, of P. Jacotin's map from Napoleon's expedition in Egypt in 1798–1801 [5], and the Survey of Egypt from 1944 [6], with agricultural lands visible on contemporary satellites, shows how rapid and tremendous the spread of agricultural lands in this area is (Figure 1).A process similar to this was also confirmed through the image interpretation of archival satellite images from the CORONA mission (from 8 November 1968 (Fore) and 29 June 1969 (Aft)) [7]) and historical images from Google Earth [8].

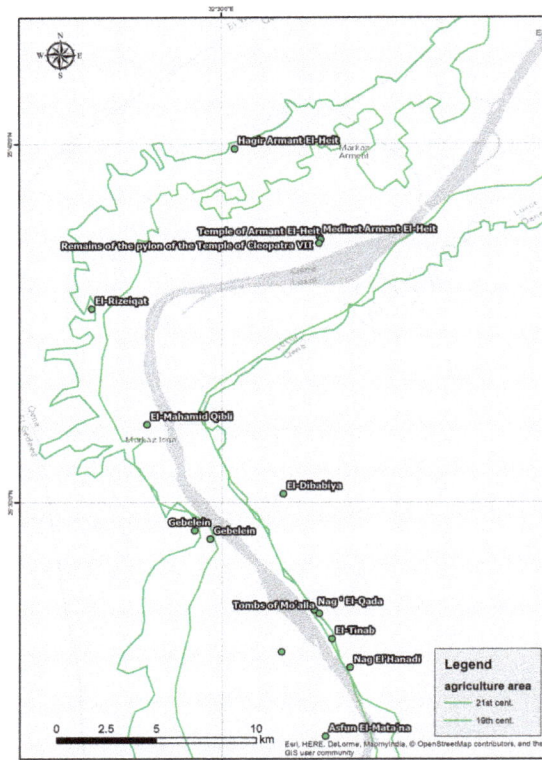

Figure 1. The spread of agricultural lands based on the comparison of P. Jacotin's map and contemporary satellite images.

The goal of this article is to present how remote sensing can help to measure the possible threats posed by the agricultural expansion of archaeological sites and consequently to help in their protection. The detection of temporal changes in land use is "the process of identifying differences in the state of an object or phenomena by observing it at different times. Essentially, it involves the ability to quantify temporal effects using multitemporal data sets" [9]. Such an approach is often used in urban growth detection, and with the help of analyzes, LANDSAT archival images, e.g., [10,11].

A study of croplands mapping and their change with time for the whole Nile Valley and Nile Delta was conducted by Xu et al. [12]. A total of 961 LANDSAT images (TM/ETM+/OLI) images 1984–2015) were used to conduct an annual frequency detection of agricultural land expansion. In their results, the authors noted a gradually increasing tendency in the "distribution of cropland dynamics from the Nile Valley towards desert". However, the article does not show in a detailed way how this dynamic looks like at the Qena bend of the Nile.

Recently, a number of papers describing change detection in the regions of the Nile Delta and Dakhla Oasis were published [13–17]. Their main focus was to present land degradation, urbanization, and the status of water resources. But the Nile Delta and Dakhla Oasis landscapes, both ancient and contemporary, are completely different from Upper Egypt's.

A similar analysis was also used to monitor urban and agricultural changes on the eastern side of the Qena Bend in Upper Egypt [18]. Abd El-Aziz processed 5 LANDSAT images (MSS-TM-ETM-OLI) from 1972, 1984, 1998, 2006, and 2013, and as an outcome, the area was split into three groups: an urban area, cultivation area, and water surfaces, with a sub pixel classification method. The result compares categorized images from different periods and highlights some areas where towns and agricultural lands have spread. However, the above described approaches do not include the possible influence of land use changes on archaeological sites, nor do they highlight areas were sites could be under threat. On the other hand, a paper published by Ahmed and Fogg [19] describes the possible impact of groundwater and the expansion of agricultural land on the archaeological sites in Luxor. One of the many methods used for creating recommendations in order to mitigate the deterioration of the archaeological sites is the supervised classification of 3 LANDSAT images (MSS-TM-ETM+) from 1982 to 2011. However, this paper does not mark which sites and how they are threatened by agricultural expansion.

Additionally, when satellite imagery analysis is used in archaeological research, it often focuses on the detection of new sites or features, e.g., [20–22], instead of a threat assessment. Just recently, a new approach of visualizing changes and their effects on archaeological sites has become a focus for research, especially for areas heavily influenced by war and looting such as the Near East, e.g., [23,24]. Still, there is a need to develop a low-budget and fast method for quantifying changes causing the destruction of archaeological sites. Such methods have the potential to support Cultural Management authorities.

2. Materials and Methods

As mentioned above, it was noted that a key threat to archaeological sites in the Upper Egypt area is the development of cultivated lands. The detection of agricultural areas' changes followed the proposed workflow below (see Figure 2).

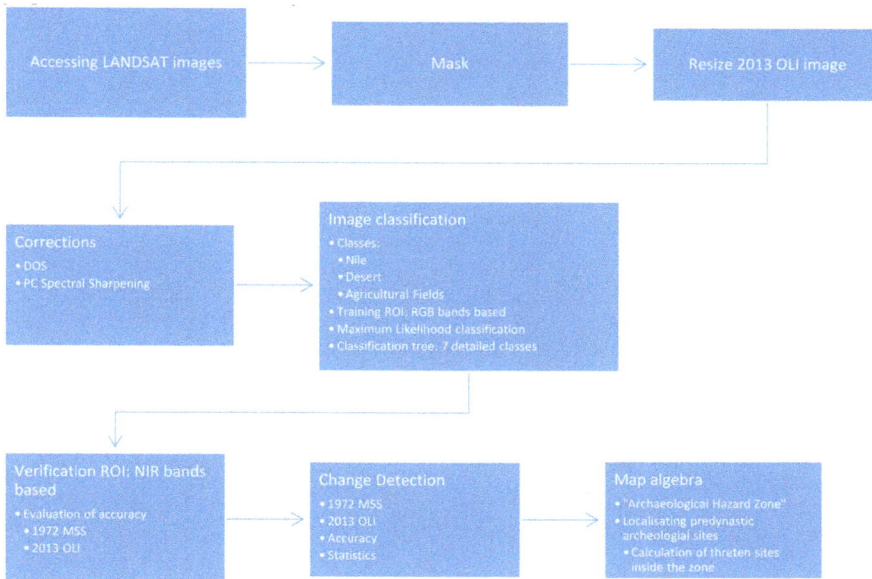

Figure 2. Analysis workflow.

The first step was to access two LANDSAT images from the United States Geological Survey (USGS) website [25]. The goal was to compare two images from the widest possible time span, which could show the landscape before and after the construction of the Aswan High Dam, but also, which could take the unstable period after the countrywide troubles in 2011 into consideration. Additionally, two images were chosen for classification because of the final goal of the analysis, which was to deduct two images and to highlight one area which in 1972 was a desert and in 2013 was covered in fields (described in the Results and Conclusion section).

Therefore, it was decided that images for one scene (path 175/row 04) would be used from Multispectral Scanners (MSS): LANDSAT 1 made on 4 October 1972 (during construction of Aswan High Dam) and from the Operational Land Imager (OLI), LANDSAT 8, from 22 April 2013, both of which were cloud free. The difference in time between the two images was 41 years.

The MSS sensor had a spectral resolution of 0.5–1.1 μm (bands from 4 to 7) and a spatial resolution of 60 m [26]. OLI had a spectral resolution of 0.43–12.5 μm (bands from 1 to 7) and a spatial resolution of 30 m [27]. The next step was to create a mask excluding high desert areas unnecessary for the current purposes of analysis. Subsequently, the image from 2013 was resized to the resolution of the image from 1972. This was necessary for the future analysis' compatibility. The next step was to process the images without the use of Radiometric Correction. Dark Object Subtraction was used for atmospheric correction. Principal Components (PC) spectra sharpening was also used to receive a better-quality image.

The following step was the classification of images. Firstly, an unsupervised, k-means classification was created, which allowed for the grouping of pixels into seven classes: Nile, fields with crops, fields without crops, high desert, desert, *wadi*, and infrastructure. However, the resulting image (Figure 3) contained too much information for the purposes of further analysis.

Figure 3. Result of the classification tree method, with seven classes.

Therefore, a new classification was created consisting of only three classes: Nile, desert, and agricultural lands. This classification focuses on detecting changes at predynastic archaeological sites located at the border of previous floodplains and the desert, marked earlier due to the threat of destruction. Training and verification of "Regions Of Interest" (ROI) were created for each image: the former with the use of Red-Green-Blue (RGB) bands, the latter with the use of Near Infra-Red (NIR) bands. First, ROI were chosen based on experience from field work in Egypt, and were supplemented with a photo interpretation of Google Earth images. Second, ROI were generated randomly as pixel-based validation samples to calculate the accuracy. Supervised Maximum Likelihood Classification (MLC) was chosen as a classification method [11]. The final step was to assess the accuracy of the classified images. Two sources of information were used: pixels classified on satellite images and verification pixels prepared earlier from NIR bands. This step resulted in the creation of two matrices describing the overall accuracy—the proportion of correctly classified pixels to general number of pixels [28]. Further steps contain analyses of the classification results and an interpretation of the detected changes.

3. Results

The results of the work flow were two classified images showing cultivation areas in 1972 and in 2013 (Figure 4). The overall accuracy of the image from 2013 is 93%, which is a good result in comparison with other works [11,12]. However, it is not possible to compare the overall accuracy for the same study area as neither of the previously mentioned papers [18,19] describe it. The chosen classification method differs from the method used by Abd-Aziz [18] for the same area and from the one used by Xu et al. [12] for the whole Nile Valley, but all of them were supervised classifications. Nevertheless, MLC is good and often used as a classification method [11] and its results were satisfactory concerning monitoring the state of archeological sites.

But a problem did appear concerning the LANDSAT MSS 1972 image: the chosen image contained noise, which can be classified as Transient Detector Failure or Band Striping [29]. The noise is visible on bands 5,6, and partly 7. It appears visually as a set of parallel lines crossing the image diagonally (Figure 5). During several consultations, some methods of removing the noise were considered, like destriping or Unstandardized Principal Components Analysis (PCA) [30], but none of them seemed to be a satisfying method for removing the noise visible on the MSS image. However, for the purposes of an analysis aimed at archaeological sites threatened by agriculture, it seems that the noise is only a minor problem.

(a)

Figure 4. *Cont.*

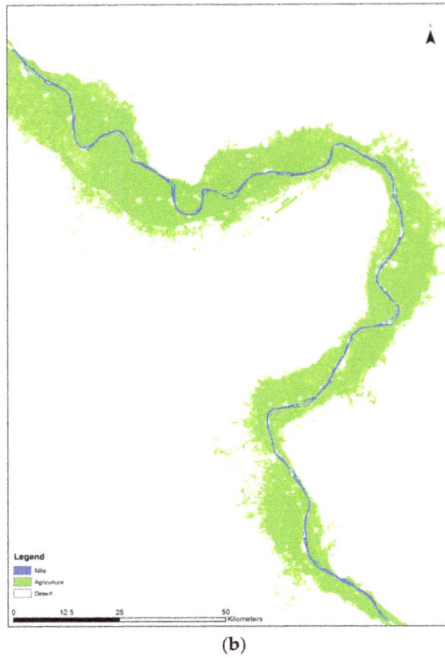

(b)

Figure 4. The result of the classification of the LANDSAT images from (a) 1972 and (b) 2013.

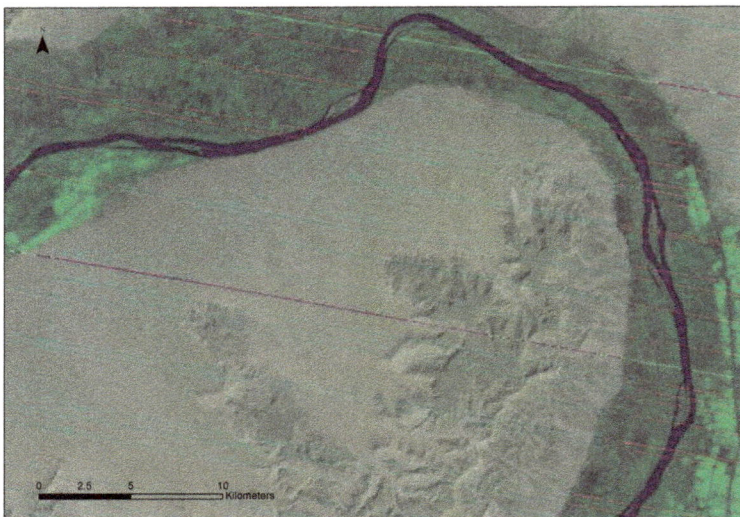

Figure 5. Noise visible on the image from 1972.

For abetter visualization of the threats, a map algebra of two classified images was generated in GIS. From the 2013 agricultural class, a 1972 agricultural class was deducted. This approach, a simple deduction of two images, gave a clear visualization of areas, in Upper Egypt, under threat by agricultural development. A so-called "Archaeological Hazard Zone" highlights the area which in 1972 was part of the desert, and in the 2013 was covered by fields (Figure 6). It appeared that, during last the 41 years, the field area has increased by 2052.65 m^2, which is an increase of 109.11%. The fragile border of the desert and agricultural lands, which have existed for a millennia, was shifted and therefore the sites located above the floodplain might be threatened. None of the papers mentioned above used map algebra and a deduction process to highlight the areas that had changed from desert to fields, nor any of them detailed, described, or analysed which sites could be threatened by those changes in the future.

Figure 6. Archaeological Hazard Zone for Upper Egypt.

4. Discussion

To fully measure how the development of agriculture in Upper Egypt threatens archaeological sites, it is necessary to locate the sites on the map. Sites were identified on Google Earth on the basis of maps and descriptions from several publications [31,32], including maps published by William Matthew Flinders Petrie [33]. A total number of 63 known sites were marked as points, and, if possible, as polygons, and saved in a *.kmz format. Subsequently, the Google Earth format was converted into a shape format, which allowed the addition of a descriptive data base. The data base includes information about the sites such as: name; Supreme Council of Antiquities inventory number; coordinates; legal status of the site; approximate size of the site (if possible); photo interpretation of the features observed through satellite images; if the site is located in the "Archaeological Hazard Zone"; field observations (to fill during survey); threats visible on the sites (to fill during survey); other comments.

The last step was to use a selection tool in GIS and to see how many and which of the sites are in the "Archaeological Hazard Zone": 35 sites might be threatened by landscape change which occurred between 1972 and 2013, which is 55% of all known sites located in the discussed area (Figure 7).

Table S1 presents all known archeological sites and their status inside the "Archeological Hazard Zone", with a verification of possible destruction through nowadays high-resolution satellite images (like Google Earth, Bing or ESRI's base maps. The oldest accessible images were from 2003). Additionally, a result of the analysis of the surroundings (not land use on the sites themselves, but around them) from a different time scope are presented, from the 1800s (based on P. Jactin's maps [5]), 1940s (based on Survey of Egypt maps [6]), 1968–1969 (based on archival images from Corona Mission [7]), and 1972 and 2013 (based on chosen LANDSAT images).

Forty-four (70%) sites' surroundings have changed from desert to agriculture, urban, or from agriculture to urban since the 1800s. Sixteen sites' (25%) landscape changed between the 1800s and 1940s and three sites (5%) between the 1940s and 1968–1969. In the case of 37 sites (59%), the change of surroundings happened between 1972 and 2013. However, in most of the cases of the sites which remain outside of the "Archaeological Hazard Zone" in both 1972 and 2013, they are located in the same surroundings: agricultural or urban. For 24 (38%) sites, a photointerpretation of high resolution images confirmed changes on the sites themselves [3] (in the case of 18 sites, the changes on the sites were not possible to estimate because they were already located in an agricultural or urban context).

An analysis of the comparison of the close surroundings of archaeological sites in Qena's bend during last 200 years shows that the landscape was already changing in the past, but was never as rapid as during last forty years.

The data described above should be verified in the field, every possibly threatened site should be visited, and the possible destruction of the sites should be examined. The "Archaeological Hazard Zone" and the location of the sites can be imported onto a GNSS (Global Navigation Satellite System) tool with a mobile GIS application with a chosen part ofthe detailed satellite image saved as geoTIFF, and it can be used during field prospection, helping to locate not only the site itself, but also zones of destruction. A similar process took place at the beginning of the Gebelein Archaeological Project. The goal of the research at the Gebelein area (Figure 8) [2,34] is to document archaeological resources and the level of threat caused by modern urbanism and agriculutre, visible on the site complex. Gebelein's (No. 45 in Table S1) surroundings have changed quite drasticly since the 1800s, from desert to urban and agricultural areas. The site is inside the "Archaeological Hazard Zone" and through the contemporary satellite images it is visiable how agricultural lands spread on the site. A photo-interpretation of archival images from Google Earth dates the biggest change to the period between 2009 and 2013 [2].

Figure 7. Archeological sites in Upper Egypt threatened by the development of agriculture.

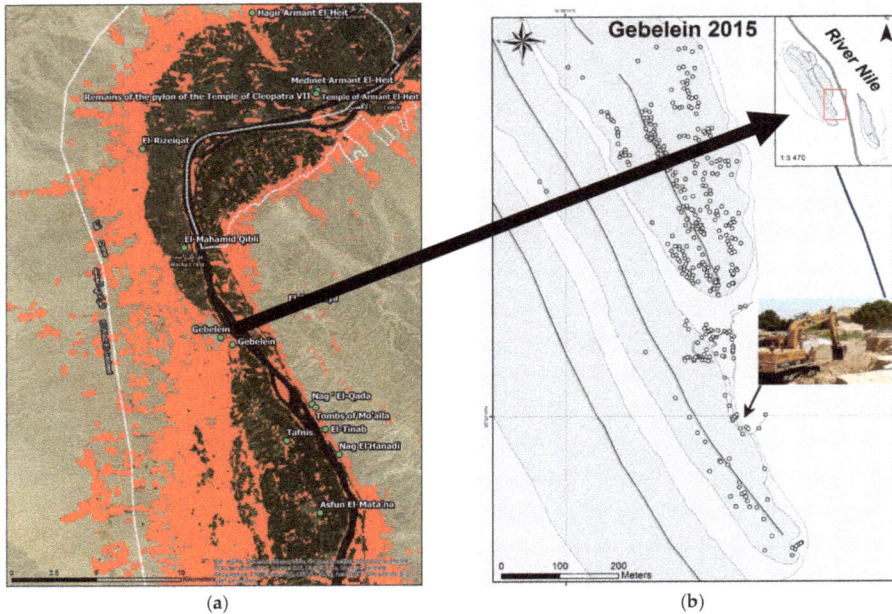

Figure 8. (a) Archaeological sites threatened by the development of agriculture in the Gebelein Region; (b)Reaserch at Gebelein has focused on the most threatened areas of the site.

5. Conclusions

The goal of the article was to describe a practical remote sensing method that can help in detecting whether archaeological sites are threatened by the expansion of agriculture. Additionally, the analysis presented the comparison archaeological sites' surroundings during the last 200 years. The result showed that the landscape in Upper Egypt changed in the past; however, during the last forty years, the change become more rapid and covering more lands. Land use around known archeological sites advanced from desert to agricultural and urban land.

This method might help Cultural Management authorities in protecting local heritage. The result of the method shows not only the sites which might be endangered, but also allows for the identification of recently threatened sites and evaluation of any threats to these sites.

The result of field prospection in Gebelein allowed not only to confirm the location of the sites inside the "Archaeological Hazard Zone", but also allowed us to determine which areas of the Gebelein site complex are the most threatened by modern development and hence which should be documented first. In the case of Gebelein, the project confirmed some recent destructions oflarge parts of the central necropolis, made by bulldozers, and the spread of the fields in valleys between the hills of Gebelein. As a result of the project, specialized maps were made, which clearly showed the scale of the destructions and the threats at Gebelein. This helped the local Inspectorate of Antiquities in Esna to take necessary steps in order to prevent further spreading of the agricultural fields and contemporary settlements. This illustrates both the threats facing sites in the area, and how targeted investigation based on the results of analyses can support and empower local authorities to take remedial action.

The method described in the article could be applied on any comparable area before archaeological field prospections and help in planning and prioritizing the archeological interventions, before sites are completely destroyed. It also allows to not only visualize the magnitude of changes in the landscape during the last 50 years, but also to quantify the number of sites which are currently under threat.

Supplementary Materials: The following are available online at www.mdpi.com/2076-3263/7/4/97/s1, Table S1: The list of known archeological sites in Qena district and time change of their surroundings.

Acknowledgments: Research for this article is part of the Gebelein Archaeological Project and was possible thanks to a scholarship of the Polish Centre of Mediterranean Archaeology, of the University of Warsaw, and the finical support of the Consultative Council for Students' Scientific Movement at the University of Warsaw, and the University of Warsaw Foundation. The author would like to express gratitude for the help of Krzysztof Misiewicz, Wojciech Ejsmond and Cezary Baka, Anastazja Stupko-Lubczyńska, Marta Grzegorek, and Daniel V. Takacs.

Conflicts of Interest: The authors declare no conflict of interest.

References

1. Kemp, B.J. *Ancient Egypt Anatomy of a Civilization*, 2nd ed.; Routledge: New York, NY, USA; Oxon, UK, 2006; pp. 31–35.

2. Ejsmond, W.; Chyla, J.M.; Baka, C. Report from field reconnaissance AT Gebelein, Khozam and El-Rizeiqat. *Pol. Archaeol. Mediterr.* **2015**, *25*, 265–274.

3. Chyla, J.M.; Ejmond, W. Wyniki Rekonesansu na Stanowiskach Archeologicznych El-Amra i Hu. In *Medjat: Egyptological Studies*; Word Press: San Francisco, CA, USA, 2016; Volume 4, pp. 13–32.

4. Hopkins, N.S. Irrigation in contemporary Egypt. In *Agriculture in Egypt: From Pharaonic to Modern Times Proceedings of the British Academy*, 1st ed.; Bowman, A.K., Rogan, E., Eds.; Oxford University Press: Oxford, UK, 1999; Volume 96, pp. 367–385. ISBN 978-0-19-726183-5.

5. Jacotin, P. *Carte Topographique de l'Égypte et de Plusieurs Parties des Pays Limitrophes, Levée Pendant L'expédition de L'armée Française, par les Ingénieurs-Géographes, les Officiers du Génie Militaire et les Ingénieurs des Ponts et ChaussesAssujettie aux Observations des Astronomes, Construite par M. Jacotin, Colonel au Corps Royal des Ingénieurs-Géographes Militaires, Gravée au Dépôt Général de la Guerre à L'echelle de 1 Millimètre pour 100 Mètres. Publiée par Ordre du Gouvernement [Document Cartographique]*, 1st ed.; C.L.F. Panckoucke: Paris, France, 1818.

6. Survey of Egypt. *Qus; Khuzam; El Shaghab; Kiman el-Matana*, 1st ed.; Department of Survey and Mines: Cairo, Egypt, 1944.

7. CORONA Atlas of the Middle East. Centre for Advanced Spatial Technologies, University of Arkansas. US Geological Survey. Available online: http://corona.cast.uark.edu/ (accessed on 20 July 2017).

8. Chyla, J. Egipt na napoleońskich mapach i z satelity. *AD REM* **2012**, *3–4*, 7–10.

9. Singh, A. Digital change detection techniques using remotely-sensed data. *Int. J. Remote Sens.* **1989**, *10*, 989–1003. [CrossRef]

10. Lu, D.; Mausel, P.; Brondizio, E.; Moran, E. Change detection techniques. *Int. J. Remote Sens.* **2004**, *25*, 2364–2407. [CrossRef]

11. Golenia, M.; Gurdak, R.; Jarocińska, A.; Mierczyk, M.; Ochtyra, A.; Zagajewski, B. Application of LANDSAT images in Urban change detection. In *Forum GIS UW*, 1st ed.; Buławka, N., Chyla, J.M., Lechnio, L., Misiewicz, K., Stępień, M., Eds.; Zakład Graficzny Uniwersytetu Warszawskiego: Warsaw, Poland, 2016; Volume III, pp. 91–101, ISBN 978-83-61376-74-3.

12. Xu, Y.; Yu, L.; Zhao, Y.; Feng, D.; Cheng, Y.; Cai, X.; Gong, P. Monitoring cropland changes along the Nile River in Egypt over past three decades (1984–2015) using remote sensing. *Int. J. Remote Sens.* **2017**, *38*, 4459–4480. [CrossRef]

13. Abd El-Kawy, O.R.; Rød, J.K.; Ismail, H.A.; Suliman, A.S. Land Use and Land Cover Change Detection in the Western Nile Delta of Egypt Using Remote Sensing Data. *Appl. Geogr.* **2011**, *31*, 483–494. [CrossRef]

14. Aboel Ghar, M.; Shalaby, A.; Tateishi, R. Agricultural Land Monitoring in the Egyptian Nile Delta Using Landsat Data. *Int.J. Environ. Stud.* **2007**, *61*, 651–657. [CrossRef]

15. El-Gammal, M.I.; Ali, R.R.; Eissa, R. Land Use Assessment of Barren Areas in Damietta Governorate, Egypt Using Remote Sensing. *Egypt. J. Basic Appl. Sci.* **2014**, *1*, 151–160. [CrossRef]

16. Hegazy, I.R.; Kaloop, M.R. Monitoring Urban Growth and Land Use Change Detection with GIS and Remote Sensing Techniques in Daqahlia Governorate Egypt. *Int. J. Sustain. Built Environ.* **2014**, *4*, 117–124. [CrossRef]

17. Kato, H.; Kimura, R.; Elbeih, S.F.; Iwasaki, E.; Zaghloul, E.A. Land Use Change and Crop Rotation Analysis of a Government Well District in Rashda Village—Dakhla Oasis, Egypt Based on Satellite Data. *Egypt. J. Remote Sens. Space Sci.* **2012**, *15*, 185–195. [CrossRef]

18. Abd El-Aziz, A.O. Monitoring and change detection along the Eastern Side of Qena Bend, Nile Valley, Egypt Using GIS and Remote Sensing. *Adv. Remote Sens.* **2013**, *2*, 276–281. [CrossRef]

19. Ahmed, A.A.; Fogg, G.E. The Impact of Groundwater and Agricultural Expansion on the Archaeological Sites at Luxor, Egypt. *J. Afr. Earth Sci.* **2014**, *95*, 93–104. [CrossRef]

20. Villa, O. Remote Sensing Applications in Archaeology. *Archeol. Calcolatori* **2011**, *22*, 147–168.

21. Campana, S. Archaeological Site Detection and Mapping: Some thoughts on differing scales of detail and archaeological 'non-visibility'. In *Seeing the Unseen—Geophysics and Landscape Archaeology*, 1st ed.; Campana, S., Piro, S., Eds.; Taylor & Francis: London, UK, 2009; pp. 5–26. ISBN 978-0-203-88955-8.

22. Lambers, K.; Zingman, I. Towards detection of archaeological objects in high-resolution remotely sensed images: The Silvretta case study. In *Archaeology in the Digital Era*, 1st ed.; Sly, T., Chrysanthi, A., Murrieta-Flores, P., Papadopoulos, C., Romanowska, I., Wheatley, D., Eds.; Amsterdam University Press: Amsterdam, The Netherlands, 2012; Volume II, pp. 781–791. ISBN 10: 9089646639.

23. Bewley, R.; Wilson, A.I.; Kennedy, D.; Mattingly, D.; Banks, R.; Bishop, M.; Bradbury, J.; Cunliffe, E.; Fradley, M.; Jennings, R.; et al. Endangered Archaeology in the Middle East and North Africa: Introducing the EAMENA Project. In *CAA 2015. Keep the Revolution Going*, 1st ed.; Campana, S., Scopigno, R., Carpentiero, G., Cirillo, M., Eds.; Archeopress Archaeology: Oxford, UK, 2016; pp. 919–932. ISBN 9781784913373.

24. Brodie, N.; Contreras, D.A. The economics of the looted archaeological site of Bab edh-Dhra: A view from Google Earth. In *All the King's Horses. Essays on the Impact of Looting and the Illicit Antiquities Trade on Our Knowledge of the Past*, 1st ed.; Lazrus, P.K., Barker, A.W., Eds.; Society for American Archaeology: Washington, DC, USA, 2012; pp. 9–24, ISBN 9780932839442.

25. NASA Global Land Cover Facility. Available online: glcf.umd.edu/data/landsat (accessed on 13 April 2013).

26. Landsat 1 History. USGS. Available online: https://landsat.usgs.gov/landsat-1-history (accessed on 20 July 2017).

27. Landsat 8 History. USGS. Available online: https://landsat.usgs.gov/landsat-8-history (accessed on 20 July 2017).

28. LANDSAT Catalogue of Known Issues. USGS. Available online: https://landsat.usgs.gov/known-issues (accessed on 20 July 2017).

29. Zagajewski, B. Ocena przydatności sieci neuronowych i danych do klasyfikacji Tatr Wysokich. *Teledetekcja Środowiska* **2010**, *43*, 57.

30. Estman, J.R. *Guide to GIS and Image Processing*, 2nd ed.; Clark University: Worcester, MA, USA, 2001; Volume 2, pp. 41–46.

31. CULTNAT. *General Map of Quena Archaeological Sites*, 1st ed.; Supreme Council of Antiquities, Library of Alexandria: Alexandria, Egypt, 2007; Volume 9–10.

32. Hendrickx, S.; van den Brink, E.C.M. Inventory of Predynastic and Early Dynastic Cemetery and Settlement Sites in the Egyptian Nile Valley. In *Egypt and the Levant: Interrelations from the 4th through the Early 3rdMillennium BCE*, 1st ed.; van den Brink, E.C.M., Levy, E., Eds.; Leicester University Press: Leicester, UK, 2000; pp. 346–393, ISBN 9780718502621.

33. Petrie, W.M.F. *Diospolis Parva: The cemeteries of Abadiyeh and Hu, 1898–1899*, 1st ed.; Offices of the Egypt Exploration Found: London, UK, 1901.

34. Ejsmond, W.; Chyla, J.M.; Witkowski, P.; Wieczorek, D.F.; Takacs, D.; Ożarek-Szilke, M.; Ordutowski, J. Comprehensive field survey at Gebelein: Preliminary results of a new method in processing data for archaeological site analysis. *Archaeol. Pol.* **2015**, *53*, 617–621.

geosciences

MDPI

Article

Using Open Access Satellite Data Alongside Ground Based Remote Sensing: An Assessment, with Case Studies from Egypt's Delta

Sarah Parcak [1,*], Gregory Mumford [1] and Chase Childs [2]

1   Department of Anthropology, The University of Alabama at Birmingham, Birmingham, AL 35294, USA;
    gmumford@uab.edu
2   Globalxplorer, Birmingham, AL 35203, USA; projects@globalxplorer.org
*   Correspondence: sparcak@uab.edu; Tel.: +1-205-996-3508

Received: 31 July 2017; Accepted: 18 September 2017; Published: 27 September 2017

Abstract: This paper will assess the most recently available open access high-resolution optical satellite data (0.3 m–0.6 m) and its detection of buried ancient features versus ground based remote sensing tools. It also discusses the importance of CORONA satellite data to evaluate landscape changes over the past 50 years surrounding sites. The study concentrates on Egypt's Nile Delta, which is threatened by rising sea and water tables and urbanization. Many ancient coastal sites will be lost in the next few decades, thus this paper emphasizes the need to map them before they disappear. It shows that high resolution satellites can sometimes provide the same general picture on ancient sites in the Egyptian Nile Delta as ground based remote sensing, with relatively sandier sedimentary and degrading tell environments, during periods of rainfall, and higher groundwater conditions. Research results also suggest potential solutions for rapid mapping of threatened Delta sites, and urge a collaborative global effort to maps them before they disappear.

Keywords: Worldview-3; Worldview-2; CORONA; magnetometer; ground penetrating radar; Nile Delta; Egypt

1. Introduction

Landscapes across the Middle East have presented many opportunities for testing a broad range of remote sensing, from ground to space-based imagery, over the past 95 years. Thanks to the first aerial photographs taken of Stonehenge by balloon in 1906, followed by reconnaissance pilots such as Father Antoine Poidebard in Syria during the 1920s, the entire field of aerial archaeology was born. Now, there is great urgency to map the rapidly disappearing landscapes of the Middle East and North Africa (MENA) region. With climate change, rising sea levels, urbanization, looting, and site destruction from civil war, for many sites, remotely sensed data is the main evidence we now have for their existence. Many RAF photos are available for Egypt from the 1930s, which are invaluable today in relation to intensive landscape changes following the construction of the Aswan High Dam. CORONA high resolution space photography from the late 1960s to early 1970s is the best data for archaeologists to examine prior to major urbanization and population growth with many discoveries at and around sites emerging from these datasets [1–3]. While archaeologists have used satellite images since the early 1980s, this technology saw limited use in the Middle East until the late 1990s to early 2000s [4,5] Since then, archaeologists working in the MENA region have applied diverse datasets to a myriad of issues, including regional site detection [6,7] and looting [8–12].

This paper shows how CORONA imagery and high-resolution satellite sensors can reveal subsurface features on sites in the Egyptian Nile Delta. It focuses on primarily visual detection methods across broad areas, where the prohibitive cost of multi-year and multi-sensor data make open sources tools like Google Earth essential. Sensors that capture images at the optimum time can reveal the layout of an entire subsurface ancient town, but the window for this data capture is narrow and can vary from year to year. This paper also shows how CORONA data helped an archaeological team to locate the enclosure wall of an ancient temple at the site of Tell Tebilla (northeast Nile Delta), where a magnetometer survey produced tomb and residential-level outlines, but it was not done at a sufficient scale and coverage to capture the presence or dimensions of the wall. Satellite data can now provide the same general picture as ground-based remote sensing tools, while the finer detail is generally missing unless 0.3 m data is available. The quality of satellite data for revealing buried features in Egypt's Nile Delta is dependent on the season, type of soil matrix at a given site (it must contain higher amounts of sand), and groundwater levels in the surrounding fields. This paper discusses the advantages and disadvantages of both satellite and ground-based remote sensing systems. Ideally, they will be used together, but this is not always possible.

2. Materials and Methods

The following section describes how both CORONA and high-resolution satellite imagery helped to map features on sites in Egypt's Nile Delta.

2.1. CORONA Imagery in Egypt

CORONA high-resolution satellite imagery is valuable to archaeologists, due to its high resolution, low cost, broad accessibility and its value in recording landscapes now built over or destroyed, especially in the Middle East. Declassified in 1995 by President Bill Clinton, the U.S. government released 860,000 images (taken between 1960 and 1972) to the public at the cost of reproduction only. The declassified intelligence imagery, code-named CORONA, ARGON and LANYARD, has helped to locate patterns in landscapes that have been changed through modern farming techniques and sites threatened by dams and construction projects. The CORONA systems included the labels KH-1, KH-2, KH-3, KH-4, KH-4A, and KH-4B, while the ARGON system was designated KH-5 and the LANYARD system KH-6. The image resolution of each system varies greatly, from one to two meters for the KH-4B system to 460 feet for the KH-1 system. The KH-4B mission series is of interest to archaeologists doing work across the Middle East, since this represents one of the primary efforts mapping the 1967 cease-fire between Egypt and Israel. In some cases, the ground resolution of this mission can reach one-two meters at nadir. In general, the KH-4A, KH-4B and KH-6 systems have the best resolution and quality, but earlier images may be useful for documenting widescale site changes. CORONA imagery is an excellent dataset for viewing longer term change in many areas of interest, and may highlight cultural features that are ephemeral today.

Numerous archaeological projects in the Middle East have made use of CORONA imagery, documenting regional landscape changes, routes to and from ancient sites, and ancient environmental features that modernization may have altered [13–15]. In Egypt, several studies have incorporated CORONA imagery in the reconstruction of palaeogeography as well as site change over time [6,16].

The initial analysis on the CORONA imagery for the Nile Delta took place in 2003, required the acquisition of the black and white CORONA negatives, and scanning them at 10 microns. The original CORONA imagery obtained in 2003 was from the KH 4B 1968 1105–2235 mission. Today, there is an invaluable map platform for Egyptologists, based at the Center for Advanced Spatial Technologies, University of Arkansas [17]. This project obtained CORONA imagery across North Africa and the Middle East, scanning and orthorectifying the data to be used in a Google Earth-like platform [18]. Some site locations in Egypt's Nile Delta are noted, but the majority are not indicated there. Given the high scanning quality of the data, this team utilized it to compare against the Google Earth imagery for landscape and site changes.

The CORONA imagery used in the paper came from the 1105-2235Fore mission (taken on 18 November 1968). An initial assessment revealed the enclosure wall of a temple wall dated to the late 4th century BC at Tell Tebilla (Figure 1) in the Northeast Delta which wasexcavated by a

Canadian project team in 2003 (Figures 2 and 3). This work was done in collaboration with Egypt's Supreme Council of Antiquities, now Ministry of Antiquities. The enclosure wall had been partially disturbed by a modern water filtration plant, so the older imagery proved invaluable for visualizing the feature for further exploration.

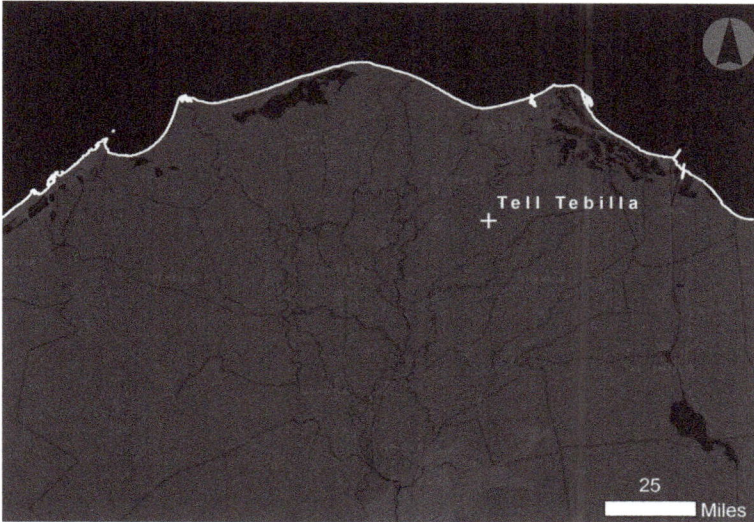

Figure 1. Location of Tell Tebilla, Eastern Delta, Egypt. Imagery courtesy Chase Childs.

Figure 2. CORONA KH-4B satellite image of Tell Tebilla, taken in 1972. Note the red arrow pointing to the enclosure wall of the Late Period (ca. 712–332 BC) temple. Imagery courtesy of the Center for Advanced Spatial Technologies, University of Arkansas/U.S. Geological Survey.

Figure 3. Excavating the eastern side of the mudbrick enclosure wall, 2003 Tell Tebilla project. The modern water plant is in the background. Imagery courtesy of S. Parcak.

An unusual subsurface rectangular feature surrounding one-quarter of the site in its northwest quadrant was identified adjacent to Tell Tebilla based on examination of the imagery. Changing contrast and brightness settings highlighted this feature, which represents an enclosure wall. The discovery proved to be particularly significant based on previous knowledge regarding traces of a destroyed temple from the same northwest quadrant. The minimum size of the Tell Tebilla temple, dating possibly as early as the New Kingdom (ca. 1550 BC–1069 BC) to early Third Intermediate Period (ca. 1069 BC–712 BC), with blocks of Ramesses II (found on the surface), had been reconstructed hypothetically from over 300 surface blocks that were photographed, measured and recorded by the Tell Tebilla team ([19] pp. 267–286). We determined the temple's exact location by speaking to Egyptian Antiquities Inspector Said el-Tahwle, who claimed that municipal digging had uncovered a 15 m × 15 m limestone block paving with column bases and a drain channel since destroyed by the construction of a water plant. The temple had likely been destroyed mostly during the Roman Period (30 BC–395 AD) (similar destruction occurs at Mendes, twelve km to the south). The extant ex-situ blocks suggest it measured a minimum of 12 m–18 m by 30 m–35 m, and was dedicated to the triad of Osiris-Isis-Horus, but also included deities such as Sobek, Duamutef, Imsety, Qebahsenuef and Hapy [19–21].

2.2. An Overview of Ground-Based Remote Sensing in Egypt

Ground-based remote sensing tools, like resistivity, magnetometry and ground-penetrating radar (GPR) surveys have helped archaeologists to identity subsurface ancient structures in Egypt (Figure 4), including temple enclosure walls, houses, tombs and palaces and have even helped to map entire cities. At tell sites in the Delta and Nile Valley, magnetometers are the ground survey tool of choice for most Egyptological expeditions. Magnetometers can, however, sometimes miss features in sandier environments. GPR surveys are the tool most often chosen for subsurface mapping efforts in the desert. They are not as greatly affected by buried metal objects compared to magnetometers, and can provide a 3D time-slice of buried features. All subsurface detection techniques have challenges in environments where modern city layers may make it more difficult to detect the more deeply buried archaeological remains.

At the sites of Mendes and Tell Tebilla, L. Pavlish conducted magnetometer surveys using a fluxgate gradiometer [22]. These techniques helped to highlight key portions of these sites for archaeological excavation. At Tell Tebilla, the magnetometry survey showed the outline of a complex (Figure 5), later found by excavators to be the remains of a structure reused as a tomb, preserved over seven meters deep. H. Becker and J. Fassbinder have done magnetometry work at Pi-Ramesses, which revealed the outlines of many structures now beneath agricultural fields [23].

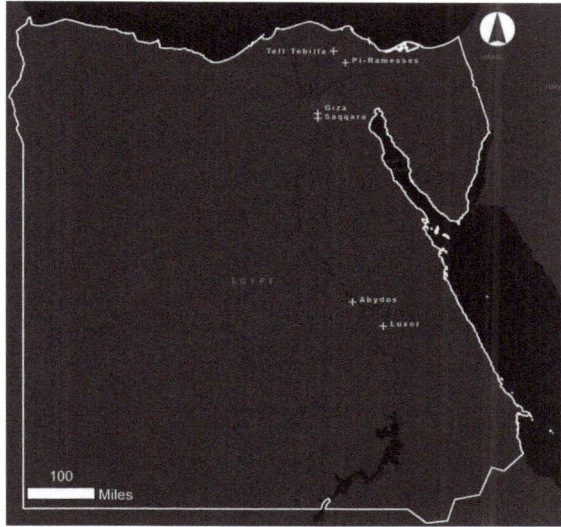

Figure 4. Map of key sites where magnetometry work has taken place. Image courtesy Chase Childs.

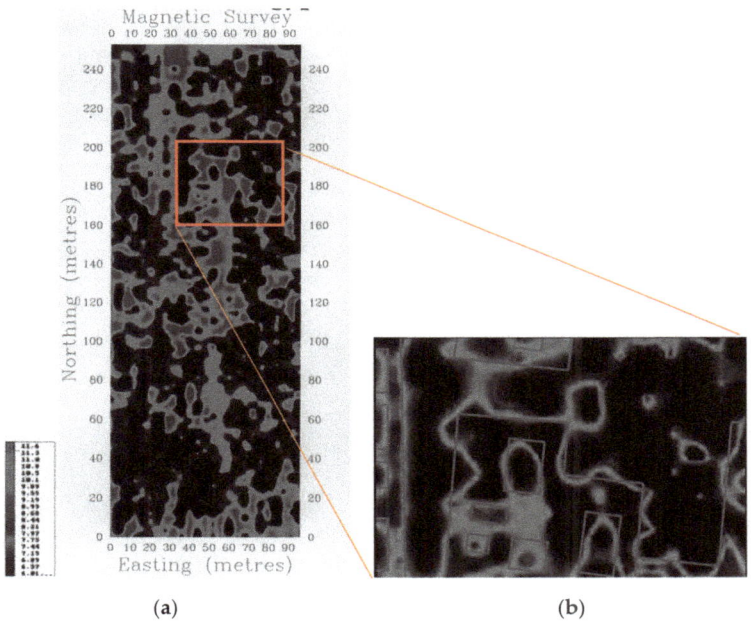

(a)   (b)

Figure 5. (a) Partial results from the 2003 magnetometer survey of Tell Tebilla indicating buried mudbrick from a housing/cemetery installation, with detail of inset; (b) Shows the location of the mudbrick walls Imagery courtesy Larry Pavlish and Greg Mumford.

In Luxor, L. Pavlish has carried out magnetic surveys over the remains of a mortuary temple equated with Amenhotep I, showing the outlines of this temple complex and related buildings. At Saqqara, I. Mathieson has revealed the outlines of an ancient settlement and nearby road [24]. At Abydos T. Herbich has found many potential Early Dynastic tombs and related enclosures [25]. In Egypt, subsurface remote sensing is clearly a standard part of excavation projects. However, bringing scientific equipment into Egypt can be problematic, with periodic challenges of clearing equipment through customs. A number of projects have had equipment sit in customs for months. Archaeologists may not be allowed to use the equipment in sensitive areas. Now, with limited budgets, it is not always feasible to include ground based remote sensing. It is also not practical to use ground based remote sensing for surveying of many sites.

This subsurface feature detection paper originated unexpectedly out of a 2014 study to examine open source methods for the detection of archaeological looting and site damage in Egypt via satellite imagery. This study [12] used both high-resolution and free access satellite data from 2002 to 2013 for a database of 1200 "sitescapes", or archaeological zones, in Egypt. The study found nearly 200,000 looting pits at 279 sites, with major spikes in looting and site destruction following the 2009 recession, and again following the 2011 Arab Spring. As our team examined Nile Delta Google Earth imagery from multiple years, we started to notice likely near-surface archaeological features at numerous sites. The Google Earth imagery represents DigitalGlobe data, which includes Worldview-1, Worldview-2, Worldview-3, Worldview-4, Quickbird-1, Quickbird-2, and Geoeye datasets. Their resolutions range from 0.3 m to 0.6 m. While the Google Earth data is true color, it can be useful in comparing data from multiple seasons. We spent additional time examining each Nile Delta site's satellite image closely, as walls and features can be ephemeral compared with looting pits. When we noted potential features, we marked the site or its components as "likely feature", and then later examined each likely site in a team effort to eliminate false positives. Since this current study is taking place in 2017, we have examined all additional Google Earth satellite data for the sites since 2014.

With a total study area of 22,000 km^2, purchasing high resolution data for 12 years of the entire Nile Delta region would have cost approximately 5.3 million US Dollars (assuming an average cost of $20.00 per sq. km). This does not include the time spent downloading and processing the data. While the data analysts were paid appropriately for their time on this project, employment represents a cost equal to 1% of the raw physical data. The visible aspect of satellite images should not be overlooked, as it can allow the creation of maps in areas where good maps do not exist or are not easily accessible, and make it possible to search for features before initiating in-depth survey season research. The main disadvantage of visual detection alone is that unclear features remain ambiguous. Purchasing the data (assuming unlimited funds and time) would have allowed for actual processing, and a higher percentage of the Nile Delta sites may have had apparent features. There are tradeoffs between cost and scale of data analysis that all researchers must consider.

3. Results

Nile Delta Remote Sensing Results

At 20 sites in the Nile Delta, an assessment of multiple years of visual site data (from varying months) revealed subsurface archaeological featuring ranging from single houses to clear outlines of settlements. While we cannot divulge their exact locations (for site protection purposes), we can describe their general region, size, observed features, vegetation coverage (if any), and general estimate of occupation time-period (some of which have previous survey data). We also relate the extent to which the sites and their landscapes have changed in a nearly 50-year period. It is noted that the spelling of specific site names in Egypt can vary widely. Much of the previous survey work is described on the Egypt Exploration Society's Nile Delta survey website [26]. The visible above-ground site size can be measured via the most recently available satellite imagery (2016 or 2017).

Abu el Humar: Located in the western Nile Delta, the site measures 300 m × 320 m, and is a lighter colored sandy mound with vegetation covering 50% of the site. A 180 m long enclosure was can be seen, with a clear southern wall and tapering eastern and western walls. The wall is most clear in an April 2002 satellite image. Given the size of the enclosure wall and other features at the site, it may be Late Period-Roman period in date (ca. 712 BC–395 AD) [26]. The village has tripled in size in almost 50 years, but the overall size of the site has not changed (Figure 6).

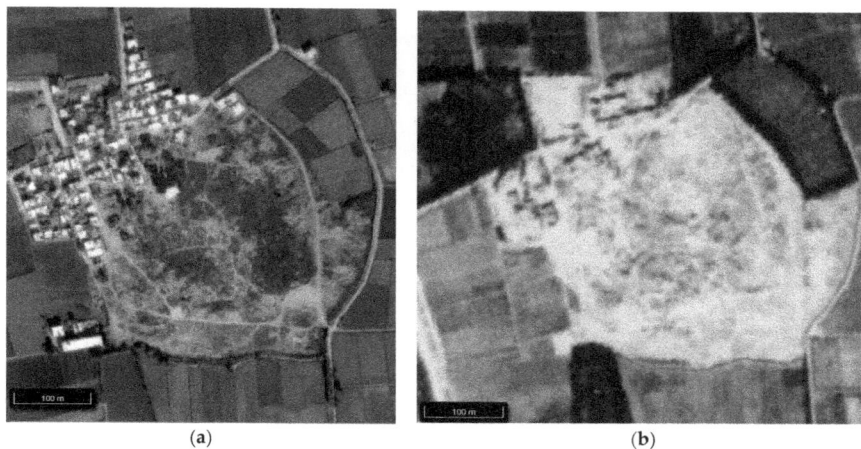

(a)   (b)

Figure 6. (a) The site of Abu el Humar. Image courtesy Google Earth; (b) 1968 CORONA image. Image courtesy of the Center for Advanced Spatial Technologies, University of Arkansas/U.S. Geological Survey.

Abu Mitawi: Located in the Central Nile Delta, the site measures 520 × 250 m, is mainly a lighter colored sandy mound, with no vegetation. The site is heavily looted along the north side, but its central and southern sections have multiple ancient subsurface structures. The southwestern part of the site has a series of larger and interconnected structures, perhaps part of an administrative complex, most clearly seen in a December 2015 image. The site is Saite Period-Roman Period (664 BC–395 AD). (Dated using ceramic evidence from previous survey work) [26] (Figure 7). The overall site seems to have shrunk since 1968, with 100-m sections to the west and east of the site now covered by modern fields.

Daba el-Qibli: Located in the western Nile Delta, the site measures 240 m × 280 m, and is a lighter colored sandy mound with limited vegetation. Multiple structures appear in a December 2014 image. The site dates from the Roman Period (30 BC–395 AD) (dated by ceramic evidence from previous survey work) [26] (Figure 8). The site has shrunk by 75% since 1968, being cut into by agricultural fields.

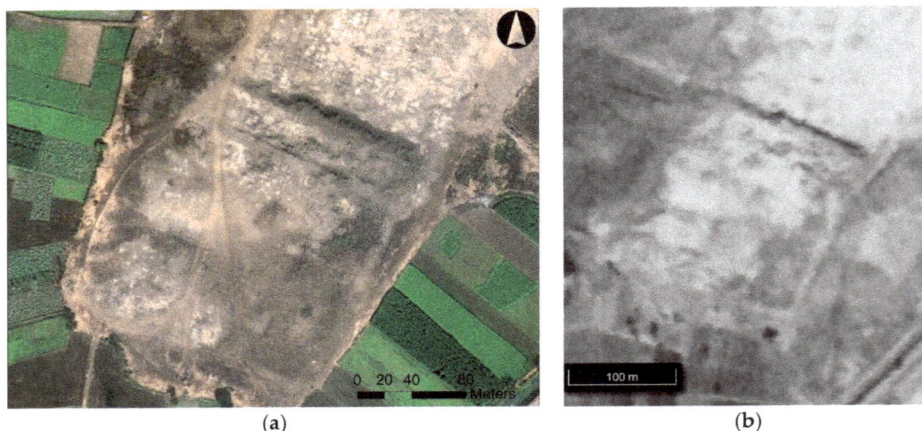

Figure 7. (**a**) The site of Abu Mitawi. Image courtesy Google Earth; (**b**) 1968 CORONA image. Image courtesy of the Center for Advanced Spatial Technologies, University of Arkansas/U.S. Geological Survey.

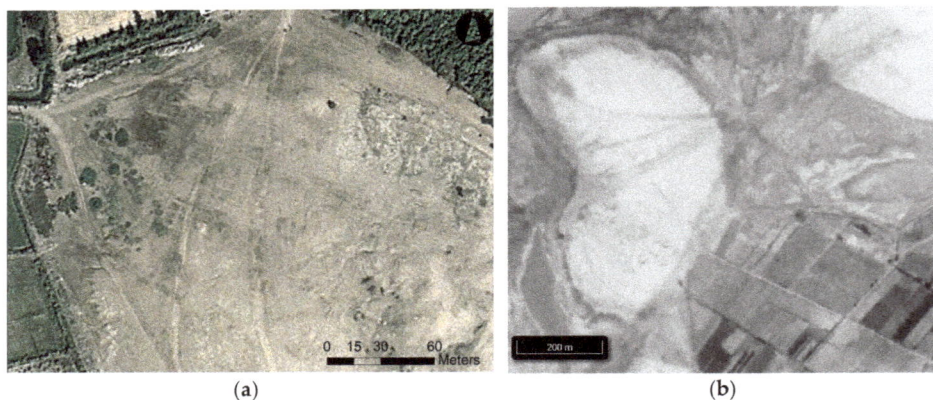

Figure 8. (**a**) The site of Daba el-Qibli. Image courtesy Google Earth; (**b**) 1968 CORONA image. Image courtesy of the Center for Advanced Spatial Technologies, University of Arkansas/U.S. Geological Survey.

Kom Biltus: Located in the western Nile Delta, the site measures 340 m × 275 m, and is covered in sparse vegetation. There are smaller features apparent in a September 2005 image. The site might have some New Kingdom (ca. 1550 BC–1069 BC) construction based on their shape (Figure 9). The southernmost part of the site has been cut into by modern agriculture since 1968.

Kom Bunduq: Located in the western Nile Delta, the site measures 240 m × 220 m, and has no vegetation. Some smaller features are visible in a November 2011 image, while others appear in the central part of the tell in a February 2013 image. The site dates to the Late Roman to Islamic Periods (ca. 395–1517 AD) [26] (Figure 10). In the 1968 image, the site appears to have a lower section (similar to sites in Syria or Iraq), now beneath agricultural fields.

Figure 9. (**a**) The site of Kom Biltus. Image courtesy Google Earth; (**b**) 1968 CORONA image. Image courtesy of the Center for Advanced Spatial Technologies, University of Arkansas/U.S. Geological Survey.

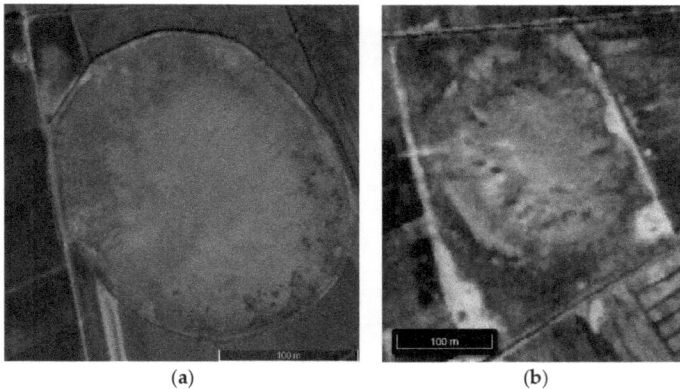

Figure 10. (**a**) The site of Kom Bunduq. Image courtesy Google Earth; (**b**) 1968 CORONA image. Image courtesy of the Center for Advanced Spatial Technologies, University of Arkansas/U.S. Geological Survey.

Kom ed-Dahaba (1): Located in the eastern Nile Delta, the site measures 550 m × 750 m, with no vegetation, and appears to be a very lighter colored sandy tell. A well-planned town can be seen in a September 2011 satellite image. The remaining town measures 360 m × 460 m. There are clear roads running N-S and E-W, with one subsurface structure in the western part of the town measuring 30 m × 40 m. The buildings in the southern part of the town seem to be upper middle class in scale, measuring 17 m × 40 m, with 8–12 rooms, while there are 10 m × 10 m 4-room houses in the north, where the roads veer to the northeast; the town does not appear to be as well planned. There is one clearer 40 m × 60 m structure in the western part of the town. There are some similarities between this town and the Third Intermediate Period (ca. 1069 BC–712 BC) town mapped at Tanis, but given the regular layout of the streets, it is probably Ptolemaic-Roman Period (332 BC–395 AD) versus New Kingdom (ca. 1550 BC–1069 BC) (Figure 11). In the 1968 imagery, the site stood along the edge of Lake Manzala, and is now surrounded by agricultural fields.

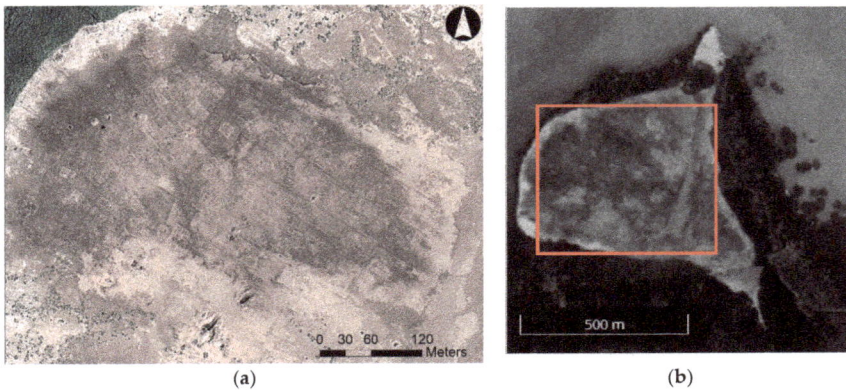

Figure 11. (**a**) The site of Kom ed Dahaba. Image courtesy Google Earth; (**b**) 1968 CORONA image. Area in image (**a**) is within the red box. Image courtesy of the Center for Advanced Spatial Technologies, University of Arkansas/U.S. Geological Survey.

Kom ed-Dahaba (2): Located in the western Nile Delta, the site measures 640 m × 220 m, and is mainly a sandy silt mound. The site is half-covered with grass and low-laying vegetation. Several rectilinear features appear in the western part of the site. They are most clear in October 2009 and 2011 images, and a late September 2009 image. The occupation time-period is unknown (Figure 12). In 1968, the site was devoid of any structures, but in the modern imagery, an installation (possibly military) covers 1/3 of the site.

Figure 12. (**a**) The site of Kom ed Dahaba. Image courtesy Google Earth; (**b**) 1968 CORONA image. Image courtesy of the Center for Advanced Spatial Technologies, University of Arkansas/U.S. Geological Survey.

Tell el-Arab: Located in the eastern Nile Delta, the site measures 340 m × 230 m, and is comprised of silt with a comparatively lighter color than other sites investigated. There is no vegetation on the site except along the edges. A clear map of a subsurface settlement is apparent in a mid-September 2011 image, measuring 140 m × 170 m, dating from the Late Roman to Coptic Periods (ca. 395 AD–620 AD) [26] (Figure 13). Modern agriculture has cut the site in half since 1968, with evidence that it was once a part of Tell el-Ghuzz to the northwest.

Figure 13. (**a**) The site of Tell el-Arab. Image courtesy Google Earth; (**b**) 1968 CORONA image. Image (**a**) is within the red box. Image courtesy of the Center for Advanced Spatial Technologies, University of Arkansas/U.S. Geological Survey.

Kom el-Arab: Located in the western Nile Delta, the site measures 560 m × 550 m, and is composed of a lighter colored sandy silt. Along its western edge, there is what appears to be a Roman Period settlement measuring 180 m × 130 m, appearing on a November satellite image. The site has Coptic-Islamic period pottery (ca. 395 AD–1517 AD) [26] (Figure 14). In 1968, the site was devoid of modern structures, and now, they cover 25% of the site, mainly in the south.

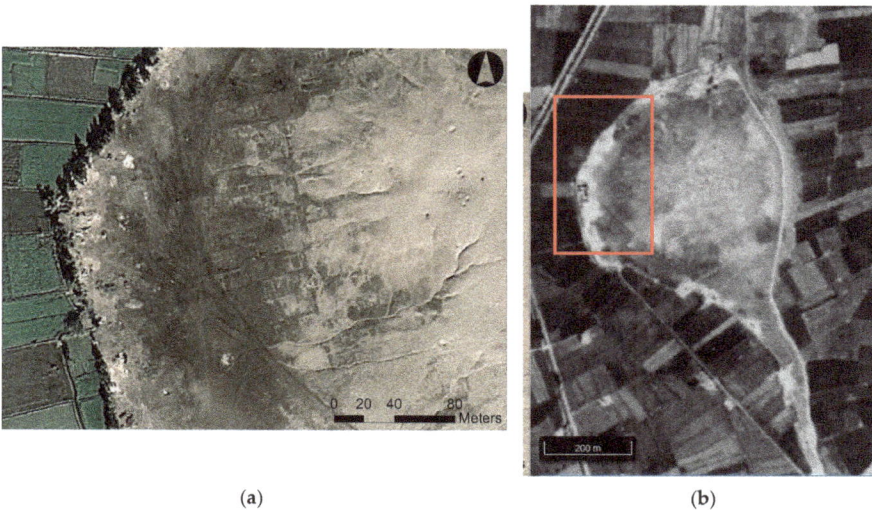

Figure 14. (**a**) The site of Kom el Arab. Image courtesy Google Earth; (**b**) 1968 CORONA image. The area of image (**a**) is in the red box. Image courtesy of the Center for Advanced Spatial Technologies, University of Arkansas/U.S. Geological Survey.

Kom el-Barud: Located in the western Nile Delta, the site measures 375 m × 250 m, composed of a lighter colored sandy silt, but 95% is covered with vegetation. On a late October 2011 image, multiple structures are evident through the vegetation. Previous survey work has shown the site dates to the

Greco-Roman and Late Roman periods (332 BC–640 AD) [26] (Figure 15). The site appears unchanged in 50 years.

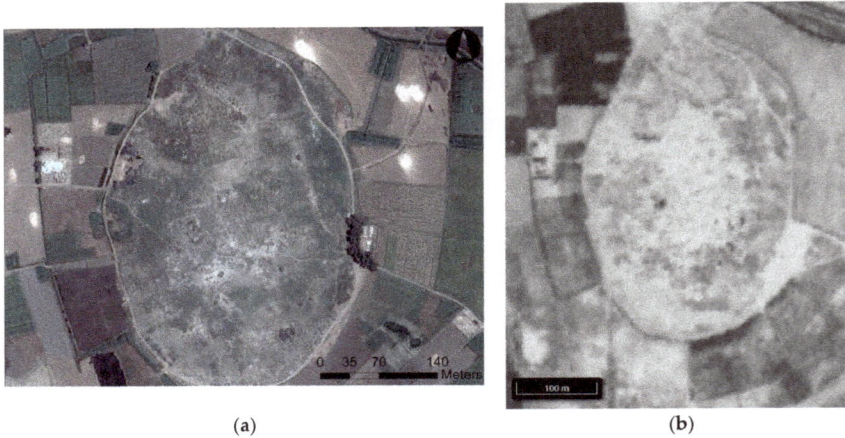

(a)　　　　　　　　　　　　　　　　　　(b)

Figure 15. (**a**) The site of Kom el-Barud. Image courtesy Google Earth; (**b**) 1968 CORONA image. Image courtesy of the Center for Advanced Spatial Technologies, University of Arkansas/U.S. Geological Survey.

Kom el-Garad: Located in the western Nile Delta, the site measures 200 m × 170 m, and is composed of a lighter colored sandy silt with virtually no vegetation. In a mid-November 2011 image, a 125 m × 135 m settlement is apparent, with multiple structures. While the habitation time-period of the site is not known, the rectilinear layout of the housing is suggestive of the Roman Period (30 BC–395 AD) (Figure 16). The site does not appear to have changed in 50 years.

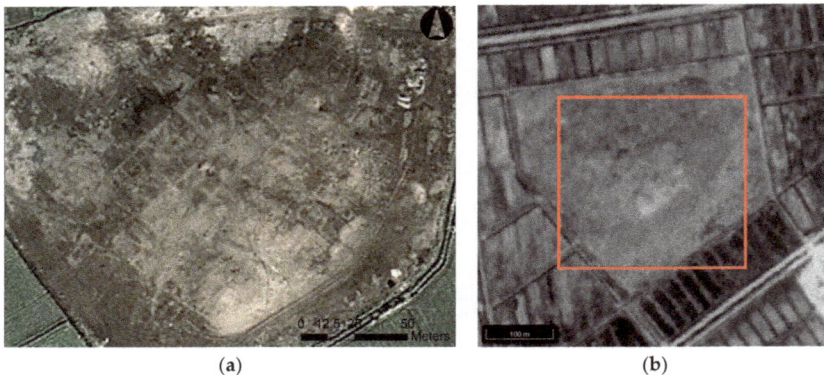

(a)　　　　　　　　　　　　　　　　　　(b)

Figure 16. (**a**) The site of Kom el-Garad. Image courtesy Google Earth; (**b**) 1968 CORONA image. The area of image (**a**) is within the red box. Image courtesy of the Center for Advanced Spatial Technologies, University of Arkansas/U.S. Geological Survey.

Tell Belim: Located in the eastern Nile Delta, the site measures 570 m × 1000 m, and contains a darker colored silt. This is a well-known site, under excavation 15 years ago by the Egypt Exploration society [27]. A great many structures, as well as the outline of a New Kingdom (ca. 1550 BC–1069 BC) temple enclosure wall [28], seen on satellite imagery and excavated by the EES, are apparent on a late July 2013 image, in what appears to be a period of inundation for the fields surrounding the site. It is curious that an image taken the very next day does not reveal as many structures. The temple enclosure is also visible on a 1935 RAF aerial photograph. The survey and excavation work date the site to the New Kingdom (ca. 1550 BC–1069 BC) and Ptolemaic-Late Roman periods (332 BC–640 AD) [26] (Figure 17). The site used to lie along the edge of Lake Manzala, and has been cut into by modern agriculture.

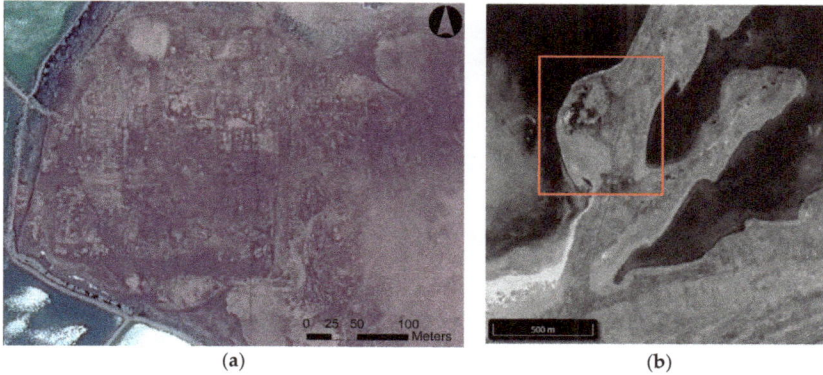

(a)   (b)

Figure 17. (a) The site of Tell Belim. Image courtesy Google Earth; (b) 1968 CORONA image. The area of image (a) is in the red box. Image courtesy of the Center for Advanced Spatial Technologies, University of Arkansas/U.S. Geological Survey.

Tell Buweib: Located in the eastern Nile Delta, the site measures 535 m × 270 m, composed of a medium colored darker silt. It is a well-known site, with excavations taking place from 2004 onwards. On an early January 2010 image, the outline of structures and a temple enclosure wall are visible. The site dates from the New Kingdom (ca. 1550 BC–1069 BC) to Late Period. (ca. 712 BC–332 BC) [23,25] (Figure 18). The site looks like it has lost 25% of its area in the last 50 years due to expanding agricultural fields.

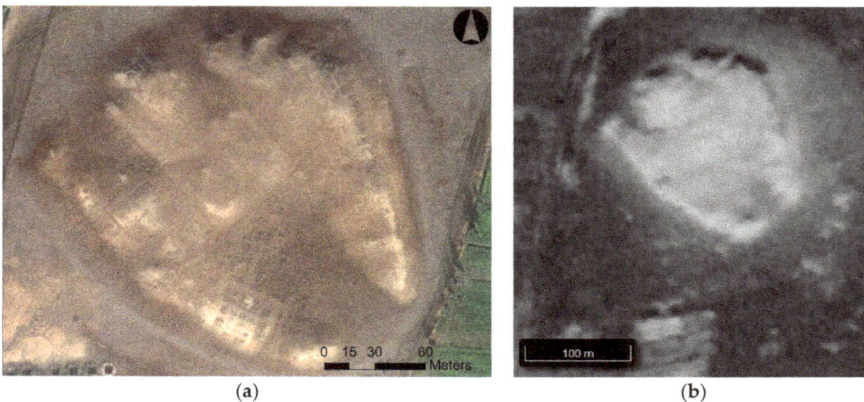

(a)   (b)

Figure 18. (a) The site of Tell Buweib. Image courtesy Google Earth; (b) 1968 CORONA image. Image courtesy of the Center for Advanced Spatial Technologies, University of Arkansas/U.S. Geological Survey.

Tell el-Damalun: Located in the western Delta, the site measures 250 m × 550 m, and is composed of a medium colored sandy silt with limited vegetation along its southern side. A mid-September 2011 image displays the outline of structures. The site date is unknown, but the buildings are suggestive of the Roman Period (30 BC–395 AD) [26] (Figure 19). 10% of the site has been lost to agricultural expansion since 1968.

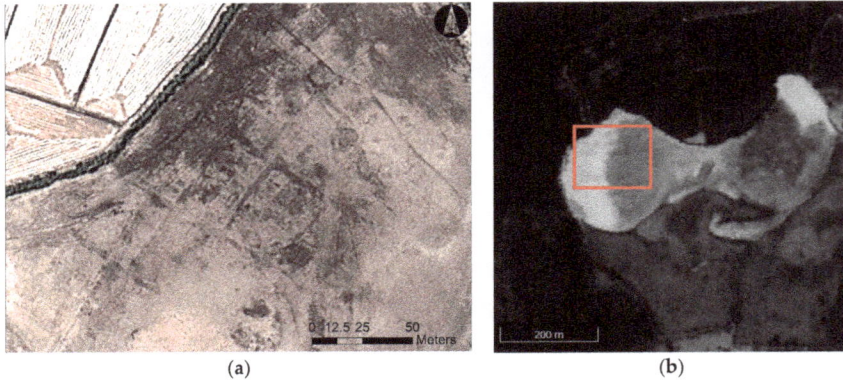

Figure 19. (**a**) The site of Tell el-Damalun. Image courtesy Google Earth; (**b**) 1968 CORONA image. The area of image (**a**) is within the red box. Image courtesy of the Center for Advanced Spatial Technologies, University of Arkansas/U.S. Geological Survey.

Tell el-Gassa: Located in the eastern Nile Delta, the site measures 340 m × 230 m, and has a medium colored sandy silt with limited vegetation along its northwestern side. In a mid-September 2011 image, the outline of structures is clear, including what appears to be a 12 m thick temple enclosure wall and the outline of the temple platform itself, measuring 15 m × 20 m. The structure could also be a fortress. There are multiple clear structures to the north of the enclosure wall. Previous survey work dates the site to the Late Roman Period (395–640 AD) [26] (Figure 20). The site was an island in 1968, located close to the shores of ancient Lake Manzala. It has now decreased in size by 50%.

Tell er-Ruhban: Located in the eastern Nile Delta, the site measured 260 m × 140 m in 2009, and is composed of a darker colored medium sandy silt. The mound measured 650 m × 350 m in 2004. In a late January 2010 image, clear rectilinear structures appear across the site. The occupation time-period is unknown (Figure 21).

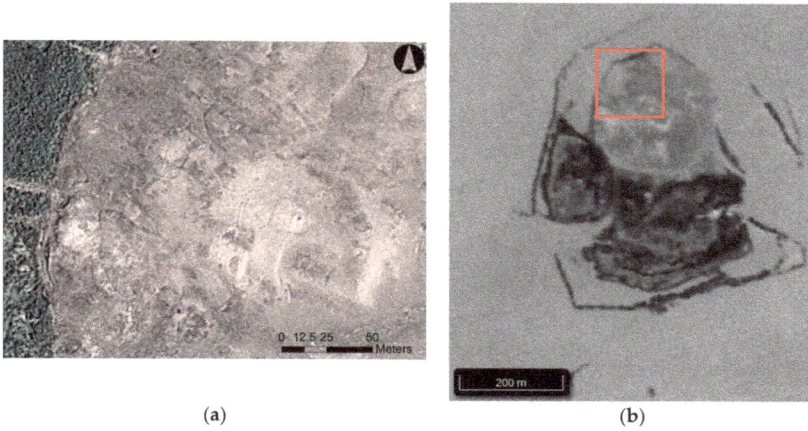

Figure 20. (**a**) The site of Tell el-Gassa Image courtesy Google Earth; (**b**) 1968 CORONA image. The area of image (**a**) is in the red box. Image courtesy of the Center for Advanced Spatial Technologies, University of Arkansas/U.S. Geological Survey.

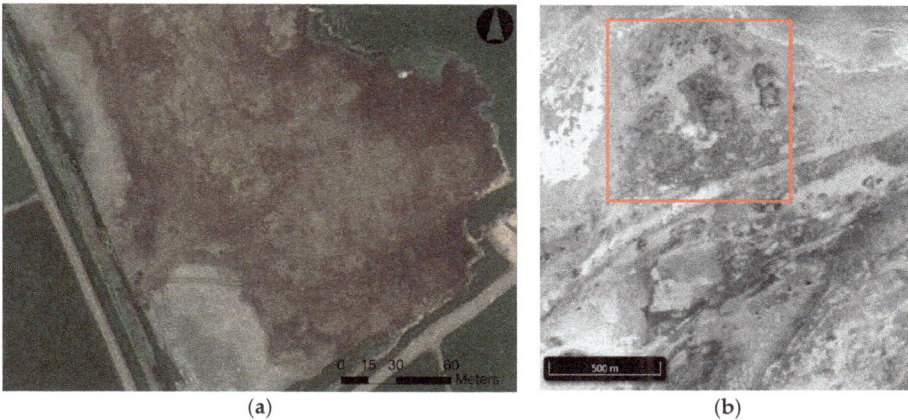

Figure 21. (**a**) The site of Tell er-Ruhban. Image courtesy Google Earth; (**b**) 1968 CORONA image. The area of image (**a**) is in the red box. Image courtesy of the Center for Advanced Spatial Technologies, University of Arkansas/U.S. Geological Survey.

Tell Mutubis: Located in the western Delta, the site measures 550 m × 500 m, and has a lighter colored sandy silt with 20% vegetation coverage. Mid-November 2011 imagery shows a series of clear rectilinear features in the southern part of the site. Previous survey work dates the site to the Greco-Roman through Islamic Periods (ca. 323 BC–1517 AD) [26] (Figure 22). The site seems unchanged in 50 years.

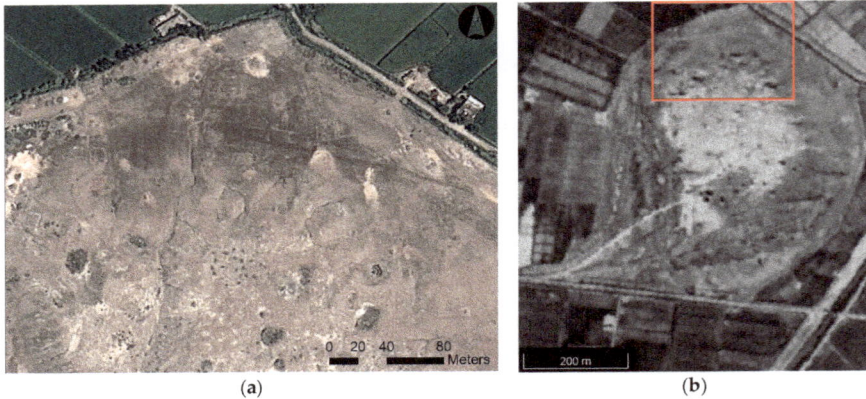

Figure 22. (**a**) The site of Tell Mutubis. Image courtesy Google Earth; (**b**) 1968 CORONA image. The area of image (**a**) is in the red box. Image courtesy of the Center for Advanced Spatial Technologies, University of Arkansas/U.S. Geological Survey.

Tukh el Qaramus: Located in the western Nile Delta, the site measures 470 m × 580 m, composed of a reddish colored sandy silt with minimal vegetation coverage. Surveys by both the Liverpool University Delta Survey and the Amsterdam Delta Survey have mapped features on the site. Multiple structures appear in an early December 2009 satellite image, including a known temple of Philip Arrhideus, and potentially a second temple in the eastern part of the site (with the same shape and orientation of the known and excavated temple). The site dates from the Ptolemaic- Roman Periods (332 BC–395 AD) based on the survey findings [26,29] (Figure 23). The modern settlement atop the site has tripled in size in 50 years.

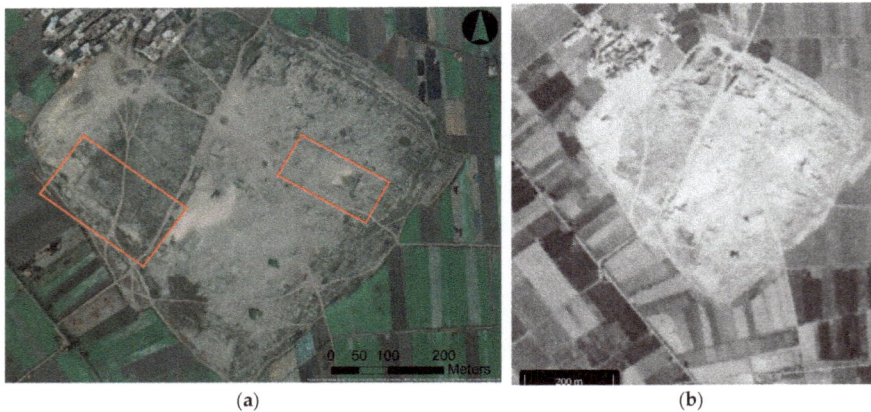

Figure 23. (**a**) The site of Tukh el Qaramus, with the temple of Philip Arrhideus in the western rectangle, and a potential temple in the eastern rectangle. Image courtesy Google Earth; (**b**) 1968 CORONA image. Image courtesy of the Center for Advanced Spatial Technologies, University of Arkansas/U.S. Geological Survey.

Umm el-Lahm: Located in the western Nile Delta, the site measures 190 m × 270 m, and is composed of a reddish colored dark silt with virtually no vegetation. In an early May image, there are faint traces of structures apparent in the central part of the mound. Previous work dates the site to the Ptolemaic-Roman periods (332 BC–395 AD) [23] (Figure 24). The site has been cut into in its southernmost corner since 1968.

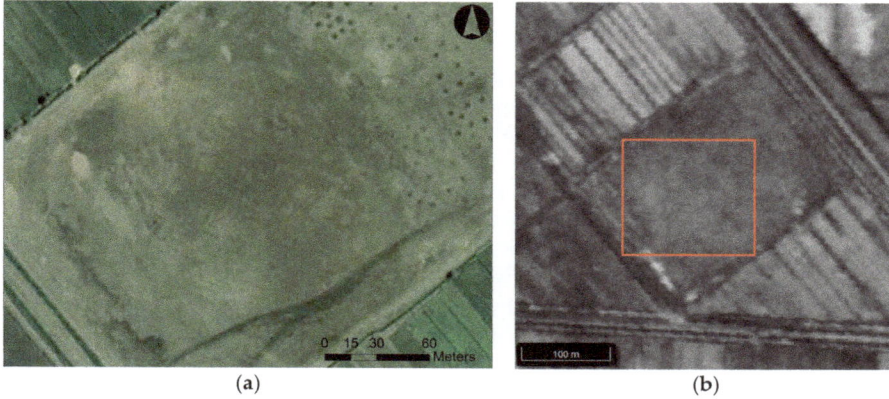

(a)

(b)

Figure 24. (a) The site of Umm el-Lahm Image courtesy Google Earth; (b) 1968 CORONA image. The area of image (a) is in the red box. Image courtesy of the Center for Advanced Spatial Technologies, University of Arkansas/U.S. Geological Survey.

Umm Gafar: Located in the western Nile Delta, the site measures 400 m × 320 m, and is composed of a medium colored sandy silt with limited vegetation along its northern side. A February 2015 image shows a well-planned settlement measuring 230 m × 280 m. Given that previous survey work has found Roman Period (30 BC–395 AD) sherds, the layout of the site may match this time-period as well [26] (Figure 25). The edges of the tell in 1968 are difficult to see, as it appears to be in a now dried up riverbed. It looks as though it has lost approximately 20% of its area, with modern buildings now on its southern side.

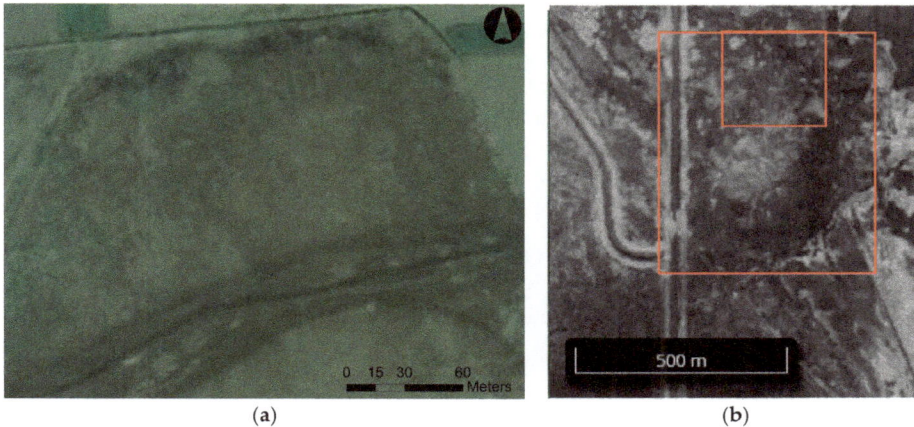

(a)

(b)

Figure 25. (a) The site of Umm Gafar Image courtesy Google Earth; (b) 1968 CORONA image. The original boundaries of the site are within the larger red box, while the area in image (a) is in the smaller box. Image courtesy of the Center for Advanced Spatial Technologies, University of Arkansas/U.S. Geological Survey.

4. Discussion

4.1. Results from the Nile Delta Satellite Imagery Assessment

To locate potential features at archaeological sites in the Egyptian Nile Delta using open-source satellite imagery, one must consider the soil matrix each site, seasonality, weather, and inundation/growing seasons. These factors are just as important as imagery processing techniques, since different images of the same site from the summer versus fall may show clear features versus almost nothing, even with enhanced processing. The results from the aforementioned 20 sites show that 15 (75%) have features appearing in the drier September-December time-period. In fact, 100% of lighter sandy sites had their features appear more clearly during these dry months. The same two images (16 September 2011; 10 November 2011) captured visible surface features for 10 separate locations. The remaining five sites with medium reddish silt on their surfaces varied slightly regarding the optimum time-period for revealing subsurface features. Three of the sites had features appear during the wetter January-February months. Of the two remaining sites, one was affected by summer inundation levels. It became clear while examining between 5 and 20 images per site (ranging from 2002 to 2016) that the specific weather conditions as well as the inundation of the fields played an essential role in feature visibility. For example, at Tell Belim, an image from 30 July 2013 had features easier to visualize versus an 1 August 2013 image. On 30 July, the fields surrounding the site were clearly inundated with water, while on the next day they did not appear to be as fully inundated. With Google Earth containing so many images across a long time-period, archaeologists can study these images of their sites, or similar sites, prior to purchasing specific imagery to determine which seasons or years might be most appropriate.

It is essential to recognize why features did not appear on some of the satellite imagery of the same locations. There are over 700 sites in the Egyptian Exploration Society Delta survey, and 1200 additional potential sites that appeared in our imagery analysis (most of which are beneath modern towns). Buried features only appear when the image is taken at the right time of year, including wetter weather conditions, with the "best" soil type (sandier soil versus siltier soil), and higher groundwater levels. While additional images have been taken in the last three years, prior to that, imagery choices can be limited, and may not be from the time of year when there was sufficient rain or inundated fields for optimate imagery analysis. On many satellite images, vegetation is dense. In some cases, the vegetation reveals buried walls, but if it is too bushy or dense, it is not possible to see anything. Heavy looting, which sadly shows up on dozens of Nile Delta sites, may also have destroyed the visible surface features. Overall, 25% of the study sites remain unchanged in 50 years. 50% of the sites have shrunk from 10 to 50% of their total area. Since these sites are a good representation of sites across the Delta, this is of great concern. 15% used to lie along Lake Manzala, which has shrunk considerably in the past 50 years. 15% have had villages on top of them double or triple in size. 10% of the sites have had other buildings constructed on them (military or private compounds).

Given these results, we can compare space based visual feature detection to ground based remote sensing. A space-based approach can be carried out in an efficient way using Google Earth if one knows the locations of sites. The main challenge is data availability. Most sites in the Delta have 10+ satellite images, with over 50% of the images being taken in the last 3–4 years. If the site's upper matrix is sandier in composition, there is no guarantee that data will emerge during a drier fall period, or that imagery might exist from when standing water surrounds the site.

Assuming there is good data from the right time of year, season, or inundation conditions, what about the quality of the data? Of the 20 sites mapped, five sites displayed features like walls or single structures (Abu el Humar, Kom Biltus, Kom Bunduq, Kom ed-Dahaba 2, Kom el-Barud), and eight sites had areas containing interconnected buildings under 1 ha. (100 m × 100 m) (Abu Mitawi, Daba el-Qibli, Tell Belim, Tell Buweib, Tell el-Damalun, Tell el-Gassa, Tell Mutubis, Tukh el-Qaramus). There were four sites that contained site outlines greater than a hectare (Tell el-Arab, Kom el-Arab, Kom el-Garad). Three sites (Kom ed-Dahaba 1, Tell er-Ruhban, Umm Gafar), yielded features across their remaining surfaces, greater than 200 m × 200 m in size. 65% of the sites examined might have shown additional features if higher resolution multispectral imagery were purchased and processed. Additional data is needed from other times of year. Typically, if subsurface features show up in visual site assessment alone, then additional buried features will appear after the application of algorithms. This multi-image assessment is indeed a crucial first step for further remote sensing work at optimum locations. 35% of the sites using visual assessment alone had enough data to begin comparative work against known excavated examples of sites from similar time periods. While it was possible to see large subsurface settlement remains in 15% of the sites, only two (Tell er-Ruhban and Kom ed-Dahaba 1) had data of the same size and quality one would expect from ground based remote sensing. Kom ed-Dahaba 1 represents the most detailed settlement map result from this research effort. Similarly, even the most detailed magnetometer maps may not be complete [30], like the gaps seen in some of the visible surface remains in the satellite imagery.

4.2. Results from the Tell Tebilla Excavations

G. Mumford, director of the Tell Tebilla project, compared a copy of the satellite image with the exposed stratigraphy along the west side of the mound to locate the enclosure wall. During the 2003 season, the excavation team explored the western side of Tell Tebilla, beginning with part of an exposed section revealed by the recent construction of a water plant. The team scraped along the edges of the *tell* and found a mud brick wall measuring 11.5 m wide, representing a southern cross-section through the enclosure wall's foundation. Ten cm of loose degraded clay and occupation debris covered the wall, which the team cleared and 100-m to the east and then 280-m north.

Overall, the structure measured 235-m east-west. The southeastern and southwestern corners were preserved, with the northeastern corner laying beneath modern rice fields. The wall is between 10.50 m and 11.50 m wide, with a series of slight buttresses on the eastern and southern exterior. We also know that a 13.50 m wide foundation trench had cut through Third Intermediate Period (ca. 1069 BC–712 BC), to Dynasty 26 mastabas (664 BC–525 BC) (Figure 18). It seems likely that the structure represents a "fort"-temple or temple enclosure of Nectanebo I/II, constructed along this branch of the Nile (Figure 26), possibly fortifying the area in expectation of an invasion by the Persians around 343 BC. It did not last for long: the Persians likely destroyed it at Mendes and elsewhere, leaving only the foundations [31,32]. Although only the temple enclosure wall's foundations remain, the degraded matrix above it had absorbed moisture in such a way as to reflect light differently from the slightly different surface layer around it, making the wall quite visible on the CORONA imagery. It must be noted that only when walls are of sufficient thickness is it possible to detect them on CORONA imagery. With a one-two meter pixel resolution, 11-m wide walls covered with ten to thirty-cm of overlaying debris will still appear quite clearly. Overall, the temple enclosure wall at Tell Tebilla measures 25% of the size of the mud brick enclosure found at Mendes, and fits in as a medium sized enclosure wall compared with other Egyptian temple enclosure walls dating to the same period [19].

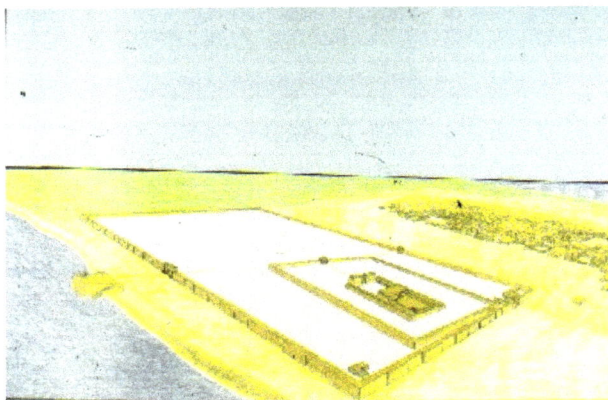

Figure 26. 3D reconstruction of the ancient site of Tell Tebilla. Imagery courtesy Greg Mumford.

4.3. Comparing Space-Based Remote Sensing Results to Ground Based Remote Sensing at Tell Tebilla

The CORONA imagery helped to reveal the temple enclosure wall at Tell Tebilla, yet only faint features are apparent on Google Earth imagery. The soil at Tebilla has higher concentrations of silt and less sand compared to the other sites in this study. The CORONA imagery only showed the large enclosure wall but no other features, likely because the other buried walls are 1 m or less in width. The magnetometer survey only caught part of the temple enclosure wall, and it was impossible to discern the exact nature of the wall from the smaller survey area. This shows one limitation of magnetometer surveys: with limited time in the field, it is possible to miss larger features. However, the magnetometer survey shows far greater detail across a site when compared with open-access imagery results. Magnetometer survey would have likely shown clear features at nearly all the tell sites described in this report. We cannot say for sure if 100% of these sites could be surveyed on the ground due to soil types, metal debris and proximity to power lines. At sites like Tell Tebilla, magnetometer survey is the best option, with a high degree of correlation between the excavated structures and the outlines seen in the magnetometer results

5. Conclusions

Overall, this is a way to target and scale ground based remote sensing and make it more effective. Ground based remote sensing can also fill in gaps that visual satellite data does not capture. Permits in Egypt are granted for regional survey, but it is rare to get permission to do ground-based remote sensing at more than one site. Satellites could help teams to target clear areas with near-surface remains. Magnetometer survey has proven effective for subsurface survey in Egypt's Delta, but may not always work given potential metal waste on sites. Would magnetometer survey work still be necessary at the three sites with greater than 200 m × 200 m of their near-surface structures visible? There were certainly gaps within and around those blocks of structures, and the detail in some areas was clearer than others. Smaller rooms (1 m or less in width) appeared blurry. The overall detail of structures seen on those three sites appeared like standard magnetometer surveys, but they only represent 15% of the sites mapped. The numbers of sites with this level of features apparent may be higher, but we await the release of satellite data from other times of year.

Every high-resolution satellite image is not on Google Earth. Once the time of year is known that works for visual subsurface feature detection from the open-access data, imagery can be purchased for a site or sites in question, even if the data is greyscale. This would be worth the $200.00–$500.00 in investment (with archive data costing between $8.00–$20.00 per sq. km) compared to the overall cost of magnetometer or other subsurface survey without any good satellite data. The cost of a specialist in the field, including plane ticket, lodging, food, cost per day of effort, and post processing, can run into thousands of dollars. It becomes a high-cost risk in relationship to typical excavation budgets and fieldwork windows to have them come out to survey a site nearly blind. The satellite data, if the conditions are right, might be able to direct the location of the ground-based remote sensing efforts.

With the varied clear threats to many known, little known, and unknown sites in Egypt, what might be done to hasten their mapping and protection? Presently, we are engaged in a collaborative training project with Egypt's Ministry of Antiquities via the Lisht Mission field school. Egyptian archaeologists are being trained in free access satellite mapping tools like Google Earth, and then are taking the data to examine potential findings in the field locally. This began in 2016, and will continue in 2017. In addition to Google Earth, students will be trained on NASA data, as well as the CAST CORONA platform. They have been, and will continue to be taught how much landscapes can change over time, affecting site preservation efforts. In the long term, it is hoped that the training and analysis done during the field school will give each participant a starting point for survey work in their own inspectorate regions. Access to good satellite maps is a challenge to all archaeological projects in countries where these data are too expensive or difficult to access. This is a point that all our field school students and local project collaborators have emphasized. Poor internet connections and accessing Google Earth are a significant challenge in rural areas, which is why we will be using WIFI boosting service BRCK [33] for our Lisht field school. The Ministry of Antiquities did have an ongoing mapping branch, called the Egyptian Antiquities Information System (EAIS) [34]. Starting as an Egyptian–Finnish project, this aimed to map sites across Egypt and create a comprehensive GIS. Presently, it is part of the MoA, but does not have the same levels of funding.

The monitoring of archaeological sites has become more critical. It does not depend on the season, but imagery does need to be taken every few months. This was not the case prior to 2013, but since then, imagery has become more frequent in the Delta region. It is now possible to see ongoing agricultural and urban encroachment, or in desert areas, illegal quarrying and cemetery construction. In the coastal Delta areas, one can map rising ground water levels.

In future, Egyptologists might consider crowdsourcing of satellite data (with protective measures ensuring site location masking) to assist with examining regional efforts to detect potential new sites, as well as tracking site changes through time using CORONA data. While the looting project took 3 people 6 intensive months, examining the entirety of Egypt's flood plain, including the Eastern Desert and Western Desert would take far longer. However, having thousands of people examine the data using crowdsourcing would allow this to be done in a matter of months. An initial concern would be access by looters and looting. Would there be any way to mask the satellite data effectively to prevent specific site identification?

We have overcome this concern with Globalxplorer (GX) [35], an online citizen science satellite imagery crowdsourcing platform that launched in January 2017. The official campaign "ended" in late April 2017, but it is still possible to participate. Over 60,000 individuals from 200 countries took part in GX, finding thousands of potential anthropogenic features in Peru (the country where the first campaign is based). Platform design played a key role in ensuring that no site identification of looting could take place. GX users, after taking a tutorial and seeing many examples of site looting, join the looting detection campaign. They are given a 300 m × 300 m satellite image clip of an undisclosed location somewhere in Peru. There are no visible maps, no GPS coordinates, and no indications of where you might be. A user can vote "looting" or "no looting" on the visible image. The same goes for the feature detection part of the campaign, where a vote is possible for "feature" or "no feature." One image might be in northern Peru, the next might be in southwestern Peru. The only people that can see actual site locations are the GX back end team, who evaluate each site seen by the crowd. Final map data is being shared only with professional Andean archaeologists and Peru's Ministry of Culture for ground truthing efforts.

A similar campaign could be done in Egypt in full collaboration with Egypt's Ministry of Antiquities. Instead of finding new sites or mapping looting, the crowd could assist with landscape changes and threats to sites via modern development. They could look at CORONA imagery alongside high resolution site data from the past 15 years. There are 1930s aerial photographs available for the Luxor and Giza regions, but no entity has yet fully digitized them. Apparently, the UK RAF has full aerial coverage of Egypt from the 1930s and 1940s, but attempts by S. Parcak to obtain these data has failed.

Egyptologists may use Google Earth imagery or DigitalGlobe data, but not realize that the time of year is an essential factor for revealing potential subsurface features. There needs to be better training for Egyptologists, or specialists working with them, on open source data and how to obtain and use satellite imagery appropriately in their projects. While many remote sensing papers appear on the archaeology of other countries in the ancient Near East, not as many papers can be seen for Egyptology, especially for survey efforts. This must change. DigitalGlobe and other satellite companies might be encouraged to collect additional data from the Middle East and North Africa regions during the wetter winter months, especially after periods of rainfall when subsurface features can appear in site imagery.

It is clear from this assessment of archaeological sites in the Delta that many of its sites are at risk from partial to total loss in the coming decades. Survey efforts by groups like the EES Delta survey and others is commendable, but they are composed of small teams of dedicated individuals. What is needed as a first step is a total geospatial assessment of the highest at-risk sites in the Egyptian Delta to focus ground-based remote sensing survey and excavation efforts. This is not unlike the multinational survey efforts around Aswan with the dam construction in the 1960s to 1970s. Resources can and should be pooled by nations for ground work to map sites as quickly as possible. The time has come for a concerted and focused effort by the entire Egyptology community, before many of these sites are either obscured from space or more fully eradicated.

Acknowledgments: Funding support for this research came from the National Geographic Society; The National Science Foundation, the Social Sciences and Humanities Research Council of Canada, and private donors. We wish to thank Egypt's Ministry of Antiquities for their ongoing support of our research, especially Minister of Antiquities Khaled el Anany, Aynman Eshmawy, Mohammed el Badaie, Mohammed Ismail, Adel Okasha, and Sharif Abd el Monaem.

Author Contributions: Sarah Parcak, Greg Mumford, and Chase Childs conceived and designed the experiments; Sarah Parcak, Chase Childs and Gregory Mumford performed the experiments; Sarah Parcak and Chase Childs analyzed the data; Gregory Mumford contributed analysis tools; Sarah Parcak, Greg Mumford, and Chase Childs wrote the paper.

Conflicts of Interest: Sarah Parcak is President of the Board of Globalxplorer. She is also an archaeology Fellow with the National Geographic Society. Chase Childs is Executive Director of Globalxplorer.

References

1. Challis, K.; Priestnall, G.; Gardner, A.; Henderson, J.; O'Hara, S. Corona remotely-sensed imagery in Dryland archaeology: The islamic city of al-Raqqa, Syria. *J. Field Archaeol.* **2002–2004**, *29*, 139–153. [CrossRef]
2. Hritz, C. Tracing settlement patterns and channel systems in Southern Mesopotamia using remote sensing. *J. Field Archaeol.* **2010**, *35*, 184–203. [CrossRef]
3. Kennedy, D. Declassified satellite photographs and archaeology in the Middle East: Case studies from Turkey. *Antiquity* **1998**, *72*, 553–561. [CrossRef]
4. Wilkinson, T. Surface collection, field walking, theory and practice, sampling theories. In *Handbook of Archaeological Sciences*; Pollard, A., Brothwell, D., Eds.; John Wiley: New York, NY, USA, 2001.
5. Stone, E. Patterns of looting in southern Iraq. *Antiquity* **2008**, *82*, 125–138. [CrossRef]
6. Parcak, S. Going, going, gone: Towards a satellite remote sensing methodology for monitoring archaeological tell sites under threat in the Middle East. *J. Field Archaeol.* **2007**, *42*, 61–83.
7. Menze, B.; Ur, J. Mapping patterns of long-term settlement in Northern Mesopotamia at a large scale. *Proc. Natl. Acad. Sci. USA* **2012**, *109*, E778–E787. [CrossRef] [PubMed]

8. Casana, J.; Panahipour, N. Notes on a disappearing past: Satellite-based monitoring of looting and damage to archaeological sites in Syria. *J. East. Mediterr. Heritage Stud.* **2012**, *2*, 128–151.

9. American Association for the Advancement of Science (AAAS). *Ancient History, Modern Destruction: Assessing the Current Status of Syria's World Heritage Sites Using High-Resolution Satellite Imagery*; AAAS: Washington, DC, USA, 2014.

10. Contreras, D.; Brodie, N. The utility of publicly-available satellite imagery for investigating looting of archaeological sites in Jordan. *J. Field Archaeol.* **2010**, *35*, 101–114. [CrossRef]

11. Parcak, S. Archaeological looting in Egypt: A geospatial View (case studies from Saqqara, Lisht, and el Hibeh). *Near East. Archaeol.* **2015**, *78*, 196–203. [CrossRef]

12. Parcak, S.; Gathings, D.; Childs, C.; Mumford, G.; Cline, E. Satellite evidence of archaeological site looting in Egypt: 2002–2013. *Antiquity* **2016**, *90*, 188–205. [CrossRef]

13. Wilkinson, T.J.; Ur, J.; Casana, J. Nucleation to dispersal: Trends in settlement patterns in the northern Fertile Crescent. In *Side-by-Side Survey: Comparative Regional Studies in the Mediterranean World*; Alcock, S., Cherry, J., Eds.; Oxbow: Oxford, UK, 2004; pp. 198–205, ISBN 1842170961.

14. Ur, J. Corona satellite photography ancient road networks: A northern Mesopotamian case study. *Antiquity* **2003**, *77*, 102–115. [CrossRef]

15. Ur, J. Sennacherib's northern Assyrian canals: New insights from satellite imagery and aerial photographs. *Iraq* **2005**, *67*, 317–345. [CrossRef]

16. Moshier, S.; El-Kalani, A. Late Bronze Age Paleogeography along the ancient ways of Horus in Northwest Sinai, Egypt. *Geoarchaeology* **2008**, *23*, 450–473. [CrossRef]

17. Corona Atlas & Referencing System. Available online: http://CORONA.cast.uark.edu/ (accessed on 19 September 2017).

18. Casana, J.; Cothren, J. The Corona Atlas Project: Orthorectification of corona satellite imagery and regional-scale archaeological exploration in the near east. In *Mapping Archaeological Landscapes from Space*; Comer, D., Harrower, M., Eds.; Springer: New York, NY, USA, 2013; pp. 33–43, ISBN 978-1-4614-6073-2.

19. Mumford, G. Reconstruction of the temple at Tell Tebilla (East Delta). In *Egypt, Israel and the Ancient Mediterranean World: Studies in Honour of Donald B. Redford*; Knoppers, G., Hirsch, A., Eds.; Brill: Leiden, The Netherlands, 2002; pp. 267–286, ISBN 9789004138445.

20. Mumford, G. Concerning the 2001 Season at Tell Tebilla (Mendesian Nome). In *The Akhenaten Temple Project Newsletter*; Pennsylvania State University: College Park, PA, USA, 2002; pp. 1–4.

21. Mumford, G. *The University of Toronto Tell Tebilla Project (Eastern Delta). The American Research Center in Egypt Annual Report, 2001*; Emory University West Campus: Atlanta, GA, USA, 2001; pp. 26–27.

22. Pavlish, L. 2004 Archaeometry at Mendes, 1990–2002. In *Egypt, Israel and the Ancient Mediterranean World: Studies in Honour of Donald B. Redford*; Knoppers, G., Hirsch, A., Eds.; Brill: Leiden, The Netherlands, 2002; pp. 61–112, ISBN 789004138445.

23. Becker, H.; Fassbinder, J. In search of Piramesses—The lost capital of Ramesses II in the Nile Delta (Egypt) by caesium magnetometry. In *Archaeological Prospection, Arbeitshefte des Bayerischen Landesamtes für Denkmalpflege*; Fassbinder, J., Irlinger, W., Petzet, M., Eds.; Bayerischen Landesamtes für Denkmalpflege: München, Germany, 1999; pp. 146–150, ISBN 387499699X.

24. Mathieson, I. The National Museums of Scotland Saqqara Survey Project, 1990–2000. In *Abusir and Saqqara in the year 2000*; Bárta, M., Krejčí, J., Eds.; Archiv Orientalni: Prague, Czech Republic, 2002; pp. 27–39, ISBN 8085425394.

25. Herbich, T. Archaeological geophysics in Egypt: The Polish contribution. *Archaeol. Pol.* **2003**, *41*, 13–56.

26. Egypt Exploration Society Delta Survey Website. Available online: http://deltasurvey.ees.ac.uk/ds-home. html (accessed on 12 July 2017).

27. Spencer, A. The exploration of Tell Belim, 1999–2002. *J. Egypt. Archaeol.* **2002**, *88*, 37–51. [CrossRef]

28. Spencer, A.; Spencer, P. A new kingdom temple in the Delta. *Egypt. Archaeol.* **2004**, *45*, 5–8.

29. Milne, J. The Tukh el-Qaramus Gold Hoard. *J. Egypt. Archaeol.* **1941**, *27*, 135–137. [CrossRef]

30. Spencer, N. *Kom Firin I: The Ramesside Temple and the Site Survey*; British Museum Research Publication 170; British Museum Press: London, UK, 2008; ISBN 978-086159-170-1.

31. Lloyd, A. The Late Period, 664–323 BC. In *Ancient Egypt: A Social History*; Trigger, B., Kemp, B., O'Connor, D., Lloyd, A., Eds.; Cambridge University Press: Cambridge, UK, 1983; pp. 279–348, ISBN 0521284279.

32. Mumford, G. A late period riverine and maritime port town and cult center at Tell Tebilla (Ro-Nefer). *J. Anc. Egypt. Interconnect.* **2013**, *5*, 38–67.

33. Connecting Africa to the Internet. Available online: https://www.brck.com/ (accessed on 19 September 2017).

34. The Egyptian Antiquities Information System. Available online: http://www.arce.org/files/resource/h/rsrc/EAIS_July_2003.pdf (accessed on 19 September 2017).

35. Globalxplorer. Available online: https://www.globalxplorer.org (accessed on 19 September 2017).

geosciences

MDPI

Article

Optical Remote Sensing Potentials for Looting Detection

Athos Agapiou * , Vasiliki Lysandrou and Diofantos G. Hadjimitsis

Remote Sensing and Geo-Environment Laboratory, Eratosthenes Research Centre, Department of Civil Engineering and Geomatics, Cyprus University of Technology, Saripolou 2-8, 3603 Limassol, Cyprus; vasiliki.lysandrou@cut.ac.cy (V.L.); d.hadjimitsis@cut.ac.cy (D.G.H.)
* Correspondence: athos.agapiou@cut.ac.cy; Tel.: +357-25-002-471

Received: 31 July 2017; Accepted: 2 October 2017; Published: 4 October 2017

Abstract: Looting of archaeological sites is illegal and considered a major anthropogenic threat for cultural heritage, entailing undesirable and irreversible damage at several levels, such as landscape disturbance, heritage destruction, and adverse social impact. In recent years, the employment of remote sensing technologies using ground-based and/or space-based sensors has assisted in dealing with this issue. Novel remote sensing techniques have tackled heritage destruction occurring in war-conflicted areas, as well as illicit archeological activity in vast areas of archaeological interest with limited surveillance. The damage performed by illegal activities, as well as the scarcity of reliable information are some of the major concerns that local stakeholders are facing today. This study discusses the potential use of remote sensing technologies based on the results obtained for the archaeological landscape of *Ayios Mnason* in Politiko village, located in Nicosia district, Cyprus. In this area, more than ten looted tombs have been recorded in the last decade, indicating small-scale, but still systematic, looting. The image analysis, including vegetation indices, fusion, automatic extraction after object-oriented classification, etc., was based on high-resolution WorldView-2 multispectral satellite imagery and RGB high-resolution aerial orthorectified images. Google Earth© images were also used to map and diachronically observe the site. The current research also discusses the potential for wider application of the presented methodology, acting as an early warning system, in an effort to establish a systematic monitoring tool for archaeological areas in Cyprus facing similar threats.

Keywords: looting; remote sensing archaeology; image analysis; satellite data; Cyprus

---

1. Introduction

Looting is considered as a major anthropogenic threat for cultural heritage due to the irreversible damage that is caused to the archaeological context and the findings themselves, often diverted into the illicit market [1,2]. Several reports can be found from all over the world indicating the size and extent of this problem [3–6]. Recent examples from the war-conflicted areas in the Middle East showcase a part of this problem [7–9].

Due to the complexity of the problem, the scientific community and local stakeholders are seeking ways to minimize the degree and the extent of looting by the exploitation of innovative technologies [10]. In this concept, Earth observation and aerial sensors are considered as important aspects of a holistic approach to eventually constraining the problem. Recent examples from both optical and passive remote sensing technologies can be found in the literature, indicating the advantages and the accuracy of the results for mapping archaeological areas that are under threat [11–14]. In some cases, Earth observation proved to be the only means of documenting the destruction made by the looters due to war conflicts [3]. In other cases, ground geophysical prospecting has also been applied, as in the case of [5] in Peru.

Even though these technologies are not capable of preventing looters, the image analysis results can be used by the local stakeholders to take all the necessary measurements for future restrictions, as well as to warn the scientific community of illegal excavations.

It should be stated that existing literature is mainly focused on the exploitation of remote sensing technologies for extended looted areas, where hundreds of looted signs are visible from space and air [3,5,6]. On the contrary, this paper aims to present small-scale looting attempts which seem to have been made in recent years in Cyprus. In addition, no scheduled flight or satellite overpass was performed to monitor the site under investigation. Therefore, the use of existing datasets captured by various sources and sensors was the only means of mapping the looting imprints. It is evident that the specific case study is limited in terms of the size of the threat, but is also bounded by the availability of existing images rarely captured from space and air. This restricted context provides a realistic case study which is appropriate to discuss the potential use of non-contact remote sensing technologies to map small-scale systematic attempts made by looters in recent years.

2. Methodology

Existing archive aerial images and satellite datasets have been exploited to meet the aims of this study. A complete list of the all of the data used is provided in Table 1. The temporal resolution of the analysis was carried out covering the last nine years (i.e., from 2008 to 2017). Aerial images included the sub-meter-resolution red-green-blue (RGB) orthophoto color composite produced in 2008 (with a spatial resolution of 0.50 m), and the latest RGB orthophoto of 2014 (with a spatial resolution of 0.20 m). Both archives were produced by the Department of Land and Surveys of Cyprus. A greyscale aerial orthophoto with a spatial resolution of 1 m taken in 1993 (and therefore prior to any looting phenomena) was used as reference. In addition, a very-high-resolution WorldView-2 multispectral satellite image taken on 20 of June 2011 was also consulted. The WorldView-2 sensor provides a high-resolution panchromatic band with a ground sampling distance (GSD) of 0.46 m (at nadir view) and eight multispectral bands with 1.84 m GSD at nadir view. The latest bands include the conventional red, green, blue, and near-infrared wavelengths, amongst other parts of the spectrum, which cover the coastal, yellow, red edge, and near-infrared wavelengths.

Table 1. Datasets used for the current study. GSD: ground sampling distance; RGB: red-green-blue.

No	Image	Date of Acquisitions	Type
1	Aerial image	1993	Greyscale (1 m pixel resolution)
2	Aerial image	2008	RGB orthophoto (50 cm pixel resolution)
3	Aerial image	2014	RGB orthophoto (20 cm pixel resolution)
4	WorldView-2	20 June 2011	Multi-spectral (1.84 m GSD for multispectral and 0.46 m at nadir view for the panchromatic image
5	Google Earth	9 June 2008	RGB
6	Google Earth	13 July 2010	RGB
7	Google Earth	20 June 2011	RGB
8	Google Earth	29 July 2012	RGB
9	Google Earth	10 November 2013	RGB
10	Google Earth	13 July 2014	RGB
11	Google Earth	16 February 2015	RGB
12	Google Earth	5 April 2015	RGB
13	Google Earth	27 April 2016	RGB

To use all possible available sources to examine looted imprints of the area, Google Earth© images have also been extracted and analyzed. The Google Earth© 3D digital globe systematically releases satellite images at high spatial resolution, which can be used for various remote sensing applications (see [15–17]). The platform provides very high-resolution natural-color (i.e., RGB) images based on existing commercial space borne sensors, such as IKONOS, QuickBird, WorldView, etc. Despite the various limitations of Google Earth© images for scientific purposes, such as the compression of the original satellite images, the loss of image quality, as well as limitations in the spectral resolution (i.e., no near-infrared band is provided), recent research demonstrated the great potential of such platforms supportive to research and providing updated information [18]. Indeed, Google Earth© images have already been used for investigation of looting phenomena in the area of Palmyra [3,19] and the ancient city of Apamea, in Syria [20].

In the case study of Politiko, looted tombs were difficult to detect directly from the aerial and satellite datasets. This is due not only to the small scale and the depth of the looted tombs (i.e., more than 3 m), but also due to the spatial resolution and the view geometry (i.e., the nadir view) of the aerial and satellite datasets. Therefore, tombs' shadows could not be used as an interpretation key as in the case of [3]. In this case, soil disturbance due to these legal activities was considered as a proxy for the looted tombs. The excavated soil was placed very close to the looted tombs, providing a homogenous spectral characteristic target compared to the surrounding non-excavated area.

The methodology followed in this study is presented in Figure 1. Nine RGB images from the Google Earth© platform between 2008 and 2017 have been extracted and interpreted in a geographical information system (GIS). To improve the photo-interpretation of these images, various histogram enhancements were applied. These included brightness and contrast adjustments for each image to enhance the looting soil disturbance against the surrounding area, which was intact and partially vegetated (see examples in Figure 3). In addition, other linear (linear percent stretch) or non-linear histogram stretches (histogram equalization) were applied for enhancement of the spectral properties of the soil disturbance, which was considered as a proxy for the looted tombs.

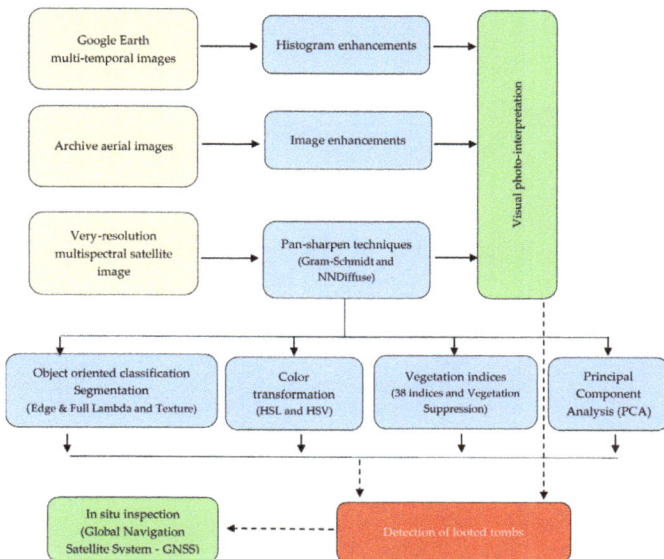

Figure 1. Overall methodology and resources used for the current study.

For the enhancement of the archive aerial orthorectified images, similar histogram adjustments have also been applied. The WorldView-2 multispectral image was also spatially improved using both Gram–Schmidt and NNDiffuse pan-sharpening algorithms. The image was then processed into various levels, including vegetation indices, vegetation suppression, orthogonal equations for the detection of crop marks [21,22], principal component analysis (PCA), and color transformations such as HSL (hue, saturation, and lightness) and HSV (hue, saturation, and value). The latest are considered as transformations of the Cartesian (cube) RGB representation. Finally, the WorldView-2 image was classified using object-oriented segmentation adjusting edge and full lambda parameters, also considering texture metrics. All of the above-mentioned image processing techniques were implemented in ENVI 5.3 (Environment for Visualizing Images, Harris Geospatial Solutions).

In addition, an in situ inspection of the site was carried out, during which the looting imprints detected through the image processing were mapped using a double differencing Global Navigation Satellite System (GNSS) and a real-time kinematic positioning technique. The vertical/horizontal combined accuracy of the in situ GNSS campaign was set to be less than 3 cm. Finally, the overall satellite and aerial image processing outputs were evaluated and cross-compared with the ground truthing investigation of the site.

3. Case Study Area

The area under investigation is in the southwestern part of the modern village of Politiko, in Nicosia District (Figure 2). In this area, looted tombs have been identified in the past, as well as in more recent years. The tombs are hewn out of the natural bedrock. Undisturbed tombs are not easily detected through aerial and/or satellite datasets since they are underground at an approximate depth of 3 m below the surface. In contrast, signs of looted tombs are more likely to be observed and identified in this manner (Figures 2 and 3).

Figure 2. Map indicating the case study area in the southwestern part of the modern village of Politiko, Nicosia District. Red dots indicate looted tombs which have been detected during the in situ investigation and mapped with GNSS (February 2016).

Figure 3. Looted tombs (February 2016).

The wider area of Politiko village consists of an intense archaeological territory which is very important for the history of Cyprus, linked to the ancient city-kingdom of Tamassos. While several archaeological missions excavated in the past or are still excavating in the area of Politiko (Politiko-Kokkinorotsos 2007: La Trobe University, Melbourne under Dr. David Frankel and Dr. Jenny Webb; Politiko–Troullia 2016: University of West Carolina Charlotte, USA under Dr. Steven Falconer and Dr. Patricia Fall, see for example [23,24]), the necropolis under investigation here has never been studied. Even though this area has been declared as an ancient monument (Scheduled B' monument) and is protected by law, the looting has not only continued but, as will be seen later, it has been augmented throughout the years.

4. Results

4.1. Aerial Orthophotos and Google Earth© Images

The investigation of the site initially started from the visual inspection of the Google Earth© images. Brightness and contrast adjustments were applied in an attempt to support the visual interpretation. Historical records from high spatial resolution images over the area of interest were examined, as shown in Figure 4. The images were imported and sorted in chronological order in a GIS environment. More specifically, the following images were extracted from the Google Earth© platform: 9 July 2008, 13 July 2010, 20 June 2011, 29 July 2012, 10 November 2013, 13 July 2014, 16 February 2015, 5 April 2015, and 27 of April 2016. Even though the looted tombs were not visible in the images, as mentioned earlier, looted areas were spotted based on the looting soil disturbance (in some instances achieved by using mechanical means). Recently disturbed terrain was clearly visible in the Google Earth© images.

Looted imprints are shown in Figure 4, in the yellow square. It is interesting to note the size, as well the systematic attempts made by the looters. The first looting activity is recorded to have taken place between 9 July 2008 and 13 July 2010 (Figure 4a,b), affecting three different areas, including more than one tomb each. In less than a year (20 June 2011; Figure 4c), a new attempt was made a few meters to the west of the previously affected northern area. The old looted areas shown in Figure 4b are partially visible now (i.e., Figure 4c) due to the vegetation growth of the area. New looting activity was captured between 29 July 2012 and 10 November 2013 (see Figure 4d,e) further to the east. Terrain disturbance was visually detected due to the characteristic white tone of the excavated soil, in contrast to the dark tone recorded by the vegetated area. The old looted areas are now difficult to spot, especially in Figure 4e. Most probably, vegetation was grown around and on top of the excavated soil, hiding the white tone of the excavated soil. It seems that the same areas were re-visited after a very short time (13 of July 2014), since a much larger disturbance has been documented at the same spots (Figure 4f). No new looting attempt was evidenced for some time (Figure 4g,h), until 2016 where a new looted imprint became visible in an image taken on the 27 April 2016 (Figure 4e).

Apart from one looted tomb in the western part of the area presented in Figure 4i, the rest of the looting marks detected in the aerial and satellite analysis have been successfully identified during the in situ investigation carried out in February 2016. In the case of the in situ documentation, the looted areas were accurately mapped. The small scale of the individual looting areas (i.e., clusters of one to three tombs each time), as well as the small size of the excavation made (approximately 1.5 m square or circle like shape trench), the detection of the looting marks is extremely difficult in case of no a priori knowledge of the area. The automatic detection of looting marks, is further hampered by the topography of the area with scattered vegetation and nude bedrock. This will be further discussed in the following section using segmentation and object-oriented analysis in the multi-spectral WorldView-2 images.

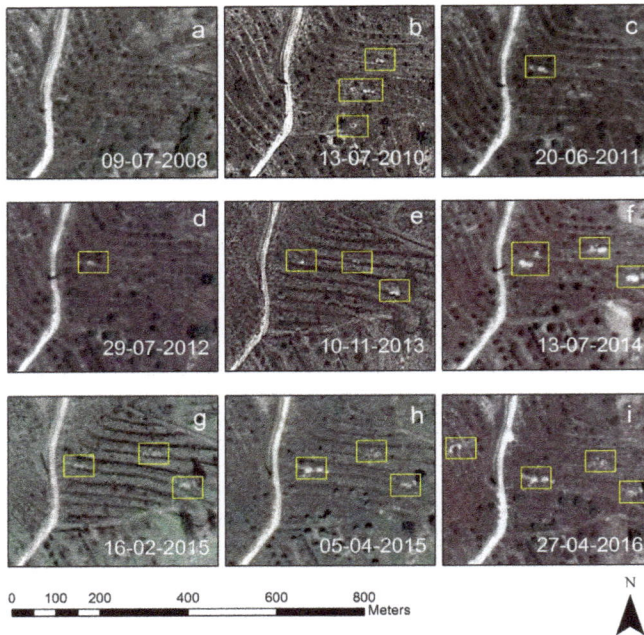

Figure 4. RGB Google Earth© images over the area of interest between the years 2008 and 2016 as follows (**a**) 9 July 2008, (**b**) 13 July 2010, (**c**) 20 June 2011, (**d**) 29 July 2012, (**e**) 10 November 2013, (**f**) 13 July 2014, (**g**) 16 February 2015, (**h**) 5 April 2015, and (**i**) 27 of April 2016. Looted tombs are indicated by the yellow squares.

Following a similar approach, photo-interpretation was carried out using the two aerial images taken in 2008 and 2014. These images were also improved using the linear percent stretch (5%) histogram enhancement technique. The earliest aerial image confirmed the results obtained from the satellite products of Google Earth©, indicating no looting attempts in the wider area of Politiko (Figure 5, bottom). Instead, at least four looting marks were spotted in the aerial image of 2014 (Figure 5, top). Looting traces indicated as b–d in Figure 5 were also recorded in the Google Earth© image (see 13 July 2014 in Figure 4f) and confirmed by the in situ inspection in February 2016. Apart from the verification of the results of the previously-elaborated images, the aerial datasets revealed a new looted tomb (see Figure 5 top,a) at the northern part of the site and approximately 100 m from other looted areas, not seen before. The interpretation of the aerial images was more efficient mainly due to the improved quality of the archive aerial datasets and the better spatial resolution. In all four cases, it was possible to identify the soil extracted from the tombs, but not the looted tombs themselves.

To proceed beyond photo-interpretation (hence, to try to detect possible changes in the funerary landscape of Politiko in a semi-automatic way), the aerial orthophotos of 1993 (single band, 1 m resolution), 2008 (RGB bands, 0.5 m resolution), and 2017 (RGB bands, 0.2 m resolution), were merged into a seven-band pseudo-color composite. In this multi-temporal image, a PCA analysis was then applied. PCA is a well-established approach to detect any significant changes. PCA analysis is a statistical tool to decompose multiple variables—as in this case study the seven-band pseudo-color composite—into principal components having orthogonality, while these components are being ranked with respect to their contribution to explaining the variances of the total seven-band image. Therefore, PCA transforms and converts high-dimensional data into linearly-uncorrelated variables (i.e., principal components).

The first two principal components (PC1 and PC2) are shown in Figure 6a,b, while a pseudo-color composite of the first three principal components (PC1–PC3) is shown in Figure 6c. The latest image (i.e., Figure 6c) was generated by displaying PC1, PC2, and PC3 into red, green, and blue bands (RGB). Looted tombs are visible in the pseudo-color composite (see the arrows in Figure 6c) because of landscape alterations.

Figure 5. Aerial RGB orthophotos taken in 2008 (**bottom**) and 2014 (**top**). Looted imprints (**a–d**) are traceable only in the latest aerial image.

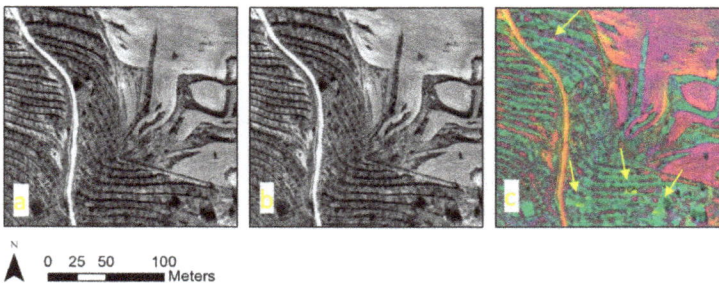

Figure 6. Principal component analysis results of the seven-band multi-temporal aerial datasets of 1993, 2008, and 2014: (**a**) Principal Component 1 (PC1); (**b**) Principal Component 2 (PC2); and (**c**) the pseudo-color composite of the first three principal components (PC1–PC3). Looted marks are indicated with arrows.

4.2. Satellite Image Processing

PCA analysis was also applied in the multi-spectral satellite image WorldView-2. The result is shown in Figure 7 (right), where a three-band pseudo-color composite is created by the first three PCs. Again, in this case, the three-band pseudo-color composite was created by displaying PC1, PC2, and PC3 into red, green, and blue bands (RGB). The looted tomb—indicated by the yellow square in this figure—was detectable after interpretation. However, it should be stressed that the identification of looting imprints in this pseudo-color composite was not a straight-forward procedure. This should be linked mainly to the spatial resolution of the multi-spectral bands (i.e., 1.84 m at nadir view). In addition, a vegetation suppression algorithm was applied in the image. The algorithm was employed, modeling the amount of vegetation per pixel, while an extended Crippen and Blom's algorithm was applied for vegetation transformation, as proposed by [25,26] based on a forced invariance approach.

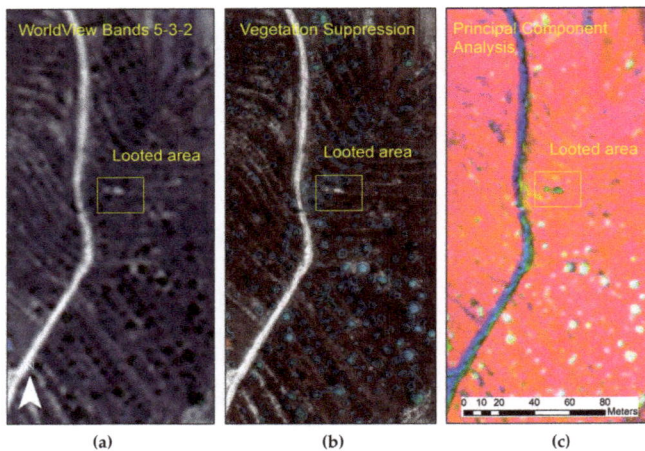

Figure 7. WorldView-2 image 5-3-2 pseudo-color composite (**a**); vegetation suppression result applied at the WorldView-2 image (**b**); and pseudo-color composite of the first three principal components (PC1–PC3) of the WorldView-2 image (**c**). The looted tomb is identified by the yellow square.

The model follows five steps to de-vegetate the bands of the satellite image. At first an atmospheric correction is applied (digital number (DN) subtraction), then a vegetation index is calculated as the simple ratio index. Following this, statistics between the DN and the vegetation index for each band are gathered and then a smooth best-fit curve to the plot is estimated. Finally, for each vegetation index level, all pixels are multiplied at that vegetation level using the smooth best-fit curves.

The model calculates the relationship of each input band with vegetation, then it decorrelates the vegetative component of the total signal on a pixel-by-pixel basis for each band. The result of the application of the vegetation suppression is shown in Figure 7 (middle). It seems that the visibility of the looted area was enhanced by this transformation compared to the initial WorldView image (Figure 7, left), since the mark is mostly surrounded by bushes and low vegetation. In addition, the specific algorithm seems to be very promising in looted areas which are fully cultivated and vegetated.

HSV and HSL color transformations results are shown in Figure 8. The color transformations were applied in the pan-sharpen WorldView-2 image after the implementation of the Gram–Schmidt and NNDiffuse pan-sharpening algorithms. Both color transformations were performed in ENVI 5.3. These two-color transformations are widely common cylindrical-coordinate representations points in an RGB color model. In this way, the intimal red, green, and blue values of this color model are transformed into new color components. In the HSV model, hue (H) defines pure color in terms of "green", "red", or "magenta", while saturation (S) defines a range from pure color (100%) to gray (0%) at a constant lightness level. Finally, value (V) refers to the brightness of the color. Similarly, HSL color transformation refers to the hue, saturation, and lightness (L) of the color.

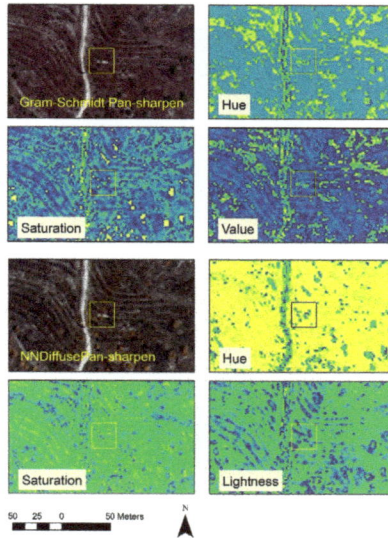

Figure 8. HSV (hue, saturation, and value) and HSL (hue, saturation, and lightness) color transformations of the WorldView-2 image after applying the Gram–Schmidt and NNDiffuse pan-sharpening algorithms. Yellow squares show the looting imprint.

Higher hue values make it easier to distinguish the looted tomb from the surrounding area, even though the overall results are not encouraging. In contrast, both pan-sharpen algorithms applied in the multi-spectral image improved the overall quality of the satellite image and the spatial resolution. The looted area became more visible, even from simple photo-interpretation.

More sophisticated algorithms have also been tested and evaluated using the WorldView-2 multispectral image. At first almost 40 different indices (mostly vegetation indices) were applied and interpreted. Table 2 provides the list of the indices applied, while the corresponding result is shown in Figure 9.

Table 2. Indices applied for the detection of looted marks in the WorldView-2 image. Promising indices are highlighted.

No.	Index	Equation	Result in Figure 9	Reference
1	Anthocyanin Reflectance Index 1	$ARI_1 = \frac{1}{p_{550}} - \frac{1}{p_{700}}$	c-I	[27]
2	Anthocyanin Reflectance Index 2	$ARI_2 = p_{800}\left[\frac{1}{p_{550}} - \frac{1}{p_{700}}\right]$	d-I	[27]
3	Atmospherically Resistant Vegetation Index	$ARVI = \frac{NIR-[Red-\gamma(Blue-Red)]}{NIR+[Red-\gamma(Blue-Red)]}$	e-I	[28]
4	Burn Area Index	$BAI = \frac{1}{(0.1-Red)^2+(0.06-NIR)^2}$	a-II	[29]
5	Difference Vegetation Index	$DVI = NIR - Red$	b-II	[30]
6	Enhanced Vegetation Index	$EVI = 2.5 \times \frac{(NIR-Red)}{(NIR+6\times Red-7.5\times Blue+1)}$	c-II	[31]

Table 2. *Cont.*

No.	Index	Equation	Result in Figure 9	Reference
7	Global Environmental Monitoring Index	$GEMI = $ $eta(1 - 0.25 \times eta) - \frac{Red - 0.125}{1 - Red}$ $eta = \frac{2(NIR^2 - Red^2) + 1.5 \times NIR + 0.5 \times Red}{NIR + Red + 0.5}$	d-II	[32]
8	Green Atmospherically-Resistant Index	$GARI = \frac{NIR - [Green - \gamma(Blue0Red)]}{NIR + [Green - \gamma(Blue0Red)]}$	e-II	[33]
9	Green Difference Vegetation Index	$GDVI = NIR - Green$	a-III	[34]
10	Green Normalized Difference Vegetation Index	$GNDVI = \frac{(NIR - Green)}{(NIR + Green)}$	b-III	[35]
11	Green Ratio Vegetation Index	$GRVI = \frac{NIR}{Green}$	c-III	[34]
12	Infrared Percentage Vegetation Index	$IPVI = \frac{NIR}{NIR + Red}$	d-III	[36]
13	Iron Oxide	$Iron\ Oxide\ Ratio = \frac{Red}{Blue}$	e-III	[37]
14	Leaf Area Index	$LAI = (3.618 \times EVI - 0.118)$	a-IV	[38]
15	Modified Chlorophyll Absorption Ratio Index	$MCARI = $ $[(p_{700} - p_{670}) - 0.2(p_{700} - p_{550})] \times$ $(\frac{p_{700}}{p_{670}})$	b-IV	[39]
16	Modified Chlorophyll Absorption Ratio Index-Improved	$MCARI2 = $ $\frac{1.5\,[2.5(p_{800} - p_{670}) - 1.3(p_{800} - p_{550})]}{\sqrt{(2 \times p_{800} + 1)^2 - (6 \times p_{800} - 5 \times \sqrt{p_{670}}) - 0.5}}$	c-IV	[40]
17	Modified Non-Linear Index	$MNLI = \frac{(NIR^2 - Red) \times (1 + L)}{NIR^2 + Red + L}$	d-IV	[41]
18	Modified Simple Ratio	$MSR = \frac{(\frac{NIR}{Red}) - 1}{(\sqrt{\frac{NIR}{Red}}) + 1}$	e-IV	[42]
19	Modified Triangular Vegetation Index	$MTVI = 1.2\,[1.2(p_{800} - p_{550}) - 2.5(p_{670} - p_{550})]$	a-V	[38]
20	Modified Triangular Vegetation Index-Improved	$MTVI2 = $ $\frac{1.5\,[1.2(p_{800} - p_{550}) - 2.5(p_{670} - p_{550})]}{\sqrt{(2 \times p_{800} + 1)^2 - (6 \times p_{800} - 5 \times \sqrt{p_{670}}) - 0.5}}$	b-V	[40]
21	Non-Linear Index	$NLI = \frac{NIR^2 - Red}{NIR^2 + Red}$	c-V	[43]
22	Normalized Difference Mud Index	$NDMI = \frac{(p_{795} - p_{990})}{(p_{795} + p_{990})}$	d-V	[44]
23	Normalized Difference Snow Index	$NDSI = \frac{(Green - SWIR1)}{(Green + SWIR1)}$	e-V	[45]
24	Normalized Difference Vegetation Index	$NDVI = \frac{(NIR - Red)}{(NIR + Red)}$	a-VI	[46]
25	Optimized Soil Adjusted Vegetation Index	$OSAVI = \frac{1.5 \times (NIR - Red)}{(NIR + Red + 0.16)}$	b-VI	[47]
26	Red Edge Position Index	Maximum derivative of reflectance in the vegetation red edge region of the spectrum in microns from 690 nm to 740 nm	c-VI	[48]
27	Renormalized Difference Vegetation Index	$RDVI = \frac{(NIR - Red)}{\sqrt{(NIR + Red)}}$	d-VI	[49]
28	Simple Ratio	$SR = \frac{NIR}{Red}$	e-VI	[50]
29	Soil Adjusted Vegetation Index	$SAVI = \frac{1.5 \times (NIR - Red)}{(NIR + Red + 0.5)}$	a-VII	[51]
30	Sum Green Index	Mean of reflectance across the 500 nm to 600 nm portion of the spectrum	b-VII	[52]
31	Transformed Chlorophyll Absorption Reflectance Index	$TCARI = $ $3[(p_{700} - p_{670}) - 0.2(p_{700} - p_{550})](\frac{p_{700}}{p_{670}})$	c-VII	[53]
32	Transformed Difference Vegetation Index	$TDVI = \sqrt{0.5 + \frac{(NIR - Red)}{(NIR + Red)}}$	d-VII	[54]
33	Visible Atmospherically Resistant Index	$VARI = \frac{Green - Red}{Green + Red - Blue}$	e-VII	[55]

Table 2. *Cont.*

No.	Index	Equation	Result in Figure 9	Reference
34	WorldView Built-Up Index	$WV - BI = \frac{(Coastal - Red\ Edge)}{(Coastal + Red\ Edge)}$	a-VIII	[56]
35	WorldView Improved Vegetative Index	$WV - VI = \frac{(NIR2 - Red)}{(NIR2 + Red)}$	b-VIII	[56]
36	WorldView New Iron Index	$WV - II = \frac{(Green \times Yellow)}{(Blue \times 1000)}$	c-VIII	[56]
37	WorldView Non-Homogeneous Feature Difference	$WV - NHFD = \frac{(Red\ Edge - Coastal)}{(Red\ Edge + Coastal)}$	d-VIII	[56]
38	WorldView Soil Index	$WV - SI = \frac{(Green - Yellow)}{(Green + Yellow)}$	e-VIII	[56]

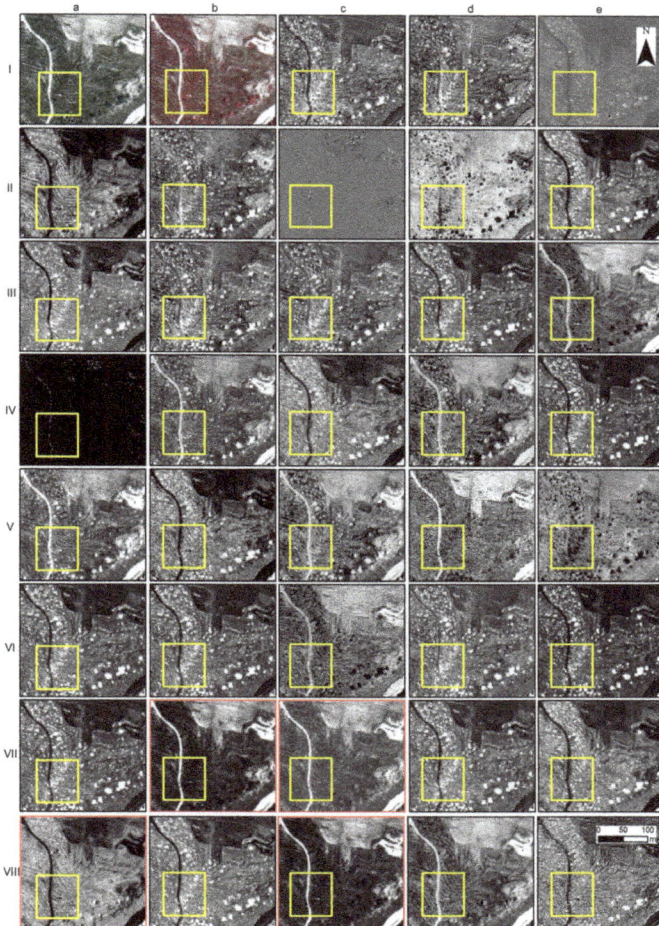

Figure 9. RGB pseudo-color composite 5-3-2 and NIR-R-G pseudo-color composite 6-5-3 are demonstrated in **a-I** and **b-I**. The rest sub-figures (**c-I**; **d-I**; .... **d-VIII** and **e-VIII**) correspond to the greyscale indices (see Table 2; total 38 indices) results applied in the WorldView-2 image. The looted area is shown in the yellow square. Images with a red outline show promising vegetation indices (i.e., **b-VII**; **c-VII**; **a-VIII**; and **c-VIII**).

The four most promising indices are the Sum Green Index (Figure 9, b-VII), the Transformed Chlorophyll Absorption Reflectance Index (Figure 9, c-VII), the WorldView Built-Up Index (Figure 9, a-VIII), and the WorldView New Iron Index (Figure 9, c-VIII). From these indices, the WorldView Built-Up Index seems to be the most promising, as far as the looted marks interpretation is concerned. The specific index is based on the spectral properties of the objects as recorded in the coastal and red-edge part of the spectrum (i.e., bands 1 and 6, respectively). The index could improve soil areas, as in the case of the looted tomb, and the earthen road in the western part of the area under investigation. The use of "non-ordinary" indices for archaeological purposes, such as the WorldView Built-Up Index, which was initially used to distinguish built-up areas, has also been reported in previous studies [57].

Other indices, including the traditional widely-applied indices, such as the Normalized Difference Vegetation Index (NDVI), performed less encouraging results as shown in Figure 9, a-VI. Some of them were demonstrated to be inappropriate as far as detecting looting imprints is concerned (i.e., the Enhanced Vegetation Index in Figure 9, c-II and the Leaf Area Index in Figure 9, a-IV). It should be mentioned that similar histogram enhancements have been equally applied in all indices.

Based upon the results from the various indices shown in Figure 9, a recent re-projection of the WorldView-2 spectral space has been implemented. This reprojection was initially developed to enhance buried archaeological remains through crop marks [21]. The WorldView-2 bands were re-projected into a new 3D orthogonal spectral space with three new axes—namely the soil component, the vegetation component, and the crop mark component, as shown in the following equations:

$$Crop\ mark_{WorldView-2} = -0.38\,\rho_{blue} - 0.71\,\rho_{green} + 0.20\,\rho_{red} - 0.56\,\rho_{NIR},$$

$$Crop\ mark_{WorldView-2} = -0.38\,\rho_{blue} - 0.71\,\rho_{green} + 0.20\,\rho_{red} - 0.56\,\rho_{NIR},$$

$$Soil_{WorldView-2} = 0.09\,\rho_{blue} + 0.27\,\rho_{green} - 0.71\,\rho_{red} - 0.65\,\rho_{NIR},$$

The results of this application are shown in Figure 10a–c, as well as the RGB pseudo-color composite (Figure 10d). The soil component (Figure 10a) enabled the enhancement of one of the looted tombs, while the vegetation component (Figure 10b) shows the three looting marks of the area. The crop mark component (Figure 10c) was less efficient, while the overall RGB pseudo-color composite (Figure 10d) improved the interpretation of the looted areas. Looted tombs in the vegetation component are detectable due to the small values (i.e., black tones of gray in Figure 10b, vegetation component) compared to the enhanced vegetated areas (i.e., white tones of gray in Figure 10b, vegetation component).

Furthermore, object-oriented classification was applied in the WorldView-2 images. An "optimum" segmentation of the image was achieved after several iterations and changes of the scale level and merge algorithms. Finally, the scale level was set to a value of 65.0, while the merge level was applied after the full lambda algorithm was set to 90.0 using the following equation (see more details in the ENVI Handbook):

$$t_{i,j} = \frac{\frac{|O_i| \cdot |O_j|}{|O_i| + |O_j|} \cdot \|u_i - u_j\|^2}{lenght(\partial(O_i, O_j))}$$

where:

$|O_i|$ is region *i* of the image,
$|O_j|$ is the area of region *i*,
u_i is the average value in region *i*,
u_j is the average value in region *j*,
$\|u_i - u_j\|^2$ is the Euclidean distance between the spectral values of regions *i* and *j*,
$lenght(\partial(O_i, O_j))$ is the length of the common boundary of $|O_i|$ and $|O_j|$.

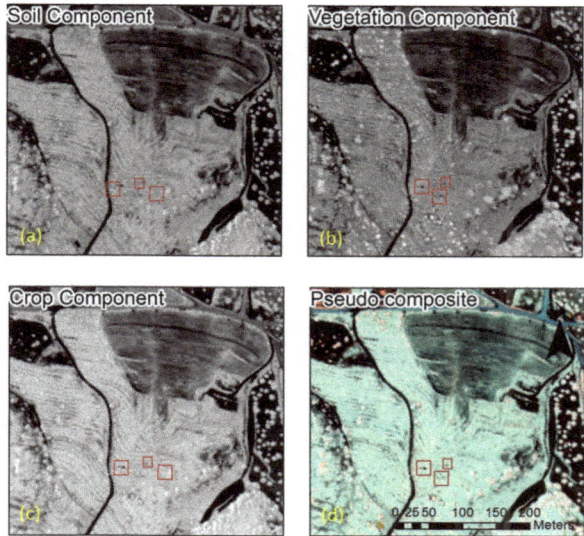

Figure 10. WorldView-2 spectral reprojection in a new 3D space: (**a**) soil component, (**b**) vegetation component, and (**c**) the crop mark component are shown in greyscale. The RGB pseudo-composite (**d**) of these three components is also shown in the lower right (soil component, vegetation component, and crop mark component reflect the red, green, and blue bands, respectively).

Full lambda algorithm was applied to merge small segments within larger ones based on a combination of spectral and spatial information, while the scale level is based on the normalized cumulative distribution function (CDF) of the pixel values in the image (see more in [58,59]). The units refer to greyscale tones. In addition, a 3 × 3 texture kernel was employed. The texture kernel refers to the spatial variation of image greyscale levels (tone) for a moving window of 3 × 3 pixels. After the segmentation of the image, rules were set for its classification. These rules included spatial attributes (areas less than 25.0), spectral properties (thresholds in coastal and red edge bands—like the WorldView Built-Up Index), and roundness parameters. The results of the object segmentation and classification are shown in Figure 11. Distinguished segments that are characterized as objects—like those presently recognized as looting marks—are shown in red, while the confirmed looted tomb is shown as a yellow square. Through this analysis, new risk-sensitive areas have been spotted in the wider area of Politiko village, while the already-known looted tomb (Figure 11, within the yellow square) was successfully detected. The false positives that were observed in the rest of the area should be linked to the similar spectral characteristics of the soil, as well as to other cultivation practices and land use properties. It is therefore evident that the automatic object-oriented approach for the extraction of looted areas is only valid to some degree in small, specific archaeological zones, and not beyond these areas. Therefore, a priori knowledge of the area under investigation is essential.

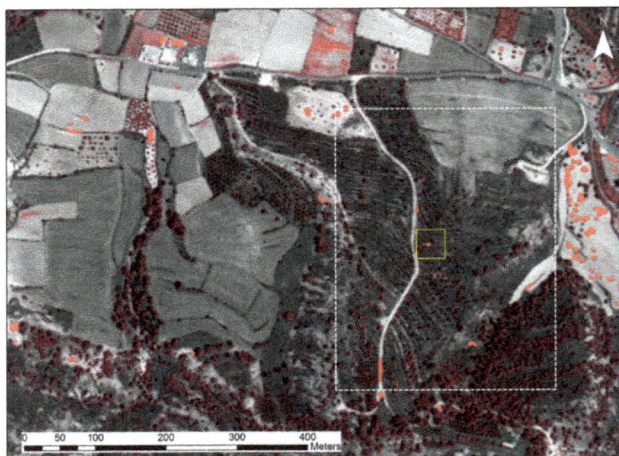

Figure 11. Object-oriented segmentation and classification of the WorldView-2 satellite image. The white rectangle shows the area of the necropolis under examination, while the yellow square shows the looting mark.

5. Conclusions

The paper aims to demonstrate the potential use of various remote sensing datasets for the detection of looting signs. Though the use of such datasets has been presented in the past by other researchers, in this example the looting signs were of a small scale (i.e., 1–3 looting attempts per year) and no schedule image was provided. Therefore, the question here was to investigate if existing datasets can be used to support local stakeholders for monitoring these threats.

Various image processing techniques have been applied to investigate the detection of small-scale looting attempts (i.e., 1–3 per year) in the wider area of Politiko village. Both archive and satellite images have been used to detect these systematic and organized events. The image analysis included archival data from the Department of Land and Surveys of Cyprus, Google Earth© images, and a very high-resolution WorldView-2 image. It should be stressed that no scheduled satellite overpass was programmed, and hence the analysis was based upon existing and available data.

The overall results demonstrated that Earth observation datasets and aerial imagery can be sufficiently used to detect looting marks in wider areas, and track the illegal excavations with high precision. The RGB-compressed images of Google Earth© are considered as a very good starting point for the interpretation of the area. These images have undergone some image histogram enhancement, namely changes in brightness and contrast, and other linear histogram enhancements. Image processing such as the vegetation indices indicated in Table 2, and spectral transformations such as PCA, orthogonal equations, HSV, etc., in multi-spectral images can further improve the final results. Automatic extraction based on object-oriented classification was also attempted in this case study, providing some interesting results. The overall interpretation of the results from the image analyses is that it is highly important to be verified with in situ inspections and ground truthing. Quantitative assessment of the overall results was not carried out due to the temporal changes of the phenomenon, as well as to the different datasets (with different spectral and spatial characteristics) used in this case study.

Areas with archaeological interest which are endangered by looting, such as the case study of *Ayios Mnason*-Politiko village, can be systematically controlled by space and aerial sensors. The establishment of such a reliable monitoring tool for local stakeholders could further act as an inhibiting factor for preventing looters.

Acknowledgments: The present communication is under the "ATHENA" project H2020-TWINN2015 of the European Commission. This project has received funding from the European Union's Horizon 2020 research and innovation programme under grant agreement No. 691936.

Author Contributions: All authors have contributed equally to the results.

Conflicts of Interest: The authors declare no conflict of interest. The founding sponsors had no role in the design of the study; in the collection, analyses, or interpretation of data; in the writing of the manuscript, and in the decision to publish the results.

References

1. *Convention on the Means of Prohibiting and Preventing the Illicit Import, Export and Transfer of Ownership of Cultural Property*; UNESCO: Paris, France, 1970.
2. UNIDROIT. *Convention on Stolen or Illegally Exported Cultural Objects*; UNIDROIT: Rome, Italy, 1995.
3. Tapete, D.; Cigna, F.; Donoghue, N.M.D. 'Looting marks' in space-borne SAR imagery: Measuring rates of archaeological looting in Apamea (Syria) with TerraSAR-X Staring Spotlight. *Remote Sens. Environ.* **2016**, *178*, 42–58. [CrossRef]
4. Chase, F.A.; Chase, Z.D.; Weishampel, F.J.; Drake, B.J.; Shrestha, L.R.; Slatton, L.C.; Awe, J.J.; Carter, E.W. Airborne LiDAR, archaeology, and the ancient Maya landscape at Caracol, Belize. *J. Archaeol. Sci.* **2011**, *38*, 387–398. [CrossRef]
5. Lasaponara, R.; Leucci, G.; Masini, N.; Persico, R. Investigating archaeological looting using satellite images and GEORADAR: The experience in Lambayeque in North Peru. *J. Archaeol. Sci.* **2014**, *42*, 216–230. [CrossRef]
6. Contreras, A.D.; Brodie, N. The utility of publicly-available satellite imagery for investigating looting of archaeological sites in Jordan. *J. Field Archaeol.* **2010**, *35*, 101–114. [CrossRef]
7. Cerra, D.; Plank, S.; Lysandrou, V.; Tian, J. Cultural heritage sites in danger—Towards automatic damage detection from space. *Remote Sens.* **2016**, *8*, 781. [CrossRef]
8. Tapete, D.; Cigna, F.; Donoghue, D.N.M.; Philip, G. Mapping changes and damages in areas of conflict: From archive C-band SAR data to new HR X-band imagery, towards the Sentinels. In Proceedings of the FRINGE Workshop 2015, European Space Agency Special Publication ESA SP-731, Frascati, Italy, 23–27 March 2015; European Space Agency: Rome, Italy, 2015; pp. 1–4.
9. Stone, E. Patterns of looting in southern Iraq. *Antiquity* **2008**, *82*, 125–138. [CrossRef]
10. Parcak, S. Archaeological looting in Egypt: A geospatial view (Case Studies from Saqqara, Lisht, andel Hibeh). *Near East. Archaeol.* **2015**, *78*, 196–203. [CrossRef]
11. Agapiou, A.; Lysandrou, V. Remote sensing archaeology: Tracking and mapping evolution in European scientific literature from 1999 to 2015. *J. Archaeol. Sci. Rep.* **2015**, *4*, 192–200. [CrossRef]
12. Agapiou, A.; Lysandrou, V.; Alexakis, D.D.; Themistocleous, K.; Cuca, B.; Argyriou, A.; Sarris, A.; Hadjimitsis, D.G. Cultural heritage management and monitoring using remote sensing data and GIS: The case study of Paphos area, Cyprus. *Comput. Environ. Urban Syst.* **2015**, *54*, 230–239. [CrossRef]
13. Deroin, J.-P.; Kheir, B.R.; Abdallah, C. Geoarchaeological remote sensing survey for cultural heritage management. Case study from Byblos (Jbail, Lebanon). *J. Cult. Herit.* **2017**, *23*, 37–43. [CrossRef]
14. Negula, D.I.; Sofronie, R.; Virsta, A.; Badea, A. Earth observation for the world cultural and natural heritage. *Agric. Agric. Sci. Procedia* **2015**, *6*, 438–445. [CrossRef]
15. Xiong, J.; Thenkabail, S.P.; Gumma, K.M.; Teluguntla, P.; Poehnelt, J.; Congalton, G.R.; Yadav, K.; Thau, D. Automated cropland mapping of continental Africa using Google Earth Engine cloud computing. *ISPRS J. Photogramm. Remote Sens.* **2017**, *126*, 225–244. [CrossRef]
16. Boardman, J. The value of Google Earth™ for erosion mapping. *Catena* **2016**, *143*, 123–127. [CrossRef]
17. Gorelick, N.; Hancher, M.; Dixon, M.; Ilyushchenko, S.; Thau, D.; Moore, R. Google earth engine: Planetary-scale geospatial analysis for everyone. *Remote Sens. Environ.* **2017**. [CrossRef]
18. Agapiou, A.; Papadopoulos, N.; Sarris, A. Detection of olive oil mill waste (OOMW) disposal areas in the island of Crete using freely distributed high resolution GeoEye's OrbView-3 and Google Earth images. *Open Geosci.* **2016**, *8*, 700–710. [CrossRef]
19. Contreras, D. Using Google Earth to Identify Site Looting in Peru: Images, Trafficking Culture. Available online: http://traffickingculture.org/data/data-google-earth/using-google-earth-to-identify-site-looting-in-peru-images-dan-contreras/ (accessed on 27 July 2017).

20. Contreras, D.; Brodie, N. Looting at Apamea Recorded via Google Earth, Trafficking Culture. Available online: http://traffickingculture.org/data/data-google-earth/looting-at-apamea-recorded-via-google-earth/ (accessed on 27 July 2017).

21. Agapiou, A. Orthogonal equations for the detection of archaeological traces de-mystified. *J. Archaeol. Sci. Rep.* **2016**. [CrossRef]

22. Agapiou, A.; Alexakis, D.D.; Sarris, A.; Hadjimitsis, D.G. Linear 3-D transformations of Landsat 5 TM satellite images for the enhancement of archaeological signatures during the phenological of crops. *Int. J. Remote Sens.* **2015**, *36*, 20–35. [CrossRef]

23. RDAC 2010, *Annual Report of the Department of Antiquities for the Year 2008, "Excavations at Politiko-Troullia"*; Department of Antiquities: Nicosia, Cyprus, 2010; p. 50.

24. RDAC 2013, *Annual Report of the Department of Antiquities for the Year 2009, "Excavations at Politiko-Troullia"*; Department of Antiquities: Nicosia, Cyprus, 2013; pp. 57–58.

25. Yu, L.; Porwal, A.; Holden, E.-J.; Dentith, C.M. Suppression of vegetation in multispectral remote sensing images. *Int. J. Remote Sens.* **2011**, *32*, 7343–7357. [CrossRef]

26. Crippen, R.E.; Blom, R.G. Unveiling the lithology of vegetated terrains in remotely sensed imagery. *Photogramm. Eng. Remote Sens.* **2001**, *67*, 935–943.

27. Gitelson, A.A.; Merzlyak, M.N.; Chivkunova, O.B. Optical properties and nondestructive estimation of anthocyanin content in plant leaves. *Photochem. Photobiol.* **2001**, *74*, 38–45. [CrossRef]

28. Kaufman, Y.J.; Tanré, D. Atmospherically resistant vegetation index (ARVI) for EOS-MODIS. *IEEE Trans. Geosci. Remote Sens.* **1992**, *30*, 261–270. [CrossRef]

29. Chuvieco, E.; Martin, P.M.; Palacios, A. Assessment of different spectral indices in the red-near-infrared spectral domain for burned land discrimination. *Remote Sens. Environ.* **2002**, *112*, 2381–2396. [CrossRef]

30. Tucker, C.J. Red and photographic infrared linear combinations for monitoring vegetation. *Remote Sens. Environ.* **1979**, *8*, 127–150. [CrossRef]

31. Huete, A.R.; Liu, H.Q.; Batchily, K.; van Leeuwen, W. A comparison of vegetation indices over a global set of TM images for EOS-MODIS. *Remote Sens. Environ.* **1997**, *59*, 440–451. [CrossRef]

32. Pinty, B.; Verstraete, M.M. GEMI: A non-linear index to monitor global vegetation from satellites. *Plant Ecol.* **1992**, *101*, 15–20. [CrossRef]

33. Gitelson, A.; Kaufman, Y.; Merzlyak, M. Use of a green channel in remote sensing of global vegetation from EOS-MODIS. *Remote Sens. Environ.* **1996**, *58*, 289–298. [CrossRef]

34. Sripada, R.P.; Heiniger, R.W.; White, J.G.; Meijer, A.D. Aerial color infrared photography for determining early in-season nitrogen requirements in corn. *Agron. J.* **2006**, *98*, 968–977. [CrossRef]

35. Gitelson, A.A.; Merzlyak, M.N. Remote Sensing of Chlorophyll Concentration in Higher Plant Leaves. *Adv. Space Res.* **1998**, *22*, 689–692. [CrossRef]

36. Crippen, R. Calculating the vegetation index faster. *Remote Sens. Environ.* **1990**, *34*, 71–73. [CrossRef]

37. Segal, D. Theoretical basis for differentiation of ferric-iron bearing minerals, using Landsat MSS Data. In Proceedings of the 2nd Thematic Conference on Remote Sensing for Exploratory Geology, Symposium for Remote Sensing of Environment, Fort Worth, TX, USA, 6–10 December 1982; pp. 949–951.

38. Boegh, E.; Soegaard, H.; Broge, N.; Hasager, C.; Jensen, N.; Schelde, K.; Thomsen, A. Airborne multi-spectral data for quantifying leaf area index, nitrogen concentration and photosynthetic efficiency in agriculture. *Remote Sens. Environ.* **2002**, *81*, 179–193. [CrossRef]

39. Daughtry, C.S.T.; Walthall, C.L.; Kim, M.S.; de Colstoun, E.B.; McMurtrey, J.E. Estimating corn leaf chlorophyll concentration from leaf and canopy reflectance. *Remote Sens. Environ.* **2000**, *74*, 229–239. [CrossRef]

40. Haboudane, D.; Miller, J.R.; Pattey, E.; Zarco-Tejada, P.J.; Strachan, I. Hyperspectral vegetation indices and novel algorithms for predicting green LAI of crop canopies: Modeling and validation in the context of precision agriculture. *Remote Sens. Environ.* **2004**, *90*, 337–352. [CrossRef]

41. Yang, Z.; Willis, P.; Mueller, R. Impact of band-ratio enhanced AWIFS image to crop classification accuracy. In Proceedings of the Pecora 17, Remote Sensing Symposium, Denver, CO, USA, 18–20 November 2008.

42. Sims, D.A.; Gamon, J.A. Relationships between leaf pigment content and spectral reflectance across a wide range of species, leaf structures and developmental stages. *Remote Sens. Environ.* **2002**, *81*, 337–354. [CrossRef]

43. Goel, N.; Qin, W. Influences of canopy architecture on relationships between various vegetation indices and LAI and Fpar: A computer simulation. *Remote Sens. Rev.* **1994**, *10*, 309–347. [CrossRef]

44. Bernstein, L.S.; Jin, X.; Gregor, B.; Adler-Golden, S. Quick atmospheric correction code: Algorithm description and recent upgrades. *Opt. Eng.* **2012**, *51*, 111719-1–111719-11. [CrossRef]
45. Hall, D.; Riggs, G.; Salomonson, V. Development of methods for mapping global snow cover using moderate resolution imaging spectroradiometer data. *Remote Sens. Environ.* **1995**, *54*, 127–140. [CrossRef]
46. Rouse, J.W.; Haas, R.H.; Schell, J.A.; Deering, D.W.; Harlan, J.C. *Monitoring the Vernal Advancements and Retrogradation (Greenwave Effect) of Nature Vegetation*; NASA/GSFC Final Report; NASA: Greenbelt, MD, USA, 1974.
47. Rondeaux, G.; Steven, M.; Baret, F. Optimization of soil-adjusted vegetation indices. *Remote Sens. Environ.* **1996**, *55*, 95–107. [CrossRef]
48. Curran, P.; Windham, W.; Gholz, H. Exploring the relationship between reflectance red edge and chlorophyll concentration in slash pine leaves. *Tree Physiol.* **1995**, *15*, 203–206. [CrossRef] [PubMed]
49. Roujean, J.L.; Breon, F.M. Estimating PAR absorbed by vegetation from bidirectional reflectance measurements. *Remote Sens. Environ.* **1995**, *51*, 375–384. [CrossRef]
50. Jordan, C.F. Derivation of leaf area index from quality of light on the forest floor. *Ecology* **1969**, *50*, 663–666. [CrossRef]
51. Huete, A. A soil-adjusted vegetation index (SAVI). *Remote Sens. Environ.* **1988**, *25*, 295–309. [CrossRef]
52. Gamon, J.A.; Surfus, J.S. Assessing leaf pigment content and activity with a reflectometer. *New Phytol.* **1999**, *143*, 105–117. [CrossRef]
53. Haboudane, D.; Miller, J.R.; Tremblay, N.; Zarco-Tejada, P.J.; Dextraze, L. Integrated narrow-band vegetation indices for prediction of crop chlorophyll content for application to precision agriculture. *Remote Sens. Environ.* **2002**, *81*, 416–426. [CrossRef]
54. Bannari, A.; Asalhi, H.; Teillet, P. Transformed difference vegetation index (TDVI) for vegetation cover mapping. In Proceedings of the IEEE International Geoscience and Remote Sensing Symposium (IGARSS '02), Toronto, ON, Canada, 24–28 June 2002; Volume 5.
55. Gitelson, A.A.; Stark, R.; Grits, U.; Rundquist, D.; Kaufman, Y.; Derry, D. Vegetation and soil lines in visible spectral space: A concept and technique for remote estimation of vegetation fraction. *Int. J. Remote Sens.* **2002**, *23*, 2537–2562. [CrossRef]
56. Wolf, A. *Using WorldView 2 Vis-NIR MSI Imagery to Support Land Mapping and Feature Extraction Using Normalized Difference Index Ratios*; DigitalGlobe: Longmont, CO, USA, 2010.
57. Agapiou, A.; Hadjimitsis, D.G.; Alexakis, D.D. Evaluation of broadband and narrowband vegetation indices for the identification of archaeological crop marks. *Remote Sens.* **2012**, *4*, 3892–3919. [CrossRef]
58. Roerdink, J.B.T.M.; Meijster, A. The watershed transform: Definitions, algorithms, and parallelization strategies. *Fundam. Inf.* **2001**, *41*, 187–228.
59. Robinson, D.J.; Redding, N.J.; Crisp, D.J. *Implementation of a Fast Algorithm for Segmenting SAR Imagery*; Scientific and Technical Report; Defense Science and Technology Organization: Victoria, Australia, 2002.

geosciences

MDPI

Article

Analysis and Processing of Nadir and Stereo VHR Pleiadés Images for 3D Mapping and Planning the Land of Nineveh, Iraqi Kurdistan

Eva Savina Malinverni [1,*] **, Roberto Pierdicca** [1] **, Carlo Alberto Bozzi** [1] **, Francesca Colosi** [2]
and Roberto Orazi [2]

[1] Department of Construction, Civil Engineering and Architecture (DICEA), Universitá Politecnica delle
    Marche, 60131 Ancona, Italy; r.pierdicca@staff.univpm.it (R.P.); carloalbertobozzi@gmail.com (C.A.B.)
[2] Institute for technologies applied to cultural heritage (ITABC), Consiglio Nazionale delle Ricerche,
    00185 Roma, Italy; francesca.colosi@itabc.cnr.it (F.C.); roberto.orazi@itabc.cnr.it (R.O.)
*   Correspondence: e.s.malinverni@staff.univpm.it; Tel.: +39-071-220-4419

Received: 4 August 2017; Accepted: 29 August 2017; Published: 6 September 2017

Abstract: The impressive hydraulic system built by the Assyrian King Sennacherib is composed by different archaeological areas, displaced along the Land of Nineveh, in Iraqi Kurdistan. The extensive project we are working on has the aim of mapping and geo-referencing any kind of documentation in order to design an archaeological-environmental park able to preserve and enhance the archaeological complex. Unfortunately, the area is failing a topographic documentation and the available cartography is not sufficient for planning and documentation purposes. The research work presented in these pages moves towards this direction, by exploiting Pleiadés Very High Resolution (VHR) images (in both nadir and stereo configuration) for an accurate mapping of the site. In more depth, Pleiadés nadir VHR images have been used to perform a pansharpening procedure used to enhance the visual interpretation of the study area, whilst stereo-pair have been processed to produce the Digital Elevation Model (DEM) of the study area. Statistical evaluations show the high accuracy of the processing and the reliability of the outputs as well. The integration of different products, at different Levels of Detail within a unique GIS environment, besides protecting, preserving and enhancing the water system of Sennacherib's, paves the way to allow the Kurdistan Regional Government to present a proposal for the admission of the archaeological complex in the UNESCO World Heritage Tentative List (WHTL).

Keywords: remote sensing; Pleiadés Satellite Images; stereo-pair; matching; DEM; GIS; archaeology

1. Introduction

With the advent of fast and agile pipelines of data collection, it has become easier and easier to gather detailed information of archaeological areas. However, some zones can be impervious to be reached due to the morphology, the wideness or, sometimes, because the war impedes systematic campaigns on site. It is well known, in fact, that a common problem for archaeologists studying the evolutions of ancient settlement is overcoming the hazardous conditions of those risky areas. Often, investigations are realized by extensive ground surveys in which insiders perform on site surveys and excavations [1] but, if not possible, an excellent opportunity is nowadays offered by high resolution satellite images, which allow one to infer useful information of an area, also whereas unreachable; in its broadest meaning in fact, archaeology might strongly benefit from the use of remote sensing (RS) techniques, since embrace methods to uncover (and map) evidence of the past [2]. RS capabilities can be exploited as complementary data (and sometimes as alternative ones) to the ground surveys, that are more accurate but are not able to provide a large scale overview of ancient settlements, especially where they are too wide and hence need a wider sight to perform investigations [3].

The applications are many and strictly dependent on the purpose of the study. Nadiral Very High Resolution (VHR) satellite images are suitable for detecting changes over time, inferring historical information about the impact of human activities over ancient sites, or monitoring the state of conservation of a site [4]. In addition, satellite platforms are nowadays providing a growing amount of data: multispectral and multi-temporal data demonstrated an adequate accuracy even for medium-scale maps (1:10,000, 1:5,000), since they hold data related to physical and environmental parameters for the detection of potential buried structures [5]. In addition, hyper-spectral sensors are the most appropriate tools for producing thematic maps (use of the soil, extraction of road system, classification of vegetation, etc.), thanks to the image classification techniques [6,7]. Moreover, some sensors can capture images in stereo mode, enabling for the creation of accurate Digital Elevation Models (DEMs), representing a great improvement to deepen the knowledge of a certain area. The unavailability of cartography, updated or in adequate scale, is a recurring problem for the archaeological research operating in urban and territorial contexts. In those areas where cartographies are inaccessible or not updated, the use of DEM and ortho-photos represents an optimal solution for the creation of topographical maps and space maps to be used as plans during archaeological field works. In this context, the automatic extraction of 3D metrical information from satellite images taken from different angles represents a turnkey. The aid of the third dimension is fundamental for understanding the morphology of a place in order to document the area and to perform planning activities.

The research work presented in this paper moves towards this direction, by exploiting Pleiadés Very High Resolution (VHR) images (in both nadir and stereo configuration) for an accurate mapping of a wide archaeological area; the activities are conducted within the framework of an International project (http://www.terradininive.com, The project is directed by Prof. Daniele Morandi Bonacossi of Udine University. The project for Sennacherib's Archaeological Park is directed by Arch. Roberto Orazi) "Land of Nineveh—Training for the enhancement of the cultural heritage of Northern Kurdistan". An overview of the territorial framework can be found in Figure 1.

In the context of this extensive project, the task is to map and geo-reference any kind of documentation in order to design an archaeological-environmental park able to preserve and enhance the vast archaeological complex. World View2 and Pleiadés VHR images have been used; for what is regarding Khinis, the area subject of this case study, the Pleiadés images allowed us to produce both a medium scale cartography, vector maps and DEM. It is important to underline that, up to now, the archaeological area is failing a topographic documentation of the region and the activity described in the paper copes with the lack of available data; in fact, the cartography was not sufficient for planning purposes, consisting of very small scale national maps (with a scale of 1:100,000), Corona and Orbview3 imagery with a very coarse radiometric and geometric resolution and DEM arising from Shuttle Radar Topography mission SRTM (approximately 30 m). The use of VHR images give a product suitable to: (i) have a first geographical view of the area; (ii) design a preliminary planning of the site; (iii) define the first limits of the core and the buffer zone and the archaeological and geological prescriptions; (iv) geo-reference those areas investigated through the archaeological survey campaigns. These elaborations gave us a precise geometric representation of the territory, and were therefore indispensable starting points for the design process. All the products of our computations are suitable for being integrated in a Geographical Information System (GIS), where most of the available data of the area can be localized and organized: technical, thematic, historical maps, archaeological survey (by close range photogrammetry or by terrestrial laser scanning) about monuments' details and niches and so on. The contribution of this paper is twofold: on one hand, we share with the research community (dealing with RS in archaeology) a mapping dataset of an unknown area, that have been studied testing several methods of image processing. In more detail, Pleiadés nadir VHR images have been used to perform a pansharpening procedure used to enhance the visual interpretation of the study area (thanks to a 50 cm resolution image for a medium scale map of the whole area), whilst stereo-pair have been processed to produce the DEM of the study area. In this last case, the panchromatic stereo-pair produces a more reliable output w.r.t. the pansharpened ones. On the other, we provide the overall design of a large archaeological and environmental park. The integration of different products at different Levels of Detail (LOD) within a unique GIS environment, besides protecting, preserving and enhancing Sennacherib's water system of Sennacherib's, paves the way to allow the Kurdistan Regional Government (KRG) to present a proposal for the admission of the archaeological complex in the UNESCO World Heritage Tentative List (WHTL).

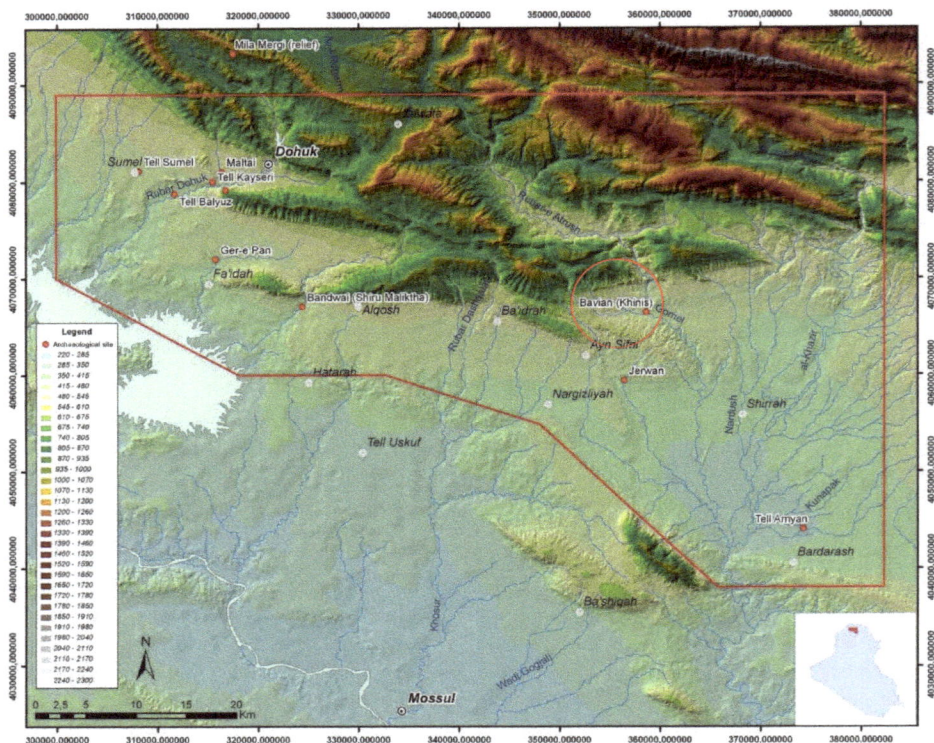

Figure 1. The borders of the "Land of Nineveh" in Iraqi Kurdistan (red line). The red circle highlights the area of Khinis, one of the areas of the Archaeological Park, subject of this case study.

The reminder of the paper is organized as follows: after the general overview already discussed, the introduction includes a brief review over the related works in the literature, besides describing in detail the study area and the reasons why our approach have been fundamental for the project. Section 2 is devoted to describe in detail the Pleiadés images processing; this section has been divided into two sub-sections to facilitate the reader to understand the different steps of the work-flow: high resolution images processing and stereo-pair computation. The output results of the process are presented in Section 3, demonstrating how the different product can be easily managed inside the GIS environment. Discussion and concluding remarks about the findings of our research are reported in Section 4.

1.1. State of the Art

Satellite remote sensing in archaeology has become a common tool of investigation, documentation and planning. The new possibility offered by remotely sensed data processing allows for understanding environmental changes and, through the development of GIS-based models and decision-support instruments, to have an enhanced decision-making process. Moreover, by using satellite RS techniques the monitoring process of archaeological sites can be efficiently supported in a reliable, repetitive, non-invasive, rapid and cost-effective way [8]. By exploiting multitemporal high-resolution satellite images, for instance, it is possible to study areas that were without documentation just few years before. Like in the work of Di Giacomo and Scaradozzi [9] where the ancient city of Ur, in southern Mesopotamia, was studied with the aid of different new satellites (QuickBird-2, Ikonos-2, WorldView-1) in comparison with older declassified spy space photos. It is also a common practice to extract 3D information from satellite dataset; the approach consists on exploiting stereo-pairs (or triplets) with an acquisition scheme insuring that every point on the ground is seen from, at least, two different points of view [10]. The approach has been used in several domains; the best choice of viewing angles for stereoscopic measurement is always a matter of compromise: a wide stereo angle provides a good geometric accuracy, but the matching of two images points may be difficult, or even impossible if the two points of view are very different. In dense urban areas, for instance, the simultaneous visibility at the street level is highly dependent on the directions and the incidence angles of the pair of images. In [11], for instance, World-View2 were used for extracting LOD for CityGML (which stands for Geography Markup Language) application of urban areas, proposing an automatic processing chain for urban modelling based on stereo images. To increase the matching between the overlapped images, sometimes the use of tri-stereo for the DEM generation is required [12]. With respect to the automation of processing large dataset, the work of Shean et al. is worth being mentioned [13]. For an open landscape instead, with moderate slopes, the use of pairs is generally sufficient [14] (depending on the purpose of the project), allowing also for detecting changes of environmental dynamics, like in the case of glacier studies [15], or volcanological applications [16]. In fact, other approaches of changes detection from stereo images are also discussed in [17], demonstrating that the production of DEMs for the change detection purposes is fundamental. In these studies, the use of 3D data at different temporal resolution allows the evaluation of changes among time.

In line with recent research trends, our work can cover different scales of representation, starting from the processing of satellite images to create the cartographic base that can be further integrated with detailed ground surveys (e.g., through Unmanned Aerial Vehicles (UAV) mapping). A similar approach in the literature can be found in [18], where the authors compute Pleiadés to map the Valley of Turu Alty (Siberia, Russia). Several articles deal with the realization of DEM highlighting the benefits of using Pleiadés sensors as the data source. As a demonstration of that, accuracy evaluations are stated in different works, using as a benchmark SRTM images at 30 m [19], as well as more accurate LiDAR data [20]. The purpose of Pleiadés is to deliver optical images of sub-metric resolution with daily access to any point on the globe (with its two satellites): a panchromatic channel with a 70 cm vertical viewing resolution and a multispectral one composed of four spectral bands (blue, green, red and near-infrared) with a 2.8 m resolution (re-sampled at 0.5 m and 2 m, respectively), image swath of about 20 km in vertical viewing, acquisition capacity, in a single pass, of a 100×100 km mosaic of images, virtually instantaneous acquisition capacity for stereoscopic pairs (and even triplets) of 20 km up to 300 km cloud-free image coverage of 2,500,000 km^2 per year. Interested readers can find more accurate features and capability of these sensors in [21], and in particular their 3D capabilities for the creation of DEM in [22]. By the way, the procedures already discussed present some bottlenecks. First of all, the stereo images need refinement and re-sampling works to make them suitable for 3D mapping purposes. The work of Cantou et al. represents an excellent example for the archaeological domain, since the image treatment depends on the material of the sites and the procedures cannot be extended to other domains [23]. Moreover, in some cases and depending on the purpose of the mapping, the resolution for the representation scale revealed insufficient [24], requiring different approaches or the integration with ground survey data. In this regard, resorting to UAV surveying can overcome the drawbacks of RS solutions (i.e., cloud coverage, resolution and so on). Recent research showed that nowadays drones are the most widespread tool for archaeological mapping campaigns and to achieve detailed and accurate documentations, but only of a small portion of a site. Just to mention some, the works in [25–27] are good examples to be reported.

The work presented in these pages ends up with the creation of an accurate cartography at a territorial scale (up to 1:5000), and it opens up to the possibility of integrating different data sources in a unique GIS, which demonstrated to represent a great advantage especially in the archaeological domain [28–31].

1.2. Description of the Study Area

The impressive hydraulic system built by the Assyrian King Sennacherib is composed by different archaeological areas. The various parts of the waterworks and celebratory works were built by the sovereign in four phases from 706 to 686 B.C. [32,33]. His purpose was to supply water to the royal palaces and gardens of Nineveh—a city neglected by his father, Sargon, but chosen by Sennacherib as his new capital—and to irrigate the land surrounding the city. Nineveh covered a larger area than all capitals before. Not only the city of Nineveh itself but the surrounding large agricultural areas that were necessary to feed the city required an enormous amount of water. To secure this water supply, Sennacherib therefore initialized his hydro-projects. They were carried out on a much higher technological level than previously known water installations. The first phase of Sennacherib's project concerned a series of canals located in the vicinity of Nineveh; these tapped water from the Khosr River near the village of Kisiri. The second phase was based on a survey made by Sennacherib himself near Mount Musri (now Jebel Bashiqah). Many springs were enlarged, 18 canals were built to feed water into the Khosr, marshes and cane-brakes were created, and a large park containing trees, fruit and animals of all kinds. The third phase concerned the so-called northern canals, which departed from the village of Rimussa (Jerrahiah); the last phase focussed on the set of canals that started in Khinis. In some parts, either at the head or along the canals, reliefs and cuneiform inscriptions were sculpted into the rock which inform us still today about the aims of King Sennacherib when ordering the construction of the canals [34]. Only two parts—Khinis and Jerwan, both belonging to the Khinis-Khosr canal—have received more detailed archaeological interventions. All other parts are known only through short-time surveys and the documentation of the findings. The project for the realization of the Archaeological Environmental Park is studying and enhancing the two last phases of the Sennacherib's irrigation system located in the Governatorate of Dohuk, in the northern part of Kurdistan region. The third construction phase is characterized by the reliefs of Maltai, situated on a very steep cliff-side overlooking the city of Dohuk, and by the Faideh bas-reliefs, situated farther south, perhaps at places where water was drawn off to irrigate the surrounding territory. Only their upper edges are visible; the rest is out of sight in the now-earth-filled canal. They are seriously threatened by cement companies operating only a few metres away. Lastly, the archaeological area of the Wadi Bandwai and of the Shiru Malikta reliefs is especially evocative, though the reliefs have largely been eroded away. The most impressive monuments belong to the fourth phase of channels, starting from the Gomel River, at the height of the Kinhis rock. Though in terms of technology and construction, the Jerwan aqueduct was the most important part of Sennacherib's water supply system [35], the Khinis monumental complex embodies the system's celebratory, ritual and artistic aspects. In both cases, their environmental impact is impressive, hence it will be vital to make sure that no extraneous elements disrupt their understanding and accessibility. For this reason, the archaeological site of Khinis and the Jerwan aqueduct are included in the boundaries of the archaeological—environmental park that the Italian Mission are planning [36]. The Khinis area seems to be of primary importance (see Figures 2 and 3).

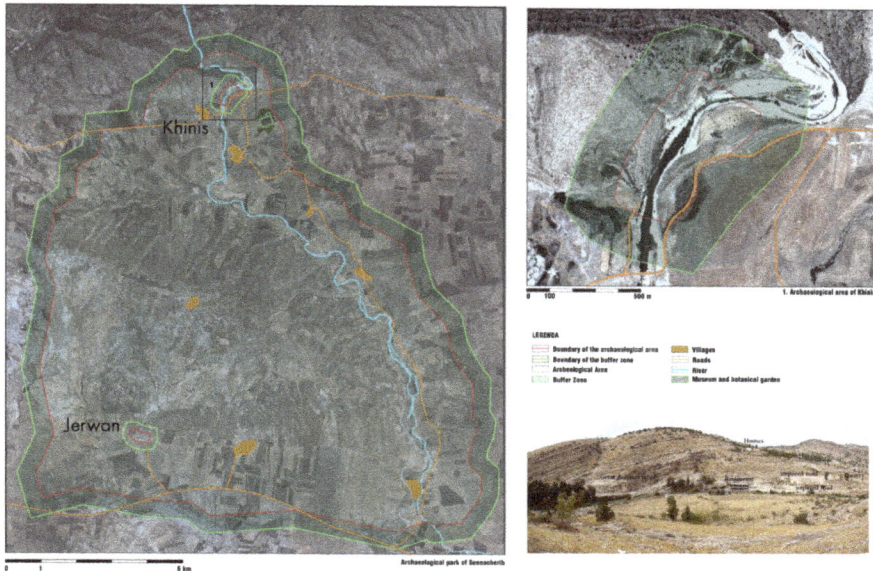

Figure 2. On the left: the borders of the entire Archaeological Park with the core (in red) buffer zones (in green) and with highlighted Khinis and Jerwan areas. The upper right side represents the borders of the Khinis area, while, in the lower right side, a picture of the area is reported.

Figure 3. A view of the Khinis' rock. On the left, the rider's panel is the great bas-relief and, on the top, many different niches with Sennacherib's inscriptions.

Though the works here date from the last phase of construction, this was actually the starting point where the canal tapped water from the Gomel River. Moreover, it stands out as a major celebratory work comprising an incredible amount of bas-reliefs, inscriptions and technical works. Blocks of stone were quarried on both sides of the Gomel and rafted downstream to land at stone wharfs; dams and leaves were built; tunnels were dug through the bedrock; ornamented niches were carved out at various heights along the whole rock wall; and artworks were created, signally the great bas-relief and the imposing monolith. The relief, approximately 12 × 14 m, showed king Sennacherib (on the sides of the panel) paying homage to the deities portrayed in the centre. In front of the relief, a huge monolith, decorated on at least two sides, marked the divide between the Gomel River and the head of the canal. One stretch of the rock face, around 300 m long, featured a great number of niches bearing inscriptions and bas-reliefs portraying Sennacherib (see Figure 3). Given this general overview of the entire archaeological site, from now on, the computations presented in this paper will be focused only on this latter area.

2. Materials and Methods: Documenting the Whole Archaeological Park

In the previous sections, a general overview of the subject area of the investigation and a brief overview of the current methods used in archaeology have been outlined. In the following, a detailed description of the steps performed to achieve a thorough topographic mapping of a portion of a landscape archaeological area will be given—in particular, the Khinis area. The work has been performed at different scales and with different products to meet the needs of archaeologists, by exploiting the following steps of the workflow:

- **Pleiadés VHR images processing**, used for a visual interpretation of the study area, together with a general overview of the entire Archaeological Park;
- **Stereo-pair processing**, used for obtaining the DEM of the area, useful to study in detail the morphology of the area of Khinis.

2.1. Pleiadé VHR Images Processing

The planning process of such wide archaeological setting cannot disregard from a first step of visual interpretation of the area. This first stage of the work consists of the use of VHR images aimed at studying the archaeological landscape and at highlighting the main areas where the cartographic features are concentrated (streets, ducts, borders, villages, and so on). In the following, the main features of the input dataset will be described and the details of the steps performed to create pansharpened pictures, together with the details about the comparison between the different methods of pansharpening adopted.

2.1.1. Dataset

The input images processed have been obtained from Pleiadés sensor. For this stage, a strip of the same region of the Iraqi Kurdistan was taken, with the simultaneous acquisition of panchromatic and multispectral, the day 01-07-2012, in Universal Transverse Mercator (UTM) planar projection, Zone 38 N, and WGS84 datum. Of course, the two set of images have two different geometric resolutions: the panchromatic ones with a ground resolution at nadir of about 0.50 m, while the multispectral ones can reach a ground resolution of 200 m. The acquisition in multispectral mode for this type of satellites is in turn subdivided in four bands of the spectrum: red, green, blue (bands of the spectrum visible to the naked eye), and near infra-red (NIR-spectrum invisible to the naked eye). The dataset, ortho-ready, is 100% cloud free and it has not been necessary to perform any radiometric calibration to the data, since there was not the presence of meaningful distortions (Figure 4a).

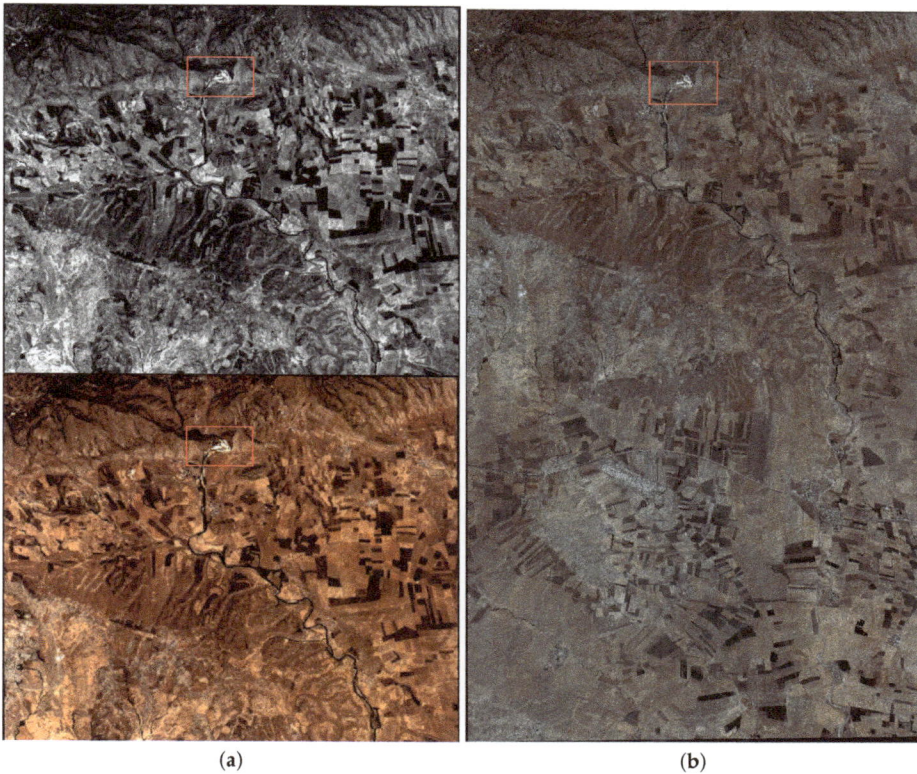

Figure 4. The input images arising from Pleiadés. Image (**a**) represents the panchromatic and multispectral original images (northern part); and (**b**) the pansharpened one. The red rectangle indicates a quarry inside the Khinis area, particularly important for the project; this subset of the image was studied due to its high brightness.

2.1.2. Images Processing

Pansharpening is a well known methodology of image fusion which combines the high resolution of panchromatic images with the lower resolution of multispectral ones. The advantage of such method is to get as a final result a coloured image of a certain area with a high resolution, optimizing the starting panchromatic one. This step is required since the images have different geometric resolution, and this does not allow their direct overlay, unless an appropriate breakdown of the pixels is carried out, and a re-sampling through dedicated algorithms. It is hence necessary to produce a pansharpened image (Figure 4b), obtaining high-resolution images (0.50 cm) in the four bands acquired by Pleiadé sensors. We therefore performed a co-registration (using the red band) to achieve a perfect overlap of the two images, with the automatic selection of the tie points and choosing a Root Mean Square (RMS) lower than 0.25 m on them. Afterwards, HSV (Hue, Saturation, Value), Principal Components (PC), Brovey and Gram–Schmidt (G–S) methods have been tested; the visual comparison exhibits that all the pansharpening algorithms have significantly improved the initial low resolution Multi-Spectral (MS) image (see Figure 5).

To validate this assumption, the first and second order statistics (namely, mean and variance) were calculated for all images to make a comparison between the original and the pansharpened ones. The comparison was done for the whole strip and for a subset of the image particularly over-exposed (see Figures 4 and 6a). The results are shown in Table 1 and can be compared with the image subset in Figure 5, which reports the four outputs from the pansharpening algorithms used. The PC algorithm obtains the maximum mean compared with the original one. The variance values are very close to each other, apart from the HSV method, which performs worse. The Gram–Schmidt method is the most balanced.

Table 1. Comparison of mean and variance for all images.

	Original	Hue, Saturation, Value	Principal Components	Brovey	Gram–Schmidt
Mean	111,721	106.473	120.683	103.687	120.496
Variance	66.107	69.795	64.667	65.506	65.090
Mean (subset)	127.834	126.668	140.149	123.126	123.364
Variance (subset)	71.078	77.913	71.787	74.438	77.037

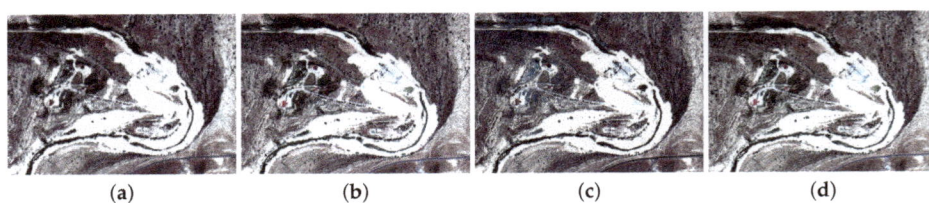

(a) (b) (c) (d)

Figure 5. A comparison of the different pansharpening algorithms. The details show the area of the quarry since it is meaningful to evaluate the performances of the test. (a) HSV; (b) PC; (c) Brovey; (d) G–S.

The final pansharpened, used is the one obtained by using the Gram–Schmidt [37] algorithm implemented in ENVI© software. The results are in line with similar research in the field [38,39]. This choice is supported with the statistics performed in a subset (row three and four of Table 1). We finally added some filters in order to achieve results closer to the real color of the area. The results of this step are shown in Figure 6. The figure demonstrates how the image have been enhanced with respect to the original one. Choosing the quarry as a demonstrative example, it is clear that, if in the initial input image, the white area was overexposed and hence not usable, after the pansharpening procedure and the application of a linear filter to the pansharpened image, the picture can be exploited for a more accurate visual interpretation.

2.2. Stereo-Pair Imagery Processing

The second stage of the work consists of the creation of the DEM of the Khinis area, starting from the satellite stereo-pair. It is indeed fundamental to recreate the morphology of this territory, with the purpose of planning all the services of the future Archaeological Park. The 3D model of the site is an instrument for controlling the urban development, for planning efficient exploitation of resources and raw materials and for studying intervention of sustainable agriculture, including the monitoring of hydrographical net and possible environmental risks. The remote sensing image observation can localize the areas where there are critical situations; this type of monitoring can be extremely important in a moment in which climatic changes are going to produce even more probable floods and drought and other consequences in the cycle of water. As in the duct fell several rests, the program of intervention requires a more detailed cartography of the area made of level curves and ortophotos in adequate scale to be managed within the GIS. The process of DEM extraction followed different tests performed to achieve the best balance between accuracy and resolution to better represent the morphology; namely, the purpose was to compare the performances of panchromatic and multispectral (then pansharpened) stereo-pairs for the creation of a DEM with 1 m of grid. To do so, we created, using the same procedure described in the previous section, the pansharpened images of the stereo-pair.

The software suite used for the processing are ERDAS IMAGINE© and ERDAS eATE© software, respectively for the photogrammetric orientation of the images and for the creation of the DEM.

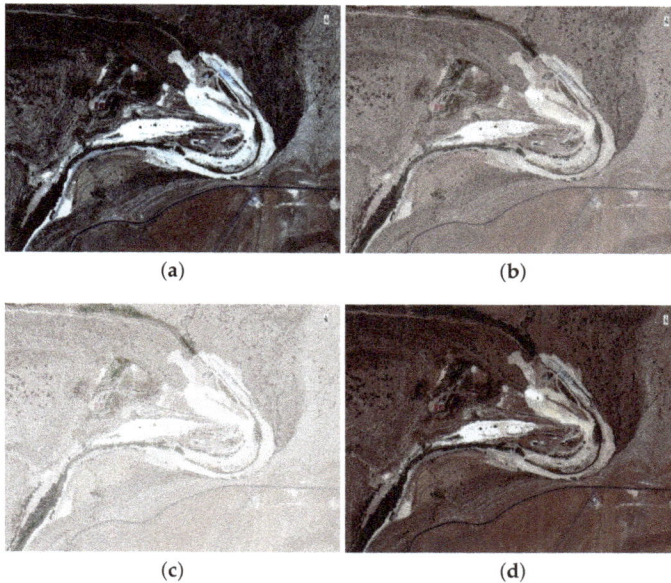

Figure 6. A comparison of the different test performed to achieve a more comprehensible result. The detail shows the area of the quarry that is overexposed and not suitable for a visual interpretation of the area. (**a**) the original picture; (**b,c**) pansharpened with failure filtering; (**d**) optimized pansharpened.

2.2.1. Dataset

As in the previous stage, the input dataset comes from Pleiadés sensor, acquired on 30 June 2013, considering in this stage the stereo-pair. In fact, thanks to the huge satellite agility obtained with control momentum gyros as actuators, the optical system delivers as well instantaneous stereo images, under different stereoscopic conditions and mosaic images, issued from along the track, thus enlarging the field of view. The dataset is composed of a couple of panchromatic images with a resolution of 0.5 m and a couple of multispectral images with a resolution of 2 m (the dataset is depicted in Figure 7).

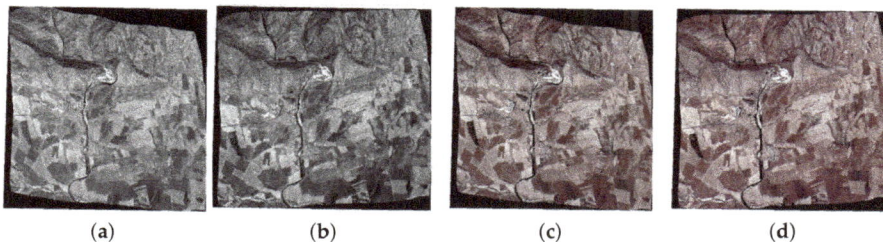

Figure 7. Stereo-pair images from Pleiadés. (**a,b**) panchromatic images; (**c,d**) multispectral images.

The dataset is even provided accompanied with information of the image capture (Rational Polynomial Coefficients—RPCs), that is, the data regarding the satellite orbit (ephemerides), platform and sensor orientation (in WGS84), and the data on optical-geometric conditions of the sensor at the time of the acquisition. ERDAS IMAGINE software uses these parameters implementing Rational Polynomial Functions (RPFs) to create a connection between image space and terrain space. After the image pyramid generation, the automatic tie point measurement has been run to provide an automatic selection of tie points to make the relative orientation. This process involves a bundle adjustment, a review of the results and, if it is necessary, a refinement of the performances to improve the quality of the solution. All these steps have been implemented for the panchromatic and pansharpened stereo-pairs, getting as a result of the triangulation accuracy, the values reported in Table 2.

Table 2. The table reports the comparison of the triangulation results between the panchromatic and pansharpened images. The first one performed using 39 tie points with a standard deviation of around 7 cm, while the second one using 37 tie points with a standard deviation of around 12 cm.

Image	Points	Iterations	Weighted Std Error (cm)	Residuals Max Value	Residuals Min Value
Panchromatic	39	2	0.067	0.454	0.001
Pansharpened	37	2	0.126	0.424	0.009

The absolute orientation of the photogrammetric model was obtained using the original RPC, since it was not possible to collect any Ground Control Point on site with Global Navigation Satellite System (GNSS) positioning instruments.

2.2.2. DEM Extraction

Given the satisfactory values of the residuals from the triangulation phase, the set of images was then ready to be used for the creation of the DEM. For its production, we used eATE (Automatic Terrain Extraction) by ERDAs, with the new LPS module for generating high resolution terrain information from multi-stereo imagery (both panchromatic and pansharpened, the latter obtained with Gram–Schmidt to work with the same resolution). Thus, we performed the automatic extraction of cloud points setting some parameters: the overlapping over 90%, the area with min and max altitude (0–800 m) and the smoothing strategy, suitable for the archaeological area. The correlation method used by eATE is the well known Normalized Cross Correlation (NCC) [40], and we set up a correlation threshold between 0.3 and 0.8, getting in output eight iterations in both cases. The comparison between the panchromatic and pansharpened are reported in Table 3.

Table 3. Synthesis of the orientation residual after the optimization.

Image	Check Points	Standard Deviation	Root Mean Square Error	Matched Points	Output Points	Matching Quality
Panchromatic	18	0.587	0.593	67362234	40156839	Excellent: 92.7% Good: 7.5%
Pansharpened	21	0.326	0.335	85901630	45195964	Excellent: 91.1% Good: 8.9%

As it can be inferred from the values of the table, the two points cloud is comparable in terms of both accuracy and points density, and it is therefore impossible to choose the point cloud to be used, and we proceeded with a geometrical comparison of the DEMs, setting up a grid of 1 m. In certain critical areas, where the matching algorithm fails on extracting correlated tie points, it is visible that the panchromatic DEM is better than the pansharpened one. As a demonstration of that, we report, in Figure 8, a comparison of the two DEMs, taking as an example the quarry area, where the computation of the 3D model generates some mistakes.

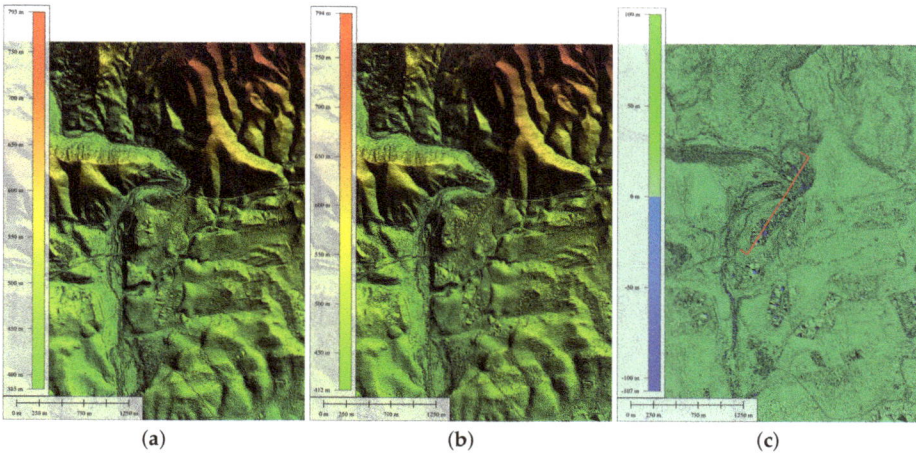

(a)   (b)   (c)

Figure 8. Comparison between the two DEMs. (**a,b**) are the panchromatic and pansharpened DEM, respectively; (**c**) is the subtraction of (**a,b**) with highlighted the zones with major defects.

For a better understanding, we also report the profile extracted from the DEM, where the approximation made for the pansharpened dataset is visible, caused by the lack of information on the rocks composing the hill (see Figure 9). In fact, the highland on the left is very coarse and the saddle on the right created by the rock is more accurate with the panchromatic DEM.

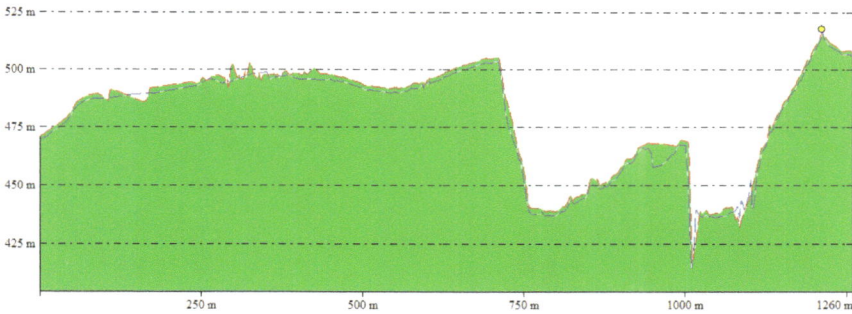

Figure 9. The profile of the two DEMs is reported. The red dashed line represents the DEM arising from the panchromatic image, while the blue dashed line represents the pansharpened one.

3. Results

Once all the steps of image processing have been performed, all the output products are suitable to be managed within the GIS. First of all, and in line with previous research conducted to evaluate the differences between the DEM production using panchromatic and pansharpened images [10], to produce the cartography of the site, we used the first one, which final DEM have been submitted to a post processing and editing phase. A 3D orthophoto has been further realized using a draping technique, by associating the DEM with the otho-ready pansharpened of the area (see Figure 10).

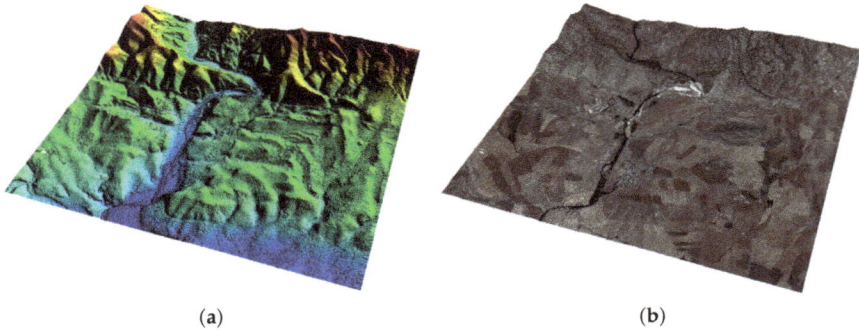

(a)   (b)

Figure 10. (**a**) 3D visualization of the DEM with color intensity visualization scaled according to the elevation; (**b**) the pansharpened image projected on the DEM.

With the consequent extraction of contour lines, with a step of 2 m, we upgraded the archaeological plans of the sites and create a vector documentation utilized for 3D reconstructions of monuments and ancient cities. Thus, the vectorization of all archaeological remains and traces visible in all multitemporal remote sensing data allowed the creation of new archaeological maps ready for input in a GIS with new data on ancient layout of sites, monuments and roads; in these maps, the contour lines were extracted from DEMs (see Figure 11).

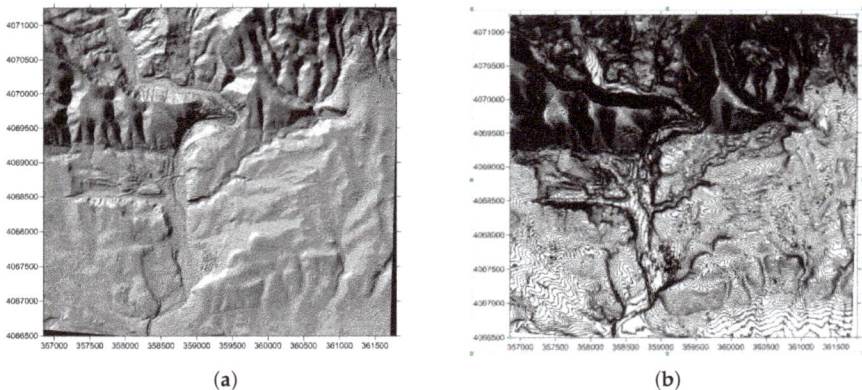

(a)   (b)

Figure 11. (**a**) shaded visualization of the DEM and (**b**) the contour lines with 2 m steps.

The produced cartographic base is fundamental to arrange the general organization of the structure, the different location of the functional areas and of the archaeological ones and the interrelation between the sites included in the core zone of the park and those located outside but belonging, according to the historical and architectural point of view, to the same archaeological complex.

4. Discussion and Conclusions

The methodology adopted for the case study proved to produce useful updating of the available cartography. The use of nadir and stereo Pleiadés images represents a great aid for the archaeological research since, against a medium expense, they provide a dataset suitable to be used at different scales of representation. Some processing is notwithstanding required. For a visual approach of such wide archaeological areas, the pansharpening procedure is required since it gives a more reliable instrument of interpretation. The Gram–Schmidt method showed the best results and, with a simple operation of filtering, we were able to produce usable maps, whereas the initial brightness did not allow any use. We cannot say the same for the creation of DEM. In fact, even if the same procedure had been used for the stereo-pair, the more reliable DEM was the one produced with the panchromatic dataset. The work presented in these pages opens up new possibilities for the Land of Nineveh, up to now almost unknown and lacking of any kind of spatial information. The purposes of the Archaeological-Environmental Park are to safeguard the cultural landscape related to the Sennacherib's irrigation complex and to offer to the local population the possibility of enjoying all its cultural, spiritual and social benefits. The region's heritage, if properly protected and enhanced, can constitute a notable driver of economic and social development by encouraging local and international tourism. The greater economic resources that can be created by new eco-friendly businesses such as those related to tourism and crafts, and the consequent development of small and medium enterprises (SMEs), can raise the society's standard of living, including in terms of physical and psychophysical health. Moreover, the arrival of foreign tourists in the region can help Iraqi Kurdistan extricate itself from the state of international isolation caused by recent warfare, and lead to new and beneficial cultural exchange and to economic progress. Regarding the boundaries of the Archaeological-Environmental Park, we must keep thinking that we have a unique monumental complex composed of different archaeological areas widespread over a vast territory. Therefore, the sites of Khinis, Jerwan, Maltai, Bandaway or other sites that will be brought to light by the future archaeological investigation have to be considered separately but also in connection with each other, since they all belong to the same complex that, for its historical, technical and artistic values, has to be preserved in its entirety. Anyway, the vastness of the archaeological complex, which is extended over a territory of almost 3.500 km^2, suggests an operative strategy tended to do so that the proposal is realistically evaluated by the UNESCO World Heritage Committee. Therefore, the boundaries of the Archaeological and Environmental Park are determined by the need to find an area small enough to be controlled, but full of important ancient structures. Under these conditions, it seems natural to think of the area that contains the bas-reliefs of Khinis and the great aqueduct of Jerwan as the basic elements of the park. As a matter of fact, we have to remember that the realization of a WHTL has to be accompanied by the management project requested by UNESCO. In this frame, the planning of an Archaeological Park will allow the creation of a decision-making structure able to manage the maintenance, the monitoring and the conservation of those sites located in its territory, but, at the same time, to start the necessary connections with the sites located outside the Park and belonging to the same system of canalization, bas-reliefs and inscriptions. A Museum Center provided with modern multimedia technologies will allow management not only of the archaeological areas of the Park, but also all those sites located at great distances that will be possible to study and visualize both in their specific location and related to the entire complex of Sennacherib. To achieve such ambitious goals, in the future, we are planning to integrate different data at different scale (drone acquisitions have been already done in different areas) within a unique and complex GIS (preliminary achievements of this phase of the work are showed in Figure 12), containing the whole graphic project of the Archaeological and Environmental Park and, at the same time, representing the managing project required by UNESCO for any site inscribed in the World Heritage List.

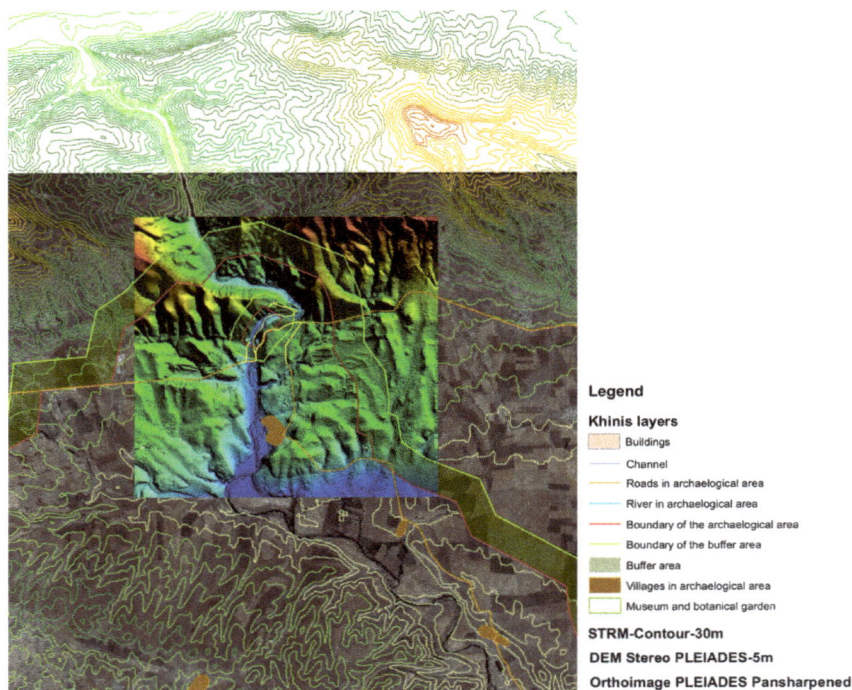

Figure 12. Topographic/archaeological layers have been stored inside a GIS environment. The management of different layers with different information are fundamental for the planning process of the area. The legend shows the different layers represented in the figure.

The GIS, which will have different levels of access for scholars and tourists, will allow the organization of the different areas of the Park and the visualization of documents (photos, videos, 3D models, AutoCAD drawings) of the archaeological structures and of the related paths to reach them, the visualization of the progress of the archaeological investigation and of the architectural and surface restoration, management of the technical structures of the park and of the tourist flow, valorisation of the site through the diffusion on the internet of a web site and of all the information on the cultural and social initiatives of the Centre. By means of mobile equipment [41], the GIS will allow tourists to move around the territory following pre-established itineraries.

Acknowledgments: The authors would like to thank Daniele Morandi Bonacossi of Udine University for involving the working team in the project. We also thank the students of Universitá Politecnica delle Marche for their work on the data processing conducted for the Master's Thesis (Carloni Soledad, Molinelli Michele, Dimitriou Anthoura, Pagnanelli Matteo). The authors also thank Planetek S.R.L. enterprise for providing satellite images and the temporary software licence.

Author Contributions: All the activities of the research work were led by R.O. The introduction section has been written by R.P. and E.S.M. (general overview and the state-of the art study) and R.O. and F.C. (description of the project historical framework). Data processing, computation and results were performed by R.P., E.S.M and C.A.B. Finally, discussion and conclusions have been outlined by R.P. and E.S.M. (technical part), and R.O. and F.C. (perspective outlook). All authors have reviewed and approved the content of this paper.

Conflicts of Interest: The authors declare no conflict of interest.

Abbreviations

The following abbreviations are used in this manuscript:

RS   Remote Sensing
DEM   Digital Elevation Model
SRTM   Shuttle Radar Topography Mission
GIS   Geographical Information System
VHR   Very High Resolution
HVS   Hue Value Saturation
G–S   Gram–Schmidt
PC   Principal Components
GCP   Ground Control Point
WHTL   World Heritage Tentative List
GNSS   Global Navigation Satellite System

References

1. Malinverni, E.; Barbaro, C.C.; Pierdicca, R.; Bozzi, C.; Tassetti, A. UAV surveying for a complete mapping and documentation of archaeological findings. The early Neolithic site of Portonovo. *Int. Arch. Photogramm. Remote Sens. Spat. Inf. Sci.* **2016**, *41*, 1149–1155.

2. Laet, V.D.; Paulissen, E.; Waelkens, M. Methods for the extraction of archaeological features from very high-resolution Ikonos-2 remote sensing imagery, Hisar (southwest Turkey). *J. Archaeol. Sci.* **2007**, *34*, 830–841.

3. Forte, M.; Campana, S. *Digital Methods and Remote Sensing in Archaeology: Archaeology in the Age of Sensing*; Springer: Berlin, Germany, 2017.

4. Hadjimitsis, D.G.; Agapiou, A.; Themistocleous, K.; Alexakis, D.D.; Sarris, A. Remote sensing for archaeological applications: Management, documentation and monitoring. In *Remote Sensing of Environment-Integrated Approaches*; Hadjimitsis, D.G., Ed.; InTech: Rijeka, Croatia, 2013.

5. Orlando, P.; Villa, B.D. Remote sensing applications in archaeology. *Arch. Calcolatori* **2011**, *22*, 147–168.

6. Castrianni, L.; Giacomo, G.D.; Ditaranto, I.; Scardozzi, G. High resolution satellite ortho-images for archaeological research: Different methods and experiences in the Near and Middle East. *Adv. Geosci.* **2010**, *24*, 97–110.

7. Malinverni, E.S.; Tassetti, A.N.; Mancini, A.; Zingaretti, P.; Frontoni, E.; Bernardini, A. Hybrid object-based approach for land use/land cover mapping using high spatial resolution imagery. *Int. J. Geogr. Inf. Sci.* **2011**, *25*, 1025–1043.

8. Parcak, S.H. *Satellite Remote Sensing for Archaeology*; Routledge: Abingdon, UK, 2009.

9. Di Giacomo, G.; Scardozzi, G. Multitemporal high-resolution satellite images for the study and monitoring of an ancient Mesopotamian city and its surrounding landscape: The case of Ur. *Int. J. Geophys.* **2012**, *2012*, doi:10.1155/2012/716296.

10. Malinverni, E. DEM automatic extraction on Rio de Janeiro from WV2 stereo pair images. *IOP Conf. Ser. Earth Environ. Sci.* **2014**, *18*, doi:10.1088/1755-1315/18/1/012022.

11. Krauß, T.; Lehner, M.; Reinartz, P. Generation of coarse 3D models of urban areas from high resolution stereo satellite images. *Int. Arch. Photogramm. Remote Sens.* **2008**, *37*, 1091–1098.

12. Perko, R.; Raggam, H.; Gutjahr, K.; Schardt, M. Advanced DTM generation from very high resolution satellite stereo images. *ISPRS Ann. Photogramm. Remote Sens. Spat. Inf. Sci.* **2015**, *2*, doi:10.5194/isprsannals-II-3-W4-165-2015.

13. Shean, D.E.; Alexandrov, O.; Moratto, Z.M.; Smith, B.E.; Joughin, I.R.; Porter, C.; Morin, P. An automated, open-source pipeline for mass production of digital elevation models (DEMs) from very-high-resolution commercial stereo satellite imagery. *ISPRS J. Photogramm. Remote Sens.* **2016**, *116*, 101–117.

14. Casana, J.; Cothren, J. Stereo analysis, DEM extraction and orthorectification of CORONA satellite imagery: Archaeological applications from the Near East. *Antiquity* **2008**, *82*, 732–749.

15. Berthier, E.; Vincent, C.; Magnússon, E.; Gunnlaugsson, Á.; Pitte, P.; Le Meur, E.; Masiokas, M.; Ruiz, L.; Pálsson, F.; Belart, J.; et al. Glacier topography and elevation changes derived from Pléiades sub-meter stereo images. *Cryosphere* **2014**, *8*, 2275–2291.

16. Bagnardi, M.; González, P.J.; Hooper, A. High-resolution digital elevation model from tri-stereo Pleiades-1 satellite imagery for lava flow volume estimates at Fogo Volcano. *Geophys. Res. Lett.* **2016**, *43*, 6267–6275.

17. Parmegiani, N.; Poscolieri, M. DEM data processing for a landscape archaeology analysis (Lake Sevan-Armenia). *Int. Arch. Photogramm. Remote Sens. Spat. Inf. Sci.* **2003**, *34*, 255–258.

18. Vandenbulcke, A.; Stal, C.; Lonneville, B.; Bourgeois, J.; De Wulf, A. Using 3D modelling in the valley of Turu Alty (Siberia, Russia) for research and conservational purposes. In Proceedings of the 8th International Congress on Archaeology, Computer Graphics, Cultural Heritage and Innovation 'ARQUEOLÓGICA 2.0', Valencia, Spain, 5–7 September 2016; Lerma, J.L., Cabrelles, M., Eds.; Universitat Politécnica de Valéncia: Valencia, Spain, 2016; pp. 408–411.

19. Nasir, S.; Iqbal, I.A.; Ali, Z.; Shahzad, A. Accuracy assessment of digital elevation model generated from pleiades tri stereo-pair. In Proceedings of the IEEE 7th International Conference on Recent Advances in Space Technologies (RAST), Istanbul, Turkey, 16–19 June 2015; pp. 193–197.

20. Sofia, G.; Bailly, J.S.; Chehata, N.; Tarolli, P.; Levavasseur, F. Comparison of Pleiades and LiDAR Digital Elevation Models for terraces detection in farmlands. *IEEE J. Sel. Top. Appl. Earth Obs. Remote Sens.* **2016**, *9*, 1567–1576.

21. Gleyzes, M.A.; Perret, L.; Kubik, P. Pleiades system architecture and main performances. *Int. Arch. Photogramm. Remote Sens. Spat. Inf. Sci.* **2012**, *39*, doi:10.5194/isprsarchives-XXXIX-B1-537-2012.

22. Bernard, M.; Decluseau, D.; Gabet, L.; Nonin, P. 3D capabilities of Pleiades satellite. *Int. Arch. Photogramm. Remote Sens. Spat. Inf. Sci.* **2012**, *39*, 553–557.

23. Cantou, J.; Maillet, G.; Flamanc, D.; Buissart, H. Preparing the use of Pleiades images for mapping purposes: Preliminary assessments at IGN-France. *Int. Arch. Photogramme. Remote Sens.* **2006**, *36*.

24. Šedina, J.; Pavelka, K.; Housarová, E. Archaeological documentation of a defunct Iraqi Town. *Int. Arch. Photogramm. Remote Sens. Spat. Inf. Sci.* **2016**, *XLI-B1*, 1031–1035

25. Asăndulesei, A. Inside a Cucuteni Settlement: Remote Sensing Techniques for Documenting an Unexplored Eneolithic Site from Northeastern Romania. *Remote Sens.* **2017**, *9*, 41, doi:10.3390/rs9010041.

26. Jiang, S.; Jiang, W.; Huang, W.; Yang, L. UAV-Based Oblique Photogrammetry for Outdoor Data Acquisition and Offsite Visual Inspection of Transmission Line. *Remote Sens.* **2017**, *9*, 278, doi:10.3390/rs9030278.

27. Sonnemann, T.F.; Ulloa Hung, J.; Hofman, C.L. Mapping Indigenous Settlement Topography in the Caribbean Using Drones. *Remote Sens.* **2016**, *8*, 791, doi:10.3390/rs8100791.

28. Harrower, M.J. Geographic Information Systems (GIS) hydrological modeling in archaeology: An example from the origins of irrigation in Southwest Arabia (Yemen). *J. Archaeol. Sci.* **2010**, *37*, 1447–1452.

29. Parcak, S.H. *GIS, Remote Sensing, and Landscape Archaeology*; Oxford Handbooks Online; Oxford University Press: New York, NY, USA, 2017.

30. Campana, S. Ikonos-2 multispectral satellite imagery to the study of archaeological landscapes: An integrated multi-sensor approach in combination with "traditional" methods. In Proceedings of the 30th Conference CAA, Heraklion, Greece, 2–6 April 2002; pp. 2–6.

31. Pierdicca, R.; Malinverni, E.; Clini, P.; Mancini, A.; Bozzi, C.; Nespeca, R. Development of a GIS environment for archaeological multipurpose applications: The Fano historic centre. In Proceedings of the XIII International Forum Le Vie dei Mercanti, Lecce, Italy, 11–13 June 2015; pp. 588–597.

32. Ur, J. Sennacherib's northern Assyrian canals: New Insights from Satellite Imagery and Aerial Photography. *Iraq* **2005**, *67*, 317–345.

33. Bagg, A.M. Irrigation in northern Mesopotamia: Water for the Assyrian Capitals (12th–7th centuries BC). *Irrig. Drain. Syst.* **2000**, *14*, 301–324.

34. Bachmann, W. *Felsreliefs in Assyrien: Bawian, Maltai und Gundük*; Hinrichs: Leipzig, Germany, 1927.

35. Jacobsen, T.; Lloyd, S. *Sennacherib's Aqueduct at Jerwan*; University of Chicago Press: Chicago, IL, USA, 1935; Volume 24.

36. Orazi, R.; Colosi, F. Archaeological Environmental Park of Sennacherib's irrigation system (Iraqi Kurdistan). In *Heritage and Technology. Mind, Knowledge, Experience, Proceedings of the Le vie dei Mercanti, XIII Forum Internazionale di Studi, Aversa, Capri, Italy, 11–13 June 2015*; Gabardella, C., Ed.; Consiglio Nazionale Delle Ricerche: Roma, Italy, 2015; pp. 2021–2030.

37. Lasaponara, R.; Masini, N. Pan-sharpening techniques to enhance archaeological marks: An overview. In *Satellite Remote Sensing*; Lasaponara, R., Masini, N., Eds.; Springer: Cham, Switzerland, 2012; pp. 87–109.

38. Lasaponara, R.; Masini, N. *Satellite Remote Sensing: A New Tool for Archaeology*; Springer: Cham, Switzerland, 2012; Volume 16.

39. Manakos, I.; Kalaitzidis, C. *Imagin [e, G] Europe: Proceedings of the 29th Symposium of the European Association of Remote Sensing Laboratories, Chania, Greece*; IOS Press: Amsterdam, The Netherlands, 2010.

40. Liua, C.; He, G.J. Auto-matching algoritmh for Remote Sensing images. In Proceedings of the 4th International Conference on GEographic Object Based Image Analysis (GEOBIA), Rio de Janeiro, Brazil, 7–9 May 2012; p. 248.

41. Pierdicca, R.; Frontoni, E.; Zingaretti, P.; Malinverni, E.S.; Colosi, F.; Orazi, R. Making visible the invisible. Augmented reality visualization for 3D reconstructions of archaeological sites. In Proceedings of the International Conference on Augmented and Virtual Reality, Ugento, Italy, 12–15 June 2016; Springer: Cham, Switzerland, 2015; pp. 25–37.

geosciences

MDPI

Article

Are We There Yet? A Review and Assessment of Archaeological Passive Airborne Optical Imaging Approaches in the Light of Landscape Archaeology

Geert J. Verhoeven

Ludwig Boltzmann Institute for Archaeological Prospection & Virtual Archaeology (LBI ArchPro),
Franz-Klein-Gasse 1, A-1190 Vienna, Austria; Geert.Verhoeven@archpro.lbg.ac.at

Received: 18 August 2017; Accepted: 11 September 2017; Published: 14 September 2017

Abstract: Archaeologists often rely on passive airborne optical remote sensing to deliver some of the core data for (European) landscape archaeology projects. Despite the many technological and theoretical evolutions that have characterised this field of archaeology, the dominant aerial photographic surveys, but also less common approaches to archaeological airborne reconnaissance, still suffer from many inherent biases imposed by sub-par sampling strategies, cost, instrument availability and post-processing issues. This paper starts with the concept of landscape (archaeology) and uses it to frame archaeological airborne remote sensing. After introducing the need for bias reduction when sampling an already distorted archaeological population and expanding on the 'theory-neutral' claim of aerial survey, the paper presents eight key characteristics that all have the potential to increase or decrease the subjectivity and bias when collecting airborne optical imagery with passive sensors. Within this setting, the paper then offers some technological-methodological reflection on the various passive airborne optical imaging solutions that landscape archaeology has come to rely upon in the past decades. In doing so, it calls into question the effectiveness and suitability of these highly subjective approaches for landscape archaeology. Finally, the paper proposes a new, more objective approach to aerial optical image acquisition with passive sensors. In the discussion, the text argues that the suggested exhaustive (or total) airborne sampling of the preserved archaeological record might transcend particular theoretical paradigms, while the data generated could span various interpretational perspectives and oppositional analytical approaches in landscape archaeology.

Keywords: aerial archaeology; aerial photography; aerial reconnaissance; airborne imaging spectroscopy; archaeological record; archaeological theory; bias; landscape archaeology; multispectral imaging; survey bias

1. Setting the Scene

Archaeology becomes more specialised by the day as it continuously integrates new subjects and methods. To properly combine and integrate these methods, reflection is needed. This reflectivity means to assess data acquisition and the grounds on which archaeological interpretations rest [1]. Although some have criticised the general lack of theoretical reflexivity in remote sensing archaeology [2], this specific discipline has seen reflection by archaeologists working in the field. They tried to critically assess how the interpretation of remotely-sensed data can shape the understanding of past cultural and physical landscapes, identify pitfalls in current and future practice and question why things are done the way they are done (e.g., [3–9]). Most papers that tackled the use of remote sensing products in the wider framework of landscape archaeology have thus approached the subject from a more interpretational angle rather than a purely methodological-technological one.

This paper tries to fill this gap. It results from the author's more general concern with the effectiveness of archaeological airborne reconnaissance (see also [10–12]). It reflects on how a more systematic approach to passive airborne optical imaging could yield data that can be used by practitioners of different archaeological theoretical viewpoints. The intention of this paper is to reflect on—and encourage a critical approach to—passive airborne optical imaging for the benefit of landscape archaeology. The text is, however, not a simplistic manifesto to blindly enforce passive optical airborne imaging. Remote sensing data can never be a mandatory or sole source to investigating the landscape. Landscape archaeology must be holistic or 'total' by building up a complementary body of data from multiple distinct survey approaches [13–15] which, when properly integrated, constitute a digital landscape [16] to explore. Obviously, some specific techniques such as pollen analyses or toponymy may have greater importance than repeated airborne surveys depending on the character of the area, the time and money available for the project and the questions being asked [17,18]. These fitness-for-purpose issues are, however, not a part of this article's core scope. The text simply assumes that if a given archaeological landscape project uses remotely sensed optical imagery, they are considered a valid data source to help answer some of the project-specific questions, despite some uncontrollable environmental and climatological biases that might characterise these data. Nevertheless, this assumption does not mean that the subjectivity, which characterises many of the controllable parameters of current passive airborne optical image acquisitions, should not be critiqued. Moreover, the author explains in the conclusion that much of the discussion about the appropriateness of existing airborne imaging methodologies is skewed exactly because of the failure to properly address and resolve the survey methodology issues that this paper tackles.

Overall, the author asks if an 'optimisation' of the dominant airborne sampling paradigms can remove many of the current inherent, but controllable, biases and as such, render the airborne data acquisition phase less subjective, and, in turn, yield a more consistent record to study our past? Thus, the article begins with an introduction to the often ambiguous concepts of landscape and landscape archaeology, followed by a short overview of archaeological remote sensing. Both sections will introduce some key concepts that are vital to evaluate the remaining parts of the paper. Starting from a discussion of the archaeological record and the obligatory samples that are acquired from it, the paper then reviews all the main passive airborne optical imaging approaches that have been used in the past decade. It identifies their strengths and weaknesses regarding eight particular key concepts to highlight their inherent sampling shortcomings and major biases. Following this, the case is made for a more objective approach to aerial optical image acquisition with passive sensors, using a geographical total coverage approach and consistent imaging in particular narrow spectral bands, combined with a large degree of computer-based automation in post-processing. Although the perspective is rather technical-methodological, landscape archaeology frames this review. Also, the article considers and comments on some of the different theoretical contexts surrounding these remote observations.

1.1. The Vogue but Vague Concepts of Landscape (Archaeology)

Depending upon research context, the definition and interpretation of the term 'landscape' changes [19–22]. This disparity in definitions of, and views on, landscape makes it sometimes very difficult to communicate clearly, certainly when the debate is focused on the archaeological study of landscapes. However, even among archaeologists, there is no consensus on the notion of landscape, simply because it remains very challenging to theoretically define concepts of space and place [23]. During the past decades, landscape has become an overused, integrated term, encapsulating a certain amount of human as well as environmental aspects of a delimited area of land. This means that landscape not only includes the combined effects of fauna and flora, geology, topography, soils, waterways and climate, but it also incorporates the cumulative cultural imprints created by settlement building, ploughing, herding, polluting and travelling [24]. This amalgam of environmental and anthropogenic processes has constantly changed in proportion and intensity across space and through time, both very often at vastly different scales. As such, landscapes can be summarised as complex, vague, messy entities that are uninterruptedly shaped through human engagement with them [25,26]; they only possess defined spatial and temporal limits when imposed by analytical procedures and intellectual traditions [27].

Notwithstanding their complexity and vagueness, 'understanding landscapes' is the central aim of landscape archaeology. This archaeological sub-discipline emerged in the mid-1970s [28] and is here roughly defined as the study of all environmental and sociocultural variables influencing the mutual interaction (both physical and perceptual) of humans with their natural environment [29,30]. This means that landscape archaeology is "about what lies beyond the site, or the edge of the excavation" [31] (p. 3), landscape being the physical reality comprising all anthropogenic features and their natural context that exist across space, collectively called 'the land'. In other words, the cultural spaces or rather diffuse human traces such as roads, land division systems or isolated artefacts in between hot spots of activity (traditional 'sites') [32]. This perspective was framed in what became known as distributional archaeology [33], non-site archaeology [34], siteless archaeology [35,36] or off-site archaeology [37].

This process of filling the gaps between the sites gradually became the practical implementation of landscape archaeology since the 1980s, providing valuable information about human exploitation of the environment. However, it also reminded archaeologists of all the possible, subtle or obvious, recursive and mutual interactions of human presence with the land through space and time as well as the strict non-physical aspect of landscape. One that im- or explicitly relates to the various layers of cultural meaning and human value [24,38,39]. Therefore, landscape archaeology is undoubtedly also about how these physical realities are perceived, culturally and socially exploited, understood and given meaning (i.e., 'land-scaped') by past and current human presence [31,40–42]. Landscapes are accepted as being polysemous: geometric, quantitative and economic, but also qualitative, subjectively perceived and socially constructed [43,44]. In other words: landscapes structure and are structured by us; landscapes move, and they move us too [26]. Finally, being the multi-temporal present product of superpositioned and intermingled temporalities, landscapes should not be metaphorically described as palimpsests. In a palimpsest, everything gets erased, but not necessarily so in a landscape. This gives landscapes their cluttered, complex and messy character [45] in which function and meaning are not separable [40].

Bringing the diverse intellectual trends in archaeological theory together with the fact that one can describe landscape only contentiously and ambiguously, has led to several incarnations of landscape archaeology. In other words, a broad array of practices, theories, and tools have emerged to produce knowledge about the physicality and meaning of landscapes in the past [46–49]. Despite this variety of approaches, much European archaeological landscape research has relied on (and still relies on) remote sensing data (although the incorporation of remote sensing methods in landscape archaeology is very variable when considered globally). This relationship should not come as a surprise. Even though they are characterised by diffuse physical and cultural boundaries, a landscape implies a meaningful place for everyday human living. Although the boundaries of these inhabiting and experiencing areas have changed through time, landscape can be considered a concept with an intermediate spatial extent, representing a human scale of lived experience for most of history [24,50]. Being larger than a complex of sites, a landscape has the ideal spatial extent to be observed from above.

1.2. Archaeological Remote Sensing

Archaeology was one of the first disciplines to use remote sensing in scientific investigations [51]. As soon as it became apparent that excavations alone did not provide sufficient clues for understanding the various aspects of past settlements and their transformation over time, 'auxiliary' approaches such as geophysical survey, surface artefact collection and airborne reconnaissance were brought in [52,53]. Although these approaches initially focused on finding 'sites' for excavation, they were instrumental in the development of landscape archaeology (see also Section 3). Today, there is a large consensus among archaeologists that remote sensing encompasses a range of useful observational techniques for discovering and registering archaeological traces and landscape patterning. Although some, mainly British and phenomenological critiques have been formulated on the airborne perspective (see [54] for an overview), remote sensing has not attracted the global and strongly outspoken theoretical controversy characteristic for related tools such as GIS [55]. Overall, non-destructively observing from above is commonly considered a valid backdrop for contemporary European archaeological landscape research [17,47,52,56–61], certainly when fully integrated with other specialised techniques such as artefact survey, geophysical and geochemical prospection, cartographic research, botany and geology. Subsequent sections will comment on the extent to which this background is theory-laden.

Although remote sensing products have already played a fundamental role in formulating explanations of past landscapes, it is still important to review their key concepts and sub-methods. In archaeology, remote sensing is a general name given to all techniques that use propagated signals to observe the Earth's surface from above. This definition rules out terrestrial geophysical methods, even though some scholars call them ground-based remote sensing [62,63]. Their widely differing characteristics allow for the classification of remote sensing techniques in diverse ways: imaging versus non-imaging, passive versus active, optical versus non-optical, and airborne versus spaceborne. Whereas passive remote sensing systems capture naturally occurring radiation such as emitted thermal energy or reflected solar radiation, active systems produce their own radiation. Airborne systems operate from within the Earth's atmosphere, whereas spaceborne systems deploy a sensor mounted onboard a spacecraft (often a satellite) that orbits the Earth. When dealing with an imaging system, the output is an image, whereas non-imaging systems can deliver sounding data or emission spectra.

Archaeological remote sensing relies mainly on the active sounding technique known as Airborne Laser Scanning (ALS) and passive air- and spaceborne imaging in the optical spectrum. However, the selection of appropriate techniques is entirely due to the nature of most (hidden) archaeological remains. Since ALS systems can record sub-canopy features, the vegetation can be digitally removed to expose the underlying terrain and model solid standing structures. Given its capability to digitise surfaces with decimetre- to centimetre-level accuracy, ALS is also able to document the geometry of small topographic undulations such as earthworks or other, even submerged features of archaeological interest [64–68]. Currently, ALS is still considered inadequate to record entirely buried archaeological residues. Since the latter can change the chemical and physical properties of the local soil matrix, they might be disclosed by growing- or ripening-induced differences in the colour or height of vegetation on top of the remains (i.e., crop or vegetation marks, exploited most by aerial survey) or distinct tonal differences in the ploughed soil (soil marks). Although advances in the radiometric correction of ALS data [69] already demonstrated their value to map reflectance differences in soils and plants [70], the amount of spatial detail in these ALS-based images can still not compete with the spatial resolution of data products from the best airborne passive imaging systems. Moreover, reflectance values are only recorded for the wavelength(s) actively emitted by the ALS system. Dedicated air- or spaceborne optical imaging devices can, therefore, more comprehensively map plough-levelled sites through the proxies of vegetation, shadow and soil marks [71].

Although optical electromagnetic radiation conventionally incorporates the ultraviolet to infrared spectrum with wavelengths between 10 nm (0.01 μm) and 1 mm (1000 μm) [72,73], remote passive imaging in the optical regime usually starts at the visible waveband (i.e., 400 nm) [74]. Together with the Near-InfraRed (NIR; 700 nm to 1100 nm) and Short Wavelength InfraRed (SWIR; 1.1 μm to 3 μm), this waveband is known as the solar-reflective spectral range because the imagery is predominantly generated by reflected solar energy (Figure 1). The neighbouring Mid Wavelength InfraRed (MWIR; 3 μm to 6 μm) range is considered an optical transition zone, as the solar-reflective behaviour slowly shifts in favour of self-emitted thermal radiation. In both the Long Wavelength InfraRed (LWIR; 6 μm to 15 μm) and Far/extreme-InfraRed (FIR; 15 μm to 1 mm), the imaging process is almost uniquely governed by the thermal electromagnetic radiation emitted by the scene objects themselves. As a result, the MWIR to FIR optical region is commonly denoted the thermal region.

When the imager uses different spectral bands or channels to capture the reflected or thermal energy, the terms 'multi-' or 'hyper-' spectral apply.

Optical electromagnetic radiation						
Division	Subdivision	Abbreviation	Cut-on (nm)	Cut-off (nm)	Division by imaging principle	
UltraViolet (UV)	Vacuum UV	VUV / UV-D*	10	200		
	Far UV	FUV / UV-C*	200	280		
	Middle-UV	MUV / UV-B	280	315		
	Near-UV	NUV / UV-A	315	400		
Visible (Vis)	Blue	B	400	500		
	Green	G	500	600	*Solar-reflective spectral region*	
	Red	R	600	700		
InfraRed (IR)	Near-IR	NIR	700	1 100		
	Short Wavelength IR	SWIR	1 100	3 000		
	Mid Wavelength IR	MWIR	3 000	6 000	*Transition zone*	*Thermal region*
	Long Wavelength IR	LWIR	6 000	15 000	*Self-emitted thermal radiation region*	
	Far/Extreme-IR	FIR	15 000	1 000 000		

Figure 1. The divisions of the optical electromagnetic spectrum (* VUV does not perfectly correspond to UV-D. While VUV runs from 10 nm to 200 nm and FUV from 200 nm to 280 nm, UV-D encompasses the 10 nm to 100 nm region and UV-C the 100 nm to 280 nm zone).

Archaeology relies often on multi-spectral satellite sensors to record archaeologically-induced reflectance differences in soil and vegetation [75–79]. It is the combination of an ever increasing spatial and spectral resolving power combined with the frequency by which they can cover large areas of interest and their inherent synoptic perspective (i.e., footage with an extensive spatial coverage) that provide many interesting opportunities for landscape archaeologists when compared to conventional aerial imagery. Besides, satellite sensors can image areas that have limited or no access due to airspace restrictions, bureaucratic red tape or other obstacles to reconnaissance. However, the comparatively limited spatial resolution of spaceborne imagery limits its utility for the discovery and detailed recording of smaller landscape features. When one needs to spatially resolve the soil and vegetation marks that result from small features (such as pits and postholes), an airborne vantage point remains still superior. As a result, satellite reconnaissance is better suited for mapping large-scale and broad landscape features such as paleochannels or detecting upstanding monuments in semi-arid environments, while airborne imaging remains the preferred approach for a detailed study of past geo-cultural activity. Hence, this paper will target only passive airborne optical imaging approaches when presenting methodological reflections and deconstruct existing practices. Moreover, the text implicitly assumes manned platforms.

1.3. Sampling the Archaeological Record

The discipline of archaeology relies by default and as a whole on very limited data, which are from an analytical viewpoint called samples in statistics [80]. In the words of Clarke: "Archaeology in essence then is the discipline with the theory and practice for the recovery of unobservable hominid behaviour patterns from indirect traces in bad samples." [81] (p. 17). With bad samples, Clarke indicates that archaeology does not just simply rely on a sample of the complete, untouched record of all tangible and intangible geo-cultural traces that ever occurred (called the population in statistics and here denoted the potential or target archaeological record). Rather, archaeology and the explanations it tries to construct, rely on a heavily sampled version of an already inherently sampled and thus imperfectly preserved archaeological record, which by itself is only a partial representation of the initial archaeological record (Figure 2).

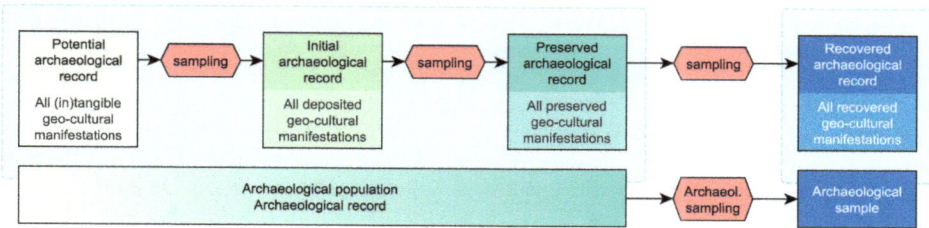

Figure 2. Explanations about the past are based on the deficient and biased archaeological data at our disposal. These data, the archaeological sample (or recovered archaeological record), are only a small subset of all present remains of past human behaviour.

In landscape archaeology, scholars research the continuous record of traces created by human and environmental activity (and their mutual interrelation) that the many post-depositional agencies in that environment (including anthropogenic disturbance) have allowed to be preserved through time. Landscapes are the framework that contain all these testimonies and patterns of human behaviour. The latter, here collectively called the preserved archaeological record, is never pristine. It constitutes only a very limited subset of all those accumulative geo-cultural manifestations and engagements that were included in the initial archaeological record, the latter being only a sample of all possible tangible and intangible human responses to cultural, demographic, environmental and social opportunities that were originally present at a certain moment in time. Depending on the view of the archaeologist, potential, initial and preserved archaeological record can be denoted the 'archaeological record' or the statistical population about which archaeologists want to draw conclusions. Cowgill, for instance, distinguished between a so-called 'physical consequences' population (the initial archaeological record) and a 'physical finds' population (a combination of the preserved and recovered archaeological record) [82]. Both the initial and preserved archaeological populations are thus residual consequences of countless spatio-temporally varying and interacting cultural and natural processes that all act as agents of sampling bias [82–84].

In the fields of metrology and engineering, bias is due to a systematic error and as such an estimate for the closeness of agreement with the unknown 'true' value [85]. Statistically speaking, bias results from an unfair or faulty sampling and is a measure of accuracy. More generally, the concept of bias often denotes the partiality or one-sidedness or unfairness of (one or more) observations, which thus renders that observation also inaccurate according to a third party. Regardless of field, highly accurate observations thus always have a small bias. In archaeology, biased data are characterised by ambiguity, unfair tendencies (systematic over or under representation) and partiality (i.e., inappropriate recovery rate). That is why the preserved archaeological record is by default a progressively biased, distorted and lacking remnant of a once functioning but unknown whole [86].

However, if the past is accessible at all, it can only be understood as a secondary construct that has to be established on the basis of this biased preserved archaeological continuous record. To access this continuum, archaeologist still must rely on finite archaeological samples to infer information about this biased population since observing the totality of a particular aspect of the preserved archaeological record is usually practically impossible or utterly expensive. Lack of consistency and errors in these archaeological sampling processes just add additional bias. It is the latter that archaeologists want to minimise, because the inadequacies in the preserved archaeological record can only be dealt with conceptually.

Since the rise of the processual paradigm, the complex and controversial topic of sampling has received increasing attention in archaeology [87–89]. Although archaeologists always sample according to both theoretical and methodological biases [90], it is generally agreed upon that large samples offer improved accuracy over small samples [91]. It is, therefore, sensible to revise any sampling strategy that knowingly and unnecessarily limits the recovered archaeological record. Only then can these unfortunate and needless constraints on the collected data be mitigated.

The consistent nature of observing is one of the main underpinnings to increase the relevance and accuracy of archaeological data gathering. In other words, only systematic recording in various dimensions can eliminate measurement bias to a large extent and render the archaeological data analytically and interpretationally useful [91,92]. However, lack of consistency is exactly what characterises the most common archaeological optical remote imaging approaches. This is the main rationale that drives this paper and its methodological-technical reflection on (and revision of) current airborne optical sampling approaches.

Finally, it remains important to stress once more the point made at the very beginning of this text. Although this paper acknowledges that topography, soil and vegetation type, land use and climate largely govern (and thus bias) the data generated by passive aerial optical imaging, the article does not provide an in-depth reflection upon these uncontrollable variables since they are inherent to the working principles of airborne passive imaging approaches. In contrast, the text pulls the variables that one can control during the image acquisition into sharp focus. This approach will also make it possible (in Section 8) to juxtapose airborne imaging with excavations and geophysical surveys and criticise some of the general reservations about the fitness-for-purpose of passive aerial optical imaging for the benefit of landscape archaeology.

1.4. Which Theory to Follow?

Reflecting on data acquisition in archaeology necessitates the contextualising of this archaeological practice. In addition, the latter requires some archaeological theory. However, this is a difficult balance and one must take care "to avoid overt theorising which loses touch with the empirical basis; to avoid swinging back too far from theoretical debate into the empirical, and starting bean-counting again" [1] (p. 65). Over the past century, but mainly during the past six decades, successive movements in archaeological theory have influenced archaeological research practices [93]. All of these differing theoretical reflections, of which most seem to originate in other disciplines, stem from the doubts archaeologists have regarding their ability to yield valid inferences about past societies. However, all possible theoretical and methodological stances and their reciprocal relationships have undergone significant changes and multiple critiques to the extent that archaeology still possesses little agreement on theoretical core concepts [94–96]. Or, as Pearce states: theory has become bricolage [97].

When going back to the above-mentioned practical level of landscape archaeology as the study of the geo-cultural spaces between (and incorporating) the sites, remote sensing is in many research contexts a valid means to accomplish this. However, remote sensing data (and the tools to interact with them) stem from the worlds of quantification and generalisation, to which the qualitative and subjective aspects of humanity defended by the broadly labelled post-processual movement, represents a partial reaction [23]. Post-processual archaeologists mainly critiqued the scientific, analytical approach to reach objective conclusions. In that light, it is unsurprising that remote sensing was generally of little interest to them. However, it is not because certain schools of thought will more readily adopt airborne optical imaging that opposing theoretical schools cannot rely on airborne observations and learn from them. Post-processualism, for instance, questioned the relationships between the observed and the object of study. From this point of view, aerial imaging has quite some notions to contribute. It also indicates that most theories relate to the interpretive framework of (landscape) archaeology. They are concerned by how information (and in a later stadium knowledge) are gained from archaeological data and why certain data are gathered. They widely disagree on the extent to which archaeological data are theory-laden, but debate only to a lesser degree the data collection procedures themselves. In contrast to what some might believe, deciding upon a specific method does not necessarily assume a particular theoretical orientation. Many archaeologists even (implicitly) believe that remote sensing is independent of any theoretical foundations, that it is 'theory-neutral' when it comes to landscape archaeology [2,98].

This aura of theoretical neutrality stems from the beginning of the 20th century and the fact that remote sensing methods were often developed outside the main theoretical reflections. However, any aerial image is an archaeological piece of data just like any other source such as a ceramic bowl. Both the photograph and the ceramic bowl are mere samples of the preserved archaeological record (see Section 1.3) and have to be the subject of a particular theory-driven interpretation, the outcome of which is affected by many underlying assumptions and pre-understandings [99]. Although the theoretical interpretational frameworks that enable the reading and narrativising of aerial images are not a part of this paper, it must be clear that the archaeological function of airborne imagery (or any remote sensing output for that matter) are far from being culturally unbiased and theoretically neutral [2,100]. Moreover, also the application of a specific method is governed by theory, since one always consciously chooses airborne remote sensing as (one of) the data collection methods to answer an (often tacit) theory-based research problem. Again, this does not mean that the choice for airborne imaging automatically assumes a particular theoretical orientation, certainly not in a time where theory seems to have become less partisan and more pragmatic.

Any aerial image only depicts a few dimensions of the contemporary, preserved archaeological record, most of it through proxies like soil, shadow and vegetation marks. The values of and relations between the individual pixels must still undergo an evidential interpretational stage (which in itself poses still many problems in archaeology [101]). The very process of passive airborne imaging should, however, be executed as exhaustively and systematically as possible (given the contemporary technological constraints of observing various dimensions). Whether the landscape is considered abstract, quantifiable, static and neutral or rather seen as a humanised, qualitative, dynamic and contextual entity [39], landscape archaeology is always about totality: it is about all kinds of people and the superposition and intermingling of all possible landscapes. This paper will, therefore, explore the need for a more holistic approach to airborne data collection in all feasible dimensions (being spatial, spectral and temporal among others). It proposes several improvements to the current dominant optical aerial imaging approaches so that the airborne data gathering stage can become of undeniable use to many possible theoretical perspectives on landscape.

In this way, the dominant form of archaeological airborne remote sensing can move beyond its traditional cultural-historical point-based approach and provide data that are deeper and more explicitly rooted within the application of landscape archaeology. Moreover, removing much of the existing selectivity from the airborne data collection phase essentially mitigates major sampling biases and largely ignores the social context and pre-understandings of the aerial archaeologist, thereby allowing optical aerial imaging to transcend theoretical perspectives and become less theory-laden. Too often, incorporating (new) digital technologies in archaeology while lacking any theoretical framework is said to be meaningless. Albeit correct to some extent, this paper hopes to show that even the best conceptual framework will not prevent any biased and incompetent use of technology (hard-plus software) and methodology. Fancy theories can still lead to the acquisition of useless data from which insupportable, meaningless and even erroneous conclusions are drawn.

In this light, the next section will present eight particular characteristics that apply to all passive optical airborne imaging approaches. Afterwards, this paper will outline the methodological-technological pros and cons of the main passive airborne optical imaging approaches applied in archaeology. Using these eight features, it can be shown that all airborne imaging techniques are—implicitly or explicitly—sub-par samplers of one or more dimensions of the preserved archaeological record, thereby introducing major bias and subjectivity into the collected data. It is important to mention once more that this assessment does not equal the often implicitly accepted (and during cultural-historical and processual times outspoken) objective and neutral nature of remotely sensed imagery. The choice of a specific study region, aircraft, flying height, pilot, imager, lens and image acquisition time are all very subjective decisions. However, subjectivity should only take its leading role after the data acquisition stage, when experience, intuition and knowledge govern data processing, visualisation, interpretation and information synthesis.

2. Capturing Multi-Dimensionality in Eight Key Characteristics

2.1. Bias Versus Cumulativety

Earlier (Section 1.3), the case was made for a consistent recording of airborne data to eliminate measurement bias as much as possible. This text also argues for an airborne approach that maximally excludes potential data partiality by exhaustive sampling. In other words, the recovery rate of the preserved archaeological record should be as high as possible. When the airborne observation does not adhere to both conditions, it is denoted biased. Not that this concept of archaeological bias strongly deviates from the metrological idea. In its strict mathematical sense, bias cannot be reduced, let alone eliminated, by acquiring more samples. Since it is the difference between the mean of many observations and the true value, a single measurement is also considered unbiased in a strict statistical sense.

However, aerial archaeological survey is a strange animal. Not only should it seek to minimise partiality by an exhaustive sampling scheme (which equals in some aspects a total coverage approach—see Section 3), it should also be executed repeatedly. Although the proponents of the empirical approach would always advocate repeatable aerial data collection sorties as they believe that more data will lead to greater veracity, even those following deductive approaches have to resort to repeated surveys if they want the aerial data collection to be effective. Even with a total approach, the visibility of (predominantly) vegetation marks in a specific part of the electromagnetic spectrum depends on numerous variables: geology, land use, plant species, soil humidity, erosion processes and weather conditions during the flight among numerous others. Since the interactions between the many conditions do often not yield conditions that are ideal for observing crop, shadow and soil mark features with a specific imaging setup, a cumulative approach to partially resolve these uncontrollable biases is imperative [102].

2.2. The Usual Suspects

Spatial, temporal, spectral and radiometric resolving power are the first four characteristics that can express how much a specific airborne data collection strategy is prone to induce bias in the imagery recorded. They stem from the fact that digital imaging sensors always represent averages of the spatial, temporal and spectral dimensions. The macroscopic space we live in is—according to Newtonian physics—four-dimensional. Any 'event' in the universe can be defined by a location (x, y, z) and a time of occurrence (t), making this event a four-dimensional entity [103]. However, in addition to this one temporal and three spatial dimensions of the universe, more or less every variable can *sensu stricto* be thought of as a dimension [104]. Just as time is a variable, the same goes for colour, temperature or the number or pottery sherds per metre squared. For remote sensing data, a sampled spectral and radiometric dimension is usually defined in addition to the temporal and spatial ones. Note that resolving power denotes the characteristic of the imaging system, while the term resolution is used when talking about the image itself.

Spatial resolving power refers to the ability of an instrument to distinguish between neighbouring objects. It is expressed as the reciprocal of spatial resolution [105]. Although the final spatial resolution of an image depends on many factors, one of the key factors is the so-called GSD or Ground Sampling Distance. Most imaging sensors have a sensor that consists of individual photosites (see Figure 3). The distance from the centre of one photosite to the centre of the neighbouring photosite is denoted the detector pitch. The GSD is the corresponding distance projected in object space, stating the horizontally or vertically measured scene distance Δx between two consecutive sample locations. Although the detector pitch is a fixed property of a sensor, it can generate images with different GSDs. Among other factors, the GSD is determined by the scene's local topography, the distance of the camera to the scene and the focal length of the lens. To counteract observational bias towards larger features, smaller GSDs are important.

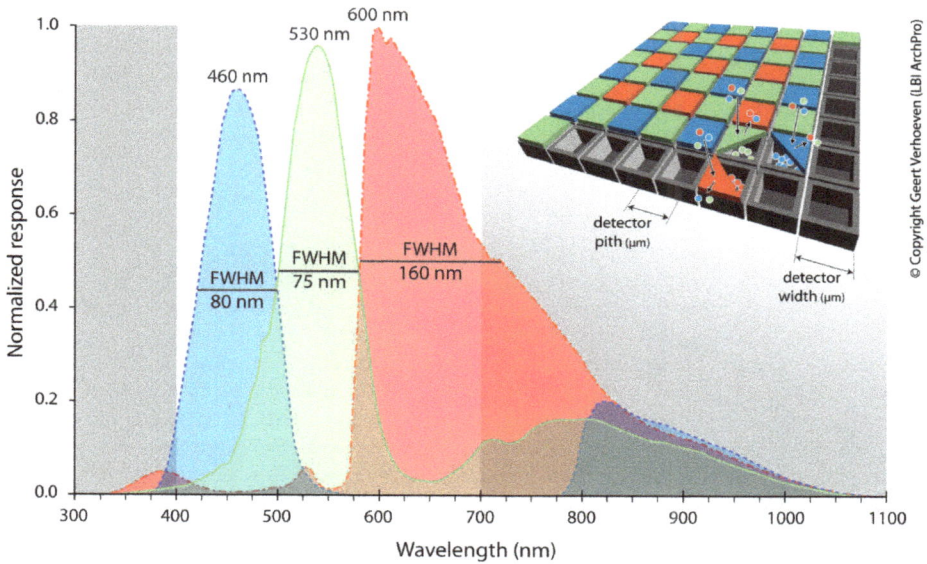

Figure 3. The working principles of a typical digital image sensor inside a photo camera. The relative spectral response curves obtained from a Nikon D200 without internal blocking filter reveal that the Blue, Green and Red channels have a FWHM of 80 nm, 75 nm and 160 nm, respectively. Note that the red channel of an unmodified camera will have a FWHM of about 50 nm since the camera will not be responsive in the invisible near-ultraviolet and near-infrared regions (both indicated in grey). With maxima at 460 nm, 530 nm and 600 nm, all bands are separated by 70 nm.

Sampling the time dimension is related to the imager's integration time or shutter speed during image acquisition. However, the temporal resolving power of an imaging system generally refers to the repeat period or the capability to view the same scene at regular intervals. This characteristic was included here since repeated airborne observations are key to build up a cumulative data record.

The spectral resolving power quantifies the spectral details that can be resolved by the imaging sensor. The Full Width at Half Maximum or FWHM is usually applied as the measure of spectral resolving power. The FWHM of a spectral band quantifies its spectral bandwidth at 50% of its peak responsitivity. Commonly, FWHM is expressed as a wavelength difference range $\Delta\lambda$ and measured in nanometre or micrometre (see Figure 3). To preserve spectral details, the spectral pass band (in terms of FWHM) should have a width smaller or comparable to the spectral details in the signal. Broad spectral bands just integrate over a too wide range. They average subtle spectral details away and create ambiguity. To sufficiently resolve small spectral features, the placement of the spectral bands—expressed as an interval in nm—is as important as the spectral bandwidth. Not being able to record the contrast exhibited by vegetation and soil marks will thus automatically lead to biased data.

Besides these three sampling processes, digitising an analogue signal also encompasses quantising the sample values to a discrete Digital/Data Number (DN). The total possible range of different DNs or quantisation values an imager can create is termed the tonal range. An image's tonal range is thus fully determined by the bit depth of this quantisation process: quantisation with N bits maps samples onto a discrete set of 2^N values. This quantisation (or radiometric sampling) provides images with a radiometric resolution. The radiometric resolving power of an imaging sensor refers to the capability of that sensor to discriminate two targets based on their emittance or reflectance, a value that is among other factors dependent on the imager's bit depth. Digital images are thus sampled and quantised representations of a scene, defined by a multi-dimensional matrix of numbers [106]. If their spatial, temporal, spectral and radiometric resolution becomes too coarse, they create ambiguity. Since all four characteristics are the result of finite interval sampling processes, a digital image will thus always induce partiality. The smaller the sampling unit becomes (i.e., the more exhaustive the sampling gets), the less biased archaeological image data will be (note again that this does not relate to the unique meaning of bias in measurement theory). Key is to minimise this partiality and ambiguity by optimising all four key characteristics.

2.3. From Availability to Processing Complexity

While an imaging sensor's resolving power and the resolution of an image are thus quoted in the spatial, spectral, temporal and radiometric measurement dimensions, this paper defines four additional but equally important properties to describe an aerial archaeological imaging solution that are not part of the usual remote sensing suspects: cost, availability, geographical bias and processing complexity.

Cost refers to the financial cost of acquiring the imaging system as well as its operational costs. Availability indicates how easy it is to have the system in operation at a certain place and a certain time. Systems that are very affordable or whose handling is straightforward, are generally easy to come by and will, therefore, be much more available than their expensive, highly specialised counterparts. When an imaging system is not available or operational funds are lacking, no data are collected or at least not when initially planned (e.g., when the conditions are favourable to record vegetation stress). Both factors might prevent the vital cumulative approach to data collection and even induce strong biases by limiting the area and the time frame in which the survey can be carried out.

The main problem with much existing aerial imagery is, however, its strong geographical, observer-related bias. The latter refers to the coverage of the scene during the data acquisition phase: is every part of the study area covered or are only selected views obtained? When airborne imagery exhibits a geographical bias, space becomes anything but neutral and remote sensing data far from a contiguous objective-like background. Since this type of bias is prevalent and of high importance, the next section covers it at length.

Finally, processing complexity encompasses all the geometric and radiometric corrections that should be executed before any of the interpretational processes can start. Highly complex post-processing chains prevent data from being considered as a relevant source in landscape archaeology, as such creating one more type of bias. Of course, one could always epistemologically argue that only archaeological research itself produces lacking data since archaeological data themselves are complete and data gaps only emerge once those samples are considered evidence of something [86]. However, since aerial image-based interpretations are used in landscape archaeology to grasp complex aspects like human behaviour, passive airborne optical imaging approach(es) should from the very start minimise data deficiencies and additional unnecessary biases. This can be obtained by maximising instrument availability and the amount of systematic observations in the various dimensions on the one hand, and minimising their operational costs, geographical bias and processing complexity on the other. The next sections will delve deeper into all individual techniques for archaeological passive airborne optical imaging, being: observer-directed aerial photography, blanket aerial photography, multispectral imaging and airborne imaging spectroscopy. All four will be assessed in the light of these eight key concepts (although less stress will be put on the radiometric and temporal resolving power of the systems given their overall good to excellent performance). A tabulated overview of these accounts can also be found in Table 1. Along the way, some thoughts on equipment- and terminology-related issues will be provided as well.

Table 1. Overview of the main passive imaging solutions in the optical domain and how they score for the eight key characteristics. +, 0 and − indicate good, neutral and bad, respectively. Scoring is specifically for archaeological purposes and established by mutually weighing all solutions.

Method	Spatial	Spectral	Temporal	Radiom.	Cost	Availability	Geo-bias	Processing
Observer-directed aerial photography	+	−	+	+	+	+	−	0
Blanket aerial photography/multi-spectral imaging	+/0	−/0	0	+	0	0	+	+
Airborne imaging spectroscopy	0	+	0	+	−	−	+	−

3. How It All Started: Observer-Directed Aerial Photography

Concerning the first phase of archaeological aerial prospection from manned platforms, much credit must go to Osbert Guy Stanhope Crawford (1886–1957). This Englishman is considered to be the inventor of scientific aerial archaeological reconnaissance, and his work in the 1920s and beyond was the basis for the future development of the discipline that became known as aerial archaeology [107–109]. Thanks to Crawford's leading role, archaeologists recognised the potential and convenience of straightforward aerial observations to unravel and study the past. His positivist perspective made Crawford consider aerial photographs as a neutral representation of objectively existing archaeological sources. Despite the processual archaeology-inspired discussions on various aspects of the method, the currently dominating practice of archaeological aerial photographic reconnaissance has not witnessed major changes since the work of Crawford. In general, aerial photographers engaged in this type of prospection acquire photographs from the cabin of a low-flying (anywhere between 300 m and higher), preferably high-wing aircraft with lift-windows to obtain a wide, unobstructed view of the scene below them. For a maximum working angle, the passenger door might be removed as well. A small- or medium-format photographic/still frame camera is hand-held and equipped with a lens that is commonly uncalibrated [110,111]. Once airborne, the archaeologist flies over targeted areas, trying to detect possible archaeologically-induced landscape features. After an archaeological or environmental feature of interest is detected, it is orbited and photographed from various positions. Although a photographer might create near vertical photographs (or simply 'verticals') during such an archaeological reconnaissance sortie, the vast majority of the photographs will be oblique in nature in that the optical axis of the imager intentionally deviates more than three degrees from the vertical to the earth's surface (Figure 4).

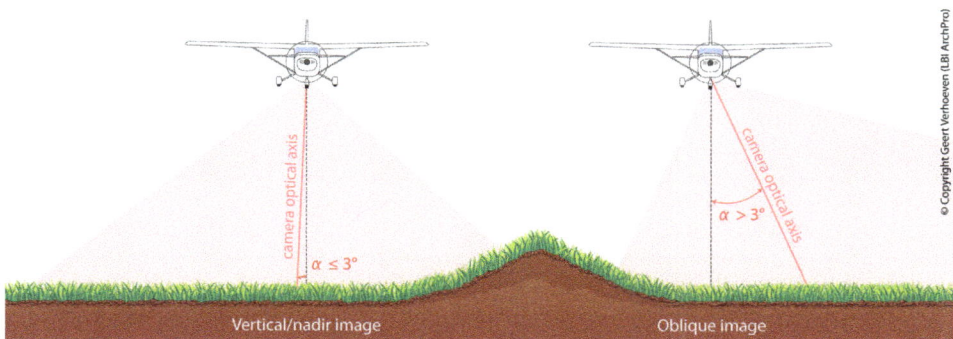

Figure 4. Vertical versus oblique aerial imaging.

Depending on the visibility of the apparent horizon, the image is then further classified as low oblique (i.e., the horizon is not included) or high oblique [112]. Sometimes, the threshold between low and high oblique photographs is also quantified at 30° [113]. Since this type of archaeological aerial prospection primarily generates oblique images (or simply 'obliques') of archaeological interest, archaeologists in Great Britain but also in other parts of Europe have used 'oblique survey' as shorthand for this type of observer-directed or observer-biased survey strategy (the latter terms coined by Palmer [114]). In the mapping and surveying community, this kind of photography is sometimes denoted free-camera oblique photography, being " . . . an enjoyable and to some extent creative form of aerial photography . . . where the pilot and photographer have the task of photographing random targets on a speculative basis" [115].

Since the 1930s, this free-camera, observer-directed aerial surveying has been the workhorse of all archaeological airborne remote sensing techniques. Its success lies mainly in its straightforward execution as well as the abundant availability and moderate cost of small aeroplanes and photo cameras. The combination of its capacity to repeatedly cover relatively large areas with ample spatial detail (typically a GSD of 5 cm to 10 cm) turns it into one of the most cost-effective methods for site discovery—or at least, that is what it is often told to be e.g., [116,117].

Despite its efficiency in certain areas and periods, the main disadvantage of this reconnaissance approach lies in its observer-directed nature which generates extremely selective, pre-interpreted [6] and thus utterly biased data that are completely dependent on an airborne photographer recognising archaeological phenomena at a relatively high air-to-ground speed. Subsurface soil disturbances that are visually imperceptible at the time of flying, or those that are simply overlooked, will not make it into a photograph. Moreover, the observer might fail to record certain features, even when looked at, because he/she is too tired or they are not part of his/her expectations or repertoire of known sites (see Cowley [18] for a more thorough discussion of this topic). Alternatively, the surveyor might simply ignore existing features and create false patterning by flying where the archaeological 'hot spots' (a.k.a. 'honey-pots') or the 'familiar stuff' can be found [4,118,119]. Aerial archaeological pioneers such as John Kenneth Sinclair St Joseph (1912–1994) are (in)famous for pursuing mainly Roman military archaeology [120,121]. To them, but also too many of their successors, taking aerial photographs is like collecting stamps. In a sense, observer-based archaeological aerial photography is still pursuing the same goals as those of the first decades of the 20th century. This site-based approach contrasts with the inherent complexity of landscapes and their internal relationships landscape archaeology tries to unravel. Also, pre-interpreting the object of study before collecting data about it renders this type of aerial observations entirely subjective.

Airborne imagery acquired during an observer-directed sortie is thus nothing but an arbitrary sample, based on the perception, interest, motivation, knowledge, expectation and logistical capabilities of the person behind a highly unoptimised data recording system called photo camera. When also taking the indispensable need for favourable interactions between all the uncontrollable variables as an additional bias into account [17,122], one can call into question the appropriateness of such haphazardly collected data to make inferences about the landscape as a whole. Yet, these biased data and the knowledge created from this century-old approach to aerial imaging continues to frame much of the research in landscape archaeology.

To counteract any of these obvious geographical biases, several scholars have already questioned this particular reconnaissance strategy over the past two decades. Although flying higher and/or with a shorter focal length can draw more of the landscape into the photograph and a second observer/photographer onboard might certainly increase the archaeological detection rate, several papers also pointed out the advantage of a geographical 'total coverage' or 'seamless' approach [114,123–129]. Moreover, since photographs that result from such a blanket survey are much better suited for automated image processing workflows, they also mitigate the prevalent bias that is created by simply not using these aerial observations.

4. Removing the Camera-Angle Delusion: Total Coverage Aerial Photography

During a total coverage or block survey sortie, photographs will predominantly be acquired with expensive, accurately calibrated, built-in (versus hand-held), gyro-stabilised and low distortion mapping frame cameras (often referred to as metric or cartographic cameras [130]) to make sure that all photographs are nadir/vertical images. These cameras are solidly housed and operated in bigger and higher flying aeroplanes. Images feature a GSD of 25 cm or smaller and are acquired in parallel strips at regular intervals, generally with a substantial frame overlap: in one flight strip, each photograph has a generally accepted degree of overlap of circa 60 ± 5% (figures to 90% can be found as well, see [113]) with the following and preceding image (longitudinal overlap). Adjacent strips have on average an overlap of 25% to 40% (lateral overlap) [115]. The camera is pointing directly down to the earth to acquire (near) nadir photographs. Because a perfect vertical is seldom achieved, an image with an angle of less than or equal to 3° is called vertical [131]. Analogous to the oblique survey approach, archaeologists began to use 'vertical survey' as shorthand for such total area coverage flying strategies. Although one could indeed argue that both flying strategies primarily deliver these types of images, this archaeological habit remains a very inappropriate way of categorising aerial survey approaches.

Just as the observer-directed flying strategy can yield vertical imagery, these aerial block surveys can deliver oblique imagery as well. This has never been more evident than in the past fifteen years, during which the geospatial community re-introduced oblique imagery for measuring applications. Although tilted views had been very popular in the 1920s to 1940s for mapping, surveillance and reconnaissance purposes [132], extracting accurate geometry from aerial images has been limited to blocks of vertical images afterwards. The recent revival of the tilted views—acquired during total coverage mapping surveys—was mainly triggered by Pictometry. Since this New York-based imaging company introduced in 2000 its PentaView capture system consisting of five cameras (of which four were taking oblique images, while the central one was directed nadir [133]), the properties of the acquired oblique imagery have been opening many eyes. As a result, a strong movement towards mapping approaches that integrated oblique images with the more traditional vertical ones was created [134,135]. This trend is now exemplified by the growing number of manufacturers that create digital multi-camera systems in which the vertical mapping camera is complemented by one or more oblique mapping cameras—for example IGI's Penta DigiCAM [136] or Microsoft's UltraCam Osprey family [137] (Figure 5).

Figure 5. The Microsoft/Vexcel UltraCam Osprey Prime II for combined nadir and oblique image capture. Image source: https://ultracam.files.wordpress.com/2014/04/ucop_2.png.

Although the current evolution and use of these systems are mainly driven by the major developments in the fields of computer vision and photogrammetry [138], the resulting imagery is also an excellent source of geo-information for many other fields. The strongest selling point of these obliques is their intuitive nature and rich content: not only do they come close to the standard human perception of objects or scenes while standing on a hill or in a high tower [139,140], they also make it easier to map façades of houses, hence providing a complete description of built-up areas [141,142]. Although parts of vertical structures such as houses, electricity pylons, trees and lamp post might indeed remain occluded in a vertical view, the information that oblique imagery delivers about vertical structures is only of minor importance in most landscape archaeology projects. The reasons for the enthusiastic revival of the slanted viewing geometry in the geospatial industry do thus not hold (or only to a minor extent and in specific cases) for most archaeological landscape projects. Moreover, nadir views provide a more than satisfying alternative to the extremes of oblique imaging for rendering shadow and colour-difference based marks [10].

From the statements above, it should be clear that the traditional archaeological remote sensing dichotomy between oblique and vertical aerial photographic reconnaissance—in which either type too often indicates a specific flying approach—is a highly inappropriate one. This terminology should be left behind in favour of observer-directed versus total coverage or block survey, both capable of generating oblique as well as vertical aerial photographs. Although the terms biased versus unbiased aerial survey are sometimes coined [114,143], their usage is also discouraged since many other biases remain. If used, one could refer to these survey strategies as being geographically unbiased—as was also done by Mills [125]. That is, geographically unbiased at least inside the survey area itself, since the specific choice of the latter is also governed by many personal and practical factors.

5. Increasing the Spectral Dimensions: VNIR Aerial Imaging

Despite many advantages, both geographically biased and unbiased aerial photography are typically characterised by major spectral shortcomings. In archaeological observer-biased aerial photography, the human visual system governs the survey working principles. However, both the human eye and a standard photo camera only create useful data streams using light (i.e., the visible electromagnetic spectrum). Although several archaeologists have at a certain point in their career experimented with false-colour or pure Near-InfraRed (NIR) airborne imaging (consider [144] for an extensive overview) or even imaging in the near-ultraviolet [145,146], the majority of aerial footage has been acquired with photographic media that were sensitised to wavelengths between approximately 400 nm and 700 nm. Although striking and revealing images have been obtained in this way, the detection of vegetation marks (and to a certain extent soil and other marks) becomes impossible in less-than-optimal circumstances. The slight differentials of height and colour in crops might simply exhibit too low contrast with the surrounding matrix to be noticed in the visible spectrum (Figure 6). As a result, the aerial observer will never orbit nor document visually imperceptible soil and crop disturbances.

Since most mapping cameras are multispectral devices that capture imagery in one NIR and three visible channels (encompassing the so-called VNIR spectrum), NIR blanket coverage generated during a block sortie tackles the issue mentioned above. However, both standard photographic solutions as well as multi-spectral imagers capture reflected solar radiation in spectrally broad wavebands. A typical colour photograph records the visible spectrum in three wide Red, Green and Blue channels (with their FWHM approximately 80 nm; Figure 3), while a NIR image generally encompasses a 200 nm waveband. From the perspective of feature recovery-rate, this is a subpar approach since particular diagnostic spectral vegetation features are often only a few nanometres wide. (Note again that this text mainly focuses on vegetation marks due to their abundance and potential to disclose very detailed morphological information about the totally buried archaeological features).

This makes acquisition of data with a high spectral resolution necessary when one seeks to assess small variations in—for example—plant physiology and minimise spectrally-induced ambiguity in the data. Vertical and oblique photography, even when executed beyond the visible waveband, thus significantly reduce the diagnostic accuracy of vegetation investigation. Despite their potentially high spatial resolving power, they spectrally undersample the at-sensor radiation and mask the spectral features that are too narrow to be distinguished [11,147]. Although more dedicated multispectral sensors like the Daedalus 1268 Airborne Thematic Mapper have been developed and used profusely in archaeological research [148–152], those cameras commonly lack the high spatial resolution advocated above. They are also ill-suited for fast deployment in the small two- to four-seat aircraft that are typically used for an archaeological aerial survey. Nevertheless, the resulting images enable the generation of radiometrically corrected orthophotographs and Digital Surface Models (DSMs).

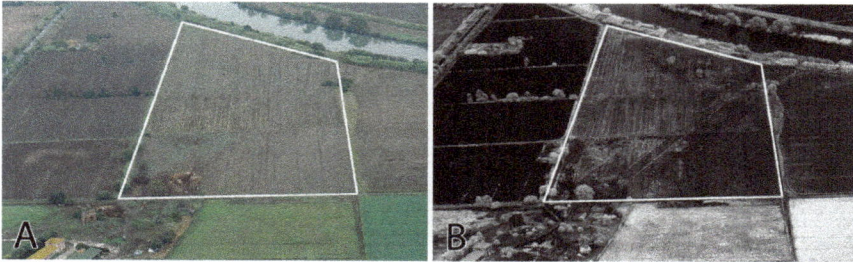

Figure 6. Visible (**A**) and NIR photograph (**B**) of the same area in *Portus*, Italy (41°46′30″N, 12°15′51″E—WGS84). Note the many features that are visible in the NIR image but go undetected in the conventional photograph.

6. Overcoming the Spectral Delusion: Airborne Imaging Spectroscopy

To resolve this spectral resolution issue, archaeologists have started to investigate Airborne Imaging Spectroscopy (AIS), also known as Airborne Hyperspectral Scanning (AHS). AIS is a passive remote sensing technique that digitises the earth's upwelling electromagnetic radiation in a multitude of small spectral bands. Compared to most multispectral sensors, hyperspectral sensors have a higher spectral resolving power (typically an FWHM of around 10 nm in the VNIR range) and capture more and (nearly) contiguous (i.e., adjacent and not overlapping) spectra [153]. Although the borderline between multi- and hyperspectral imaging is rather discipline dependent, ten to twelve spectral bands is often taken as the threshold [154]. In other words: tens to hundreds of small bands of electromagnetic radiant energy are captured per pixel location (Figure 7). Through a combination of all spectral data acquired from a particular spatial location, every pixel of the final image holds the spectral signature of the material that was sampled at that specific location.

AIS is typically used to image entire areas without having geographical data gaps. So far, AIS has been used in several archaeological projects, for example [150,155–168]. The success rate regarding archaeological subsurface structure detection is, however, variable and less effective applications seem to be connected with the lower spatial resolutions of the acquired datasets (in most cases the GSD ranges from 1 m to 4 m). Although the current generation of AIS sensors easily enables a GSD below 50 cm when flown low and slow enough, the visible image detail is still not comparable with conventional three-band aerial photographs.

Despite better spectral resolution and improving spatial resolution, working with AIS data necessitates a certain amount of expert knowledge and specialised software. Assuming that the image geometry and georeferencing can be handled correctly—which is far from a trivial task—one needs to deal with the data quality and dimensional overload by denoising procedures and intelligent data mining approaches. Even though specific archaeological tools such as the MATLAB-based ARCTIS toolbox [169] have been developed to answer these needs, the issues mentioned above are still a limiting (i.e., biasing) factor for the widespread application of AIS in archaeology. Finally, the cost of an AIS flight still places it out of reach of most institutions, despite the steadily decreasing prices due to increased demand in non-archaeological communities. If, however, hyperspectral data with a high spatial resolution are used in combination with powerful processing algorithms, AIS does certainly have a clear potential in areas where vegetation marks do not appear clearly in broad-band multi-spectral setups (e.g., [158,168]).

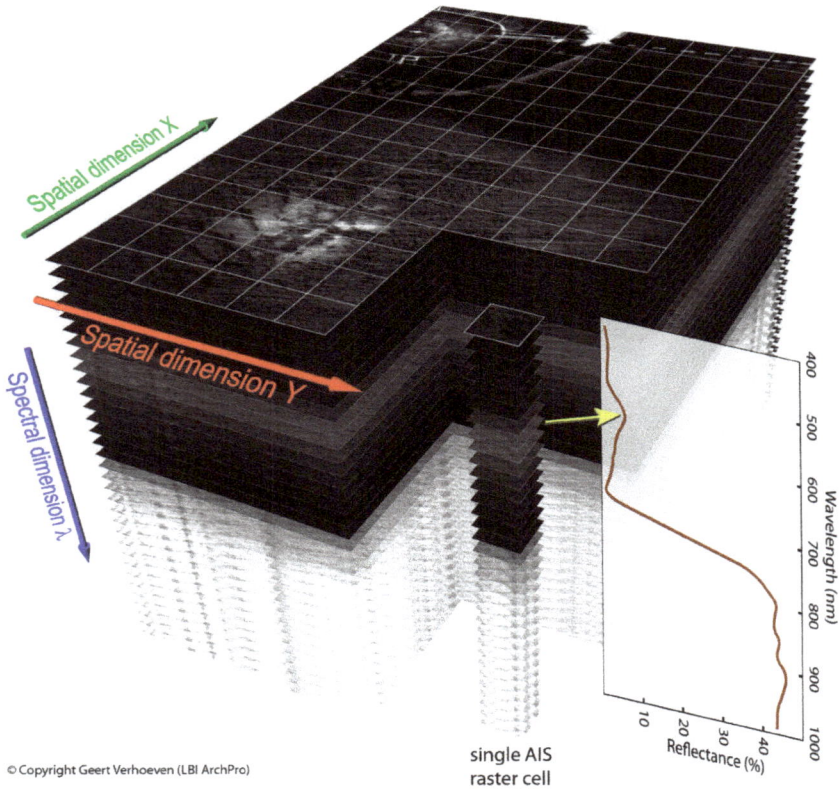

Figure 7. Visualisation of an airborne imaging spectroscopy data cube.

7. Take the Best, Leave the Rest. A Discussion

After this reflexive approach, it thus seems that, despite a hundred years of technological evolutions and methodological developments, many observational shortcomings still hamper passive airborne optical imaging to truly benefit the many flavours of landscape archaeology. This results in a wide variety of biases, both in the data and the subsequent usage of them:

- spaceborne data, consistently acquired over extended areas and often in invisible wavebands, might tackle the observer-directed and visible-radiation-limited biases. However, the data are less (or not at all) suited for the discovery and detailed recording of small archaeological features, as the spatial resolving power of the sensors exceeds one metre in all but a few panchromatic cases (a panchromatic image is a greyscale image created by one spectral band that is sensitive to more or less all ('pan') wavelengths (or colours, hence 'chroma') of visible electromagnetic energy). Moreover, the spectral bands of older spaceborne imagers (i.e., those whose products are freely available) are generally too broad or misplaced spectrally to truly detect the plant stress that governs vegetation marks [170];

- conventional airborne photographic imaging approaches are dominantly observer-directed, creating a strong geographical bias. They also lack the spectral resolving capabilities that are needed to digitise subtle reflectance features;
- existing multispectral solutions are often limited to four broad bands while the instrumentation is expensive and impossible to easily (de)mount into a light aircraft;
- hyperspectral imaging sensors do acquire data in narrow wavebands and are usually flown with a total coverage strategy in mind, but the combination of affordability, availability, data complexity, moderate temporal resolution and generally lower spatial resolving power also significantly restrict its frequent use (even of existing data) in archaeological research. Various biases and a lack of cumulative data are the result;

These remarks are not to diminish the importance of data collection strategies used so far, but the tools and knowledge are now here to eliminate several of these shortcomings. At the turn of the 21st century, Bewley and Rączkowski stated that "in examining future developments it is becoming clear that we must hold on to the best of the old and embrace the best of the new" [171] (p. 4). From the statements above, it seems that a more technologically adept archaeological airborne reconnaissance system would combine the cost-effectiveness (from a hardware point of view), availability and operating flexibility as well as the spatial and temporal resolving power of the observed-directed photographic approach. In addition, it should allow a geographically unbiased total coverage in a selection of narrow visible and invisible spectral wavebands that are suitable for detecting vegetation and soil marks.

In the author's view, the last two characteristics are crucial if the aerial archaeological community wants to evolve to an improved airborne data acquisition that aims at a more consistent record of our past. The next sections outline a proposal to that end. In short, the text further details why a rather dense airborne sampling of various dimensions of the preserved archaeological record could naturally span several perspectives and oppositional analytical approaches that are commonplace in landscape archaeology. As such, this observational (i.e., data gathering) phase might to a large extent transcend major theoretical paradigms. This proposal does not have to stay a purely hypothetical one, as Verhoeven and Sevara [12] have recently shown how this could be materialised and put into practice.

7.1. Geographically Unbiased and Vertical

From a technological point of view, geographical total coverage aerial imaging is not novel at all. However, it becomes innovative in archaeology as soon as it is executed with the same intensity as observer-directed aerial surveys. When total coverage passive aerial optical imaging would be performed by default, the uniformity of the aerial archaeological data sets would benefit enormously. Passive airborne optical image collections would be freed of (most of) their inherent weaknesses such as the observer's experience, a priori knowledge and many other methodological biases [4,18], although Wilson claims them to be unavoidable features of aerial reconnaissance [119].

The acceptance and implementation of such a strategy can, however, only be fruitful if one distances themselves from the common approaches by which block surveys are executed. It still holds true that most archaeological features appear on vertical photographs (acquired during a block survey) through what has been denoted the serendipity effect [172]. This is principally because high-level vertical block surveys have generally been executed to acquire basic material for (orthophoto)map generation rather than for archaeological purposes. Even then, much valuable information has been extracted from such blanket photography. Also, most of the usual arguments to not fly (or even use) verticals from a block survey sortie (see [117,173–175]) have been countered on many occasions by various authors [10,114,176,177].

Moreover, it is not written in stone that a total area coverage should only be flown at mid-day at high-altitudes by professional survey companies that generate true nadir photographs with overly expensive equipment. It is perfectly feasible to generate verticals using the small, two- to four-seater aeroplanes which are typically used in archaeological aerial prospection by modifying the door or mounting the camera externally (e.g., [178–181]). Fixing a camera to the side of the aircraft for taking vertical photographs was already very common during the First World War [51]. Nowadays, Unmanned Aerial Systems (UASs) will also predominantly acquire vertical still imagery in a total area style with abundant image overlap. The reasons for this are pretty obvious. First, vertical, block-covered footage offers an advantage in mapping, as the induced geometrical distortions are much smaller than those embedded in oblique footage [182]. Second, the illumination changes across the image frame are smaller and the scale is approximately constant throughout [139]. Since the data are by default captured with a high overlap, they are also perfectly suited to create three-dimensional DSMs (see Figure 8B). When acquired from operating heights that surpass allowable UAS altitudes, the high spatial resolution and broad coverage of blanket vertical imaging make them relevant for a holistic view of the landscape, to map extensive geo-archaeological traces as well as for the primary discovery of small, individual archaeological features. Finally, even low-slanting sunlight is not an absolute necessity anymore for recording earthworks or differences in vegetation height. Most orthorectification approaches now automatically generate DSMs [183,184]. These can be used to artificially produce shadow marks in a fast and (depending on the extent of the data set) even interactive way (Figure 8B) [185], while many additional techniques exist to highlight small topographic differences further (see Figure 8C) [186].

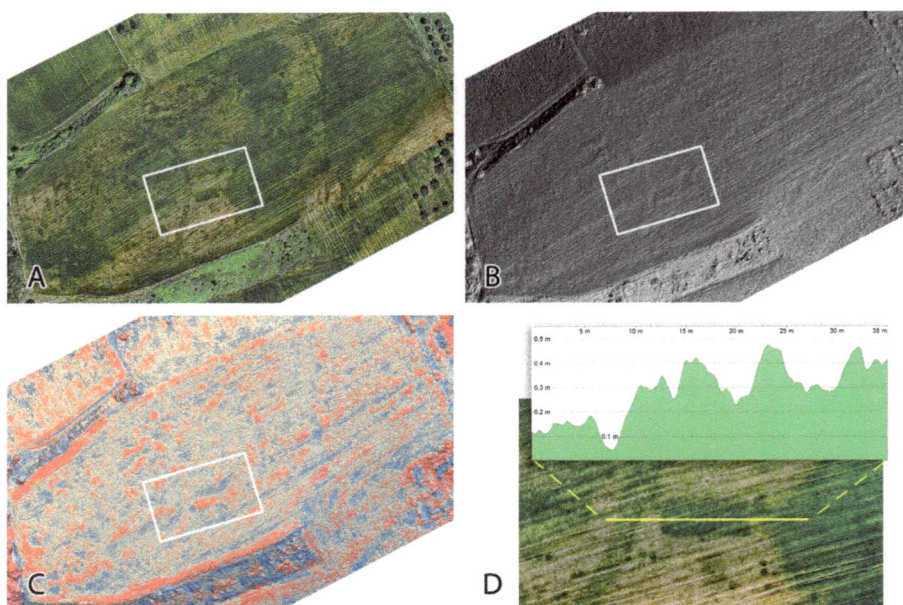

Figure 8. (**A**) Orthophotograph of the Montarice hilltop (Marche, Italy; 43°25′18″N, 13°39′33″E—WGS84) showing many vegetation marks. Although no shadow marks are visible, the next three visualisations clearly indicate differences in vegetation height; (**B**) a DSM with artificially created shadow marks; (**C**) a local relief model (annulus kernel, (45 pixels, 75 pixels)) brings out the tiny differences in vegetation height; (**D**) a zoom of the same scene (its extent is indicated in **A**–**C**) with the height of the crops extracted along the uninterrupted yellow line after computing a canopy height model. More details are mentioned in [186].

Of course, this does not mean that observer-directed flights in favourable light conditions should be discarded. In between total coverage approaches, one can certainly think of a few targeted observed-led sorties to acquire additional image material. However, observer-directed flights as a means to better understand the recorded features or the area under study might make subsequent interpretations of landscape organisation problematic if one assumes that past people also had a similar 'aerial (over)view' concept of the landscape. In contrast, the photographs that are possible with observer-based obliques are also still valid if one flies for artistic satisfaction or documenting standing remains (certainly when those are to be found environments such as deserts; e.g., [187]). However, it is the author's firm conviction that these flights should be rather the exception than the rule.

Though Campana argues to "supplement oblique aerial survey with some form of 'total' recording at those times when archaeological visibility is at its best" [188] (p. 12), he arguably makes a false assumption that archaeological aerial archaeological photographic reconnaissance cannot escape its inherent subjectivity and selectivity. On the contrary, it does not take all that much effort to replace the observer led approach with a reconnaissance strategy involving cumulative total coverage surveys, with or without oblique camera observation angles, executed from the same light aircraft commonly used for archaeological airborne prospection. Of course, a certain amount of subjectivity will always exist (such as choice of pilot, aircraft, flying height, time of data acquisition), but the main data selectivity actors are not that hard to weed out. Maybe one of the key arguments in realising the necessary change in archaeological mindset lies in the fact that information about soil type, hydrology, land use and vegetation cover—which is all extractable from continuous airborne coverage—can even help to detect bias in other data sources [189].

7.2. Multispectral and Portable

It is also deemed of the utmost importance to acquire spectral data in a handful of small, vegetation and soil mark-sensitive spectral bands as well as in the very common broad wavebands of the visible spectrum, since the interpretation of the former will greatly benefit from the latter [11]. Ideally, imagery could be acquired in the green peak at 550 nm (30 nm FWHM), the red edge at 705 nm (30 nm FWHM) and the NIR shoulder at 820 nm (100 nm FWHM). Those three spectral bands have repeatedly been shown to be strongly related to biophysical changes caused by plant stress [190–193]. They are, therefore, well-suited for assessing crop (and by extension vegetation) characteristics, while the combination of the 820 nm NIR band and the 705 nm Red Edge band has already proven its potential in archaeological prospection [156,194,195]. Additionally, the potential information captured in the NIR band is not solely restricted to vegetation marks. Since water heavily absorbs NIR [196] and existing soil moisture differences are often characteristic of soil marks [197,198], discerning the latter becomes easier in the NIR as compared to the visible range. Moreover, all three bands exhibit relatively small anisotropic reflectance effects [199]. This approach also omits the overload of highly correlated spectral data that is characteristic for conventional AIS approaches, while a high spatial resolution can be combined with a high spectral resolution (be it in less spectral bands). To measure the incoming visible spectrum, a normal state-of the-art camera can be flown alongside.

When the imaging solution is also transportable and easy to mount on a variety of light airplanes (e.g., Cessna 152, Cessna 162 Skycatcher, Cessna 172 Skyhawk and Cessna 182 Skylane), repetitive, cost-effective and total coverage airborne survey will become commonplace, offering local researchers the chance to have a high-end imaging system flown at very reasonable costs.

7.3. The Processing and Interpretation Back-End

Although it might seem that the need for the proposed passive optical imaging approach is paramount regardless of the data processing and interpretation pipeline, such drastic developments can not only stay limited to the acquisition phase. The predominant single-imaged based workflows are just too slow and cumbersome to deal with the large amounts of aerial imagery that are constantly generated. As a result, millions of aerial photographs are simply stored in archives where they are at risk of loss or obscurity and their archaeological information cannot (or will not) be exploited efficiently (contributing to another form of bias).

The only solution is to leave the current post-acquisition workflows behind in favour of processing and interpretation pipelines that can deal with a multitude of images at once, thereby saving on the demands for skill, money and time. Given the current failure to do so, the latter three are also often quoted as reasons to stay away from total area coverage [117]. As a result, new georeferencing approaches, radiometry processing algorithms, management structures and interpretation strategies must be conceptualised, tested and implemented, not only to maximise the usability of aerial optical imagery for landscape archaeology but to simply deal with the large amounts of airborne data that will result from the proposed image acquisition approach. Although it is sometimes said that studying a sample with great care might yield more information than studying the complete population [80], advances in data storage, computational power, machine vision and deep learning start to render this argument more or less invalid for these image collections. When the processing and interpretational chains are properly set-up (see also [12,200,201]), dealing with several thousands of images should not be much more of a burden than dealing with a few hundreds of them.

Although the technology is available, realising all these aspects will not come easy because the major challenge lies in the creation of a new methodological, analytical and interpretational mindset. Since this paper does not want to suggest that a blind collection and accumulation of aerial data is the way to go, this new data collection strategy must go hand in hand with (the already partly ongoing) renewed practical and theoretical reflections on the detection, identification, typological and chronological classification, perception and interpretation of landscape features [18,54,68,101,168,202–207]. These considerations remain vital because archaeology still falls short of realising the full potential that digital data have to offer [208]. However, instead of looking for the optimal solution according to one dogmatic paradigm, it is best to accept that these aerial imaging products also have their limits. As such, this paper advocates the vision that archaeologists should try to combine the complementary strengths of different theoretical and analytical methodologies while sticking to a healthy amount of source-criticism and scepticism. Only then can the combination of computing power and this new digital imagery provide new archaeological insights, regardless of the theoretical framework.

8. Conclusions

Archaeologists need to reflect on the ever-expanding array of specialised topics and methods to continue their integration and to combine their different strands of thinking. In most of the cases, archaeologists try to build data by systematically observing the preserved archaeological record. The combination of any pre-existing intellectual perspective and the research problem ultimately—but often unconsciously—determine which observations are made. Finally, the data collection tool that is used determines the exact nature of these observations. This paper has taken a reflexive approach to how archaeologists with different (but often unknown) theoretical convictions have been acquiring passive optical imagery from the air. Starting from the concept of landscape archaeology, the principles of archaeological remote sensing and the need for bias reduction when sampling the archaeological record, the author has tried to assess all passive airborne optical imaging approaches in terms of eight essential characteristics, which all bear the potential to increase or decrease the subjectivity and bias in the acquired image set.

Although archaeological optical imagery is never an 'objective' or theory-neutral dataset, it has been shown that the amount of data favouritism and partiality can easily be significantly reduced. This reflection is not to diminish the importance of data collection strategies used so far. Even though the observer-led passive airborne imaging approach of the early years was entirely appropriate, as "the imperative then was to record as many previously unknown monuments as quickly as possible" [209] (p. 65), this essentially site-based methodology seems nowadays at odds with many current core conceptions of landscape archaeology. However, also the conventional multi- and hyperspectral imaging approaches have proven to be inherently flawed, be it by their sampling design, cost, instrument availability or processing complexity. As a result, they also cannot prevent highly biased data collection and usage.

Can any form of meaningful insight about the past be obtained at all if biased remote sensing methods continue to generate some of the key datasets in landscape archaeology? How can one argument in a coherent way if the data at hand do not allow for a proper evaluation of the argumentation? Instead of considering possible answers to these questions, this paper started from an intellectually-integrative approach to landscape archaeology to propose a new technological-methodological data acquisition strategy. This attempt is based on assumptions that (1) any lack of understanding all inherent airborne data flaws will be prone to erroneous and unsupportable conclusions and that (2) remote sensing will become ever more integral to the process of understanding the human past and that (3) to "extract the most complete understanding of the settlement dynamics of an extensive territory requires the application of special field methods" [52] (p. 23).

One might say that the attention for the method and the need to exhaustively and cumulatively collect imagery exemplifies a positivistic, empirical approach to archaeology. However, the need to optimise data collection in various dimensions does not hamper its interpretation. Although this paper advocates in true processual style a standardisation of practice and aims at systematic, repeatable imaging surveys with better sampling characteristics, it does not advocate a separation between observation and interpretation, between analytical and interpretive archaeologies. The attention given to the analytical method of aerial survey does not prevent an interpretative approach to the data, nor does it mean that one will automatically better know the past or that cognitive objectivity is assumed.

This ties directly into the question of the fitness-for-purpose of aerial imaging in landscape archaeology and the seemingly logical follow-up argument that there is a need for an initial broad-brush landscape characterisation approach to determine those portions of the study area for which passive airborne imaging has archaeological potential [210]. First, a major part of this argument stems from the fact that broadband photography in the visible spectrum has always been the *de facto* way of aerial imaging. This approach does indeed often fail in less-than-optimal circumstances (see Section 5). Despite the fact that Taylor already wrote forty years ago that "Continuous reconnaissance, using new techniques, over areas normally regarded as poor or unrewarding is a basic need for future field archaeology" [211] (p. 140), unrewarding or 'unresponsive' zones have seldom been targeted with more specific imaging approaches such as those presented in this paper.

Second, this type of criticism regularly contrasts aerial optical imaging with geophysical approaches and excavation techniques (e.g., [17]), both of which are often considered to deliver uniform data sets that provide almost absolute knowledge about the archaeology hidden within the soil matrix (excavations even more so that geophysics [212]). However, the chemical and physical properties of the soil with its embedded archaeological residues as well as the sensor characteristics both limit all three approaches to landscape investigation. It is a well-known but often neglected fact that excavations might fail to uncover archaeological features that are revealed by both remote sensing and geophysics [213,214]. Although this situation might cause serious incomprehension among many archaeologists, it is not difficult to understand that a geophysically measured contrast (such as an increase in magnetic susceptibility) does not have to translate into a colour difference (which is the contrast on which the human visible system or a standard photographic camera relies).

On the other hand, geophysical sensors might be unable to record archaeological features in certain conditions. For instance, high topsoil salinity or wet and clayish conditions can render ground-penetrating radar almost useless. As a result, archaeological features might show up in aerial photographs but not in geophysical datasets [215,216]. Since both excavations and geophysical methods can only recover some specific aspects of the preserved archaeological record, their data are thus by default also incomplete. While excavations and geophysical prospection usually have a better reputation regarding their archaeological effectiveness and data consistency, they thus also suffer from the uncontrollable biases that characterise passive airborne optical imaging.

Even though the number of data representativity issues might indeed be greater for airborne imaging, one should never consider the absence of any archaeological evidence as proof of archaeological absence for any of these three methods (or many other prospection methods for that matter). Despite the inherent biases in their recovered archaeological record, both excavation and geophysical survey oriented archaeologists generally accept that dense sampling of large, continuous areas is key to properly understand what is going on at a site or landscape level [217–219]. This paper followed the same reasoning. Only when the sampling is done properly, one can combine the aerial imagery with colluvial, alluvial and aeolian records [149] or other data sources to assess and truly understand the biases imposed by topography, environmental factors and land uses. Otherwise, any assessment of fitness-for-purpose of the aerial imagery for answering particular landscape archaeology questions or any broad-brush landscape characterisation to matching survey methodology to local context seems the wrong way around.

Since the proposed approach has the potential to remove many of the major and controllable data gathering biases, the resulting image sets should enable holders of rival theories to agree on the fact that these data are archaeologically useable. Finally, imagery collected in such a non- (or less) selective way can also be of use to subsequent studies that might have largely different scopes. Or it can facilitate data examination with oppositional analytical and interpretative approaches. In this way, it is hoped that we can build, in combination with many other data sets, potentially novel, more robust and richer understandings of the landscapes that are diachronically and synchronically constituted by the reciprocal relationships between human and their environment.

Acknowledgments: The author would like to thank Seta Štuhec and three anonymous reviewers for their feedback on the first draft of this paper. Not all the members of the author's host institution, LBI ArchPro, necessarily share the opinions expressed in this article which are very much those of the author. The Ludwig Boltzmann Institute for Archaeological Prospection and Virtual Archaeology (archpro.lbg.ac.at) is based on the international cooperation of the Ludwig Boltzmann Gesellschaft (A), the University of Vienna (A), the Vienna University of Technology (A), ZAMG, the Austrian Central Institute for Meteorology and Geodynamics (A), the Province of Lower Austria (A), Airborne Technologies (A), 7 reasons (A), the Austrian Academy of Sciences (A), the Austrian Archaeological Institute (A), RGZM, the Roman-Germanic Central Museum Mainz (D), the National Historical Museums—Contract Archaeology Service (S), the University of Birmingham (GB), the Vestfold County Council (N) and NIKU, the Norwegian Institute for Cultural Heritage Research (N).

Author Contributions: Geert Verhoeven conceived and wrote the whole paper. He also created all illustrations.

Conflicts of Interest: The author declares no conflict of interest.

References

1. Gramsch, A. Theory in Central European Archaeology: Dead or alive? In *The Death of Archaeological Theory*; Bintliff, J.L., Pearce, M., Eds.; Oxbow: Oxford, UK, 2011; pp. 48–71.
2. Rączkowski, W. *Archeologia Lotnicza. Metoda Wobec Teorii*; Wydawnictwo Naukowe Uniwersytetu im. Adama Mickiewicza: Poznań, Poland, 2002.
3. Brophy, K.; Cowley, D.C. (Eds.) *From the Air. Understanding Aerial Archaeology*; Tempus: Stroud, UK, 2005.
4. Cowley, D.C. What kind of gaps? Some approaches to understanding bias in remote sensing data. *Archeol. Aerea* **2013**, *7*, 76–88.
5. Hanson, W.S. The Future of Aerial Archaeology (or Are Algorithms the Answer?). In *Remote Sensing for Archaeology and Cultural Heritage Management, Proceedings of the 1st International EARSeL Workshop, CNR, Rome, 30 September–4 October 2008*; Lasaponara, R., Masini, N., Eds.; Aracne: Rome, Italy, 2008; pp. 47–50.
6. Rączkowski, W. Beyond the technology: Do we need 'meta-aerial archaeology'? In *Aerial Archaeology: Developing Future Practice*; Bewley, R.H., Rączkowski, W., Eds.; IOS Press: Amsterdam, The Netherlands, 2002; pp. 311–327.
7. Rączkowski, W. Why interpretation?: Chairman's Piece. *AARGnews* **2009**, *39*, 5–8.
8. Rączkowski, W. Towards integration: Two prospection methods and some thoughts. In *From Space to Place, Proceedings of the 2nd International Conference on Remote Sensing in Archaeology, CNR, Rome, Italy, 4–7 December 2006*; Campana, S., Forte, M., Eds.; Archaeopress: Oxford, UK, 2006; pp. 203–206.

9. Campana, S. Sensing Ruralscapes. Third-Wave Archaeological Survey in the Mediterranean Area. In *Digital Methods and Remote Sensing in Archaeology: Archaeology in the Age of Sensing*; Forte, M., Campana, S., Eds.; Springer: Cham, Switzerland, 2016; pp. 113–145.

10. Verhoeven, G.J.J. BRDF and Its Impact on Aerial Archaeological Photography. *Archaeol. Prospect.* **2017**, *24*, 133–140. [CrossRef]

11. Verhoeven, G.J.J. Beyond Conventional Boundaries. New Technologies, Methodologies, and Procedures for the Benefit of Aerial Archaeological Data Acquisition and Analysis. Ph.D. Dissertation, Ghent University, Zelzate, Belgium, 2009.

12. Verhoeven, G.J.J.; Sevara, C. Trying to Break New Ground in Aerial Archaeology. *Remote Sens.* **2016**, *8*, 918. [CrossRef]

13. Powlesland, D. Why bother? Large scale geomagnetic survey and the quest for "Real Archaeology". In *Seeing the Unseen: Geophysics and Landscape Archaeology*; Campana, S., Piro, S., Eds.; CRC Press/Balkema: Boca Raton, FL, USA; Leiden, The Netherlands, 2009; pp. 167–182.

14. Campana, S. 'Total Archaeology' to reduce the need for Rescue Archaeology: The BREBEMI Project (Italy). In *Remote Sensing for Archaeological Heritage Management, Proceedings of the 11th EAC Heritage Management Symposium, Reykjavík, Iceland, 25–27 March 2010*; Cowley, D.C., Ed.; Europae Archaeologia Consilium (EAC), Association Internationale sans But Lucratif (AISBL): Brussels, Belgium, 2011; pp. 33–41.

15. Taylor, C.C. Total Archaeology or studies in the history of the landscape. In *Landscapes and Documents*; Rogers, A., Rowley, T., Eds.; Bedford Square Press of the National Council of Social Service: London, UK, 1974; pp. 15–26.

16. Lock, G. *Using Computers in Archaeology: Towards Virtual Pasts*; Routledge: London, UK; New York, NY, USA, 2003.

17. Cheetham, P.N. Noninvasive Subsurface Mapping Techniques, Satellite and Aerial Imagery in Landscape Archaeology. In *Handbook of Landscape Archaeology*, 1st ed.; David, B., Thomas, J., Eds.; Left Coast: Walnut Creek, CA, USA, 2010; pp. 562–582.

18. Cowley, D.C. What Do the Patterns Mean? Archaeological Distributions and Bias in Survey Data. In *Digital Methods and Remote Sensing in Archaeology: Archaeology in the Age of Sensing*; Forte, M., Campana, S., Eds.; Springer: Cham, Switzerland, 2016; pp. 147–170.

19. Rodaway, P. *Sensuous Geographies: Body, Sense and Place*; Routledge: New York, NY, USA, 1994.

20. Wylie, J. *Landscape*; Routledge: Abingdon, UK, 2007.

21. Bender, B. (Ed.) *Landscape. Politics and Perspectives*; Berg: Providence, RI, USA; Oxford, UK, 1993.

22. Bruns, D.; Kühne, O.; Schönwald, A.; Theile, S. (Eds.) *Landscape Culture—Culturing Landscapes. The Differentiated Construction of Landscapes*; Springer: Wiesbaden, Germany, 2015.

23. Witcher, R.E. GIS and Landscapes of Perception. In *Geographical Information Systems and Landscape Archaeology*; Gillings, M., Mattingly, D.J., van Dalen, J., Eds.; Oxbow Books: Oxford, UK, 1999; pp. 13–22.

24. Denham, T. Landscape Archaeology. In *Encyclopedia of Geoarchaeology*; Gilbert, A.S., Goldberg, P., Holliday, V.T., Mandel, R.D., Sternberg, R.S., Eds.; Springer: Dordrecht, The Netherlands, 2016; pp. 464–468.

25. Mlekuž, D. Messy landscapes: Lidar and the practices of landscaping. In *Interpreting Archaeological Topography: 3D Data, Visualisation and Observation*; Opitz, R.S., Cowley, D.C., Eds.; Oxbow Books: Oxford, UK; Oakville, ON, Canada, 2013; pp. 88–99.

26. Mlekuž, D. Messy landscapes manifesto. *AARGnews* **2012**, *44*, 22–23.

27. Darvill, T. Pathways to a Panoramic Past: A Brief History of Landscape Archaeology in Europe. In *Handbook of Landscape Archaeology*, 1st ed.; David, B., Thomas, J., Eds.; Left Coast: Walnut Creek, CA, USA, 2010; pp. 60–76.

28. Aston, M.; Rowley, T. *Landscape Archaeology. An Introduction to Fieldwork Techniques on Post-Roman Landscapes*; David and Charles: Newton Abbot, UK, 1974.

29. Yamin, R.; Metheny, K.B. *Landscape Archaeology. Reading and Interpreting the American Historical Landscape*; University of Tennessee Press: Knoxville, TN, USA, 1996.

30. Ingold, T. The temporality of the landscape. *World Archaeol.* **1993**, *25*, 152–174. [CrossRef]

31. Johnson, M. *Ideas of Landscape*; Blackwell: Oxford, UK, 2007.

32. Knapp, A.B.; Ashmore, W. Archaeological Landscapes: Constructed, Conceptualized, Ideational. In *Archaeologies of Landscape: Contemporary Perspectives*; Ashmore, W., Knapp, A.B., Eds.; Blackwell: Malden, MA, USA, 1999; pp. 1–30.

33. Ebert, J.I. *Distributional Archaeology*, 1st ed.; University of New Mexico Press: Albuquerque, NM, USA, 1992.
34. Thomas, D.H. Nonsite Sampling in Archaeology: Up the Creek Without a Site? In *Sampling in Archaeology*; Mueller, J.W., Ed.; University of Arizona Press: Tucson, AZ, USA, 1975; pp. 61–81.
35. Dunnell, R.C. The Notion Site. In *Space, Time, and Archaeological Landscapes*; Rossignol, J., Wandsnider, L., Eds.; Springer: Boston, MA, USA, 1992; pp. 21–41.
36. Dunnell, R.C.; Dancey, W.S. The Siteless Survey: A regional Scale Data Collection Strategy. In *Advances in Archaeological Method and Theory*; Schiffer, M.B., Ed.; Academia Press: New York, NY, USA, 1983; Volume 6, pp. 267–287.
37. Foley, R. Off-site archaeology: An alternative approach for the short-sited. In *Pattern of the Past: Studies in Honour of David Clarke*; Hodder, I., Isaac, G., Hammond, N., Eds.; Cambridge University Press: Cambridge, UK, 1981; pp. 157–183.
38. Bradley, R. *An Archaeology of Natural Places*; Routledge: Abingdon, UK, 2000.
39. Tilley, C.Y. *A Phenomenology of Landscape. Places, Paths, and Monuments*; Berg: Oxford, UK, 1994.
40. Gramsch, A. Landscape Archaeology: Of Making and Seeing. *J. Eur. Archaeol.* **2013**, *4*, 19–38. [CrossRef]
41. Herring, P. The past informs the future; Landscape archaeology and historic landscape characterisation in the UK. In *Landscape Archaeology between Art and Science: From a Multi- to an Interdisciplinary Approach*; Kluiving, S.J., Gutmann-Bond, E.B., Eds.; Amsterdam University Press: Amsterdam, The Netherlands, 2012; pp. 485–501.
42. Thomas, J. Archaeology, Landscape, and Dwelling. In *Handbook of Landscape Archaeology*, 1st ed.; David, B., Thomas, J., Eds.; Left Coast: Walnut Creek, CA, USA, 2010; pp. 300–306.
43. Bender, B. Introduction: Landscape—Meaning and action. In *Landscape: Politics and Perspectives*; Bender, B., Ed.; Berg: Providence, RI, USA; Oxford, UK, 1993; pp. 1–18.
44. Boaz, J.S.; Uleberg, E. The potential of GIS-based studies of Iron Age cultural landscapes in eastern Norway. In *Archaeology and Geographical Information Systems: A European Perspective*; Lock, G.R., Stančič, Z., Eds.; Taylor & Francis: London, UK, 1995; pp. 249–259.
45. Mlekuž, D. Skin Deep: LiDAR and Good Practice of Landscape Archaeology. In *Good Practice in Archaeological Diagnostics: Non-Invasive Survey of Complex Archaeological Sites*; Corsi, C., Slapšak, B., Vermeulen, F., Eds.; Springer: Cham, Switzerland, 2013; pp. 113–129.
46. David, B.; Thomas, J. (Eds.) *Handbook of Landscape Archaeology*, 1st ed.; Left Coast: Walnut Creek, CA, USA, 2010.
47. Doneus, M. *Die Hinterlassene Landschaft—Prospektion und Interpretation in der Landschaftsarchäologie*; Verlag der Österreichischen Akademie der Wissenschaften: Wien, Austria, 2013.
48. Ashmore, W.; Knapp, A.B. (Eds.) *Archaeologies of Landscape: Contemporary Perspectives*; Blackwell: Malden, MA, USA, 1999.
49. Novaković, P. *Osvajanje Prostora. Razvoj Prostorske in Krajinske Arheologije*; Filozofska Fakulteta: Ljubljana, Slovenia, 2003.
50. Denham, T. Environmental Archaeology: Interpreting Practices-in-the-Landscape through Geoarchaeology. In *Handbook of Landscape Archaeology*, 1st ed.; David, B., Thomas, J., Eds.; Left Coast: Walnut Creek, CA, USA, 2010; pp. 468–481.
51. Barber, M. *A History of Aerial Photography and Archaeology. Mata Hari's Glass Eye and Other Stories*; English Heritage: Swindon, UK, 2011.
52. Gojda, M. (Ed.) *Ancient Landscape, Settlement Dynamics and Non-Destructive Archaeology. Czech Research Project 1997–2002 (Dávnověká Krajina a Sídla ve Světle Nedestruktivní Archeologie: Český Výzkumný Projekt 1997–2002)*; Academia: Prague, Czech Republic, 2004.
53. Sever, T.L. Remote Sensing Methods. In *Science and Technology in Historic Preservation*; Williamson, R.A., Nickens, P.R., Eds.; Springer: New York, NY, USA, 2000; pp. 21–51.
54. Millican, K. The Outside Inside: Combining Aerial Photographs, Cropmarks and Landscape Experience. *J. Archaeol. Method Theory* **2012**, *19*, 548–563. [CrossRef]
55. Hu, D. Advancing Theory? Landscape Archaeology and Geographical Information Systems. *Pap. Inst. Archaeol.* **2011**, *21*, 80–90. [CrossRef]
56. Pasquinucci, M.; Trément, F. (Eds.) *Non-Destructive Techniques Applied to Landscape Archaeology*; Oxbow Books: Oxford, UK, 2000.

57. Frachetti, M. Digital archaeology and the scalar structure of pastoral landscapes: Modeling mobile societies of prehistoric central Aasia. In *Digital Archaeology: Bridging Method and Theory*; Daly, P.T., Evans, T.L., Eds.; Routledge: London, UK, 2006; pp. 128–147.

58. Bowden, M. (Ed.) *Unravelling the Landscape. An Inquisitive Approach to Archaeology*; The History Press: Stroud, UK, 2013.

59. Haupt, P. *Landschaftsarchäologie. Eine Einführung*; WBG (Wissenschaftliche Buchgesellschaf): Darmstadt, Germany, 2012.

60. Campana, S. Towards mapping the archaeological continuum. New perspectives and current limitationsin Planning-Led-Archaeology in Italy. In *Looking to the Future, Caring for the Past: Preventive Archaeology in Theory and Practice*; Boschi, F., Ed.; Bononia University Press: Bologna, Italy, 2016; pp. 27–40.

61. Vermeulen, F. Aerial survey in an Italian landscape: From archaeological site-detection and monitoring to prevention and management. In *Looking to the Future, Caring for the Past: Preventive Archaeology in Theory and Practice*; Boschi, F., Ed.; Bononia University Press: Bologna, Italy, 2016; pp. 135–146.

62. Johnson, J.K. (Ed.) *Remote Sensing in Archaeology. An Explicitly North American Perspective*; University of Alabama Press: Tuscaloosa, AL, USA, 2006.

63. David, A. Finding Sites. In *Archaeology in Practice: A Student Guide to Archaeological Analyses*; Balme, J., Paterson, A., Eds.; Blackwell Publishing: Malden, MA, USA, 2006; pp. 1–38.

64. Crutchley, S.; Crow, P. *The Light Fantastic. Using Airborne Lidar in Archaeological Survey*, 1st ed.; English Heritage: Swindon, UK, 2010.

65. Doneus, M.; Briese, C.; Fera, M.; Janner, M. Archaeological prospection of forested areas using full-waveform airborne laser scanning. *J. Archaeol. Sci.* **2008**, *35*, 882–893. [CrossRef]

66. Doneus, M.; Doneus, N.; Briese, C.; Pregesbauer, M.; Mandlburger, G.; Verhoeven, G.J.J. Airborne Laser Bathymetry—Detecting and recording submerged archaeological sites from the air. *J. Archaeol. Sci.* **2013**, *40*, 2136–2151. [CrossRef]

67. Doneus, M.; Briese, C. Airborne Laser Scanning in forested areas—Potential and limitations of an archaeological prospection technique. In *Remote Sensing for Archaeological Heritage Management, Proceedings of the 11th EAC Heritage Management Symposium, Reykjavík, Iceland, 25–27 March 2010*; Cowley, D.C., Ed.; Europae Archaeologia Consilium (EAC), Association Internationale sans But Lucratif (AISBL): Brussels, Belgium, 2011; pp. 59–76.

68. Opitz, R.S.; Cowley, D.C. (Eds.) *Interpreting Archaeological Topography: 3D Data, Visualisation and Observation*; Oxbow Books: Oxford, UK; Oakville, ON, Canada, 2013.

69. Briese, C.; Pfennigbauer, M.; Lehner, H.; Ullrich, A.; Wagner, W.; Pfeifer, N. Radiometric calibration of multi-wavelength airborne laser scanning data. In Proceedings of the XXII ISPRS Congress, Technical Commission VII. Imaging a Sustainable Future, Melbourne, Australia, 25 August–1 September 2012; Shortis, M.R., Wagner, W., Hyyppä, J., Eds.; International Society for Photogrammetry and Remote Sensing (ISPRS): Melbourne, Australia, 2012; pp. 335–340.

70. Briese, C.; Doneus, M.; Verhoeven, G.J.J. Radiometric calibration of ALS data for archaeological Interpretation. In *Archaeological Prospection, Proceedings of the 10th International Conference on Archaeological Prospection, Vienna, Austria, 29 May–2 June 2013*; Neubauer, W., Trinks, I., Salisbury, R.B., Einwögerer, C., Eds.; Austrian Academy of Sciences: Vienna, Austria, 2013; pp. 427–429.

71. Wilson, D.R. *Air Photo Interpretation for Archaeologists*, 2nd ed.; Tempus: Stroud, UK, 2000.

72. Palmer, J.M.; Grant, B.G. *The Art of Radiometry*; SPIE Press: Bellingham, DC, USA, 2010.

73. Ohno, Y. Basic concepts in photometry, radiometry and colorimetry. In *Handbook of Optoelectronics*; Dakin, J.P., Brown, R.G.W., Eds.; Taylor & Francis: Boca Raton, FL, USA, 2006; pp. 287–305.

74. Schowengerdt, R.A. *Remote Sensing. Models and Methods for Image Processing*, 3rd ed.; Academic Press: Burlington, ON, Canada, 2007.

75. Lasaponara, R.; Masini, N. (Eds.) *Satellite Remote Sensing. A New Tool for Archaeology*; Springer: Dordrecht, The Netherlands, 2012.

76. Parcak, S.H. *Satellite Remote Sensing for Archaeology*; Routledge: London, UK; New York, NY, USA, 2009.

77. Comer, D.C.; Harrower, M.J. (Eds.) *Mapping Archaeological Landscapes from Space*; Springer: New York, NY, USA, 2013.

78. Hanson, W.S.; Oltean, I.A. (Eds.) *Archaeology from Historical Aerial and Satellite Archives*; Springer: New York, NY, USA, 2013.

79. Scardozzi, G. An introduction to satellite remote sensing in archaeology: State of the art, methods, and applications. In *Looking to the Future, Caring for the Past: Preventive Archaeology in Theory and Practice*; Boschi, F., Ed.; Bononia University Press: Bologna, Italy, 2016; pp. 217–239.

80. Drennan, R.D. *Statistics for Archaeologists. A Common Sense Approach*, 2nd ed.; Springer: Dordrecht, The Netherlands, 2009.

81. Clarke, D.L. Archaeology: The loss of innocence. *Antiquity* **1973**, *47*, 6–18. [CrossRef]

82. Cowgill, G.L. Some Sampling and Reliability Problems in Archaeology. In *Archéologie et Calculateurs: Problèmes Sémiologiques et Mathématiques, Proceedings of the Colloques Internationaux du Centre National de la Recherche Scientifique: Sciences Humaines, Marseille, France, 7–12 April 1969*; Gardin, J.-C., Ed.; Centre National de la Récherche Scientifique: Paris, France, 1970; pp. 161–175.

83. Collins, M.B. Sources of bias in processual data: An appraisal. In *Sampling in Archaeology*; Mueller, J.W., Ed.; University of Arizona Press: Tucson, AZ, USA, 1975; pp. 26–32.

84. Schiffer, M.B. *Formation Processes of the Archaeological Record*; University of Utah Press: Salt Lake City, UT, USA, 1996.

85. Grabe, M. *Measurement Uncertainties in Science and Technology*; Springer: Cham, Germany, 2014.

86. Nativ, A. No Compensation Needed: On Archaeology and the Archaeological. *J. Archaeol. Method Theory* **2017**, *24*, 659–675. [CrossRef]

87. Cherry, J.F.; Gamble, C.; Shennan, S. (Eds.) *Sampling in Contemporary British Archaeology*; British Archaeological Reports (B.A.R.): Oxford, UK, 1978.

88. Orton, C.R. *Sampling in Archaeology*; Cambridge University Press: Cambridge, UK, 2000.

89. Mueller, J.W. (Ed.) *Sampling in Archaeology*; University of Arizona Press: Tucson, AZ, USA, 1975.

90. Schiffer, M.B. Toward the Identification of Formation Processes. *Am. Antiq.* **1983**, *48*, 675–706. [CrossRef]

91. VanPool, T.L.; Leonard, R.D. *Quantitative Analysis in Archaeology*; Wiley-Blackwell: Chichester, UK, 2011.

92. Lyman, R.L.; VanPool, T.L. Metric Data in Archaeology: A Study of Intra-analyst and Inter-analyst Variation. *Am. Antiq.* **2009**, *74*, 485–504. [CrossRef]

93. Bentley, R.A.; Maschner, H.D.G.; Chippindale, C. (Eds.) *Handbook of Archaeological Theories*; AltaMira Press: Lanham, MD, USA, 2008.

94. Wallace, S. *Contradictions of Archaeological Theory. Engaging Critical Realism and Archaeological Theory*; Routledge: Abingdon, UK; New York, NY, USA, 2011.

95. Bintliff, J.L. The Death of Archaeological Theory? In *The Death of Archaeological Theory*; Bintliff, J.L., Pearce, M., Eds.; Oxbow: Oxford, UK, 2011; pp. 7–22.

96. Johnson, M.H. On the nature of theoretical archaeology and archaeological theory. *Archaeol. Dialogues* **2006**, *13*, 117–132. [CrossRef]

97. Pearce, M. Have Rumours of the 'Death of Theory' been Exaggerated? In *The Death of Archaeological Theory*; Bintliff, J.L., Pearce, M., Eds.; Oxbow: Oxford, UK, 2011; pp. 80–89.

98. Rączkowski, W. Aerial Archaeology. In *Encyclopedia of Global Archaeology*; Smith, C., Ed.; Springer: New York, NY, USA, 2014; pp. 33–38.

99. Hodder, I. *The Archaeological Process. An Introduction*; Blackwell: Oxford, UK, 1999.

100. Banaszek, Ł. *Przeszłe Krajobrazy w Chmurze Punktów*; Wydawnictwo Naukowe Uniwersytetu im. Adama Mickiewicza: Poznań, Poland, 2015.

101. Chapman, R.; Wylie, A. *Evidential Reasoning in Archaeology*; Bloomsbury Academic: London, UK; New York, NY, USA, 2016.

102. Gojda, M. The Archaeology of Lowlands: A Few Remarks on the Methodology of Aerial Survey. In *Landscape Ideologies*; Meier, T., Ed.; Archaeolingua Alapítvány: Budapest, Hungary, 2006; pp. 117–123.

103. Muller, R.A. *Now. The Physics of Time*, 1st ed.; W.W. Norton & Company: New York, NY, USA, 2016.

104. Stewart, I. One hundred and one dimensions. *New Sci.* **1995**, *148*, 28–31.

105. Born, M.; Wolf, E. *Principles of Optics. Electromagnetic Theory of Propagation, Interference and Diffraction of Light*, 7th ed.; Cambridge University Press: Cambridge, UK, 1999.

106. Verhoeven, G.J.J. Basics of photography for cultural heritage imaging. In *3D Recording, Documentation and Management of Cultural Heritage*; Stylianidis, E., Remondino, F., Eds.; Whittles Publishing: Caithness, UK, 2016; pp. 127–251.

107. Crawford, O.G.S. *Air Survey and Archaeology*; Ordnance Survey: Southampton, UK, 1924.

108. Crawford, O.G.S.; Keiller, A. *Wessex from the Air*; Oxford University Press: Oxford, UK, 1928.

109. Crawford, O.G.S. Air Photographs of the Middle East: A Paper Read at the Evening Meeting of the Society on 18 March 1929. *Geogr. J.* **1929**, *73*, 497–509. [CrossRef]

110. Wilson, D.R. Photographic Techniques in the Air. In *Aerial Reconnaissance for Archaeology*; Wilson, D.R., Ed.; Council for British Archaeology: London, UK, 1975; pp. 12–31.

111. Crawshaw, A. Oblique Aerial Photography-Aircraft, Cameras and Films. In Proceedings of the Luftbildarchäologie in Ost- und Mitteleuropa/Aerial Archaeoloy in Eastern and Central Europe: Internationales Symposium, Kleinmachnow, Germany, 26–30 September 1994; Kunow, J., Ed.; Verlag Brandenburgisches Landesmuseum für Ur- und Frühgeschichte: Potsdam, Germany, 1995; pp. 67–76.

112. Harman, W.E., Jr.; Miller, R.H.; Park, W.S.; Webb, J.P. Aerial photography. In *Manual of Photogrammetry*, 3rd ed.; Thompson, M.M., Eller, R.C., Radlinski, W.A., Speert, J.L., Eds.; American Society of Photogrammetry: Falls Church, VA, USA, 1966; Volume I, pp. 195–242.

113. Schneider, S. *Luftbild und Luftbildinterpretation*; Walter de Gruyter: Berlin, Germany; New York, NY, USA, 1974.

114. Palmer, R. If they used their own photographs they would not take them like that. In *From the Air: Understanding Aerial Archaeology*; Brophy, K., Cowley, D.C., Eds.; Tempus: Stroud, UK, 2005; pp. 94–116.

115. Read, R.E.; Graham, R. *Manual of Aerial Survey. Primary Data Acquisition*; CRC Press/Whittles Publishing: Boca Raton, FL, USA, 2002.

116. The British Academy. *Aerial Survey for Archaeology. Report of a British Academy Working Party 1999*; The British Academy: London, UK, 2001.

117. Musson, C.; Palmer, R.; Campana, S. *Flights into the Past. Aerial Photography, Photo Interpretation and Mapping for Archaeology*; AARG—ArcLand: Heidelberg, Germany, 2013.

118. Palmer, R. Aerial archaeology and sampling. In *Sampling in Contemporary British Archaeology*; Cherry, J.F., Gamble, C., Shennan, S., Eds.; British Archaeological Reports (B.A.R.): Oxford, UK, 1978; pp. 129–148.

119. Wilson, D.R. Bias in aerial reconnaissance. In *From the Air: Understanding Aerial Archaeology*; Brophy, K., Cowley, D.C., Eds.; Tempus: Stroud, UK, 2005; pp. 64–72.

120. St Joseph, J.K.S. Air Reconnaissance of Roman Scotland, 1939-75. *Glasg. Archaeol. J.* **1976**, *4*, 1–28. [CrossRef]

121. Jones, R.H. The advantages of bias in Roman studies. In *From the Air: Understanding Aerial Archaeology*; Brophy, K., Cowley, D.C., Eds.; Tempus: Stroud, UK, 2005; pp. 86–93.

122. Gould, R.A. Archaeological Survey by Air: A Case from the Australian Desert. *J. Field Archaeol.* **1987**, *14*, 431–443. [CrossRef]

123. Coleman, S. Taking Advantage: Vertical Aerial Photographs Commissioned for Local Authorities. In *Populating Clay Landscapes*; Mills, J., Palmer, R., Eds.; Tempus: Stroud, UK, 2007; pp. 28–33.

124. Doneus, M. Vertical and Oblique Photographs. *AARGnews* **2000**, *20*, 33–39.

125. Mills, J. Bias and the World of the Vertical Aerial Photograph. In *From the Air: Understanding Aerial Archaeology*; Brophy, K., Cowley, D.C., Eds.; Tempus: Stroud, UK, 2005; pp. 117–126.

126. Palmer, R. Seventy-Five Years v. Ninety Minutes: Implications of the 1996 Bedfordshire Vertical Aerial Survey on our Perceptions of Clayland Archaeology. In *Populating Clay Landscapes*; Mills, J., Palmer, R., Eds.; Tempus: Stroud, UK, 2007; pp. 88–103.

127. Palmer, R. Editorial. *AARGnews* **1996**, *13*, 3. [CrossRef]

128. Cowley, D.C. A case study in the analysis of patterns of aerial reconnaissance in a lowland area of southwest Scotland. *Archaeol. Prospect.* **2002**, *9*, 255–265. [CrossRef]

129. Doneus, M. On the archaeological use of vertical photographs. *AARGnews* **1997**, *15*, 23–27.

130. Slater, P.N.; Doyle, F.J.; Fritz, N.L.; Welch, R. Photographic systems for remote sensing. In *Manual of Remote Sensing: Volume 1: Theory, Instruments and Techniques*, 2nd ed.; Colwell, R.N., Simonett, D.S., Ulaby, F.T., Eds.; American Society of Photogrammetry: Falls Church, VA, USA, 1983; pp. 231–291.

131. Estes, J.E.; Hajic, E.J.; Tinney, L.R.; Carver, L.G.; Cosentino, M.J.; Mertz, F.C.; Pazner, M.I.; Ritter, L.R.; Sailer, C.T.; Stow, D.A.; et al. Fundamentals of Image Analyis: Analysis of Visible and Thermal Infrared Data. In *Manual of Remote Sensing: Theory, Instruments and Techniques*, 2nd ed.; Colwell, R.N., Simonett, D.S., Ulaby, F.T., Eds.; American Society of Photogrammetry: Falls Church, VA, USA, 1983; Volume 1, pp. 987–1124.

132. Spurr, S.H. *Photogrammetry and Photo-Interpretation. With a Section on Applications to Forestry*, 2nd ed.; The Ronald Press Company: New York, NY, USA, 1960.

133. Lemmens, M. Digital Oblique Aerial Cameras (1): A Survey of Features and Systems. *GIM Int.* **2014**, *28*, 20–21, 23–25.

134. Nurminen, K. Oblique aerial photographs—An "old-new" data source. *Photogramm. J. Finl.* **2015**, *24*, 1–19. [CrossRef]
135. Remondino, F.; Gerke, M. Oblique Aerial Imagery—A Review. In *Photogrammetric Week 2015*; Fritsch, D., Ed.; Wichmann/VDE Verlag: Belin/Offenbach, Germany, 2015; pp. 75–83.
136. IGI mbH. Penta DigiCAM. Available online: http://www.igi.eu/penta-digicam.html (accessed on 4 March 2016).
137. Microsoft. UltraCam Osprey Prime II/Prime Lite. Available online: https://www.microsoft.com/en-us/Ultracam/UltraCamOsprey.aspx (accessed on 4 March 2016).
138. Rupnik, E.; Nex, F.; Toschi, I.; Remondino, F. Aerial multi-camera systems: Accuracy and block triangulation issues. *ISPRS J. Photogramm. Remote Sens.* **2015**, *101*, 233–246. [CrossRef]
139. Paine, D.P.; Kiser, J.D. *Aerial Photography and Image Interpretation*, 3rd ed.; Wiley: Hoboken, NJ, USA, 2012.
140. General Staff. *Notes on the Interpretation of Aeroplane Photographs*; Revised March 1917; General Staff War Office: London, UK, 1917.
141. Haala, N.; Rothermel, M.; Cavegn, S. Extracting 3D urban models from oblique aerial images. In Proceedings of the 2015 Joint Urban Remote Sensing Event (JURSE), Lausanne, Switzerland, 30 March–1 April 2015; IEEE: Piscataway, NJ, USA, 2015; pp. 1–4.
142. Xiao, J.; Gerke, M.; Vosselman, G. Building extraction from oblique airborne imagery based on robust façade detection. *ISPRS J. Photogramm. Remote Sens.* **2012**, *68*, 56–68. [CrossRef]
143. Brophy, K.; Cowley, D.C. From the Air—An introduction. In *From the Air: Understanding Aerial Archaeology*; Brophy, K., Cowley, D.C., Eds.; Tempus: Stroud, UK, 2005; pp. 11–23.
144. Verhoeven, G.J.J. Near-Infrared Aerial Crop Mark Archaeology: From its Historical Use to Current Digital Implementations. *J. Archaeol. Method Theory* **2012**, *19*, 132–160. [CrossRef]
145. Verhoeven, G.J.J. Exploring the Edges of the Unseen: An Attempt to Digital Aerial UV Photography. In *Remote Sensing for Archaeology and Cultural Heritage Management, Proceedings of the 1st International EARSeL Workshop, CNR, Rome, 30 September–4 October 2008*; Lasaponara, R., Masini, N., Eds.; Arracne: Rome, Italy, 2008; pp. 79–83.
146. Verhoeven, G.J.J.; Schmitt, K.D. An attempt to push back frontiers—Digital near-ultraviolet aerial archaeology. *J. Archaeol. Sci.* **2010**, *37*, 833–845. [CrossRef]
147. Carter, G.A.; Miller, R.L. Early detection of plant stress by digital imaging within narrow stress-sensitive wavebands. *Remote Sens. Environ.* **1994**, *50*, 295–302. [CrossRef]
148. Donoghue, D.N.M. Multispectral Remote Sensing for Archaeology. In *Remote Sensing in Archaeology: XI Ciclo di Lezioni Sulla Ricerca Applicata in Archeologia, Certosa di Pontignano (Siena), 6–11 Dicembre 1999*; Campana, S., Forte, M., Eds.; All'Insegna del Giglio: Firenze, Italy, 2001; pp. 181–192.
149. Powlesland, D.; Lyall, J.; Hopkinson, G.; Donoghue, D.; Beck, M.; Harte, A.; Stott, D. Beneath the sand: Remote sensing, archaeology, aggregates and sustainability: A case study from Heslerton, the Vale of Pickering, North Yorkshire, UK. *Archaeol. Prospect.* **2006**, *13*, 291–299. [CrossRef]
150. Challis, K.; Kincey, M.; Howard, A.J. Airborne remote sensing of valley floor geoarchaeology using Daedalus ATM and CASI. *Archaeol. Prospect.* **2009**, *16*, 17–33. [CrossRef]
151. Aqdus, S.A.; Drummond, J.; Hanson, W.S. Discovering Archaeological Cropmarks: A Hyperspectral Approach. *Int. Arch. Photogramm. Remote Sens. Spat. Inf. Sci.* **2008**, *37*, 361–365.
152. Winterbottom, S.J.; Dawson, T. Airborne multi-spectral prospection for buried archaeology in mobile sand dominated systems. *Archaeol. Prospect.* **2005**, *12*, 205–219. [CrossRef]
153. Van der Meer, F.D.; de Jong, S.M. (Eds.) *Imaging Spectrometry. Basic Principles and Prospective Applications*; Kluwer Academic Publishers: Dordrecht, The Netherlands, 2001.
154. Shrestha, R.; Mansouri, A.; Hardeberg, J.Y. Multispectral imaging using a stereo camera: Concept, design and assessment. *EURASIP J. Adv. Signal Process.* **2011**, *57*, 1–15. [CrossRef]
155. Aqdus, S.A.; Hanson, W.S.; Drummond, J. The potential of hyperspectral and multi-spectral imagery to enhance archaeological cropmark detection: A comparative study. *J. Archaeol. Sci.* **2012**, *39*, 1915–1924. [CrossRef]
156. Bennett, R.; Welham, K.; Hill, R.A.; Ford, A.L.J. The Application of Vegetation Indices for the Prospection of Archaeological Features in Grass-dominated Environments. *Archaeol. Prospect.* **2012**, *19*, 209–218. [CrossRef]
157. Cavalli, R.M.; Licciardi, G.A.; Chanussot, J. Detection of Anomalies Produced by Buried Archaeological Structures Using Nonlinear Principal Component Analysis Applied to Airborne Hyperspectral Image. *IEEE J. Sel. Top. Appl. Earth Obs. Remote Sens.* **2012**, *5*, 1–12. [CrossRef]

158. Verhoeven, G.J.J.; Doneus, M.; Atzberger, C.; Wess, M.; Ruš, M.; Pregesbauer, M.; Briese, C. New approaches for archaeological feature extraction of airborne imaging spectroscopy data. In *Archaeological Prospection, Proceedings of the 10th International Conference on Archaeological Prospection, Vienna, Austria, 29 May–2 June 2013*; Neubauer, W., Trinks, I., Salisbury, R.B., Einwögerer, C., Eds.; Austrian Academy of Sciences: Vienna, Austria, 2013; pp. 13–15.

159. Barnes, I. Aerial remote-sensing techniques used in the management of archaeological monuments on the British Army's Salisbury Plain Training Area, Wiltshire, UK. *Archaeol. Prospect.* **2003**, *10*, 83–90. [CrossRef]

160. Bassani, C.; Cavalli, R.M.; Goffredo, R.; Palombo, A.; Pascucci, S.; Pignatti, S. Specific spectral bands for different land cover contexts to improve the efficiency of remote sensing archaeological prospection: The Arpi case study. *J. Cult. Herit.* **2009**, *10*, e41–e48. [CrossRef]

161. Coren, F.; Visintini, D.; Prearo, G.; Sterzai, P. Integrating LiDAR Intensity Measures and Hyperspectral Data for Extracting of Cultural Heritage. In Proceedings of the Italy–Canada 2005 Workshop on 3D Digital imaging and Modeling: Applications of Heritage, Industry, Medicine and Land, Padua, Italy, 17–18 March 2005.

162. Emmolo, D.; Franco, V.; Lo Brutto, M.; Orlando, P.; Villa, B. Hyperspectral Techniques and GIS for Archaeological Investigation. In Proceedings of the ISPRS 2004 Commission IV—Geo-Imagery Bridging Continents. XXth ISPRS Congress, Istanbul, Turkey, 12–23 July 2004; Altan, O., Ed.; ISPRS: Istanbul, Turkey, 2004.

163. Forte, E.; Pipan, M.; Sugan, M. Integrated Geophysical Study of Archaeological Sites in the Aquileia Area. In Proceedings of the 1st Workshop on The New Technologies for Aquileia (NTA-2011), Aquileia, Italy, 2 May 2011; Roberto, V., Ed.; Department of Mathematics and Computer Science, University of Udine: Udine, Italy, 2011.

164. Merola, P.; Allegrini, A.; Bajocco, S. Hyperspectral MIVIS data to investigate the Lilybaeum (Marsala) Archaeological Park. In Proceedings of the Remote Sensing for Environmental Monitoring, GIS Applications, and Geology V, Bruges, Belgium, 19 September 2005; Ehlers, M., Michel, U., Eds.; Society of Photo-Optical Instrumentation Engineers (SPIE): Bellingham, WA, USA, 2005; pp. 212–222.

165. Pietrapertosa, C.; Vellico, M.; Sterzai, P.; Coren, F. Remote Sensing Applied to the Detection of Archaeological Buried Structures in the Aquileia Site. In Proceedings of the 27° Convegno Nazionale GNGTS—2008, Trieste, Italy, 8 October 2008; pp. 368–372.

166. Traviglia, A. Integration of MIVIS Hyperspectral Remotely Sensed Data and Geographical Information Systems to Study Ancient Landscapes: The Aquileia Case Study. *Agri Centuriati* **2005**, *2*, 139–170.

167. White, D.A. AVIRIS and Archaeology in Southern Arizona. In *AVIRIS Proceedings*; NASA Jet Propulsion Laboratory: Pasadena, CA, USA, 2003.

168. Doneus, M.; Verhoeven, G.J.J.; Atzberger, C.; Wess, M.; Ruš, M. New ways to extract archaeological information from hyperspectral pixels. *J. Archaeol. Sci.* **2014**, *52*, 84–96. [CrossRef]

169. Atzberger, C.; Wess, M.; Doneus, M.; Verhoeven, G.J.J. ARCTIS—A MATLAB® Toolbox for Archaeological Imaging Spectroscopy. *Remote Sens.* **2014**, *6*, 8617–8638. [CrossRef]

170. Carter, G.A. Ratios of leaf reflectances in narrow wavebands as indicators of plant stress. *Int. J. Remote Sens.* **1994**, *15*, 697–703. [CrossRef]

171. Bewley, R.; Rączkowski, W. Past achievements and prospects for the future development of aerial archaeology: An introduction. In *Aerial Archaeology: Developing Future Practice*; Bewley, R.H., Rączkowski, W., Eds.; IOS Press: Amsterdam, The Netherlands, 2002; pp. 1–8.

172. Brugioni, D.A. The Serendipity Effect of Aerial Reconnaissance. *Interdiscip. Sci. Rev.* **1989**, *14*, 16–28. [CrossRef]

173. Crawshaw, A. Letter. *AARGnews* **1997**, *14*, 59.

174. Gates, T. Recording upland landscapes: A personal account from Northumberland. In *From the Air: Understanding Aerial Archaeology*; Brophy, K., Cowley, D.C., Eds.; Tempus: Stroud, UK, 2005; pp. 127–140.

175. Wilson, D.R. Vertical versus oblique photography. *AARGnews* **2005**, *20* (Suppl. S1), 32–34.

176. Doody, M. Medium altitude aerial photographic survey in East Limerick and West Tipperary. *J. Irish Archaeol.* **2001**, *10*, 13–24.

177. Doneus, M. Vom Luftbild zur Karte. In *Aus der Luft-Bilder Unserer Geschichte: Luftbildarchäologie in Zentraleuropa: Katalog zur Ausstellung*; Oexle, J., Ed.; Landesamt für Archäologie mit Landesmuseum für Vorgeschichte: Dresden, Germany, 1997; pp. 38–45.

178. Warner, W.S.; Graham, R.W.; Read, R.E. *Small Format Aerial Photography*; Whittles Publishing: Caithness, UK, 1996.

179. Graham, R.; Koh, A. *Digital Aerial Survey. Theory and Practice*; CRC Press/Whittles Publishing: Boca Raton, FL, USA, 2002.

180. Bäumker, M.; Brechtken, R.; Heimes, F.-J.; Richter, T. Practical Experiences with a High-Precision Stabilised Camera Platform Based on INS/(D)GPS. In Proceedings of the First North American Symposium on Small Format Aerial Photography, Cloquet, MN, USA, 14–17 October 1997; American Society for Photogrammetry and Remote Sensing: Bethesda, MD, USA, 1997; pp. 45–54.

181. Mills, J.P.; Newton, I.; Graham, R.W. Aerial Photography for Survey Purposes with a High Resolution, Small Format, Digital Camera. *Photogramm. Rec.* **1996**, *15*, 575–587. [CrossRef]

182. Imhof, R.K.; Doolittle, R.C. Mapping from oblique photographs. In *Manual of Photogrammetry*, 3rd ed.; Thompson, M.M., Eller, R.C., Radlinski, W.A., Speert, J.L., Eds.; American Society of Photogrammetry: Falls Church, VA, USA, 1966; Volume II, pp. 875–917.

183. Verhoeven, G.J.J.; Doneus, M.; Briese, C.; Vermeulen, F. Mapping by matching: A computer vision-based approach to fast and accurate georeferencing of archaeological aerial photographs. *J. Archaeol. Sci.* **2012**, *39*, 2060–2070. [CrossRef]

184. Verhoeven, G.J.J.; Sevara, C.; Karel, W.; Ressl, C.; Doneus, M.; Briese, C. Undistorting the past: New techniques for orthorectification of archaeological aerial frame imagery. In *Good Practice in Archaeological Diagnostics: Non-Invasive Survey of Complex Archaeological Sites*; Corsi, C., Slapšak, B., Vermeulen, F., Eds.; Springer: Cham, Switzerland, 2013; pp. 31–67.

185. Verhoeven, G.J.J. Mesh Is More—Using All Geometric Dimensions for the Archaeological Analysis and Interpretative Mapping of 3D Surfaces. *J. Archaeol. Method Theory* **2016**, 1–35. [CrossRef]

186. Verhoeven, G.J.J.; Vermeulen, F. Engaging with the Canopy: Multi-Dimensional Vegetation Mark Visualisation Using Archived Aerial Images. *Remote Sens.* **2016**, *8*, 752. [CrossRef]

187. Kennedy, D.L.; Bewley, R.H. *Ancient Jordan from the Air*; Council for British Research in the Levant: London, UK, 2004.

188. Campana, S. Archaeological site detection and mapping: Some thoughts on differing scales of detail and archaeological 'non-visibility'. In *Seeing the Unseen: Geophysics and Landscape Archaeology*; Campana, S., Piro, S., Eds.; CRC Press/Balkema: Boca Raton, FL, USA; Leiden, The Netherlands, 2009; pp. 5–26.

189. Deravignone, L.; Blankholm, H.P.; Pizziolo, G. Predictive Modeling and Artificial Neural Networks (ANN): From Model to Survey. In *Mathematics and Archaeology*; Barceló, J.A., Bogdanovic, I., Eds.; CRC Press: Boca Raton, FL, USA, 2015; pp. 335–351.

190. Buschmann, C.; Nagel, E. In vivo spectroscopy and internal optics of leaves as basis for remote sensing of vegetation. *Int. J. Remote Sens.* **1993**, *14*, 711–722. [CrossRef]

191. Datt, B. Visible/near infrared reflectance and chlorophyll content in Eucalyptus leaves. *Int. J. Remote Sens.* **1999**, *20*, 2741–2759. [CrossRef]

192. Gitelson, A.A.; Merzlyak, M.N. Quantitative estimation of chlorophyll-*a* using reflectance spectra: Experiments with autumn chestnut and maple leaves. *J. Photochem. Photobiol. B Biol.* **1994**, *22*, 247–252. [CrossRef]

193. Carter, G.A.; Estep, L.; Muttiah, R.S. General spectral characteristics of leaf reflectance responses to plant stress and their manifestation at the landscape scale. In *From Laboratory Spectroscopy to Remotely Sensed Spectra of Terrestrial Ecosystems*; Muttiah, R.S., Ed.; Kluwer Academic Publishers: Dordrecht, The Netherlands, 2002; pp. 271–293.

194. Agapiou, A.; Hadjimitsis, D.G.; Georgopoulos, A.; Sarris, A.; Alexakis, D.D. Towards an Archaeological Index: Identification of the Spectral Regions of Stress Vegetation due to Buried Archaeological Remains. In *Progress in Cultural Heritage Preservation, Proceedings of the 4th International Conference, EuroMed 2012, Lemessos, Cyprus, 29 October–3 November 2012*; Ioannides, M., Fritsch, D., Leissner, J., Davies, R., Remondino, F., Caffo, R., Eds.; Springer: Berlin/Heidelberg, Germany, 2012; pp. 129–138.

195. Verhoeven, G.J.J.; Doneus, M. Balancing on the Borderline—A Low-cost Approach to Visualize the Red-edge Shift for the Benefit of Aerial Archaeology. *Archaeol. Prospect.* **2011**, *18*, 267–278. [CrossRef]

196. Curcio, J.A.; Petty, C.C. The Near-Infrared Absorption Spectrum of Liquid Water. *J. Opt. Soc. Am.* **1951**, *41*, 302–304. [CrossRef]

197. Avery, T.E.; Lyons, T.R. *Remote Sensing. Aerial and Terrestrial Photography for Archeologists*; Cultural Resources Management Division: Washington, DC, USA, 1981.

198. Jones, R.J.A.; Evans, R. Soil and Crop Marks in the Recognition of Archaeological Sites by Air Photography. In *Aerial Reconnaissance for Archaeology*; Wilson, D.R., Ed.; Council for British Archaeology: London, UK, 1975; pp. 1–11.

199. Sandmeier, S.R.; Itten, K.I. A Field Goniometer System (FIGOS) for Acquisition of Hyperspectral BRDF Data. *IEEE Trans. Geosci. Remote Sens.* **1999**, *37*, 978–986. [CrossRef]

200. Barceló, J.A. *Computational Intelligence in Archaeology*; Information Science Reference: Hershey, PA, USA, 2009.

201. McCoy, M.D.; Ladefoged, T.N. New Developments in the Use of Spatial Technology in Archaeology. *J. Archaeol. Res.* **2009**, *17*, 263–295. [CrossRef]

202. Traviglia, A.; Cowley, D.C.; Lambers, K. Finding common ground: Human and computer vision in archaeological prospection. *AARGnews* **2016**, *53*, 11–24.

203. Bennett, R.; Cowley, D.C.; De Laet, V. The data explosion: Tackling the taboo of automatic feature recognition in airborne survey data. *Antiquity* **2014**, *88*, 896–905. [CrossRef]

204. Cowley, D.C. In with the new, out with the old? Auto-extraction for remote sensing archaeology. In Proceedings of the Remote Sensing of the Ocean, Sea Ice, Coastal Waters, and Large Water Regions 2012 Conference, Edinburgh, UK, 26–27 September 2012; Bostater, C.R., Jr., Mertikas, S.P., Neyt, X., Nichol, C., Cowley, D.C., Bruyant, J.-P., Eds.; SPIE: Bellingham, DC, USA, 2012; p. 853206.

205. Lambers, K.; Zingman, I. Towards Detection of Archaeological Objects in High-Resolution Remotely Sensed Images: The Silvretta Case Study. In *Archaeology in the Digital Era. Volume II: E-Papers from the 40th Conference on Computer Applications and Quantitative Methods in Archaeology, Southampton, 26–30 March 2012*; Earl, G.P., Sly, T., Chrysanthi, A., Murrieta-Flores, P., Papadopoulos, C., Romanowska, I., Wheatley, D., Eds.; Amsterdam University Press: Amsterdam, The Netherlands, 2013; pp. 781–791.

206. Hanson, W.S. The future of aerial archaeology in Europe. *Photo Interprét. Eur. J. Appl. Remote Sens.* **2010**, *46*, 3–11.

207. Sevara, C.; Pregesbauer, M.; Doneus, M.; Verhoeven, G.J.J.; Trinks, I. Pixel versus object—A comparison of strategies for the semi-automated mapping of archaeological features using airborne laser scanning data. *J. Archaeol. Sci.* **2016**, *5*, 485–498. [CrossRef]

208. Llobera, M. Working the digital: Some thoughts from landscape archaeology. In *Material Evidence: Learning from Archaeological Practice*; Chapman, R., Wylie, A., Eds.; Routledge: Abingdon, UK; New York, NY, USA, 2015; pp. 173–188.

209. Cowley, D.C. Creating the Cropmark Archaeological Record in East Lothian, South-East Scotland. In *Prehistory without Borders: The Prehistoric Archaeology of the Tyne-Forth Region*; Crellin, R., Fowler, C., Tipping, R., Eds.; Oxbow Books: Oxford, UK; Havertown, PA, USA, 2016; pp. 59–70.

210. Cowley, D.C. Remote sensing for archaeology and heritage management—Site discovery, interpretation and registration. In *Remote Sensing for Archaeological Heritage Management, Proceedings of the 11th EAC Heritage Management Symposium, Reykjavík, Iceland, 25–27 March 2010*; Cowley, D.C., Ed.; Europae Archaeologia Consilium (EAC), Association Internationale sans But Lucratif (AISBL): Brussels, Belgium, 2011; pp. 43–55.

211. Taylor, C.C. Aerial photography and the field archaeologist. In *Aerial Reconnaissance for Archaeology*; Wilson, D.R., Ed.; Council for British Archaeology: London, UK, 1975; pp. 136–141.

212. Löcker, K.; Kucera, M.; Trinks, I.; Neubauer, W. Successfully falsified . . . on epistomological problems of archaeological excavations and geophysical surveys. *Archaeol. Pol.* **2015**, *53*, 222–224.

213. Seren, S.; Trinks, I.; Hinterleitner, A.; Neubauer, W. The anomaly that wasn't there—On the visibility of archaeological prospection anomalies and their causative structures in the subsurface. In *Archaeological Prospection, Proceedings of the 10th International Conference on Archaeological Prospection, Vienna, Austria, 29 May–2 June 2013*; Neubauer, W., Trinks, I., Salisbury, R.B., Einwögerer, C., Eds.; Austrian Academy of Sciences: Vienna, Austria, 2013; pp. 252–254.

214. Fassbinder, J.W.E.; Irlinger, W.; Schleifer, N.; Stanjek, H. Methodische Untersuchungen zur Magnetometerprospektion: Das frühmittelalterliche Gräberfeld von Alburg, Stadt Straubing, Niederbayern. In *Das archäologische Jahr in Bayern, 1998*; Bayerisches Landesamt für Denkmalpflege, Ed.; Konrad Theiss Verlag: Darmstadt, Germany, 1999; pp. 112–114.

215. Filzwieser, R.; Olesen, L.H.; Neubauer, W.; Trinks, I.; Mauritsen, E.S.; Schneidhofer, P.; Nau, E.; Gabler, M. Large-scale geophysical archaeological prospection pilot study at Viking Age and medieval sites in west Jutland, Denmark. *Archaeol. Prospect.* **2017**. [CrossRef]

216. Filzwieser, R.; Olesen, L.H.; Verhoeven, G.J.J.; Mauritsen, E.S.; Neubauer, W.; Trinks, I.; Nowak, M.; Nowak, R.; Schneidhofer, P.; Nau, E.; et al. Integration of Complementary Archaeological Prospection Data from a Late Iron Age Settlement at Vesterager—Denmark. *J. Archaeol. Method Theory* **2017**, *24*, 1–21. [CrossRef]

217. David, A.; Linford, N.; Linford, P.; Martin, L.; Payne, A.; Jones, D.M. *Geophysical Survey in Archaeological Field Evaluation*; English Heritage: Swindon, UK, 2008.

218. Roskams, S. *Excavation*; Cambridge University Press: Cambridge, UK, 2001.

219. Wallner, M.; Löcker, K.; Neubauer, W.; Doneus, M.; Jansa, V.; Verhoeven, G.J.J.; Trinks, I.; Seren, S.; Gugl, C.; Humer, F. ArchPro Carnuntum Project Large-scale non-invasive archaeological prospection of the Roman town of Carnuntum. *Archaeol. Pol.* **2015**, *53*, 400–403.

Article

SAR Imaging of Archaeological Sites on Intertidal Flats in the German Wadden Sea

Martin Gade [1,*], Jörn Kohlus [2] and Cornelia Kost [3]

[1] Universität Hamburg, Institut für Meereskunde, Bundesstraße 53, 20146 Hamburg, Germany
[2] Landesbetrieb für Küstenschutz, Nationalpark und Meeresschutz Schleswig-Holstein (LKN), Nationalparkverwaltung, 25832 Tönning, Germany; joern.kohlus@lkn.landsh.de
[3] Wördemanns Weg 23a, 22527 Hamburg, Germany; info@cornelia-kost.de
* Correspondence: martin.gade@uni-hamburg.de; Tel.: +49-40-42838-5450

Received: 4 August 2017; Accepted: 2 October 2017; Published: 13 October 2017

Abstract: We show that high-resolution space-borne synthetic aperture radar (SAR) imagery with pixel sizes smaller than 1 m^2 can be used to complement archaeological surveys on intertidal flats. After major storm surges in the 14th and 17th centuries ("*Grote Mandrenke*"), vast areas on the German North Sea coast were lost to the sea. Areas of settlements and historical farmland were buried under sediments for centuries, but when the surface layer is driven away under the action of wind, currents, and waves, they appear again on the Wadden Sea surface. However, frequent flooding and erosion of the intertidal flats make any archaeological monitoring a difficult task, so that remote sensing techniques appear to be an efficient and cost-effective instrument for any archaeological surveillance of that area. Space-borne SAR images clearly show remains of farmhouse foundations and of former systems of ditches, dating back to the times before the "*Grote Mandrenke*". In particular, the very high-resolution acquisition ("staring spotlight") mode of the TerraSAR/TanDEM-X satellites allows detecting various kinds of remains of historical land use at high precision. Moreover, SARs working at lower microwave frequencies (e.g., that on Radarsat-2) may complement archaeological surveys of historical cultural traces, some of which have been unknown so far.

Keywords: storm surge; cultural traces; German Wadden Sea; intertidal flats; synthetic aperture radar; TerraSAR-X; St. Marcellus Flood; Burchardi Flood

1. Introduction

In the Medieval Period, the German North Sea coastline looked very different from how it looks today: the North Frisian Islands did not yet exist, but were part of the so-called "*Uthlande*" (outer lands) that belonged to, or were connected with, the mainland [1]. Vast areas along the coast were dominated by swamps, marshes, and swamp forests, which often made any settlements difficult or impossible. From the 11th and 12th centuries on, however, many settlements on the German North Sea coast appeared. In those settlements houses were often built on dwelling mounds and were protected by small "summer dikes", which could provide protection against high water only during summer, when storms are rare. Later on, coastal protection was improved through a system of "winter dikes" that were strong enough to prevent the marsh lands from frequent flooding [2]. Systems of drainage ditches were built to remove the water from the farmlands, thereby allowing for any kind of agriculture.

Salt extraction from salty peat, which was found and mined all along the North Sea coast, became an important economical factor, but the peat mining also caused a decrease in surface level in the protected (diked) areas [2,3]. Together with the draining the extensive land use caused the land surface to be below the mean high tide level at many places.

In mid-January 1362, severe westerly storms hit the Northern German coast, and on January 16 an immense storm tide flooded the coast, causing the small dikes to break at many places and killing a great number of people and cattle. This major storm surge is known in history as the *Saint Marcellus'* *Flood* or *"Grote Mandrenke"* ("great drowning of men") and caused that vast areas were lost to the sea. Those areas haven't been diked since then (compare the left and middle panels of Figure 1). Thereafter, it took a long time until new dikes were built to protect the remaining marsh land. The new farmland was characterized by wider plots of land, the dikes enclosed larger polders than in the centuries before, and farmhouses on terps were connected by narrow lanes.

Figure 1. Changes in the German North Sea coastline during the past 700 years (note the upper left part of each panel), after [1]. The left panel shows vast coastal wetlands that were lost after the first *"Grote Mandrenke"* in 1362. In the middle panel, the horseshoe-shaped island of Strand (white arrow) can be seen, whose major, central parts got lost after the second *"Grote Mandrenke"* in 1634.

Another major storm surge occurred on 11 October 1634, again, causing death of people and cattle, after the dikes had broken at several places. This second *"Grote Mandrenke"* (also known as *Burchardi Flood*) hit the area of North Frisia in an economically weak period, after the plague had caused many deaths only about 30 years before. The island of Strand, a horseshoe-shaped island in the center of the North Frisian coast (see the white arrow in the middle panel of Figure 1), was flooded and eventually broke up into pieces (compare the middle and right panels of Figure 1) and farmland, farms, and whole villages were destroyed. Consequently, the *Burchardi Flood* is still the most-known storm surge in history in the area of the North Frisian Wadden Sea.

Over the following centuries, great parts of this former agricultural area have been buried under muddy and sandy sediments, which nowadays form the German Wadden Sea. This area is being flooded, and falls dry, once during each tidal cycle, thereby making archaeological excavations very difficult. However, under the permanent action of the tidal forces the muddy and sandy marine sediments are partly driven away, and traces of former peat digging, drainage systems, and settlements appear again on the surface of the intertidal flats [3]. Since those areas are difficult to reach, and thus to observe from ground, airborne sensors have proven to be advantageous for a systematic observation of those historic places [4,5]. Airborne surveillance, however, is cost-intensive and the use of unmanned aerial vehicles (UAV) is prohibited in the National Park (to which the study area belongs). Therefore, high-resolution space-borne sensors are an alternative source of data that can be used by archaeologists for their frequent surveillance of that area [6,7].

Spaceborne synthetic aperture radar (SAR) imagery has been used for archaeological studies at various places worldwide since the 1980s [8,9]. SAR sensors image the Earth's surface at high spatial resolution and independent of daylight and weather conditions, thereby making dedicated acquisitions easier. Particularly their all-weather capabilities can make them advantageous over optical sensors in areas where those do not perform effectively [10]. In this respect, SAR-based archaeological studies have been focusing mainly on mountainous, arid, and forested regions [11–14], partly using SAR data of very high resolution [15,16]. In order to complement those studies, here, we introduce a new domain for SAR-driven archaeology.

2. Materials and Methods

We already used SAR images to derive surface characteristics of exposed intertidal flats and to detect bivalve beds [17]. During our analyses of SAR imagery of intertidal flats north of Pellworm we discovered bright and dark linear features that could not be of natural origin, but rather be manifestations of anthropogenic structures dating back to the times before the "*Grote Mandrenke*". These findings called for further studies that were based on high-resolution SAR imagery.

2.1. Region of Interest

A map of the study area is shown in Figure 2, with the location of the SAR image of Figure 3 inserted. This area lies in the center of the North Frisian Wadden Sea, i.e., in an area that was most affected by the major storm surges in the 14th and 17th centuries (Figure 1). Remains of the former island of Strand are Pellworm and Nordstrand, the latter meanwhile being attached to the mainland after long-standing land reclamation. In the area of interest traces of former land use have been observed for some decades [3,18].

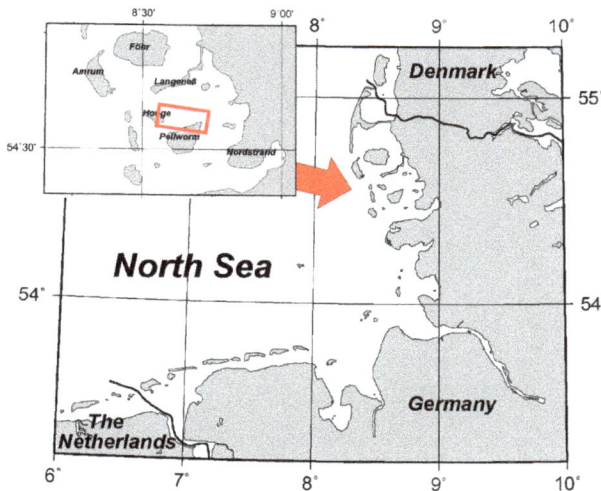

Figure 2. Area of interest on the German North Sea coast, north off Pellworm island, i.e., at the location of the former island of Strand. The red rectangle in the close-up on the upper left marks the location of the SAR image shown in Figure 3.

A TerraSAR-X image (11.6 km × 5.2 km) of the area of interest, acquired on 12 December 2012 (at 05:33 UTC, 18 minutes after low tide), is shown in Figure 3. The islands of Pellworm and Hooge can be seen in the lower and left parts of the image, respectively. Two red squares mark (1.0 km × 1.0 km) areas, in which remains of former land use were found.

Because of the low wind speed during that image acquisition (4 m/s), tidal channels and creeks show up dark: the radar backscattering from the water surface and moist soil mainly depends on its surface roughness; therefore, a flat water surface at low wind speeds causes low radar backscatter and, thus, dark image areas. Bright features in the right image half mark the rough edges of tidal creeks and dry, sandy sediments [19]; however, they are not of interest herein.

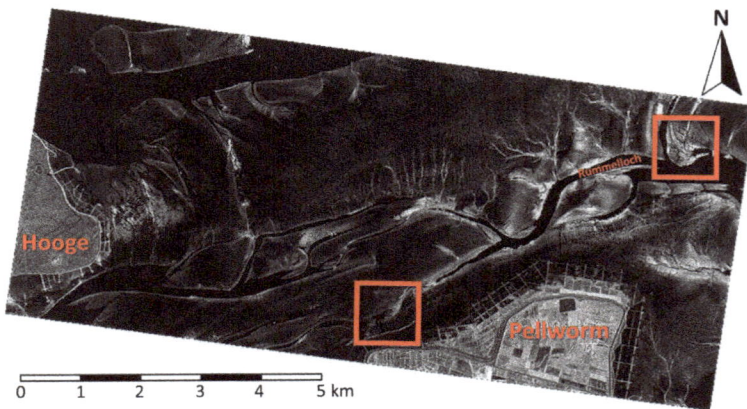

Figure 3. TerraSAR-X image of the area of interest, north of Pellworm and east of Hooge, acquired on 12 December 2012. The red squares denote two areas that were observed in greater detail. © DLR 2012.

The locations of all (known) cultural traces are depicted in Figure 4, where the light-green background color marks the islands of Pellworm (bottom) and Hooge (left) and the yellow background color the former island of Strand (Figure 1). Several of those traces are meanwhile lost, particularly around the small island (so-called Hallig) of Südfall, because of the strong erosion by the tidal currents, wind, and waves. The red rectangle, again, marks the location of the SAR image shown in Figure 3. Four dark blue arrows mark those locations, where the dikes broke during the second *"Grote Mandrenke"* in 1634.

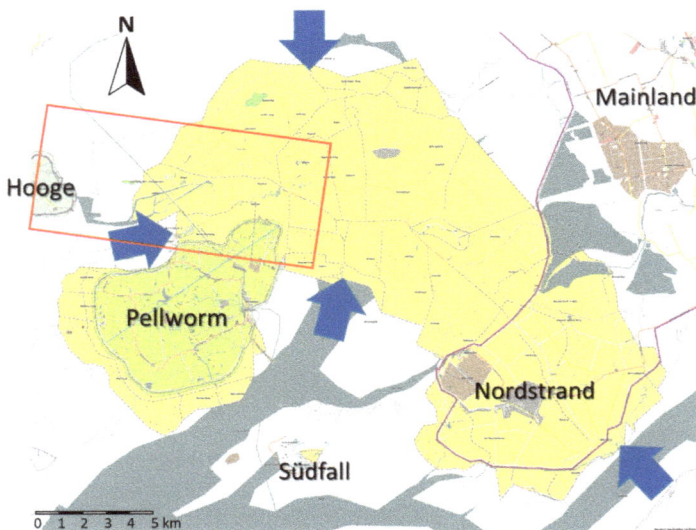

Figure 4. Map showing in yellow the former Island of Strand, and in green today's islands Pellworm and Hooge. The mainland and Nordstrand are located on the right. Locations of residuals of former landuse, along with roads, dikes, and names of former settlements are also inserted. The blue arrows denote those main locations, at which the dikes broke in 1634, and the red rectangle denotes the locations of the SAR image shown in Figure 3. Map: C. Kost.

Many lines mark the (known) locations of former dikes and lanes, as derived from historical maps of that area. Within our study area, i.e., within the red rectangle, groups of parallel lines denote known locations of former drainage systems. The systems consisted of a mesh of ditches, whose remains have become visible after they appeared again on the surface of the intertidal flats.

2.2. Synthetic Aperture Radar Data

A total of 19 SAR images acquired between 2012 and 2017 by the German TerraSAR/TanDEM-X and the Canadian Radarsat-2 satellites in high-resolution modes were available for our analyses of SAR signatures of historical land use. The pixel sizes of all images are on the order of 1 m^2, or even below, thereby allowing for the detection of fine structures such as remains of narrow ditches, settlements, etc. All images were calibrated and geocoded, and the speckle noise was reduced applying a Lee filter [20]. Of particular use was the new "staring spotlight" mode of the TerraSAR/TanDEM-X missions, with pixel sizes reaching down to 0.05 m^2 (Table 1) and spatial resolutions down to 0.5 m. This made it possible to detect fine structures even after speckle filtering.

SAR data from the European ENVISAT ASAR [21] and the Japanese ALOS PALSAR [22] missions have been successfully used for archaeological prospection and monitoring. However, since the pixel sizes of those images are on the order of 10 to 100 m^2, they did not allow studying the fine structures that are of interest herein. Table 1 gives an overview of all SAR images used, including the acquisition times and modes, pixel sizes, and the times of the closest low tide.

Table 1. SAR data used in the present study.

Sensor	Date/Time	Low Tide	Mode [1]	Pixel Size
TerraSAR-X	16 June 2012, 17:10 UTC	16:47 UTC	Spotlight	1.00 m × 1.00 m
TerraSAR-X	30 July 2012, 17:10 UTC	16:04 UTC	Hires SL	1.00 m × 1.00 m
TerraSAR-X	15 August 2012, 17:19 UTC	17:29 UTC	Hires SL	1.00 m × 1.00 m
Radarsat-2	28 October 2012, 05:44 UTC	05:39 UTC	Ultra Fine	1.00 m × 1.00 m
TerraSAR-X	28 October 2012, 05:50 UTC	05:39 UTC	Spotlight	1.00 m × 1.00 m
TerraSAR-X	25 November 2012, 05:42 UTC	04:13 UTC	Hires SL	0.75 m × 0.75 m
TerraSAR-X	11 December 2012, 05:50 UTC	04:09 UTC	Hires SL	0.75 m × 0.75 m
TerraSAR-X	12 December 2012, 05:33 UTC	05:15 UTC	Hires SL	0.50 m × 0.50 m
TanDEM-X	25 December 2012, 17:19 UTC	16:48 UTC	Hires SL	0.50 m × 0.50 m
TanDEM-X	25 April 2013, 17:19 UTC	18:42 UTC	Hires SL	1.00 m × 1.00 m
TanDEM-X	6 May 2013, 17:19 UTC	16:22 UTC	Hires SL	1.00 m × 1.00 m
TerraSAR-X	7 May 2013, 17:02 UTC	17:15 UTC	Hires SL	1.00 m × 1.00 m
TerraSAR-X	25 May 2013, 05:50 UTC	06:38 UTC	Hires SL	1.00 m × 1.00 m
TanDEM-X	4 October 2013, 05:50 UTC	06:14 UTC	Hires SL	1.00 m × 1.00 m
TanDEM-X	19 November 2014, 17:02 UTC	16:35 UTC	Staring SL	0.26 m × 0.26 m
TanDEM-X	21 November 2014, 05:42 UTC	05:42 UTC	Staring SL	0.22 m × 0.22 m
TanDEM-X	20 January 2015, 05:51 UTC	06:27 UTC	Staring SL	0.28 m × 0.28 m
TerraSAR-X	18 July 2016, 05:42 UTC	05:32 UTC	Staring SL	0.22 m × 0.22 m
TerraSAR-X	22 June 2017, 17:02 UTC	17:13 UTC	Staring SL	0.26 m × 0.26 m

[1] Hires SL = High Resolution Spotlight; Staring SL = Staring Spotlight.

Intertidal flats are highly morphodynamic, and when the muddy and sandy marine sediments of the flats' upper layer are moved away, banks of peat, old clay, and remains of farmland and settlements appear again on the dry-fallen surface. Moreover, the deposition of fine sediments along those morphologically harder structures can increase the local contrast, thereby making them (better) visible in aerial and satellite imagery. Analyzing the high-resolution SAR imagery we found at several places fine linear structures, which are clearly anthropogenic.

3. Examples of Cultural Traces Found on SAR Imagery

The two SAR images in Figure 5 were acquired on the same day, 28 October 2012, only 7 mins apart, by Radarsat-2 (upper panel) and TerraSAR-X (lower panel), both with pixel sizes of 1 m × 1 m. They show the same 1.7 km × 1.3 km area, north of Pellworm, containing the lower left red square in Figure 3. The diagonal parallel lines in the images' centers are manifestations of remains of historical ditches (compare with Figure 4 as well) that were buried for centuries under the sandy sediment.

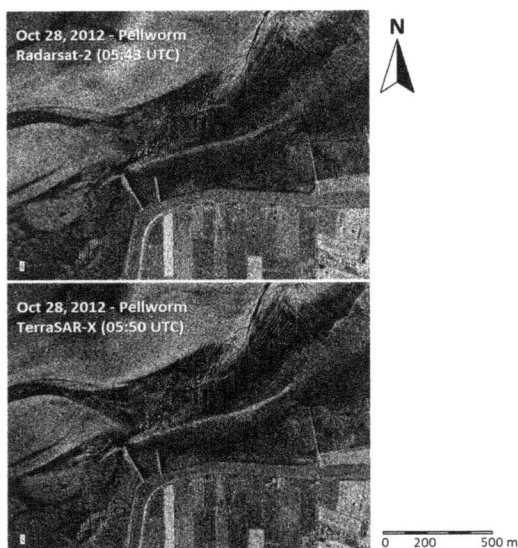

Figure 5. Radarsat-2 (upper) and TerraSAR-X (lower) SAR images of the same (1.7 km × 1.3 km) spot, north off Pellworm, acquired on 28 October 2012. Parallel diagonal lines in the images' centers are remains of historical ditches. Radarsat-2 data and products © MacDonald, Dettwiler and Associates Ltd. 2012—All Rights Reserved; TerraSAR-X data © DLR 2012.

These SAR image examples demonstrate that cultural traces on exposed intertidal flats can be detected by space-borne SAR, if the spatial resolution is high enough. We also note that the two SAR systems operate at different microwave frequencies (Radarsat-2: 5.41 GHz; TerraSAR-X: 9.65 GHz), but show similar image contrasts caused by the cultural traces. The radar backscattering depends on the surface roughness, whose scales are of the same order than the radar wavelength (some centimeters). Therefore, we can conclude that the roughness caused by the cultural traces must be of scales of several centimeters, too (see below).

3.1. Remains of Settlements in "Bupsee"

A series of 1 km × 1 km details of TerraSAR/TanDEM-X images is included in Figure 6. The SAR images show the same spot, north-east of Pellworm, corresponding to the upper right red square in Figure 3. In the lower part of each panel the tidal channel "Rummelloch" can be delineated, as well as a smaller tidal channel entering from the north. The intertidal flat in the panels' centers had fallen dry to different extent, depending on the water level at the time of image acquisition. A rectangular feature, which appears in some of the panels (but not in all of them), is marked by the red arrow in the lower left panel. This feature was observed in high-resolution TerraSAR-X imagery first, and it was studied in greater detail using data acquired in the 'staring spotlight' mode (lower right panel of Figure 6).

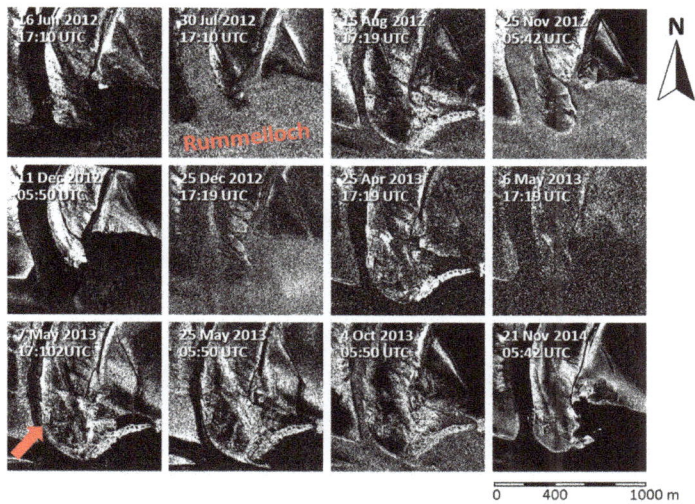

Figure 6. Examples of (1 km × 1 km) TerraSAR-X imagery of the area of interest marked by the right square in Figure 3. The red arrow in the lower left panel denotes cultural traces, whose imprints can be seen on some of the SAR images, and best on the 'staring spotlight' scene on the lower right. © DLR 2012, 2013, 2014.

Figure 7 shows an aerial photograph of the same intertidal flat, taken at low tide on 29 July 2009 (i.e., some years before the SAR images were acquired). Here, it is obvious that the above-mentioned linear structures originate from foundations of former constructions. In addition, several dark spots are visible, which originate from former wells, pits, cisterns, etc. Those spots can also be found as bright spots on SAR imagery (Figure 8) if the spatial resolution is high enough. Note that in SAR images of lower resolutions they could easily be confused with the speckle noise typical for SAR imagery. Three arrows are included in the aerial photograph for better comparison with the SAR images shown in Figure 8. Also visible in the photograph is the sandy sediment, under which those structures were buried for long, and which was driven apart by currents, wind, and waves.

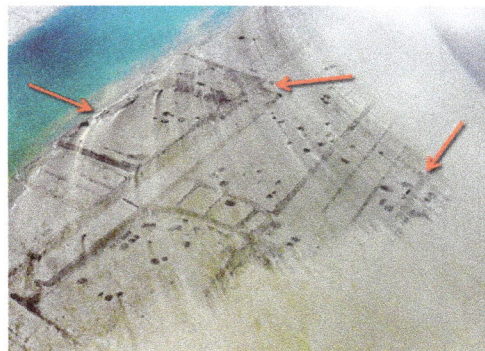

Figure 7. Aerial photograph of exposed intertidal flats north of Pellworm, taken in July 2009. Residuals of former settlements, close to a tidal creek (upper left), can be clearly seen. The arrows were included for easier comparison with the SAR images shown in Figure 8. Photograph: B. Hälterlein, LKN.

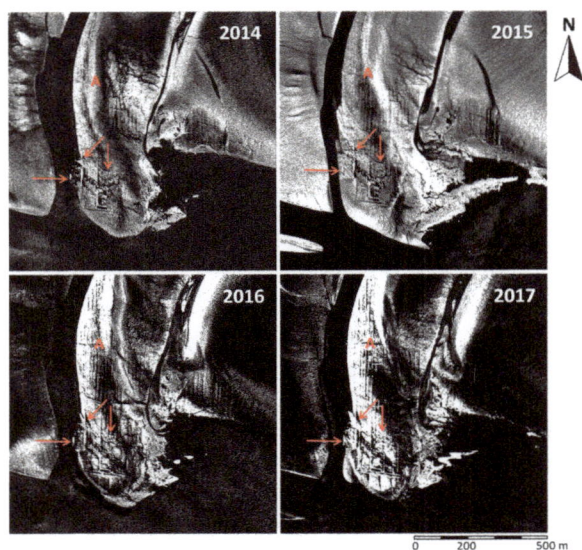

Figure 8. Subsections (1000 m × 1000 m) of four TerraSAR-X/TanDEM-X staring spotlight scenes acquired on (**upper left**) 19 November 2014, (**upper right**) 20 January 2015, (**lower left**) 18 July 2016, and (**lower right**) 22 June 2017. The linear structures are cultural traces. The letter (A) is introduced for convenience (see text), the arrows are inserted for intercomparison with Figure 7. © DLR 2014, 2015.

Figure 8 contains 1000 m × 1000 m details of four TerraSAR-X/TanDEM-X images acquired in staring spotlight mode on 19 November 2014, at 17:02 UTC (upper left; 27 minutes after low tide), on 20 January 2015, at 05:51 UTC (upper right; 36 minutes before low tide), on 18 July 2016, at 05:42 UTC (lower left; 10 minutes after low tide), and on 22 June 2017, at 17:02 UTC (lower right; 11 minutes before low tide). The location of these 1 km^2 details is the same as in Figure 6 (upper right red square in Figure 3).

The very fine pixel sizes of 28 cm × 28 cm and below allow imaging of remains of historical land use (e.g., houses, ditches, lanes), which usually are too narrow to be delineated on SAR imagery of conventional resolution, with pixel sizes on the order of 10 m × 10 m. Clearly visible in all panels of Figure 8 are linear, partly rectangular structures in the lower left image centers, close to the tidal creek (note the red arrows and compare with Figure 7). A close comparison of the aerial photograph (Figure 7) with the high-resolution SAR images (Figure 8) also reveals that, during the approx. Five years between the acquisitions, parts of the exposed remains were already lost, due to the permanent erosion, sedimentation, and morphological changes. However, the comparison of the four panels of Figure 8 also reveals that those rectangular structures are expanding eastward, i.e., that an increasing amount of those traces is appearing at the surface. This example demonstrates that erosion may erase, but may also produce, the traces on the surface that are visible on SAR imagery.

Groups of parallel vertical lines can be seen at different locations in all image centers. The mean distance between those parallel lines is about 15 m, thus indicating that they are remains of former ditches and drainage channels [18]. The letter (A) marks the same location in all panels; note that the spatial extent of the parallel vertical lines surrounding this letter has been increasing, due to the erosion of the upper sediment layers. A closer comparison with the map of all known cultural traces (see below) also reveals that these are structures that had not been known so far.

The above comparisons illustrate the strong morphodynamics on intertidal flats, but also the need for a frequent monitoring of those archaeological sites.

3.2. Remains of Drainage Systems and Lanes in "Waldhusen"

A series of 12 TerraSAR/TanDEM-X images (1 km × 1 km details, corresponding to the lower left red square in Figure 3) is shown in Figure 9. The red arrows mark the same diagonal parallel lines that were shown in Figure 5 and that are due to former ditches and lanes. However, different environmental conditions, namely different water levels and wind speeds during the image acquisitions, may strongly reduce the radar contrast of those structures, thereby making it difficult to extract them from SAR images. Also note that the brightness of the exposed intertidal flats varies, depending not only on the type of sediment [19], but also on the environmental conditions [23]. These variations, however, are not of interest herein.

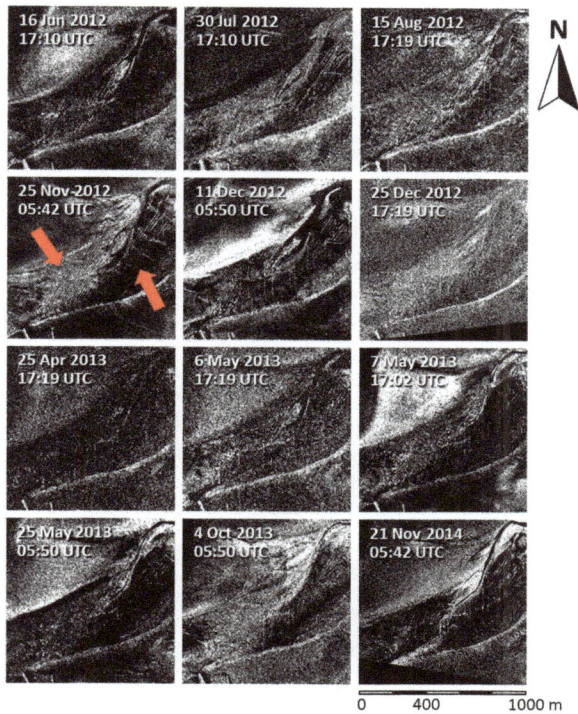

Figure 9. Examples of TerraSAR-X imagery of the area of interest marked by the left square in Figure 3. The red arrows mark several diagonal bright lines that can be identified on all SAR images, and that are due to traces of old drainage systems. They are best visible on the "staring spotlight" scene on the lower right. © DLR 2012, 2013, 2014.

Figure 10 shows two photographs taken during low tide on 15 June 2017 in the study area. Dark parallel structures (marked by the red arrows) on the sand and mud flats can be delineated, which correspond to those seen on the SAR images. Note the differences in sediment composition of the linear structures and the surrounding sandy flats. The ditch remains are marked by denser (harder) sediment causing higher surface roughness, which in turn results in higher radar backscattering. Also note that those lines show higher concentration of bivalves and vegetation (sea grass). Also visible is remnant water in between those structures and in small tidal pools scattered over the study area. This water effectively flattens the surface seen by the SAR (at low wind speeds), thereby leading to a stronger contrast between the parallel bright lines and the dark areas in between (Figure 11).

Figure 10. Two photographs of traces of historical ditches in the area of interest, taken at low tide during a field excursion in June 2017. The red arrows in both panels mark manifestations of the (former) ditches' rims that can be seen on SAR imagery. Photographs: M. Gade.

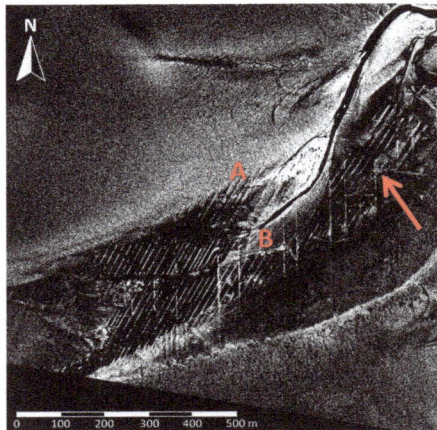

Figure 11. Subsection (1000 m × 1000 m) of a TanDEM-X staring-spotlight scene acquired on 21 November 2014, north of Pellworm Island. Letters (A) and (B) are described in the text. The photographs shown in Figure 10 were taken in the area marked by the red arrow. © DLR 2014.

An example of very high resolution SAR imagery of the same site (corresponding to the lower right panel of Figure 9) is shown in Figure 11. Inserted is a red arrow that marks the location, at which the photographs of Figure 10 were taken. This small (1000 m × 1000 m) section of a TanDEM-X staring spotlight scene was acquired on 21 November 2014, at 05:42 UTC (low tide) and shows many bright and dark parallel lines all over the image center. The distance of those lines is between 10 m and 20 m, again, indicating that they are remains of a former mesh of draws and ditches built for the drainage of the farmland [18]. However, we also note that, once the exposed space in between is partly filled with sandy sediments, some of the lines may also appear dark (seen in the image center of Figure 11, below letter (A)).

A part of the same area, north of the red arrow in Figure 11, is shown on the photograph in Figure 12 that was taken from a vantage point on Pellworm's northern dike at low tide on 15 May 2009. During that time, easterly winds were driving the water out of the study area and caused very low water levels. Under such favorable environmental conditions the parallel features could be easily observed from land. We also note, however, that the prevailing westerly winds in that area, along with the greater distance of most of the remains from the coast, usually make such land-based observations very difficult.

Figure 12. Photograph of the study area north of the red arrow in Figure 11. The picture was taken at low tide on 15 May 2009, when easterly winds caused a very low water level. Photograph: M. Gade.

Figure 13 shows a reconstruction of a historical lane, with ditches on either sides, which can be found on the intertidal flats north of Pellworm and which may cause structures like those observed in the SAR imagery (below letter (B) in Figure 11). Residuals of fossil farmland structures, mostly of ditches, but also of lanes or dikes, cannot be observed through their relief of less than 10 cm height. Instead, it is the sediments on the lost pastures that are different from those in the linear structures of ditches. Typical Wadden sediments on the flat sand banks consist of marine fine sand, which had been the basic compound of the old marsh land and which is still a major part of the coastal environment on the German North Sea coast. In contrast, the surface of the fossil ditches is different: in their center pillow-like sediments can be found, while the ditch edges are often stabilized by fossil roots and other plant material connected with the sediment (Figure 10). This causes narrow ridges of only 10 cm to 30 cm width, which can still be found today and which show up on SAR imagery, if its spatial resolution is high enough.

Figure 13. Reconstruction of a historical lane with ditches on either side. Photograph and sketch: J. Kohlus.

4. Comparison with Existing Data

A detail of the map in Figure 4 is shown in Figure 14. Two red squares mark the areas investigated herein in greater detail, "Waldhusen" (lower left) and "Bupsee" (upper right), where remains of former ditches and lanes and of former settlements, respectively, have been found on SAR imagery.

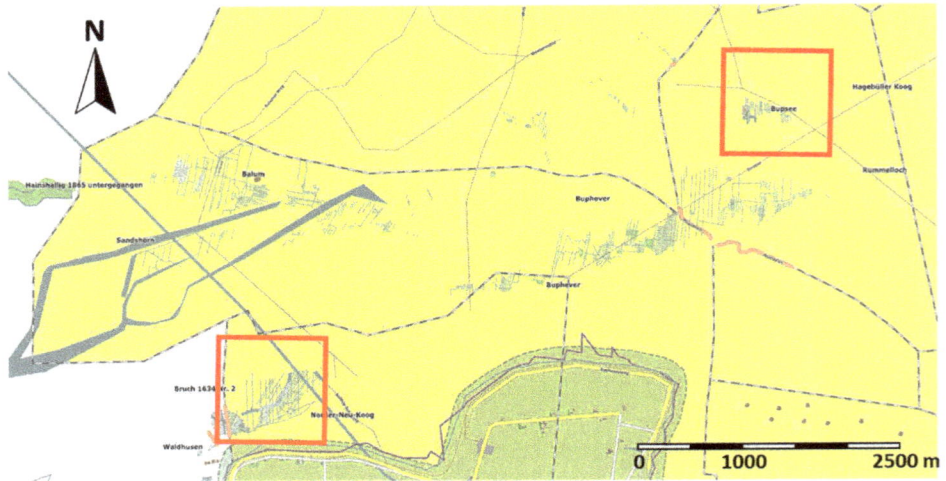

Figure 14. Detail of the map shown in Figure 4. Two red squares denote the locations of the SAR images shown in Figure 6 and in Figure 9, respectively. Map: C. Kost.

In the area "Waldhusen" (lower left square) parallel linear features on the map correspond well with the SAR observations; however, new features were found in the area "Bupsee" (upper right square). Here, more vertical lines (i.e., north-southerly oriented structures) were found on the SAR images (Figure 8) than known before [18]. This is a clear indication that SAR imagery can be used to update und to improve maps of remains of historical land use.

We also note, however, that the SAR did not capture all features that are included in the map: the extent of the linear features on the SAR imagery shown in Figure 9 is smaller than in the lower left square on the map. Again, this proves the strong morphodynamic changes in that area, which may lead to an appearance, but also to a disappearance, of exposed cultural traces.

5. Conclusions

We have demonstrated that high-resolution SAR imagery can be used to complement archaeological surveys on intertidal flats on the German North Sea coast. The new high-resolution TerraSAR-X acquisition mode ("staring spotlight") allows detecting various kinds of remains of historical land use, some of which have been unknown so far. Signatures of remains of both former settlements and former systems of ditches and of peat cutting, dating back to periods before major storm surges in the 14th and 17th centuries, can be found on high-resolution TerraSAR/TanDEM-X images. In this respect, best results were obtained (i.e., strongest and clearest signatures were found) when SAR images acquired in the new "staring spotlight" mode were used, with pixel sizes on the order of $0.1\ \mathrm{m}^2$.

In many cases the observed signatures of former ditches are due to different sediment types, which in turn are due to the actual ditch morphology. The presence of fossil roots and other organic material results in denser and harder sediments, which may be directly sensed by space-borne SAR, or which may cause additional sedimentation (i.e., a deposition of sandy sediments) that can be seen on SAR imagery. We also note that different sediments may cause different biological productivity, and are therefore often marked by benthic organisms causing different surface roughness patterns. Also, bivalves tend to settle on those sediments, thereby causing higher surface roughness sensed by the SAR.

SAR sensors are advantageous over optical sensors, because their use is independent of daylight and weather conditions. This is a particular advantage for the studies presented herein, since data acquisitions of the study area always have to be made at low water levels. Moreover, acquisitions made by spaceborne SAR sensors allow a frequent surveillance of the study areas, which is required in a strongly morphodynamic environment.

Further research will see thorough analyses of more SAR imagery of the area of interest, but also of other locations further north (close to the island of Langeneß), where more cultural traces were reported before. Automated detection of those remains that manifest in linear structures on SAR imagery seems possible through an application of Fourier and Principal Component Analyses as well as Hough Transformations, but only if the image resolution is high enough.

Acknowledgments: Parts of this research received funding from the German Ministry of Economy (BMWi) under contract 50 EE 0817, RS2 and TSX data were provided by the Canadian Space Agency (CSA) under contract 5077 and by the German Aerospace Center (DLR) under contract COA0118, respectively. RADARSAT is an official mark of the Canadian Space Agency.

Author Contributions: Martin Gade designed and performed all SAR image analyses and wrote most parts of the text. Jörn Kohlus initiated this research and provided a-priori knowledge about the site and cultural traces. Cornelia Kost provided knowledge about all known cultural traces in the study area. All authors participated in the data interpretation and in the final text editing.

Conflicts of Interest: The authors declare no conflict of interest. The funding sponsors had no role in the design of the study; in the collection, analyses, or interpretation of data; in the writing of the manuscript, and in the decision to publish the results.

References

1. Behre, K.-H. *Landschaftsgeschichte Norddeutschlands: Umwelt und Siedlung von der Steinzeit bis zur Gegenwart*; Wachholtz: Neumünster, Germany, 2009; p. 308.

2. Meier, D. Landschaftsentwicklung und historische Nutzung der Nordseeküste. In *Warnsignale aus Nordsee & Wattenmeer. Eine Aktuelle Umweltbilanz*; Lozán, J.L., Rachor, E., Reise, K., Sündermann, J., Westernhagen, V., Eds.; Wissenschaftliche Auswertungen: Hamburg, Germany, 2002; pp. 161–166.

3. Bantelmann, A. Die Landschaftsentwicklung an der schleswig-holsteinischen Westküste, dargestellt am Beispiel Nordfriesland. *Die Küste* **1966**, *14*, 1–182.

4. Gade, M.; Kohlus, J.; Mertens, C. Archaeological Surveys on the German North Sea Coast Using High-Resolution Synthetic Aperture Radar Data. In Proceedings of the 37th ISRSE Symposium, Tshwane, South Africa, 8–12 May 2017.

5. Gade, M.; Kohlus, J. SAR Imaging of Archaeological Sites on Dry-Fallen Intertidal Flats in the German Wadden Sea. In Proceedings of the 2015 International Geoscience and Remote Sensing Symposium (IGARSS), Milan, Italy, 26–31 July 2015.

6. Gade, M.; Melchionna, S.; Stelzer, K.; Kohlus, J. Multi-Frequency SAR Data Help Improving the Monitoring of Intertidal Flats on the German North Sea Coast. *Estuar. Coast. Shelf Sci.* **2014**. [CrossRef]

7. Gade, M.; Kohlus, J. After the Great Floods: SAR-Driven Archaeology on Exposed Intertidal Flats. In Proceedings of the ESA Living Planet Symposium 2016, Prague, Czech Republic, 9–13 May 2016.

8. Lasaponare, R.; Masini, N. Satellite Synthetic Aperture Radar in Archaeology and Cultural landscape: An Overview. *Archaeol. Prospect.* **2013**, *20*, 71–78. [CrossRef]

9. Chen, F.; Lasaponara, R.; Masini, N. An overview of satellite synthetic aperture radar remote sensing in archaeology: From site detection to monitoring. *J. Cult. Herit.* **2017**, *23*, 5–11. [CrossRef]

10. Tapete, D.; Cigna, F. Trends and perspectives of space-borne SAR remote sensing for archaeological landscape and cultural heritage applications. *J. Archaeol. Sci. Rep.* **2016**, *14*, 716–726. [CrossRef]

11. Chen, F.; Jiang, A.; Tang, P.; Yang, R.; Zhou, W.; Wang, H.; Lu, X.; Balz, T. Multi-scale synthetic aperture radar remote sensing for archaeological prospection in Han Hangu Pass, Xin'an, China. *Remote Sens. Lett.* **2017**, *8*, 38–47. [CrossRef]

12. Stewart, C.; Lasaponara, R.; Schiavon, G. ALOS PALSAR Analysis of the Archaeological Site of Pelusium. *Archaeol. Prospect.* **2013**, *20*, 109–116. [CrossRef]

13. Linck, R.; Busche, T.; Buckreuss, S.; Fassbinder, J.W.E.; Seren, S. Possibilities of Archaeological Prospection by High-Resolution X-band Satellite Radar—A Case Study from Syria. *Archaeol. Prospect.* **2013**, *20*, 97–108. [CrossRef]

14. Moore, E.; Freeman, T.; Hensley, S. Spaceborne and Airborne Radar at Angkor: Introducing New Technology to the Ancient Site. In *Remote Sensing in Archaeology. Interdisciplinary Contributions to Archaeology*; Wiseman, J., El-Baz, F., Eds.; Springer: New York, NY, USA, 2006; pp. 185–216.

15. Tapete, D.; Cigna, F.; Donoghue, N.M. 'Looting marks' in space-borne SAR imagery: Measuring rates of archaeological looting in Apamea (Syria) with TerraSAR-X Staring Spotlight. *Remote Sens. Environ.* **2016**, *178*, 42–58. [CrossRef]

16. Balz, T.; Caspari, G.; Fu, B.; Liao, M. Discernibility of Burial Mounds in High-Resolution X-Band SAR images for Archaeological Prospections in the Altai Mountains. *Remote Sens.* **2016**, *8*, 817. [CrossRef]

17. Müller, G.; Stelzer, K.; Smollich, S.; Gade, M.; Melchionna, S.; Kemme, L.; Geißler, J.; Millat, G.; Reimers, C.; Kohlus, J.; et al. Remotely sensing the German Wadden Sea—A new approach to address national and international environmental legislation. *Environ. Monit. Assess.* **2016**, *188*, 1–17. [CrossRef] [PubMed]

18. Bantelmann, A. *Die Landschaftsentwicklung an der schleswig-holsteinischen Westküste*; Wachholtz: Neumünster, Germany, 1967; p. 97.

19. Gade, M.; Alpers, W.; Melsheimer, C.; Tanck, G. Classification of sediments on exposed tidal flats in the German Bight using multi-frequency radar data. *Remote Sens. Environ.* **2008**, *112*, 1603–1613. [CrossRef]

20. Lee, J.S. Digital Image Enhancement and Filtering by Use of Local Statistics. *IEEE Trans. Pattern Anal. Mach. Intell.* **1980**, *2*, 165–168. [CrossRef] [PubMed]

21. Tapete, D.; Cigna, F.; Masini, N.; Lasaponara, R. Prospection and Monitoring of the Archaeological Heritage of Nasca, Peru, with ENVISAT ASAR. *Archaeol. Prospect.* **2013**, *20*, 133–147. [CrossRef]

22. Dore, N.; Patruno, J.; Pottier, E.; Crespi, M. New Research in Polarimetric SAR Technique for Archaeological Purposes using ALOS PALSAR Data. *Archaeol. Prospect.* **2013**, *20*, 79–87. [CrossRef]

23. Gade, M.; Wang, W.; Kemme, L. On the Imaging of Exposed Intertidal Flats by Dual-Polarization Synthetic Aperture Radar. *Remote Sens. Environ.* **2017**. under review.

geosciences

MDPI

Article

SARchaeology—Detecting Palaeochannels Based on High Resolution Radar Data and Their Impact of Changes in the Settlement Pattern in Cilicia (Turkey)

Susanne Rutishauser [1,*], Stefan Erasmi [2], Ralph Rosenbauer [1] and Ralf Buchbach [2]

1   Institut für Archäologische Wissenschaften, Universität Bern, Abt. Vorderasiatische Archäologie,
    Länggass-Strasse 10, 3012 Bern, Switzerland; rr@kun.de
2   Geographisches Institut, Georg-August-Universität Göttingen, Abt. Kartographie, GIS & Fernerkundung,
    Goldschmidtstr. 5, 37077 Göttingen, Germany; serasmi@uni-goettingen.de (S.E.); ralf@buchbach.com (R.B.)
*   Correspondence: rutishauser@iaw.unibe.ch; Tel.: +41-31-6318992

Received: 22 July 2017; Accepted: 16 October 2017; Published: 24 October 2017

Abstract: The fertile alluvial plain of Cilicia is bordered by the Taurus and Amanus mountain ranges to the west, north and east and the Mediterranean Sea to the south. Since the Neolithic Period, Plain Cilicia was an important interface between Anatolia and the Levant. The alluvial plain is dominated by three rivers: Tarsus, Seyhan and Ceyhan. The avulsion history of the lower course of the rivers Seyhan and Ceyhan during the Holocene remains an unresolved issue. The knowledge about how former river courses have changed is essential for the identification of ancient toponyms with archaeological sites. The analysis of silted up riverbeds based on high-resolution digital elevation models (TanDEM-X) and historic satellite imagery (CORONA) in this paper provide the first indications for the reconstruction of former river channels. Further evidence is given by the evaluation of the settlement patterns from 3rd to 1st millennium BC.

Keywords: Cilicia; Seyhan; Ceyhan; Puruna; avulsion; TanDEM-X; CORONA; remote sensing; archaeology

---

1. Introduction

The changes of river courses, which have had a strong influence on the development of sites and is important for the localization of sites has not been investigated in Cilicia in detail so far. Only the development of river deltas and the changes of the river Tarsus have been studied in detail [1–4]. For the identification of toponyms known from written sources with archaeological sites, precise knowledge about the changes of river courses is essential. The aim of this paper is a first look at the development of the Seyhan and Ceyhan rivers during the late Holocene. Due to the lack of available coring data, this will be done by the analysis of remote sensing data in combination with a study of changes of the settlement cluster in Plain Cilicia. In southern Mesopotamia, studies about ancient watercourses on which all communication moved and along which all settlement ranged itself were undertaken already in the 1960s and 1970s on the basis of archaeological data and remote sensing data (LANDSAT) [5]. Since then, much more data has become available e.g., CORONA and TanDEM-X. In this study, these archives will be analyzed for the region of Plain Cilicia.

Plain Cilicia (gr. Kilikia Pedias, lat. Cilicia Campestris) covers an area of approximately 8000 km^2 in the northeastern corner of the Mediterranean Sea. Like almost no other landscape in Asia Minor, this plain is defined by strong topographic contrasts: a large fertile plain surrounded by steep mountain ranges. The Middle Taurus Range (turk. Orta Toroslar), which exceeds 3700 m in elevation, borders the landscape to the west and north. In the east, the settlement cluster is confined by the Amanos (turk. Nur Dağları) that separates Plain Cilicia from inner Syria and reaches elevations of up to 2200 m. The Gulf of Iskenderun and the Gulf of Mersin form its southern boundary (Figure 1).

Natural passes through the mountains still form the main routes of access to the lowlands. The "Cilician Gates" (turk. Gülek Boğazı) at an elevation of 1290 m connect Anatolia with the plain and serve as the main passageway for railroad tracks and major vehicular roads up until this day. In the westernmost part of Plain Cilicia, forming the transition to Rough Cilicia, the Göksu Valley stretches north–south from the Konya Plain to coastal Silifke and represented an important connection to Anatolia in antiquity [6] (p. 88). The Bahçe Pass ("Amanian Gates") in the East leads over the Amanus Range and serves as a passage between Cilicia and the Islahiye Plain. Further south, the Orontes Valley can be reached over the Belen Pass ("Syrian Gates"). Access from the north is facilitated by several tributary valleys of the Seyhan and Ceyhan rivers. Late Bronze and Iron Age rock reliefs as well as inscriptions like Hanyeri, İmamkulu, Taşçı and Fraktın [7], mark their significance in antiquity. The plain itself is divided into a western (Çukurova) and an eastern (Yukarıova) settlement cluster by a natural border formed by the foothills of the Middle Taurus Range and the Misis Mountains. The approximately 760 m high Misis Mountains [8] (pp. 305–307) extend northwards from the western end of the Gulf of Iskenderun (Figure 1). The average height of the Çukurova is about 15 m above sea level. The height of the Yukarıova varies from 20–100 m above sea level.

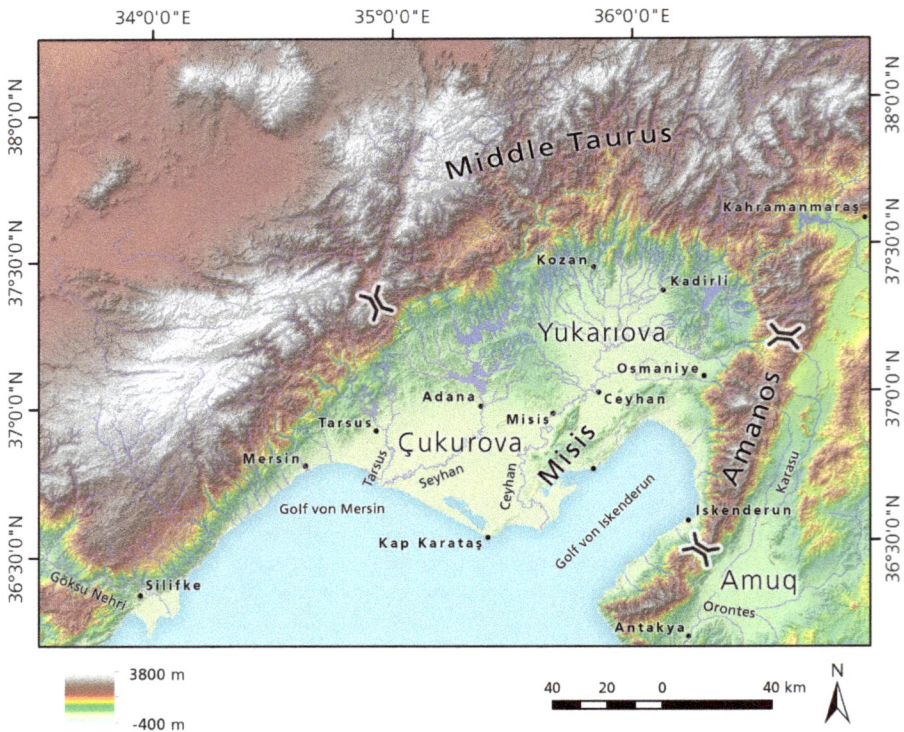

Figure 1. Map of Plain Cilicia and its neighboring regions. Topographic data based on SRTM (Shuttle Radar Topographic Mission, U.S. Geological Survey, 3 Arc Second scene) [9] © Susanne Rutishauser.

Three main rivers, originating in the Taurus Mountains, cross the lowlands and discharge into the Mediterranean Sea: Berdan or Tarsus Çayı (gr. *Kydnos*), Seyhan (hitt. *Šamri*/*Sapara*, gr. *Saros*) and Ceyhan (hitt. *Puruna*, gr. *Pyramos*) [10] (p. 6). These rivers form the deltaic complex that occupies the Adana Basin (Figure 2). The Adana Basin represents the onshore extension of the marine Cilicia Basin [11] (p. 123). The 6000 m thick sedimentary succession of the Adana Basin is spanning Miocene to Recent [12] (p. 190) and is built by alluvial variable grainsize deposits [13] (p. 135). All rivers are braided in their upper reaches with channel gradients ranging from 1:40 to 1:400, while the channel gradients decrease to >1:2000 in the lower 35–110 km of their courses, causing them to meander [14] (p. 57). The major tributaries of the 560 km long Seyhan are the Göksu, Zamantı, Çakıt and Körkün Rivers [15] (p. 647). Prior to construction of the first dam in 1954 the Seyhan River Delta expanded its area on average by 2.8 ha per year [16] (p. 64). The dam greatly reduced sedimentation in the delta [17] (p. 5). The 510 km long Ceyhan River drains today into the Gulf of Iskenderun and formed a more than 100 km^2 large river delta. Paleoenvironmental studies of the Iskenderun Bay show a muddy sedimentation, which coincides with a major delta progradation of the Ceyhan around 2140 BP [18] (p. 20). Therefore, a drainage into the Mediterranean Sea at Kap Karataş during the Iron Age is generally accepted [18] (p. 20).

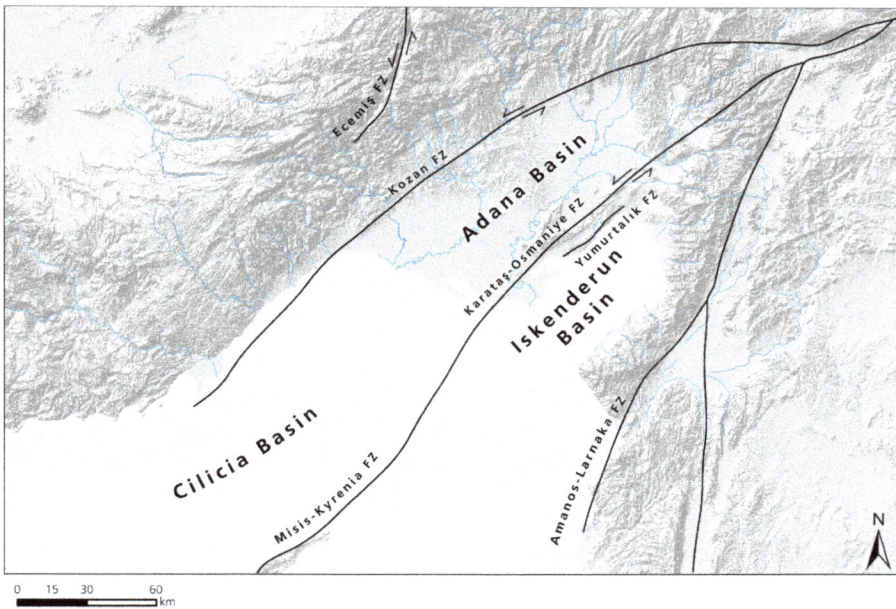

Figure 2. Morphological and simplified tectonic map of the region with the locations of major Neogene basins and the main fault zones (FZ) (after [19] Figure 1), © Susanne Rutishauser.

The course of the coastline underwent major changes. There are a few studies for a reconstruction of the palaeoshorelines available [2,4,14,17,18,20–23]. A sea level high stand around 6000–5000 BP is generally accepted, followed by a significant regression of the sea level of several meters as well as a slight rise to the modern datum [24]. A mid-Holocene coastline about 5–10 km inland from the present-day coastline is assumed [1,4].

Also from a historical perspective, Plain Cilicia is defined by contrasts between independence and foreign influence up to annexation. During the Chalcolithic Period, there is strong evidence for connections to the Amuq (Figure 1) and other Levantine and north Mesopotamian regions [25]. At the beginning of the Bronze Age, end of the 4th millennium BC, an alteration is visible by a change in the ceramic assemblage and a first orientation to the Anatolian plateau in the north [26] (pp. 141–142), [27]. However, there are still parallels to the Amuq region and the Northern Levant. At the beginning of the 2nd millennium BC, Syro-Cilician Pottery is predominant [28] (p. 11). When Cilicia became the Kingdom Kizzuwatna independent, mid 2nd millennium BC, there is a sudden change in the material evidence, with a strong orientation to the Anatolian plateau visible. From mid-15th to the beginning of the 12th century BC, Kizzuwatna stood under Hittite or Mittani supremacy [28] (p. 36). At the transition from the Late Bronze to the Early Iron Age Plain Cilicia became again independent as the kingdom Ḫiyawa/Que [29]. During the 8th century BC Plain Cilicia was integrated into the Assyrian Empire [30] (pp. 7–24).

The knowledge of the material culture of Kizzuwatna and Ḫiyawa/Que is still fragmentary. Most information come from excavations of the main Bronze and Iron Age sites in the plain. These are from West to East: Mersin-Yumuktepe, Tarsus-Gözlükule, Adana-Tepebağ, Sirkeli Höyük, Tatarlı Höyük and Kinet Höyük (see Figure 3).

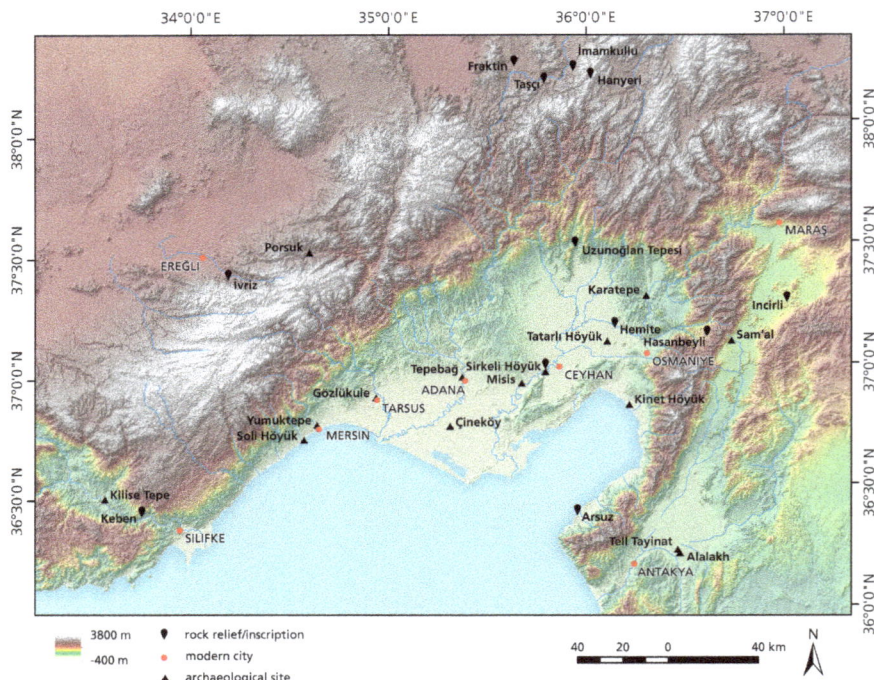

Figure 3. Map of Plain Cilicia and its neighboring regions with the most important Bronze and Iron Age sites as well as modern cities. Topographic data based on SRTM [9] © Susanne Rutishauser.

Since no archive has been discovered in Cilicia so far, only little is known about the ruling dynasties and the political organization for the Late Bronze and Early Iron Age. Today, the Çukurova is an important agriculture center with high productivity. A dozen archaeological surveys give some evidence about the distribution of settlements in the Çukurova and the Yukarıova. The settlement pattern has not changed fundamentally over the time. Mainly during the Bronze Age and in the Roman Period there is a consolidation of settlement visible. A lack of evidence exist in the foothills of the Taurus Mountains north of Adana as well as in an area between the rivers Tarsus and Seyhan (Figure 4).

The historian and geographer Abū l-Fidā' (1273–1331) reports [31] (p. 465) that the river Seyhan and Ceyhan join after Misis: "Le Syhan, au-dessous d'Adana et de Messyssa, se joint au Djyhan; tous deux ne forment plus qu'un seul fleuve, qui se jette dans la mer entre Ayas et Tharse" [32] (p. 62f).

The localization of the important classical site *Mallos* is not clear. It is known that the site must have been close to the river mouth of the Pyramos, the ancient Ceyhan. Forlanini suggest that the site was known in the Late Bronze Age as the harbor town *MLWM*. He identifies the site at Domuztepe (Seton-Williams No. 74 [26]), situated east of the river Ceyhan, 20 km northeast of the Kap Karataş [10] (p. 12).

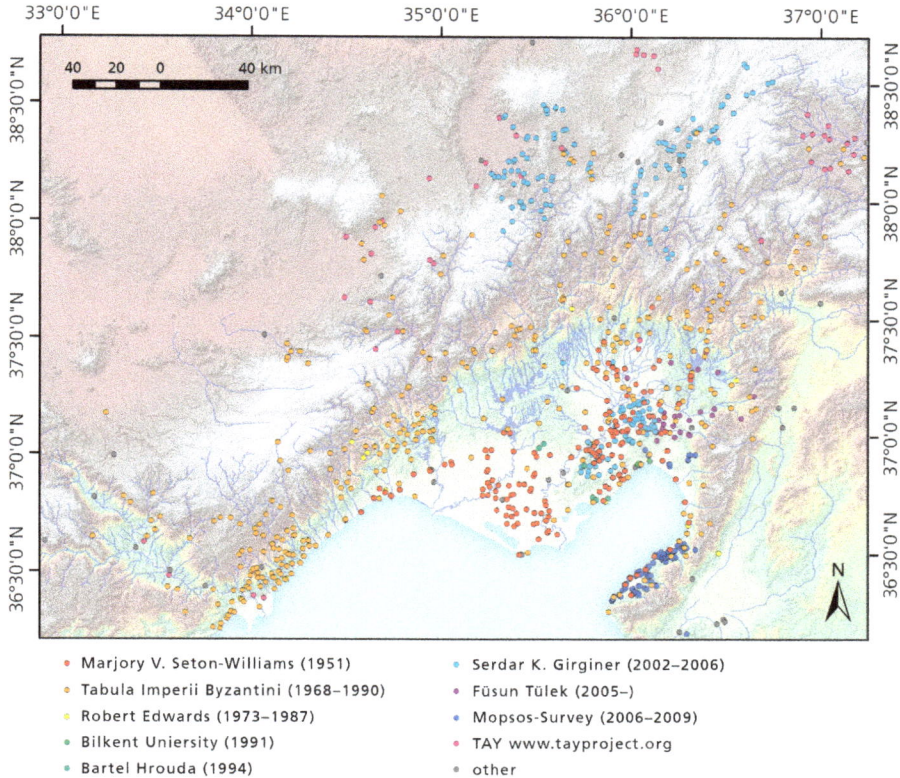

Figure 4. Distribution of archaeological sites investigated during different surveys in the Cilician Plain © Susanne Rutishauser.

2. Materials and Methods

2.1. Settlement Database

To analyze changes in the settlement cluster of Cilicia, a database with the focus on Bronze and Iron Age sites was built. Each site has been documented with its coordinates and its occupation phases (spanning from Neolithic to Ottoman Period). Because less than 3% of the sites have been investigated by excavation it was not possible to provide precise information regarding settlement size during any particular period. Most information for the settlement database was provided by a dozen surveys conducted in the area (Figure 4).

Hetty Goldman (Bryn Mawr Expedition to Cilicia) conducted the first systematic survey in Plain Cilicia in 1934 [33], one year later she started to excavate the site Tarsus-Gözlükule (Figure 3). Excavations were conducted in 1935–1939, 1947–1949 and resumed by Asli Özyar in 2001. The site can securely be identified with Hittite *Tarša*, know in Neo-Assyrian sources as *Tarzu* and as *Tarsos/Tarsus* in Classical Times [10] (p. 4). The Neilson Expedition (1936–1939, 1946–1947) under the direction of John Garstang started to excavate the site Mersin-Yumuktepe (Figure 3) with a sondage in its first year of the survey. The work was resumed by Veli Sevin and Isabella Caneva in 1993. The site might be identified with Hittite *Ellipra*, Neo-Assyrian *Illupru* [10] (p. 4).

Tepebağ Höyük, in the center of the present city of Adana, was first excavated with a small sondage in 1936 by the Museum Adana. Recently, excavations have been resumed first by the Çukurova University and later by the Museum of Adana. It can be identified with hitt. *Adaniya* and classical *Antioch on the Saros* [29] (pp. 191–195), [34] (p. 154). To gain information about routes through the Taurus Helmuth Theodor Bossert and Bahadir Alkım conducted a survey in the Upper Ceyhan valley from 1945–1952 and in 1955 [35]. They discovered the late Hittite fortress Karatepe-Arslantaş (hitt. *Azatiwataya*) and started to excavate there in 1946 [36–38]. The first and still most comprehensive survey of Plain Cilicia was directed by Marjory Veronica Seton-Williams in 1951 [26]. An important site at the eastern border of the Çukurova, Yakapınar-Misis, was excavated from 1955–1959 by Ludwig Budde and Helmuth Theodor Bossert. In 2000, the work was resumed by Giovanni Salmeri and Ana Lucia D'Agata after conducting a survey in the area. The site might be identified with *Zunnabara* mentioned in Hittite sources [10] (p. 6). In classical time its name was *Mopsuhestia* and for a short time *Seleucia ad Pyramum* [34]. Over 400 sites mainly dating from the Classical to the Byzantine Period were investigated by Friedrich Hild and Hansgerd Hellenkemper and published in volume five of the Tabula Imperii Byzantini [34]. Since 1990 Mustafa H. Sayar is conducting an epigraphic survey in the entire region. Marie-Henriette Gates in 1991 and later Ann Killebrew and Gunnar Lehmann (2006–2009) were conducting a survey (Mopsos Survey) along the eastern coast of the Gulf of Iskenderun [39,40]. From 1992–2011 excavations at the harbor site Kinet Höyük under the direction of Marie-Henriette Gates were undertaken. The site can be identified with Hittite *Izziya* and classical *Issos* [41]. Only 15 km upstream of Yakapınar-Misis is the site Sirkeli Höyük located. Excavations started in 1936–1937 by John Garstang as part of the Neilson Expedition. Later Excavations were directed by Barthel Hrouda 1992–1996 and Horst Ehringhaus (1997). The work was resumed in 2006 by Mirko Novák, Ekin Kozal and Deniz Yaşin Meier [42]. There is strong probability to identify Sirkeli Höyük with *Kummanni* known from Hittite religious texts [10]. A neighboring cult city to Kummanni was the site *Lawazantiya*/akkad. *Lusanda*. The site of Tatarlı Höyük is a strong candidate for its identification. Excavations have started in 2007 by K. Serdar Girginer [43].

Information from written sources have been added to the database with up to 1000 sites when available.

2.2. CORONA Satellite Imagery

A total of 19% of the sites in the settlement database are not visible on modern satellite imagery anymore and 76% are damaged. However, many of these archaeological sites are still visible on historical satellite imagery. Declassified in 1995, CORONA satellite images (Figure 5) acquired by the United States from 1958–1972 [44] have been used in addition to Digital Elevation Models to map palaeochannels in Plain Cilicia. The CORONA images are valuable because they were acquired before the intensification of the agriculture and the growth of urban areas. Some of the sites were destroyed due to the erecting of dams [45,46]. Images from different missions with the camera systems KH-4A and KH-4B were analyzed (Figure 5). The images have an average spatial resolution of 1.8 m [47] (p. 119). For images of the mission DS1107, orthorectified images of the CORONA Atlas of the Middle East were used (Center for Advanced Spatial Technologies, University of Arkansas/U.S. Geological Survey; [48]). To remove geometric distortion of the images from further missions (DS1043, DS1101, DS1103, DS1105, DS1109) the CORONA software written by Irwin Scollar was used [49]. The georeferencing was done with ArcGIS v. 10.2 (3rd Order Polynomial). The identification of archaeological sites and geomorphic features on CORONA imagery was performed visually.

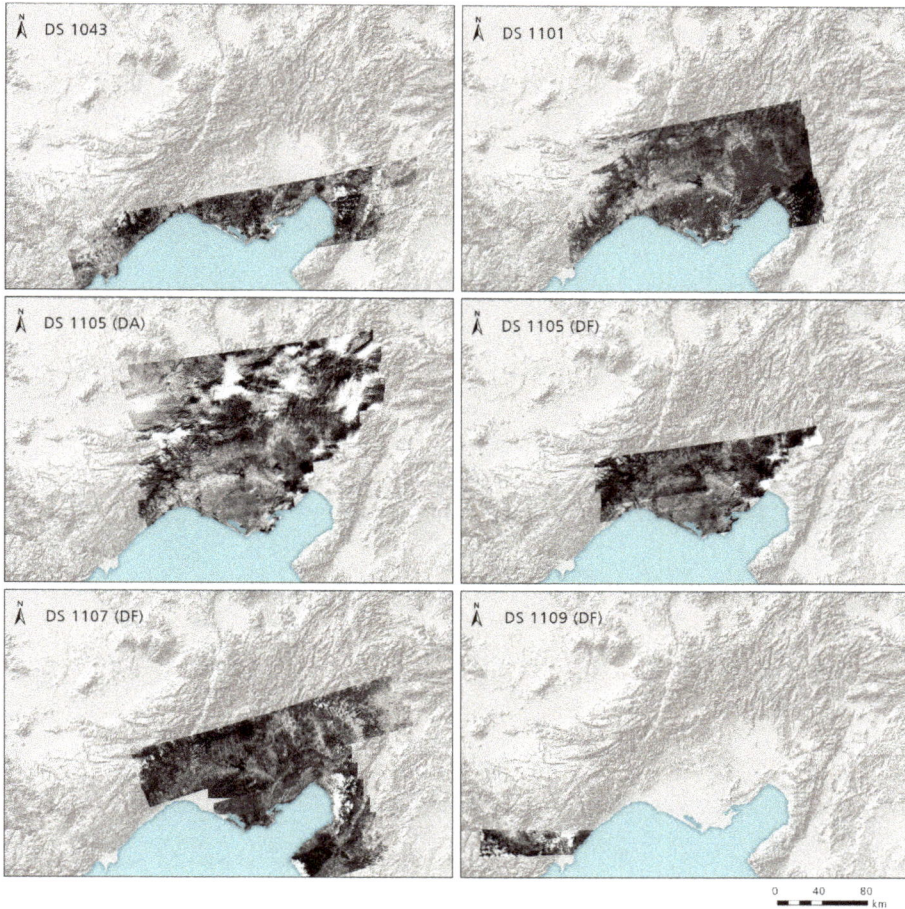

Figure 5. Orthorectified CORONA scenes (U.S. Geological Survey, CORONA Satellite Photographs) analyzed for this project © Susanne Rutishauser.

2.3. TanDEM-X: TerraSAR-X Add-on for Digital Elevation Measurement

The TanDEM-X system is a space borne radar interferometer that is based on two satellites (TerraSAR-X 3) orbiting in close formation [50]. The high resolution X-band SARs (9.65 GHz) can be operated in different imaging modes (High Resolution Spotlight, Stripmap, ScanSAR). For this study, two TanDEM-X acquisitions were made possible in two different imaging modes at horizontal polarization. In Stripmap (SM) mode the course of the Ceyhan River was covered at an azimuth resolution of 2.54 m. The High Resolution Spotlight (HS) mode enabled the registration of a smaller area of interest (approx. 5×5 km^2) at a nominal resolution of 1.13 m in azimuth (see [51] for a detailed description of image processing). In addition to TanDEM-X, other freely available DEM products from the shuttle radar topography mission (SRTM C-band version 4.1 with 90 m resolution, SRTM X-band with 30 m resolution) as well as the ASTER Global DEM (V.2) were used to generate a high resolution DEM-mosaic (Figure 6). The global spatial resolution is 5 m. 55% of the area are covered by TanDEM-X data divided into 8 spotlight and 22 stripmap scenes. The accuracy of the height value is 4.65 m on average with a standard deviation of 3.82 m [45]. The TanDEM-X DEM mosaic provided the basis for the extraction of cross sections of height values (profiles) for the detailed analysis of the topography across riverbeds and paleochannels (Figure 6).

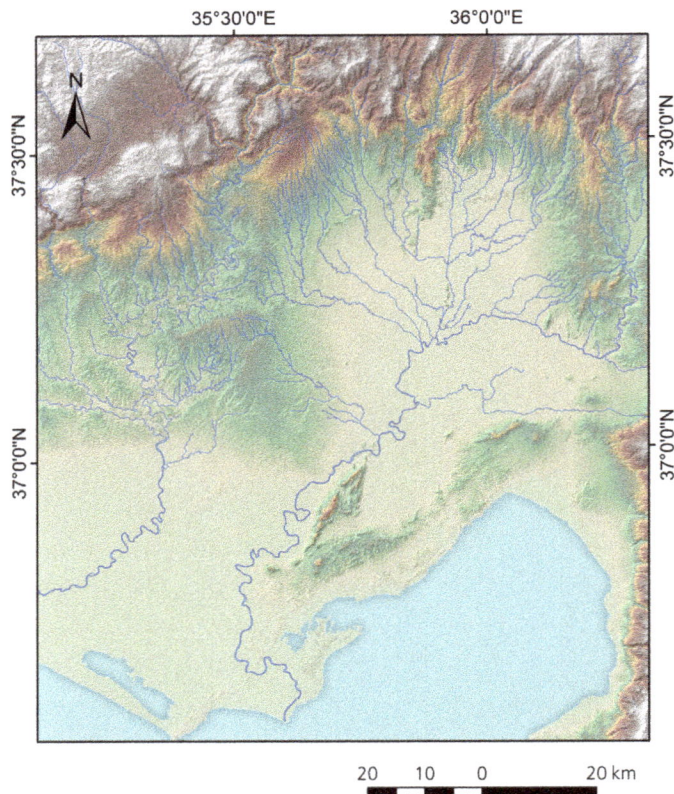

Figure 6. DEM-Mosaic based on TanDEM-X, Aster GDEM and SRTM data with modern river courses © Susanne Rutishauser.

In addition, VHR (very high resolution) satellite images in GoogleEarth and ArcGIS were visually analyzed.

3. Results

3.1. Palaeochannels and Relict Canals

Palaeochannels are sections of abandoned channels [52]. They become fully or partly infilled by abandoned channel accretion [53] (p. 136). Palaeochannels are most commonly preserved as exposed cross-sections, abandoned surface forms, or buried channels [54]. They were mapped based on the analysis of geomorphic features visible in CORONA satellite imagery from the 1960s, high resolution DEM-data and freely available VHR satellite images (Figure 7, palaeochannels in blue and green color).

Relict canals are abandoned canals for irrigation. They can be traced by their pattern in modern fields and the remains of ancient levees [55] (p. 289) (Figure 7, relict canals in pink color).

Since most of the palaeochannels were found between today's river course of Seyhan and Ceyhan, this area will be discussed here.

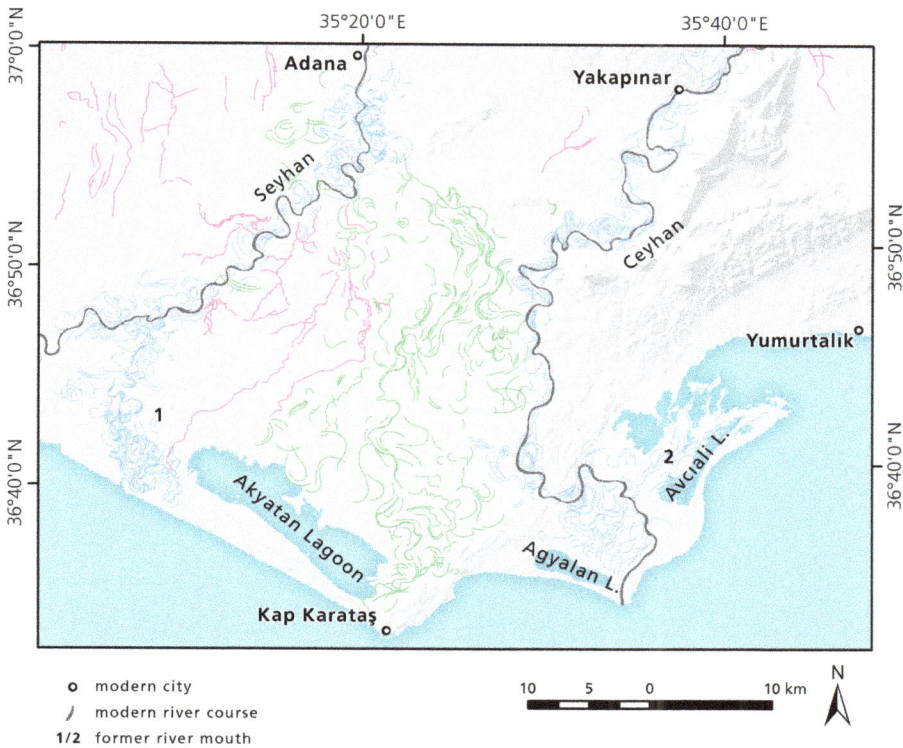

Figure 7. Palaeochannels (blue, green) and relict canals (pink) in the Çukurova digitized based on TanDEM-X, CORONA and VHR-data © Susanne Rutishauser.

Not all palaeochannels are visible on aerial or satellite imagery. Older channels are buried beneath more recent sediments. However, the mapped palaeochannels show different morphologies. The idealized river course corresponds to a sine curve [56] (p. 147) (Figure 8).

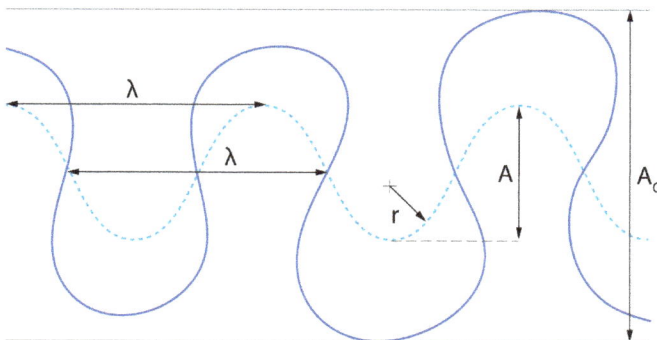

Figure 8. The dotted line represents an idealized river course, the continuous line a meandering river course. λ: meander wavelength, A: meander amplitude, r: meander radius, A_G: meander belt © Susanne Rutishauser.

The sinuosity ratio gives an indication how much a river meanders by measuring the length of a channel reach and dividing this by the straight line distance [53] (p. 138). Rivers in a flat plain follow a winding path where the meander amplitude (A) is higher than the meander wavelength (λ) (Figure 8). This can be seen exemplary in the Çukurova about one kilometer east and west of the rivers Seyhan and Ceyhan. The blue palaeochannels (Figure 7) show the most recent abandoned channels. The former Seyhan river mouth, active until 1000 AD, is clearly visible with a high sinuosity ratio (Figure 7, No. 1) [1,21]. A more recent alteration of the Ceyhan River seems to have occurred at the beginning of the 20th century. From the Middle Ages until 1935, the Ceyhan River mouth diverted eastwards [2] (Figure 7, No. 2), thus its present-day location is situated west of where it had been in medieval times [15] (p. 656).

Less meandering structures with a lower sinuosity ratio show the green palaeochannels (Figure 7). They are located west of the river Ceyhan but belong to a former river course of the Seyhan as DEM-data show. A main channel of this former river course can be traced up to 7 km before the coastline northeast of the Akyatan Lagoon (Figure 9).

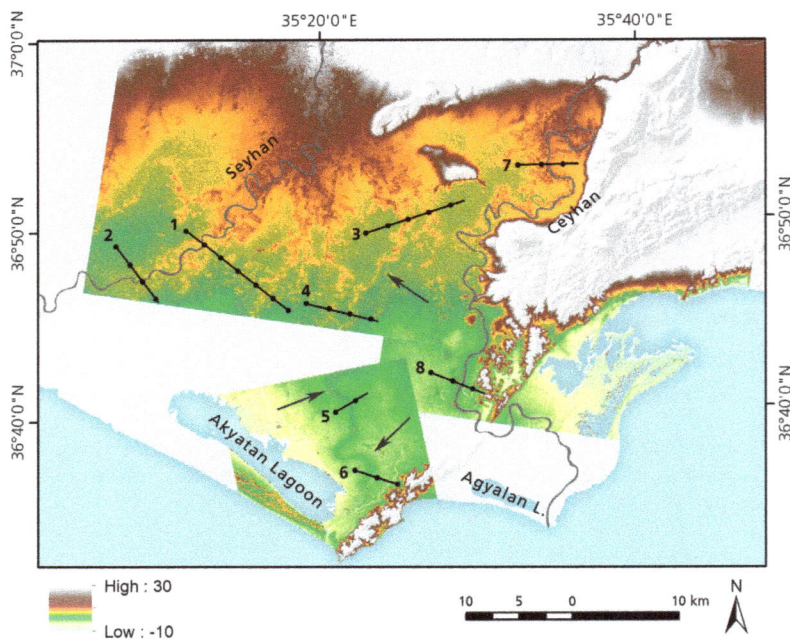

Figure 9. High-resolution digital elevation model based on TanDEM-X data (Stripmap) of the Çukurova. The grey arrows point to a large palaeochannel. The location of the cross sections is represented by black lines. The dots mark a distance of 2000 m © Susanne Rutishauser.

The data of the high resolution TanDEM-X DEM allow the generation of cross sections (Figure 10). The large palaeochannel is visible as a positive height anomaly, which is formed by natural levees. Deposition of fine sand, silt and clay alongside and within the channels results in the aggradation of levees which gradually raise the rivers until they flow several meters above the plain on low ridges. As cross sections of the modern Seyhan and Ceyhan rivers show, their riverbed is located 2–4 m above the plain as well (Figure 10).

The construction of canals for irrigation results in levees as well. In addition, channel excavation and cleaning operations result in the up-cast of clean-out banks alongside the canal [57] (p. 415), [58]

(p. 339). Compared to palaeochannels, (relict) canals show a less meandering course with a low sinuosity ratio. The course of presumed canals was mapped in pink (Figure 7). The spoil banks are visible on optical imagery as CORONA representing white lines bordered the canal (Figure 11). Relict canals itself can be identified as darkened depressions, filled with moist sediments [59]. Water flowing in active canals has a dark signature as well (Figure 11).

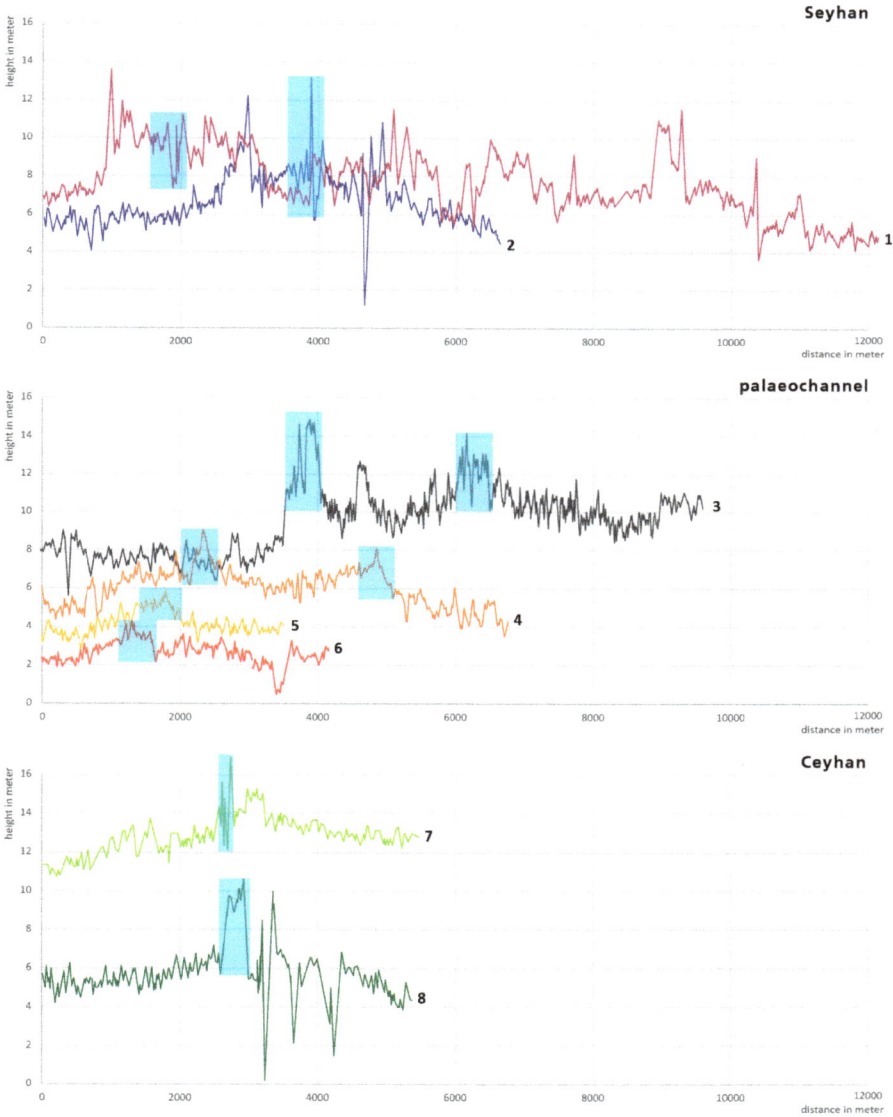

Figure 10. Cross sections of height values from the TanDEM-X DEM for the rivers Seyhan and Ceyhan and the large palaeochannel. For the location, see Figure 9. The blue rectangles mark the location of the riverbed resp. palaeochannel © Susanne Rutishauser.

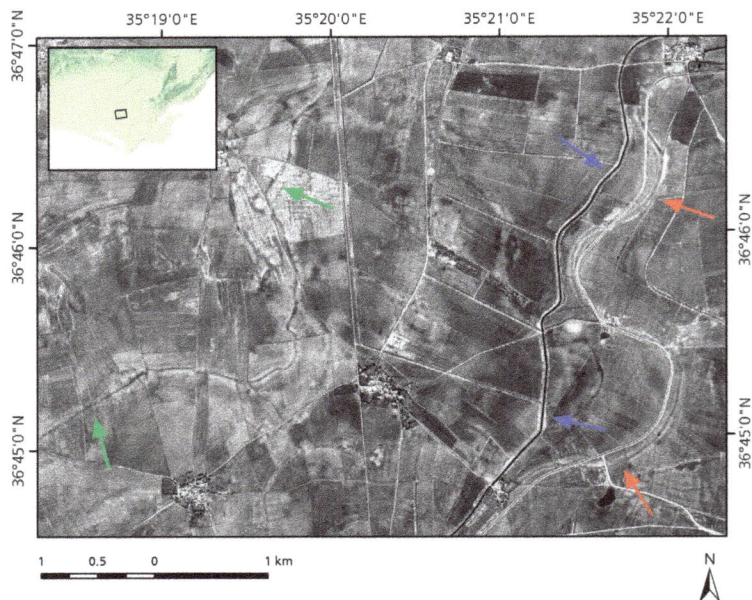

Figure 11. Detail of a CORONA scene (DS1105-2267, DA003, 20.11.1968). The course of the large palaeochannel visible on the DEM is marked with red arrows. A modern canal is following its course. The blue arrows points to a recent canal from the 1960s. The green arrow point to a relict canal © Susanne Rutishauser.

3.2. Distribution of Archaeological Sites in the Area

Even though for precise dating of palaeochannels and relict canals coring and the analysis of sediment is necessary, the location of archaeological sites give some evidence about the dating of these relict canals and palaeochannels (Figure 12).

First, a relation between the age of an archaeological site and its distance to the modern coastline can be observed. The older a site the farther it is situated from the sea. In this area, the closest chalcolithic sites are in distance of 18–20 km to the shoreline. These are Dervisli (Seton-Williams No. 135 [26]) and Domuz III (Seton-Williams No. 102 [26]). While the closest Bronze Age site Gavur Köy (Seton-Williams No. 119 [26]) is only 13 km from today's coastline located. The known Iron Age sites are located in a similar distance to the shoreline. Sites dating to the Roman Period are located right behind the Akyatan Laggon and at the Kap Karataş (Magarsos [60]). These settlement distributions give an idea about shifts of the coastline.

In the area of the large palaeochannel, there is only one site (Domuz III, Seton-Williams No. 102 [26]) dating to the Chalcolithic period known. However, several sites dating to Hellenistic and Roman period are situated close to the palaeochannels (Figure 12, mapped in green). We assume the danger of flooding during the Chalcolithic Period affected this distribution. The accumulation of sediments and the ideal slope for irrigation were the preconditions for sites in later periods. Sites on natural levee ridges were first occupied after these landforms were already developed and no longer subject to flooding [61].

Several sites dating mainly from Bronze Age to the Roman Period are situated along relict canals (mapped in pink). Therefore, the canals have to be in use since the Bronze Age.

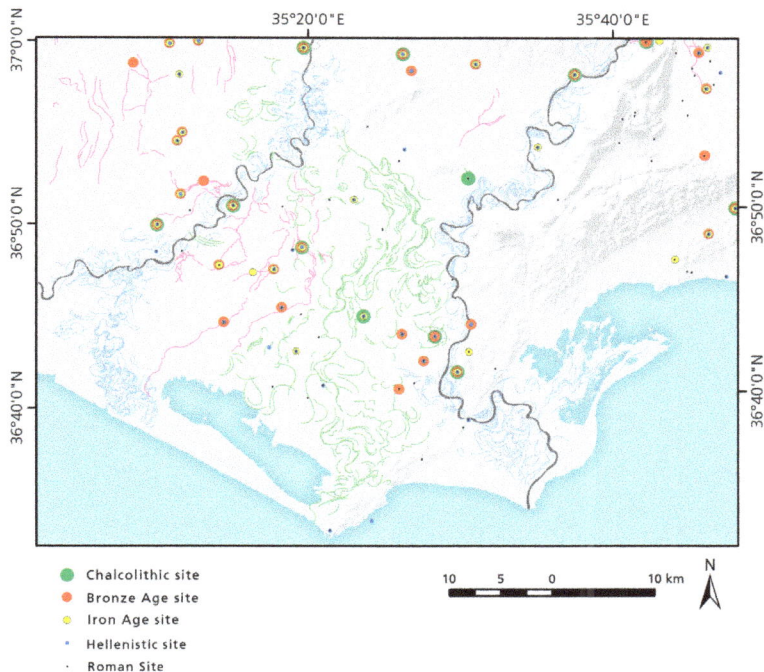

Figure 12. Sites dating from the Chalcolithic to the Roman Period in the Çukurova as well as palaeochannels and relict canals © Susanne Rutishauser.

4. Discussion

4.1. Dating of Palaeochannels and the Relict Canals

A soil survey of the Çukurova University [62] show that the palaeochannels mapped in Figure 7 correspond to late Holocene river terraces while the surrounding plain was built during the middle Holocene (Figure 13). The late Holocene river terraces spread in the area between today's river course of Seyhan and Ceyhan.

Additionally, the mapping of former river delta soils give some evidence of former location of the coastline (Figure 13). Fluvisols and Vertisols of the delta reach up to 20 km inland.

Based on the data of the soil survey of the Çukurova University (Figure 13) and the presence of Bronze Age sites along relict canals (Figure 12) we propose that the river Seyhan followed already in the Bronze Age (3rd and 2nd millennium BC) a western course like today (Figure 14). The intake of the relict canals must have been located west of the supposed former course of the large palaeochannel (mapped in green, Figure 12). Therefore, the large palaeochannel must date before the Bronze Age, possibly in the Neolithic (9th–7th millennium BC) or Chalcolithic Period (6th–4th millennium BC). However, a shared river mouth in previous times of former Seyhan and Ceyhan rivers—as authors in Antiquity like Pseudo-Scylax or in Medieval times like Abū l-Fidā' reported [31] (p. 465)—is conceivable for the time when the large palaeochannel was active.

The Late Bronze Age harbor town *MLWM* could have been a river harbor situated east of the river Ceyhan likely identified with the archaeological site Domuztepe [10] (p. 12). Figure 14 shows a preliminary reconstruction of the palaeo-landscape from the mid to the late Holocene.

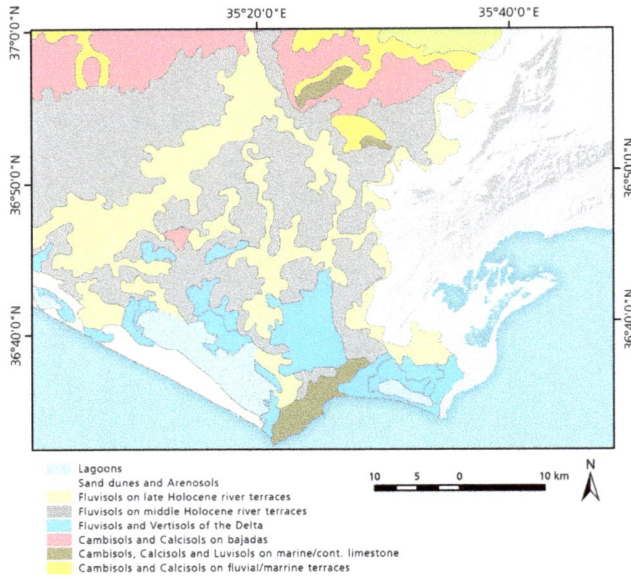

Figure 13. Soil map of Çukurova (data collection [62]), © Susanne Rutishauser.

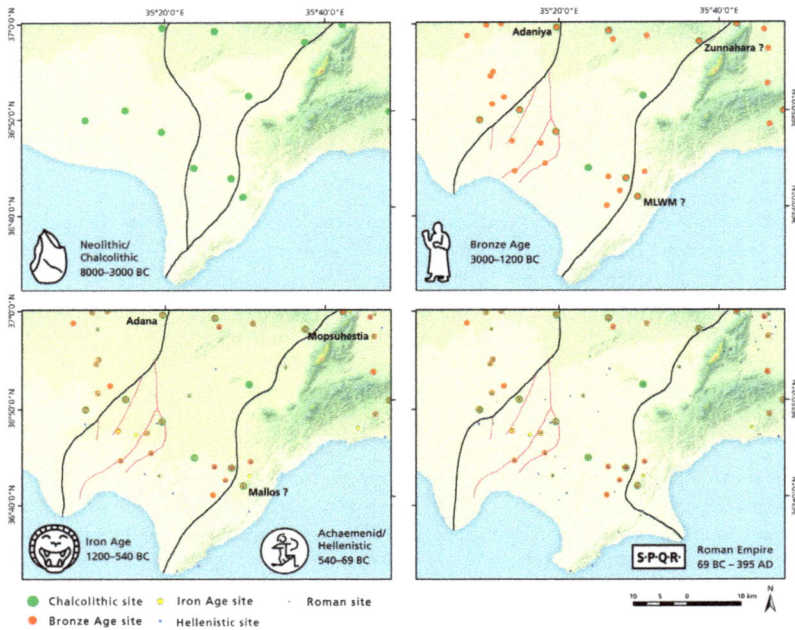

Figure 14. Preliminary reconstruction of the palaeo-landscape of the eastern Çukurova. The schematic black lines represent former river courses of Seyhan and Ceyhan, the pink lines show relict canals, © Susanne Rutishauser.

Today, several modern sites are located along a modern canal following the western relict canal (Figure 15, No. 1). However, there are no modern sites located along the eastern relict canal (Figure 15, No. 2). Therefore, the eastern relict canal must have been in use during previous periods. Since several sites dating to the Bronze Age are located along these relict canals it is possible that they were in use from the Bronze Age to Roman Times.

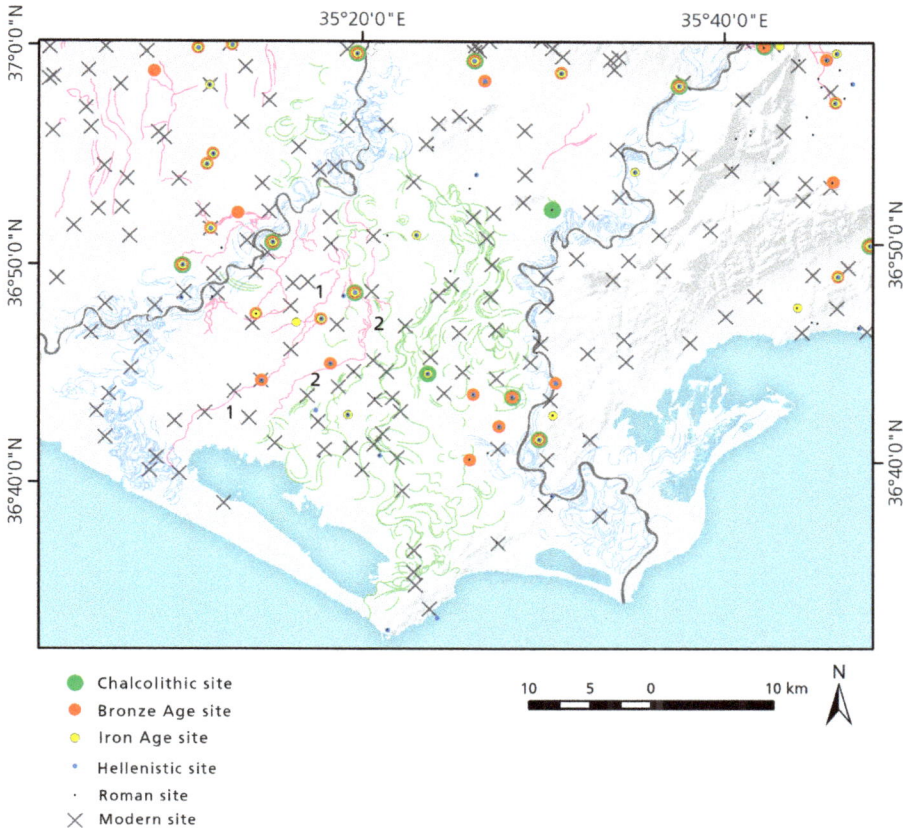

Figure 15. Location of archaeological and modern sites in the Çukurova. Modern river courses (grey), palaeochannels (blue, green) and relict canals (pink) in the Çukurova © Susanne Rutishauser.

4.2. Irrigation System

The average annual rainfall in the Çukurova is over 750 mm/year [63]. This rate is high enough for rainfed agriculture. Supplementary irrigation systems increase the productivity of rainfed areas. Due to its fertile alluvial soils and mild climate, the Çukurova produce high crop yields twice a year. This makes the region an important contributor to the gross value of Turkeys agricultural crop production of 5% in 1995 [63]. In antiquity the region was famous for agricultural products [34] (pp. 109–111).

A large limestone statue of a male bearded figure standing on a basalt basis showing a chariot pulled by two bulls was found 1997 in Çineköy (Figure 3), a village in the middle of the Çukurova [64]. The male figure represents a god, who must be identified with the Storm-god Tarhunzas. A bilingual inscription in Luwian and Phoenician is engraved in the monument [64–66]. The more than 2.50 m high monument is shown today in the Museum Adana. The inscription is formed of two main parts:

an introduction dedicated to the self-presentation of the king and a longer section describing the king's achievements [66]. The king praise himself: "I caused the plain of Hiyawa to prosper" [66] (p. 187). He named the plain also as "land of the rivers" [64] (p. 972). There is no large Late Bronze Age site in vicinity known which would explain the location of such monument. However, the find spot is located at a relict canal. Therefore, it might by possible that the monument was erected at an artificial canal. Additional irrigation must have used already during the Late Bronze and Iron Age.

The overlay of the modern canal network for irrigation with mapped relict canals and palaeochannels show that the course of modern canals corresponds to relict canals and palaeochannels (Figure 16). Today, most canals in the Çukurova are constructed with concrete slabs. Many canals have a meandering course even though a straight channel network would be more effective and reduce the number of necessary channels. Thus, the ideal slope of the palaeochannels was followed when constructing modern canals.

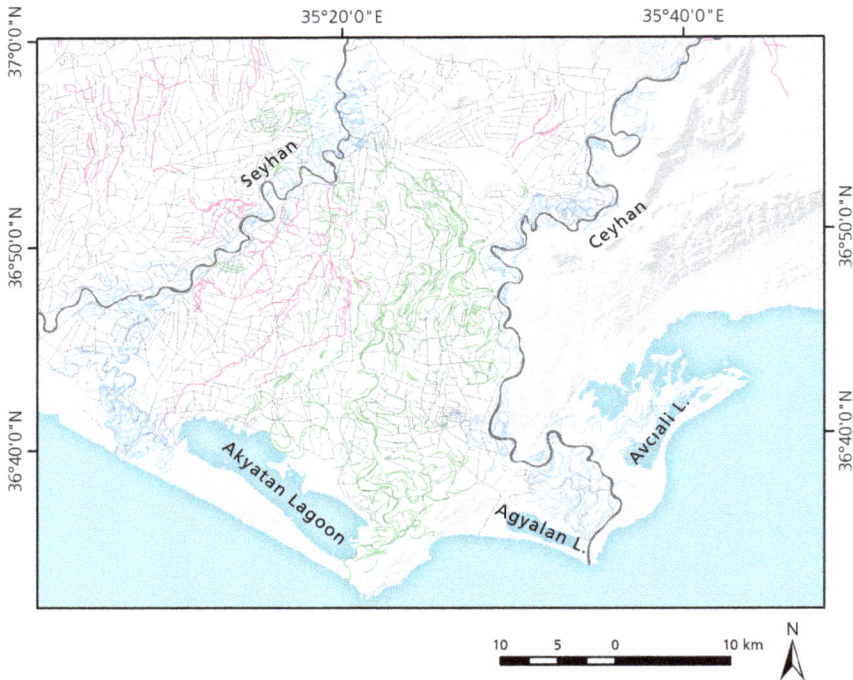

Figure 16. Modern river courses (grey), palaeochannels (blue, green) and relict canals (pink) in the Çukurova. The network of thin grey lines shows the modern network for canals for irrigation (mapped based on freely available VHR data in ArcGIS) © Susanne Rutishauser.

4.3. The Advantage of High Resolution DEM Data for Archaeology

Only a few archaeological sites are published with their coordinates. Therefore, we needed additional information for the precise localization of the sites for the settlement database. The CORONA imagery was an important source of information. Still, small sites were not always visible on this historic satellite imagery. For such cases, the TanDEM-X based height model was an important addition. A low elevated hill can be still visible in a height model while it is not visible anymore on optical data.

Large palaeochannels can be detected easily and even smaller palaeochannels can be uncovered with a stretched color ramp. High resolution DEM-data helps not only to detect features. It is also possible to take precise measurements. The data allows generating cross section to make small differences in height of 1 m visible. Nevertheless, it does not replace ground truthing.

5. Conclusions

The combined use of remote sensing data, soil surveys and archaeological data are first indications for the reconstruction of ancient river courses, when there is only limited coring available. Based on the available data we assume a more eastern course of the river Seyhan during the early or middle Holocene/the Neolithic Period. A major avulsion took place during the late Neolithic or Chalcolithic Period. This shift of the river Seyhan to the west must have an influence to the distribution of sites. In comparison to southern Mesopotamia, where major shifts of rivers were leading to depopulation of large areas, the consequences in the Çukurova might have been less drastic. While irrigation was essential for agriculture in southern Mesopotamia farmers in the Çukurova could still rely on rainfed agriculture for their subsistence. The influence of channel avulsion on settlement must be further investigated.

Today, the landscape of the Çukurova is structured from a bird's eye view by the agricultural area as well as modern river beds, irrigation canals and channel avulsion. Despite the meandering character of palaeochannels modern irrigation canals still follow their course. The location of archaeological sites along both relict canals and modern irrigation canals demonstrate that irrigation canals are in use—maybe not continuous—over a very long period. The finding of such a big monument as the Çineköy statue close to a relict canal gives further evidence for the use of irrigation in this area at least since the Late Bronze Age.

However, the limitations of this study can only be overcome by further investigations. The analysis of drilling cores along transects to get information about the past shifts of the coastline and the sedimentation history of the Ceyhan and Seyhan rivers is necessary.

Acknowledgments: All TanDEM-X data were kindly provided by German Aerospace Center (DLR) within the project "The potential of Tandem-X DEM data for archaeological prospection—case studies in Plain Cilicia, Turkey" (project ID: OTHER0340).

Author Contributions: Susanne Rutishauser and Ralph Rosenbauer designed the concept for the case study. Ralf Buchbach and Stefan Erasmi performed the processing of the TanDEM-X data. Susanne Rutishauser analyzed the data and wrote the paper. All authors contributed to the discussion of the results.

Conflicts of Interest: The authors declare no conflict of interest.

References

1. Gürbüz, K. An example of river course changes on a delta plain: Seyhan Delta (Cukurova plain, southern Turkey). *Geol. J.* **1999**, *34*, 211–222. [CrossRef]
2. Erol, O. Ceyhan Deltasinin Jeomorfolojik Evrimi: Geomorphological Evolution of the Ceyhan River Delta: Eastern Mediterranean Coast of Turkey. *Aegean Geogr. J.* **2003**, *12*, 58–81.
3. Mahmoud, Y.; Masson, F.; Meghraoui, M.; Cakir, Z.; Alchalbi, A.; Yavasoglu, H.; Yönlü, O.; Daoud, M.; Ergintav, S.; Inan, S. Kinematic study at the junction of the East Anatolian fault and the Dead Sea fault from GPS measurements. *J. Geodyn.* **2013**, *67*, 30–39. [CrossRef]
4. Öner, E.; Hocaoğlu, B.; Uncu, L. Palaeogeographical Surveys around the Mound of Gözlükule (Tarsus). In *Field Seasons 2001–2003 of the Tarsus-Gözlükule Interdisciplinary Research Project*; Özyar, A., Ed.; Ege Yayınları: İstanbul, Turkey, 2005; pp. 69–82.
5. Adams, R.M. *Heartland of Cities. Surveys of Ancient Settlement and Land Use on the Central Floodplain of the Euphrates*; The University of Chicago Press: Chicago, IL, USA, 1981.
6. Newhard, J.M.L.; Levine, N.; Rutherford, A. Least-Cost Pathway Analysis and Inter-Regional Interaction in the Göksu Valley, Turkey. *Anatol. Stud.* **2008**, *58*, 87–102. [CrossRef]

7. Ehringhaus, H.; Starke, F. *Götter, Herrscher, Inschriften. Die Felsreliefs der Hethitischen Grossreichszeit in der Türkei*; Philipp von Zabern: Mainz, Germany, 2005.
8. Schiettecatte, J.P. Geology of the Misis Mountains. In *Geology and History of Turkey*; Campbell, A.S., Ed.; Petroleum Exploration Society of Lybia: Tripoli, Libya, 1971; pp. 305–315.
9. Jarvis, A.; Reuter, H.I.; Nelson, A.; Guevara, E. CGIAR-CSI SRTM 90 m Database. Hole-filled SRTM for the Globe Version 4. 2008. Available online: http://srtm.csi.cgiar.org (accessed on 18 October 2017).
10. Forlanini, M. How to infer ancient roads and itineraries from heterogeneous Hittite texts: The case of the Cilician (Kizzuwatnean) road system. *Riv. Stor. Ambient. Cult. Vicino Oriente Antico* **2013**, *10*, 1–34.
11. Aksu, A.E.; Calon, T.J.; Hall, C.J.; Mansfield, S.; Yaşar, D. The Cilicia–Adana basin complex, Eastern Mediterranean: Neogene evolution of an active fore-arc basin in an obliquely convergent margin. *Mar. Geol.* **2005**, *221*, 121–159. [CrossRef]
12. Burton-Ferguson, R.; Aksu, A.E.; Hall, C.J. Seismic stratigraphy and structural evolution of the Adana Basin, eastern Mediterranean. *Mar. Geol.* **2005**, *221*, 189–222. [CrossRef]
13. Isola, I.; Bini, M.; Ribolini, A.; Zanchetta, G.; D'Agata, A.L. Geomorphology of the Ceyhan River lower plain (Adana Region, Turkey). *J. Maps* **2017**, *13*, 133–141. [CrossRef]
14. Aksu, A.E.; Ulug, A.; Piper, D.; Konuk, Y.T.; Turgut, S. Quaternary sedimentary history of Adana, Cilicia and Iskenderun Basins: Northeast Mediterranean Sea. *Mar. Geol.* **1992**, *104*, 55–71. [CrossRef]
15. Akbulut, N.; Bayarı, S.; Akbulut, A.; Şahin, Y. Rivers of Turkey. In *Rivers of Europe*; Tockner, K., Ed.; Elsevier: Amsterdam, The Netherlands, 2009; pp. 643–672.
16. Bal, Y.; Çetin, H.; Demirkol, C. An update on the coastline changes and evolution of the Seyhan and Ceyhan Deltas in the Northeast Mediterranean, Turkey. In Proceedings of the International Symposium on Geology and Environment, GEONEV '97, Istanbul, Turkey, 1–5 September 1997; p. 64.
17. Çetin, H.; Bal, Y.; Demirkol, C. Engineering and Environmental Effects of Coastline Changes in Turkey, Northeastern Mediterranean. *Environ. Eng. Geosci.* **1999**, *5*, 1–16.
18. Spezzaferri, S.; Basso, D.; Koral, H. Holocene palaeoceanographic evolution of the Iskenderun Bay, South-Eastern Turkey, as a response to river mouth diversions and human impact. *Mediterr. Mar. Sci.* **2000**, *1*, 19–43. [CrossRef]
19. Walsh-Kennedy, S.; Aksu, A.E.; Hall, C.J.; Hiscott, R.N.; Yaltırakb, C.; Çifçic, G. Source to sink: The development of the latest Messinian to Pliocene-Quaternary Cilicia and Adana Basins and their linkages with the onland Mut basin, eastern Mediterranean. *Tectonophysics* **2014**, *622*, 1–21. [CrossRef]
20. Bal, Y.; Kelling, G.; Kapur, S.; Akça, E.; Çetin, H.; Erol, O. An improved method for determination of Holocene coastline changes aroun two ancient settlements in southern Anatolia: A geoarchaeological approach to historical land degradation studies. *Land Degrad. Dev.* **2003**, *14*, 363–376. [CrossRef]
21. Evans, G. The recent sedimentation of Turkey and the adjacent Mediterranean and Black Seas: A review. In *Geology and History of Turkey*; Campbell, A.S., Ed.; Petroleum Exploration Society of Lybia: Tripoli, Turkey, 1971; pp. 385–406.
22. Evans, G. Recent coastal sedimentation: A review. In *Marine Archaeology, Proceedings of the 23rd Symposium of the Colston Research Society, Bristol, UK, 4–8 April 1971*; Blackman, D.J., Ed.; Butterworths: London, UK, 1973; pp. 89–112.
23. Kuleli, T. Quantitative analysis of shoreline changes at the Mediterranean Coast in Turkey. *Environ. Monit. Assess.* **2010**, *167*, 387–397. [CrossRef] [PubMed]
24. Kelletat, D. A Holocene Sea Level Curve for the Eastern Mediterranean from Multiple Indicators. *Z. Geomorphol.* **2005**, *137*, 1–9.
25. Steadman, S.R. Prehistoric sites on the cilician coastal plain: Chalcolithic and early bronze age pottery from the 1991 Bilkent University Survey. *Anatol. Stud.* **1994**, *44*, 85–103. [CrossRef]
26. Seton-Williams, M.V. Cilician Survey. *Anatol. Stud.* **1954**, *4*, 121–174. [CrossRef]
27. Steadman, S.R. Isolation or Interaction: Prehistoric Cilicia and the Fourth Millenium Uruk Expansion. *J. Mediterr. Archaeol.* **1996**, *9*, 131–165. [CrossRef]
28. Jean, E. Sociétés et Pouvoirs en Cilicie au 2nd Millénaire av. J.-C. Ph.D. Dissertation, Univeristé Paris I Sorbonne, Paris, France, 2010.
29. Hawkins, J.D. *Que, A. Reallexikon der Assyriologie*; De Gruyter: Berlin, Germany, 2006–2008; pp. 191–195.
30. Kaufman, S.A. The phoenician Inscription of the Incirli Trilingual: A tentative Reconstruction and Translation. *Maarav* **2007**, *14*, 7–26.

31. Langlois, V. *Voyage dans la Cilicie et dans les Montagnes du Taurus*; Benjamin Duprat: Paris, France, 1861.

32. Abūāl-Fidā', I.; Reinaud, J.T. *Géographie d'Aboulféda*; Paris Impr. Royale: Paris, France, 1848–1883.

33. Goldman, H. Preliminary Expedition to Cilicia 1934, and Excavations at Gözlü Kule, Tarsus 1935. *Am. J. Archaeol.* **1935**, *39*, 526–549. [CrossRef]

34. Hild, F.; Hellenkemper, H. *Kilikien und Isaurien*; Verl. der Österreichischen Akademie der Wissenschaften: Wien, Austria, 1990.

35. Bossert, H.T.; Alkım, U.B. *Karatepe. Kadirli and Its Environments/Second Preliminary Report*; Üniversite Basimevi: Istanbul, Turkey, 1947.

36. Çambel, H. *Karatepe-Aslantaş. The Inscriptions: Facsimile Edition*; Walter de Gruyter: Berlin, Germany, 1999.

37. Çambel, H. Karatepe-Aslantas, a many-sided Project. In *Istanbul University's Contributions to Archaeology in Turkey (1932–2000)*; Belli, O., Ed.; Istanbul University Rectorate Research Fund: Istanbul, Turkey, 2001; pp. 195–203.

38. Sicker-Akman, M. Der Fürstensitz der späthethitischen Burganlage Karatepe-Arslantaş. *Istanb. Mitt.* **2001**, *50*, 131–142.

39. Özgen, I.; Gates, M.-H. Report on the Bilkent University Archaeological Survey in Cilicia and the Abū āl-Fidārthern Hatay: August 1991. *Araştirma Sonuçlari Toplantisi* **1992**, *10*, 387–394.

40. Lehmann, G.; Killebrew, A.E.; Gates, M.-H.; Halpern, B. The Mopsos Project: The 2004 Season of Archaeological Survey in the bay of Iskenderun, Eastern Cilicia. *Araştirma Sonuçlari Toplantisi* **2005**, *23*, 79–87.

41. Gates, M.-H. The Hittite Seaport Izziya at Late Bronze Age Kinet Höyük (Cilicia). *Near East. Archaeol.* **2013**, *76*, 223–234. [CrossRef]

42. Kozal, E.; Novák, M. Sirkeli Höyük: A Bronze and Iron Age Urban Settlement in Plain Cilicia. *Der Anschnitt* **2013**, *25*, 229–238.

43. Girginer, K.S. 2013 Yılı Tatarlı Höyük Kazısı. *Kazı Sonuçları Toplantısı* **2015**, *36*, 431–446.

44. Day, D.A.; Logsdon, J.M.; Latell, B. *Eye in the Sky. The Story of the Corona Spy Satellites*; Smithsonian Institution Press: Washington, DC, USA, 1998.

45. Ur, J. Spying on the Past: Declassified Intelligence Satellite Photographs and Near Eastern Landscapes. *Near East. Archaeol.* **2013**, *76*, 28–36.

46. Casana, J. A Landscape Context for Paleoethnobotany: The Contribution of Aerial and Satellite Remote Sensing. In *Method and Theory in Paleoethnobotany*; University Press of Colorado: Boulder, CO, USA, 2015; pp. 315–335.

47. Galiatsatos, N. *Assessment of the CORONA Series of Satellite Imagery for Landscape Archaeology: A Case Study from the Orontes Valley, Syria*; University of Durham: Durham, UK, 2004.

48. Casana, J.; Cothren, J. The CORONA Atlas Project: Orthorectification of CORONA Satellite Imagery and Regional-Scale Archaeological Exploration in the Near East. In *Mapping Archaeological Landscapes from Space*; Comer, D.C., Harrower, M.J., Eds.; Springer: New York, NY, USA, 2013; pp. 33–43.

49. Scollar, I.; Galiatsatos, N.; Mugnier, C. Mapping from CORONA: Geometric Distortion in KH4 Images. *Photogramm. Eng. Remote Sens.* **2016**, *82*, 7–13. [CrossRef]

50. Werninghaus, R.; Buckreuss, S. The TerraSAR-X Mission and System Design. *IEEE Trans. Geosci. Remote Sens.* **2010**, *48*, 606–614. [CrossRef]

51. Erasmi, S.; Rosenbauer, R.; Buchbach, R.; Busche, T.; Rutishauser, S. Evaluating the Quality and Accuracy of TanDEM-X Digital Elevation Models at Archaeological Sites in the Cilician Plain, Turkey. *Remote Sens.* **2014**, *6*, 9475–9493. [CrossRef]

52. Kumar, V. Palaeo-Channel. In *Encyclopedia of Snow, Ice and Glaciers*; Singh, V.P., Ed.; Springer: Dordrecht, The Netherlands, 2011; p. 803.

53. Charlton, R. *Fundamentals of Fluvial Geomorphology*; Routledge: London, UK, 2008.

54. Knighton, A.D. *Fluvial Forms and Processes: A New Perspective*; Les Presses de l'Université de Montréal; Érudit: Montreal, QC, Canada, 1998.

55. Wiseman, J.; El-Baz, F. *Remote Sensing in Archaeology*; Springer: New York, NY, USA, 2007.

56. Zepp, H. *Geomorphologie. Eine Einführung*; Schöningh: Stuttgart, Germany, 2004.

57. Hritz, C.; Wilkinson, T.J. Using Shuttle Radar Topography to map ancient water channels in Mesopotamia. *Antiquity* **2006**, *80*, 414–424.

58. Rayne, L.E. Water and territorial empires. Doctoral thesis, Durham University, Durham, UK, 2014.

59. Casana, J.; Cothren, J.; Kalayci, T. Swords into Ploughshares: Archaeological Applications of CORONA Satellite Imagery in the Near East. *Int. Archaeol.* **2012**. [CrossRef]
60. Rosenbauer, R. Topographisch-urbanistischer Survey des Ruinengeländes am Kap Karatas/Türkei. Vorbericht zur ersten Kampagne 2006. *Hefte Archäol. Semin. Univ. Bern* **2007**, *20*, 107–119.
61. Saucier, R.T. *Geomorphology and Quaternary Geologic History of the Lower Mississippi Valley*; U.S. Army Engineer Waterways Experiment Station: Vicksburg, MS, USA, 1994.
62. Dinç, U.; Senol, S.; Cangir, C.; Dinç, A.O.; Akça, E.; Dingil, M.; Oztekin, E.; Kapur, B.; Kapur, S. Soil Survey and Soil Database of Turkey. *Eur. Soil Bur. Res. Rep.* **2005**, *9*, 371–375.
63. Alphan, H.; Yilmaz, K.T. Monitoring environmental changes in the Mediterranean coastal landscape: The case of Cukurova, Turkey. *Environ. Manag.* **2005**, *35*, 607–619. [CrossRef] [PubMed]
64. Tekoglu, R.; Lemaire, A. La bilingue royale louvito-phénicienne de Çineköy. *C. R. Séances l'Acad. Inscr. Belles-Lett.* **2000**, *144*, 961–1007. [CrossRef]
65. Lanfranchi, G.B. The Luwian-Phoenician Bilingual of Cineköy and the Annexation of Cilicia to the Assyrian Empire. In *Von Sumer bis Homer: Festschrift für Manfred Schretter zum 60. Geburtstag am 25. Februar 2004*; Rollinger, R., Schretter, M.K., Eds.; Ugarit-Verlag: Münster, Germany, 2005; pp. 481–496.
66. Lanfranchi, G.B. The Luwian-Phoenician bilinguals of ÇINEKÖY and KARATEPE: An ideological dialogue. In *Getrennte Wege?: Kommunikation, Raum und Wahrnehmung in der Alten Welt*; Rollinger, R., Luther, A., Wiesehöfer, J., Eds.; Verlag Antike: Frankfurt am Main, Germany, 2007; pp. 179–217.

geosciences

MDPI

Article

Detecting Landscape Disturbance at the Nasca Lines Using SAR Data Collected from Airborne and Satellite Platforms

Douglas C. Comer [1,*], Bruce D. Chapman [2] and Jacob A. Comer [1]

[1] Cultural Site Research and Management Foundation, 2113 Saint Paul Street, Baltimore, MD 21218, USA; jcomer@culturalsite.com

[2] Jet Propulsion Laboratory, California Institute of Technology, 4800 Oak Grove Drive, Pasadena, CA 91109, USA; bruce.d.chapman@jpl.nasa.gov

[*] Correspondence: dcomer@culturalsite.com; Tel.: +1-410-244-6320

Received: 1 August 2017; Accepted: 12 October 2017; Published: 16 October 2017

Abstract: We used synthetic aperture radar (SAR) data collected over Peru's Lines and Geoglyphs of Nasca and Palpa World Heritage Site to detect and measure landscape disturbance threatening world-renowned archaeological features and ecosystems. We employed algorithms to calculate correlations between pairs of SAR returns, collected at different times, and generate correlation images. Landscape disturbances even on the scale of pedestrian travel are discernible in correlation images generated from airborne, L-band SAR. Correlation images derived from C-band SAR data collected by the European Space Agency's Sentinel-1 satellites also provide detailed landscape change information. Because the two Sentinel-1 satellites together have a repeat pass interval that can be as short as six days, products derived from their data can not only provide information on the location and degree of ground disturbance, but also identify a time window of about one to three weeks during which disturbance must have occurred. For Sentinel-1, this does not depend on collecting data in fine-beam modes, which generally sacrifice the size of the area covered for a higher spatial resolution. We also report on pixel value stretching for a visual analysis of SAR data, quantitative assessment of landscape disturbance, and statistical testing for significant landscape change.

Keywords: nasca lines; UAVSAR; Sentinel-1; SAR interferometry (InSAR); disturbance; world heritage; archaeology; paracas; heritage management; *pampas*; geoglyphs

1. Introduction

1.1. Applications of SAR to Archaeology: The State of the Art

The use of a synthetic aperture radar (SAR) for archaeological research and management began as early as the 1980s, when paleo-riverine networks and soils altered by agriculture provided clues to the locations of ancient sites in desert environments [1–3]. By the first decade of the twenty-first century, the direct detection of archaeological sites with SAR data collected by aerial platforms, notably AirSAR and GeoSAR [4], sometimes in concert with multispectral data [5–7], had been demonstrated.

Recent years have seen a surge in interest in using satellite SAR data for research and the management of archaeological sites and landscapes. This is partly due to the profusion of SAR data being collected: satellite SAR instruments can cover much greater areas than airborne SAR, and satellite SAR can now be obtained for, essentially, the entire world. In this project, we used data collected in Sentinel-1's Interferometric Wide (IW) Swath mode, which covers a strip 250 km wide. This data is collected continually, and there is no need to place a special order or pay for it, making it eminently practical for the constant monitoring of large areas.

Satellite SAR data acquired in fine-beam or spotlight modes can be expensive or difficult to acquire, but it can also achieve pixel sizes as small as those generated from some kinds of optical satellite data. Even casual users of satellite imagery are familiar with the usefulness of high-resolution optical data, which facilitates the visual identification and analysis of features on the ground. The color images in Google Earth, for example, can be pan-sharpened to produce 65 cm pixels (though in certain regions, pixel size can be 2.5 m or even 15 m). For-profit companies offer imagery with even greater resolution: DigitalGlobe advertises that imagery generated from its WorldView-3 satellite has a pan-sharpened pixel size of 31 cm and an accuracy better than 5 m.

In comparison, however, TerraSAR-X satellites collect X-band data that can produce 16 m pixels when operating in ScanSAR mode over a scene 100 km by 150 km, and 0.24 m pixels in Staring Spotlight mode over an area that varies from 2.5 to 2.8 km in azimuth and 4.6 to 7.5 km in range [8]. ALOS PALSAR-2, COSMO-SkyMed, and Radarsat-2 can also provide imagery with pixels smaller than 1 m. As with high-resolution optical data, one can often discern targets of interest in high-resolution SAR amplitude imagery. Further, SAR data, unlike optical data, can be collected regardless of clouds or solar illumination. SAR's indifference to these conditions is a great advantage over spectral data, particularly in parts of the world where cloud cover is common, and it is obviously valuable for change detection.

In some cases, such as when the features under consideration are characterized more by their spectral reflectance than their physical shape, SAR may be no more useful than spectral data for detection or monitoring [9,10]. In others, SAR's sensitivity to structure makes it a clearly superior tool for site and feature detection. For example, at Pachacamac, in Peru, COSMO SkyMed X-band high-resolution imagery (with a pixel size of 3 m) was used to detect earthen structures that are only partially above ground; visually, these structures would be difficult to distinguish from the surrounding landscape [11]. An interferometric analysis of SAR data (InSAR) has been used to detect and measure deformation of the Earth's surface, including that produced by seismic or volcanic activity or by groundwater depletion. Depending on the application, SAR can be used to measure such deformation at centimeter scales [12]. Surface deformation that threatens the ancient venue for the Olympics has been monitored with Differential Interferometric SAR (DInSAR) [13]. High-resolution TerraSAR-X data is being used to monitor ground subsidence in the ancient quarters of Mexico City [14]. The movement of the monuments at Angkor, along with changes in surface elevation, has been monitored with TerraSAR/TanDEM-X data [15].

For general discussions of the technical aspects of SAR as they relate to archaeology, we direct the reader to other works [16,17]. Several studies have applied satellite SAR data analysis to our area of interest (which offers a particularly suitable environment for SAR research, as we describe in the next section), demonstrating the technology's usefulness in archaeological prospection [18,19] and research on the regional environment and its influences on human activity [20]. Satellite spectral data has been used for similar purposes [21,22].

This area has also been the focus of several studies on the use of SAR for monitoring anthropogenic and natural changes to the landscape [23–25]. It is on these applications of SAR that, with this project, we aimed to build a model by using satellite SAR to detect landscape disturbance on a finer scale than any of the studies just referenced; by proving the concept that high-temporal-resolution satellite SAR data can provide a continual stream of current, relevant information for protected area management and research; and by demonstrating that this data provides the basis for a quantitative assessment and the statistical testing of landscape disturbance in specific areas at small scales. In so doing, we demonstrate some advantages of satellite SAR over other potential monitoring methods, such as ground survey or change detection with high-resolution spectral data.

1.2. The Contribution of Correlation Images Generated with UAVSAR and Sentinel-1 Data

We demonstrate the capacity of repeat-pass InSAR to monitor small-scale landscape change, and more specifically to measure land-surface change caused by, for example, flooding, mining, erosion, and even vehicular and pedestrian traffic. The detection of landscape disturbance at the structural, and even micro-structural, level described here is different from the many applications of SAR data to archaeological landscape monitoring that have focused on ground deformation [14], large-scale erosion [25], and, more recently, the movements of monuments [15].

Our demonstration area is the Lines and Geoglyphs of Nasca and Palpa World Heritage Site (hereafter referred to as the Nasca Lines World Heritage Site), in southern Peru (see Figure 1). An analysis of L-band SAR data collected from an airborne platform over the Nasca Lines World Heritage Site demonstrates the data's utility in monitoring the landscape and its archaeological features. Section 2 begins with an account of that analysis, which follows from work published in 2015 [26]. With the results of that analysis in mind, we hypothesized that satellite SAR data, specifically C-band Sentinel-1 data, acquired at different times over the same area, could detect locations where the landscape had been changed by relatively small-scale disturbances, such as those listed above.

These disturbances are a constant concern at the Nasca Lines World Heritage Site, which was designated to recognize and protect the fragile arrangements of stones and soil left on the area's desert floor by ancient civilizations (first the Paracas people, then the Nasca) between approximately 700 BCE and 700 CE. Viewed from above, these arrangements take a variety of forms: geometric shapes, animals (such as the famous Hummingbird glyph), plants, humans, and mythological beings [27,28]. Despite the passage of centuries, the forms have been largely preserved. In cases where the glyphs' constituent stones or soil have been disturbed, further damage may be more likely due to the increased vulnerability to wind and water erosion [29,30].

This area was an ideal test site for our application of SAR to the precise detection of landscape disturbance. The Nasca Lines World Heritage Site is in the Peruvian-Chilean Desert, one of the most arid places on Earth. In the flat areas, or *pampas*, where the lines and geoglyphs are found, vegetation is almost absent, except in stream and river valleys that run through the desert on their way to the Pacific Ocean. This was helpful because SAR easily detects seasonal vegetative change and water, which manifest in the data similarly to landscape change and can therefore mask the disturbances with which we were concerned.

Villages in this area, both modern and ancient, are located along the rivers and streams that cut through the *pampas*. Rapid population increase in extant villages, many of which are becoming cities, as well as increasing tourism, have encouraged human activities that have degraded some of the lines and glyphs; during fieldwork for this project, for example, Ministerio de Cultura personnel pointed out a geometric glyph on which people from a nearby, expanding village play soccer, using the glyph's edges as boundaries. Monitoring destructive activities using traditional, on-the-ground methods is challenging because of the vast area that the lines and glyphs occupy (the Nasca Lines World Heritage Site covers over 73,000 hectares) and the impracticality of preventing access to it. As we show in the Results, the advantages of satellite SAR, and particularly Sentinel-1, that we described in Section 1.1, include high temporal resolution, wide area coverage, and sensitivity to structural change that can help circumvent this challenge.

We calculated the InSAR coherence, or correlation, for a pair of data sets collected by the National Aeronautics and Space Administration (NASA)/Jet Propulsion Laboratory (JPL) Uninhabited Aerial Vehicle Synthetic Aperture Radar (UAVSAR) platform on 19 March 2013 and 23 March 2015. In addition to the general observations that arose from analyzing this data and encouraged our further experimentation with Sentinel-1 data, an incident that occurred in the December of 2014, between the two UAVSAR flights, was of particular interest. Persons affiliated with Greenpeace, a non-governmental organization with an environmental focus, placed panels on the ground to form the message "Time for change! The future is renewable, Greenpeace" very near one of the most famous and pristine Nasca geoglyphs, the Hummingbird. The place and means of disturbance were well documented: Greenpeace posted a video on the internet that showed its team's approach to the glyph, the placement of the message, and aerial photos of the glyph and message together [31]. Because of the geoglyphs' extreme fragility, access to this area had been prohibited for decades before the incident (though this did not prevent unpermitted access). Greenpeace's activity offered an opportunity to determine, with UAVSAR, SAR data's utility in detecting, evaluating, and testing for landscape disturbance. We then followed a similar process in analyzing pairs of data sets collected by the Sentinel-1 satellites. Sentinel-1's data collection interval of six to twenty-four days made such an analysis particularly attractive, because it suggested the potential to use satellite SAR data to monitor landscape change, and therefore some threats to cultural and natural resources, over large areas and with a relatively high temporal resolution.

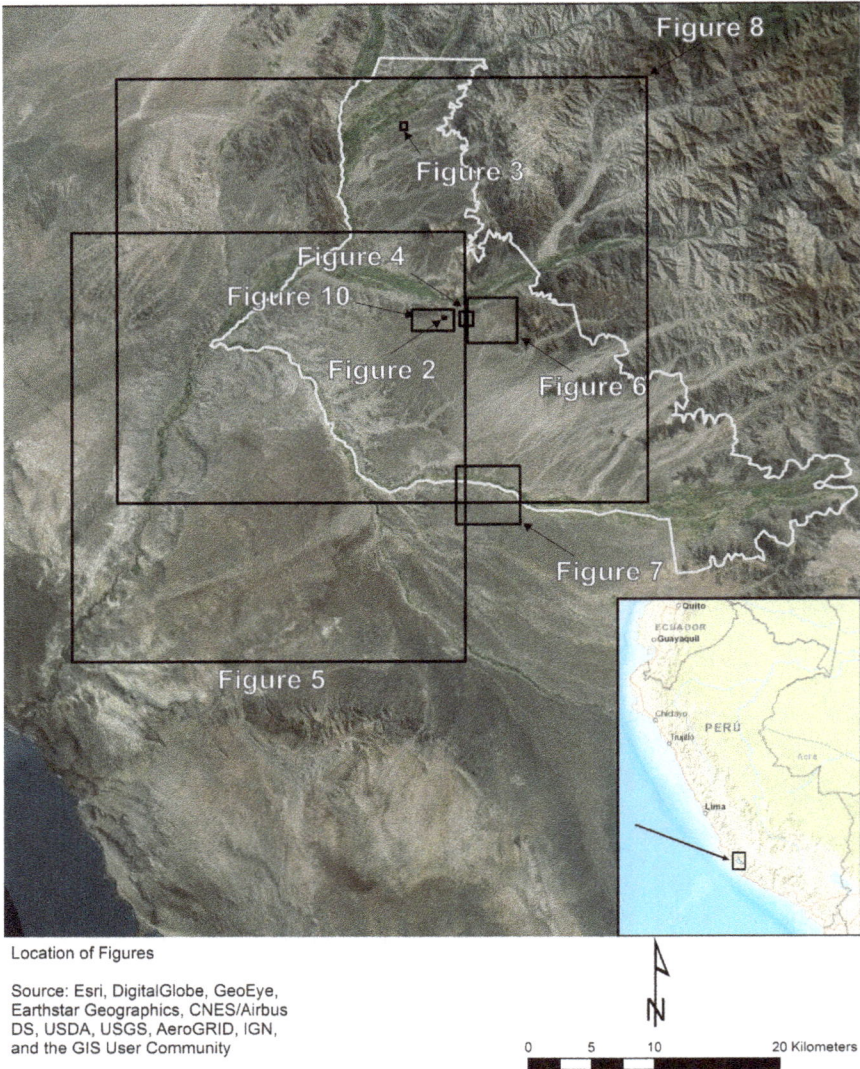

Figure 1. A map showing the Nasca Lines World Heritage Site and the locations of subsequent figures.

2. Materials and Methods

Acquiring the L-Band UAVSAR Data and Producing the Initial Correlation Images

UAVSAR is a NASA/JPL L-band airborne SAR platform, funded by NASA to conduct research for its Science Mission Directorate, that has been in operation since 2009. UAVSAR is designed to collect fully polarimetric SAR data at a spatial resolution of about two meters from an aircraft that can fly near-exact-repeat flight lines for interferometry research and applications [26].

In March 2013, UAVSAR data were collected opportunistically over the Nasca Lines, when the UAVSAR platform flew over them during a deployment to image volcanoes in Chile, simply to see how the lines and geoglyphs might appear in airborne SAR imagery. A second UAVSAR data set was gathered in March 2015 to see if the production of a very high-resolution correlation image could allow for the detection and measurement of disturbance to the geoglyphs, and to the Hummingbird geoglyph in particular. After examination of the correlation image, it became apparent that UAVSAR data could be used not only to evaluate damage to the Hummingbird geoglyph and the area immediately around it, but also to detect other ways in which the Nasca Lines had been disturbed and, more importantly, could be further disturbed in the future [26].

The UAVSAR has a wavelength of 23 cm and flies aboard the NASA Armstrong Gulfstream C-20A aircraft. This aircraft has been modified to fly extremely precise planned flight paths, such that it can repeat a previously flown path and only rarely diverge from it by more than five meters over hundreds of kilometers of flight. This allows the data from two or more UAVSAR flights to be made into interferograms or correlation images with a near-zero baseline between observations, using the technique called InSAR. With the UAVSAR datasets collected in March 2013 and March 2015, we generated a high-resolution correlation image that displays the degree to which SAR returns obtained in the first deployment of UAVSAR are statistically different from those obtained in the second [26].

The UAVSAR data were processed as "Single Look Complex" (SLC) image files. Each pixel in these files contains both the magnitude and phase of the SAR backscatter. As described below, the correlation image is produced from the SLC image files and has an output ground pixel spacing of about two meters.

To generate a correlation image that was sensitive to ground disturbance, we co-registered each pixel in the SLC images from the two UAVSAR flights. We then obtained a calibrated interferometric correlation γ_{cal} using Equation (1) [32], as follows:

$$\gamma_{cal} = \frac{|< E_1 E_2^* >|}{\sqrt{< E_1 E_1^* >< E_2 E_2^* >}} \left(1 + SNR^{-1}\right) \tag{1}$$

where E1 and E2 are the SLC estimates of the calibrated complex electric field for the two data takes and the ensemble average is over a given number of statistical looks. The * indicates the complex conjugate. The SNR is the signal-to-noise ratio, where the noise is given by an empirically derived quadratic fit to the noise-equivalent, radar-backscattered power of the image. γ_{cal} can vary between 0 (complete decorrelation or, equivalently, no correlation between the two images) and 1 (no decorrelation or, equivalently, complete correlation between the images). Normalization by the SNR can result in some areas having a correlation greater than 1, particularly in those areas that are shadowed by topographic features. For Sentinel-1 data, the correlation product can be obtained using the standard approaches implemented by the European Space Agency's (ESA's) Sentinel Application Platform (SNAP) [33]. For the UAVSAR data, Equation (1) was coded in the interactive display language (IDL).

As mentioned in the Introduction, one general property of the correlation data product is that vegetated areas have much lower values than areas without vegetation. Open water also decorrelates, because the scattering of the radio waves is highly variable over water. Atmospheric conditions can interfere with SAR collection, but we did not observe any kilometer-scale effects on the coherence from tropospheric water content, which is often visible in InSAR interferometric phase measurements.

The desert *pampas* of the Nasca Lines World Heritage Site are almost completely lifeless and static, and therefore, absent of any changes to the ground surface, the correlation in these areas should be close to 1. In areas with steep slopes, however, there is often decorrelation caused by erosion: in such places, a small movement of soil or stones can be augmented by gravity, resulting in greater decorrelation than would have been caused by a similar movement on flat ground. In both the UAVSAR and Sentinel-1 data, we found that such downslope erosion is common in the Nasca Lines World Heritage Site. (In applications to similar phenomena at larger scales, satellite SAR has been used to study landslides [34]).

The correlation products developed from SAR data can be visualized as gray-scale images, with pixel values ranging from 0 (black) to 255 (white). Lesser correlation (greater decorrelation) is indicated by darker pixel values. That is, for dry areas without vegetation, we can generally infer from the correlation images that the greater the landscape change over the two-year period between UAVSAR flights, the greater the difference between the returns from the two data takes, the lower the correlation (the greater the decorrelation), and the darker that area appears in the correlation images. This decorrelation might be caused by natural disturbances, such as erosion by flooding, or by anthropogenic disturbances, such as urban or agricultural development or surface disturbance by vehicles.

In the UAVSAR correlation image, to make decorrelation at the Greenpeace protest site more visually evident, we assigned a color scale to the correlation values, making areas of no or very low decorrelation green and areas of high decorrelation red (see Figure 2). Yellow (medium-low decorrelation) and orange (medium-high decorrelation) pertain to values between the extremes. To soften the effects of noise and of minor phenomena not pertinent to substantial landscape change, we modified the color scale with a stretching technique that assigns values farther than one standard deviation from the mean of the most extreme possible values (that is, values that are greater than one standard deviation from the mean in the positive direction take the value 255, and values greater than one standard deviation from the mean in the negative direction take 0) and squares values that are within one standard deviation of the mean. This made the minor landscape changes that were unlikely to be associated with real threats to cultural resources much less salient; it had a similar effect on most, if not all, of the noise inherent to the SAR instrument. It also highlighted the areas of greatest and least change. Red areas in the colorized image can therefore safely be considered locations where real and substantial terrain disturbance has occurred, and areas shaded yellow to orange are candidates for further investigation to determine whether there are ongoing natural or human disturbances to the landscape.

There are many options for modifying color tables and scales when visualizing results like these; we explore other options in subsequent figures. The correlation data products we generated, however, also allow for other forms of analysis. To quantitatively assess and verify the disturbances that are apparent in the colorized images, we used tests for statistically significant differences among sets of correlation pixel values. As reported elsewhere [26], the UAVSAR correlation image clearly reveals disturbance where Greenpeace placed its message, along the route the activists used to approach the Hummingbird, and in the areas where they placed and prepared their equipment. The details and results of the application of a Wilcoxon rank-sum test to corroborate visual evidence of disturbance at the Hummingbird glyph with the correlation images is discussed in Section 3.

Other locations where landscape disturbance occurred during the two-year window between UAVSAR flights can also be seen in correlation images. On-the-ground examination of these areas indicated that terrain disturbance had resulted from several kinds of ongoing activity. Among these were illegal mining operations, unsustainable agricultural practices, the looting of archaeological sites, and rain events associated with the El Niño–Southern Oscillation (ENSO) cycle. Personnel from JPL/NASA, the CSRM Foundation, and the Ministerio de Cultura Management System for the Cultural and Territorial Heritage of Nasca and Palpa went into the field several times from January through March 2016 to verify these and other apparent disturbances. The goal of these field sessions was to corroborate or correct our interpretations of the correlation data, including the location, causes, and intensity of disturbance. Data collection methods included photography, note-taking, and accessing and editing our GIS using mobile devices running the ArcGIS Collector app. This field verification was indispensable to drawing reliable conclusions about the location, degree, and cause of landscape disturbance within and around the World Heritage Site.

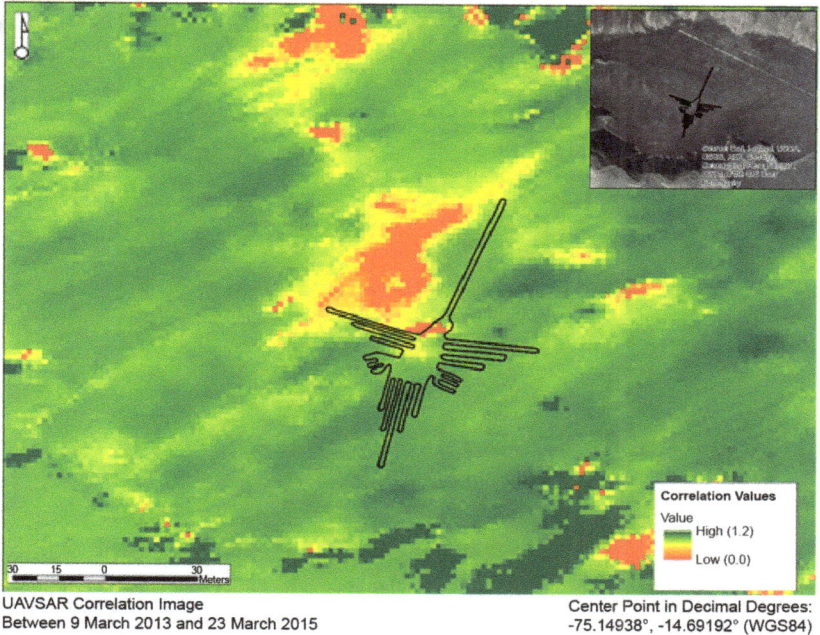

UAVSAR Correlation Image
Between 9 March 2013 and 23 March 2015

Center Point in Decimal Degrees:
-75.14938°, -14.69192° (WGS84)

Inset source credits:
Esri, i-cubed, USDA, USGS, AEX, GeoEye, Getmapping, Aerogrid, IGN, IGP, and the GIS User Community

Figure 2. A correlation image of the Hummingbird glyph (outlined) after applying the colorized stretch (the inset shows an optical satellite image of the same area). Highly decorrelated areas are in red. These areas include those at which Greenpeace activists placed their message and prepared their materials, as well as the trail they took to reach the Hummingbird (extending southward from the top of the image). The other decorrelated areas correspond to steep slopes (visible in the inset) leading down from the *pampas*; we found decorrelation in such areas to be common, because of gravity pulling sediment downslope.

Results suggesting that the UAVSAR data detected landscape change and provided insight into the degrees and causes of change are discussed further in the next section. Having acquired these results, we built upon our analysis of the UAVSAR data by applying a similar analytical process to C-band SAR data gathered by the Sentinel-1 satellites, aiming to assess the utility of that data in pursuing the kinds of monitoring tasks we had demonstrated with UAVSAR.

Sentinel-1A, launched on 3 April 2014, carries a C-band SAR instrument that transmits and receives radiation at wavelengths of about five centimeters. It and its identical companion, Sentinel-1B, which was launched on 25 April 2016, are part of the European Union's Copernicus Earth observation program. They can operate in four imaging modes: Interferometric Wide (IW), Extra Wide (EW), Strip Map (SM), and Wave (WV). The first two modes are regularly used to collect data over much of the Earth's land surface. The IW mode is commonly applied over land and typically used to cover the Nasca Lines World Heritage Site. It provides a single-look spatial resolution of better than 20 m over a strip 250 km wide [33]. Interferometric Sentinel-1 data is being used widely for the rapid imaging of surface displacements caused by earthquakes, subsidence, and landslides [35], as well as other applications at various spatial scales.

3. Results

3.1. Field Verification of UAVSAR and Sentinel-1 Correlation Images

3.1.1. UAVSAR

Our observations on the ground indicate that even small surface disturbances are visible as slightly decorrelated areas in the UAVSAR-derived correlation images. For example, minor erosion caused by low volumes of water moving slowly across the ground surface, which was documented at the time of its occurrence by Ministerio de Cultura personnel, appears as light gray areas in the gray-scale images. Heavily traveled roads and pedestrian trails are black in the gray-scale images, as are steep hillsides where substantial volumes of soil and rock moved downslope [26].

During field observations, it was evident that landscape disturbance caused by human activities, such as walking or driving across the *pampas*, are threatening the preservation of the Nasca lines and geoglyphs and degrading the Rio Grande de Nasca drainage. We also found that different forms of landscape disturbance can be distinguished by their appearances in the UAVSAR correlation image (see Figure 3). The data, and therefore the image, are highly precise: even minor terrain disturbances were detected, and the degree of decorrelation in the image is obviously associated with the intensity of disturbance on the ground.

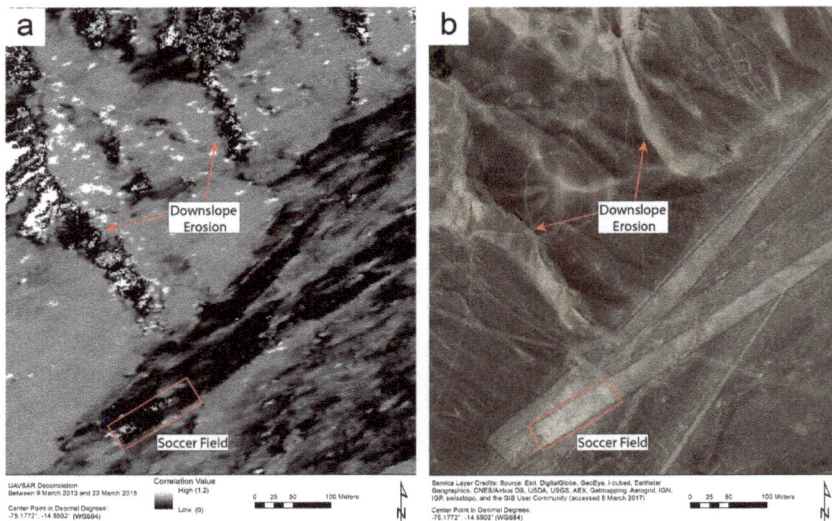

Figure 3. The correlation image in (**a**) reveals landscape disturbance in the area shown in the satellite image in (**b**). According to Ministerio de Cultura personnel, pedestrian traffic and informal games of soccer are common in and around the geometric glyphs at the bottom and right of (**b**); such activity manifests as smooth, intense decorrelation. The steep and rugged terrain in the top left of the image has been disturbed by foot traffic and encroaching development from the river valley just to the north of the image. As these images attest, we found that disturbance tends to be greater in sloped areas, where gravity augments small movements of rocks or soil. This downslope erosion results in uneven, speckled decorrelation.

Among the more remarkable applications of the UAVSAR correlation images is in revealing the unpaved paths and roads that were traveled between data takes. Figure 4 shows disturbance to a dirt road leading from the Pan-American Highway in the east to a power line running across the *pampas* in the west. In satellite imagery of the same area, there is little to distinguish this road from other, nearby lines on the ground, which may be geoglyphs, trails, or other dirt roads (during fieldwork, the features labeled as geoglyphs at the bottom of Figure 4b were positively identified). The correlation image, however, clearly shows that this particular road was disturbed. Also visible in Figure 4a are signs of interference with SAR waves by vehicles on the Pan-American Highway. These so-called sensor artifacts have no association with features on the ground.

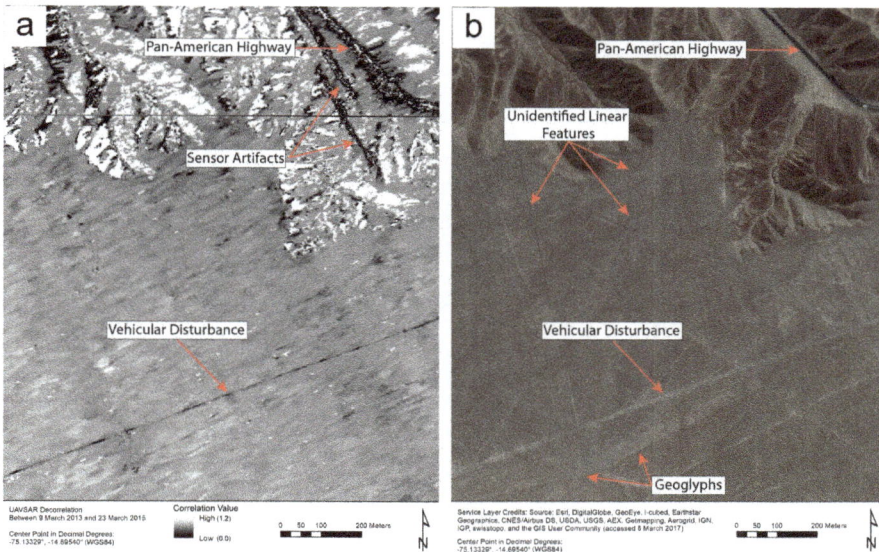

Figure 4. The correlation image in (**a**) indicates that only one of the linear features visible in (**b**)—some of which may be not trails or roads, but geoglyphs—was substantially disturbed between UAVSAR flights (without ground verification, it is difficult to confidently classify the unidentified linear features). Sensor artifacts, visible in (**a**), are due to interference with radar waves by fast-moving vehicles on the highway and do not correspond to any physical features or changes on the ground.

3.1.2. Sentinel-1

Despite its lower spatial resolution, the Sentinel-1 data proved useful in similar ways. Figure 5 illustrates the difference in utility between a Sentinel-1 amplitude image and a correlation image derived from two acquisitions forty-two days apart. While the amplitude image shows similar backscatter features for very different kinds of ground cover and structure, the correlation image specifically differentiates three land cover classes from the dry *pampas* landscape: natural vegetated areas, cultivated vegetated areas, and sand dunes. All these areas were decorrelated after forty-two days.

Satellite SAR data, specifically TerraSAR-X [25] and ERS-2 [23] data, has been used to identify erosion and sedimentation in the Nasca area at small scales, but the disturbance revealed here by Sentinel-1 is yet more subtle. Figure 6 shows the area along the Pan-American Highway near an observation tower. From the tower, visitors can view some of the nearby lines and geoglyphs on both sides of the road.

Panel 6a is a correlation image generated from Sentinel-1 data takes on 16 May 2015 and 18 December 2016. The Pan-American Highway appears as a dark line that runs from the southeast to the northwest (in [24], the highway manifests similarly in a correlation image derived from ENVISAT-ASAR data). Panel 6b is a pan-sharpened WorldView-2 optical satellite image. In it, the observation tower and the area where vehicles were driven off the highway are visible, but this single image cannot tell us when the vehicle tracks were made. The decorrelation in Figure 6a, however, tells us that the disturbance occurred between 16 May 2015 and 18 December 2016.

The green dots in Figure 6a,b show where the photo in Figure 6c was taken. This picture shows where vehicles were driven off the paved highway near the observation tower. Their tracks are clearly visible and indicate that some of the vehicles were driven some distance past the highway's eastern edge. The correlation image in Figure 6a shows these disturbed areas as dark patches of pixels. Other decorrelation in Figure 6a is very near the geoglyph called the Hands, which is visible at the right of the photo in Figure 6c.

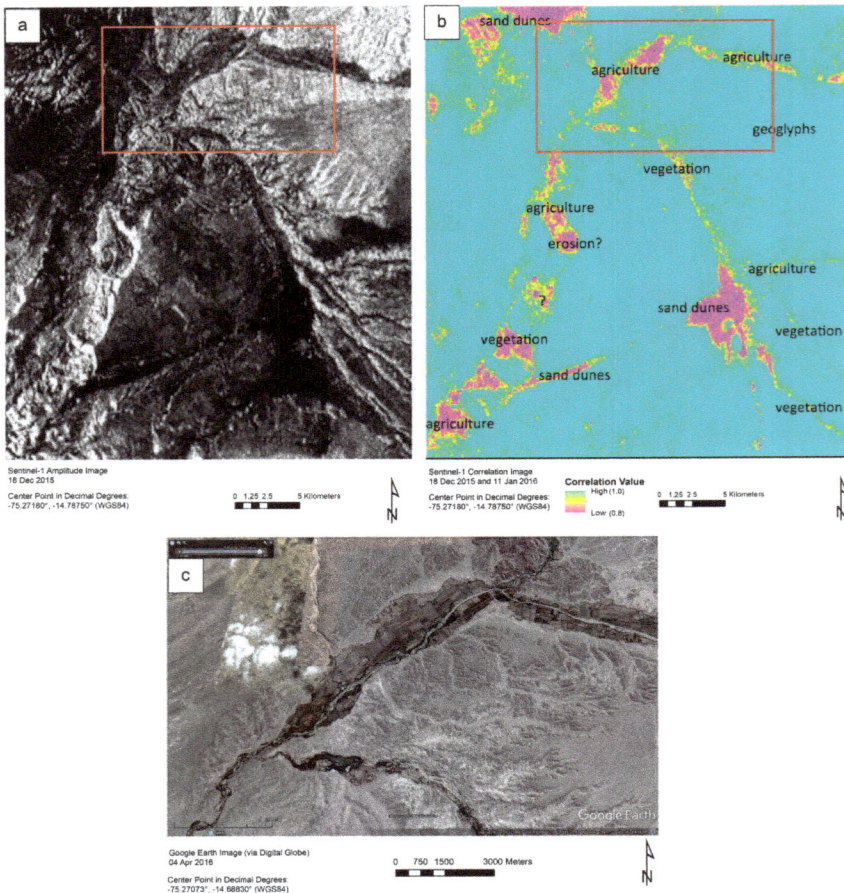

Figure 5. Panel (**a**) shows a Sentinel-1 amplitude image created using data collected on 18 December 2015. While many large landscape features are discernible here, the correlation image in (**b**); derived from data collected on 18 December 2015 and 11 January 2016, provides more information. In this image, we can differentiate among three types of land cover: natural vegetation, cultivated vegetation, and sand dunes. Panel (**c**) shows a pan-sharpened WorldView-2 image in which the agricultural areas demarcated by the box in (**b**) are visible.

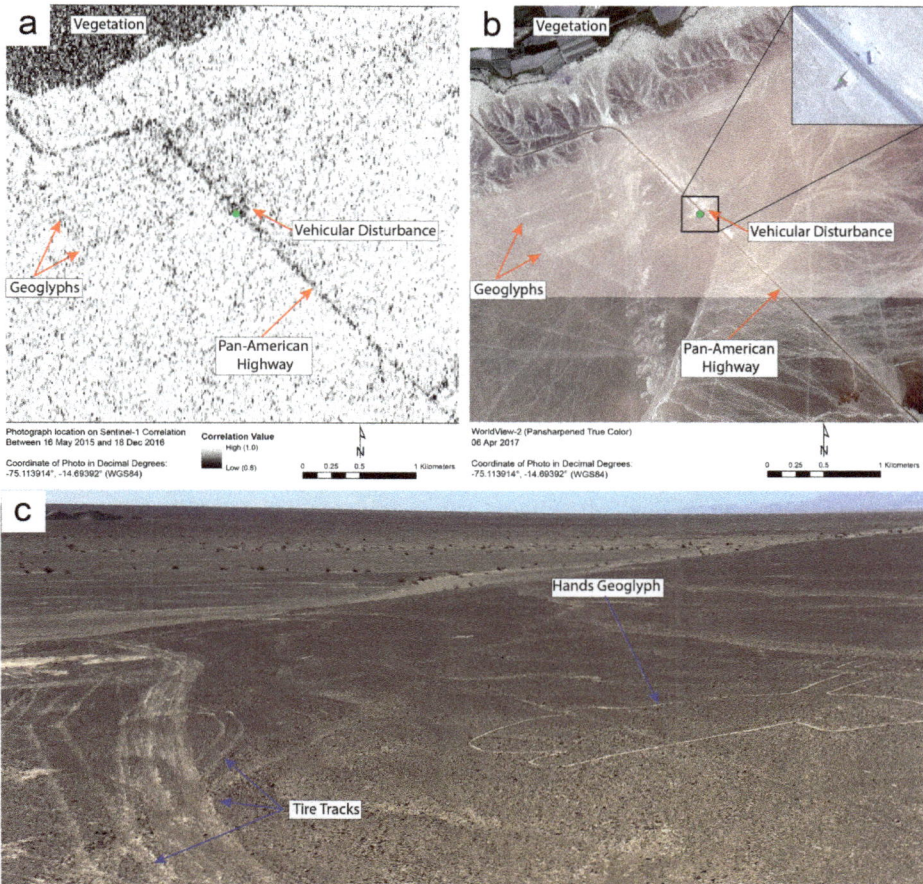

Figure 6. Panel (**a**) reveals slight disturbance to geoglyphs west of the Pan-American Highway and heavier disturbance, caused by vehicles, near the observation tower. These geoglyphs and the tower are visible in (**b**). In both (**a,b**), the tower is marked by the green dot, which is where the photo in (**c**) was taken. This photo shows vehicle tracks off the east side of the highway, running very near the Hands geoglyph. In (**a**), these tracks manifest as a patch of dark pixels near the green dot, and the highway appears as a dark line running southeast-northwest.

Figure 7 provides similar ground-verified evidence of disturbance. Panel Figure 7a contains a correlation image of Cahuachi, a major ceremonial center of the Nasca culture during the early centuries of the Common Era, generated using data collected by Sentinel-1 on 16 May 2015 and 18 December 2016. As in the other correlation images, dark pixels indicate decorrelation. Heavily vegetated areas, such as those along the Rio Grande de Nasca (which runs across the image from top left to center right), appear as masses of dark pixels. The green dot in Figure 7a,b, which shows a pan-sharpened WorldView-2 image, indicates the intersection of an unpaved road and a pedestrian trail where the photo in Figure 7c was taken. Both the road and the trail are visible in Figure 7b,c. They also show up as lines of dark pixels in Figure 7a, which suggests they were used between 16 May 2015 and 18 December 2016. Satellite SAR has been used to detect ground disturbance and soil moisture changes at Cahuachi, but not, apparently, at such a high spatial resolution [20]; past studies have not, as far as we are aware, revealed evidence of vehicle or foot traffic along unpaved routes near the site.

Figures 6 and 7 are examples of our evidence that support the idea described at the end of Section 1.2: that Sentinel-1 SAR data, with its high temporal resolution, is useful for monitoring landscape change, and threats to cultural and natural resources in particular, in areas like the Nasca Lines. This data allows us not only to identify the location and measure the intensity of disturbance, but also to infer a usefully small time window within which the disturbance must have occurred. By continually generating Sentinel-1 correlation images, researchers or site managers can gather helpful information on where disturbance is happening, what kind of disturbance it might be, and when it happened. For example, looting is a major problem in and near Cahuachi; correlation images derived from Sentinel-1 data can show which paths or roads leading to the site have been used, or whether new looting pits have been dug, every six to twenty-four days.

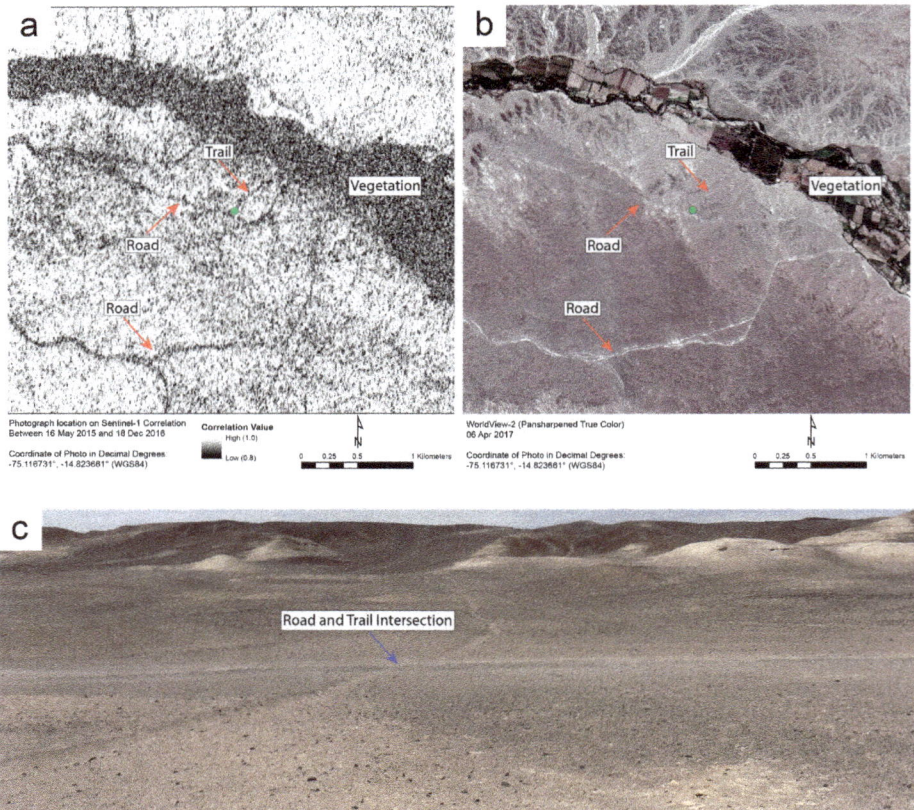

Figure 7. Roads and trails near the archaeological site of Cahuachi that were used between 16 May 2015 and 18 December 2016 are visible in the Sentinel-1 correlation image in (**a**) as lines of dark pixels. In (**b**), the roads and trails in the same area are visible in a pan-sharpened WorldView-2 satellite image. Panel (**c**) is a photo taken at the juncture of the road and trail near the green dot in both (**a**,**b**).

The correlation change image in Figure 8a was produced by computing pixel value ratios from two Sentinel-1 correlation images. Each of these two correlation images was derived from a pair of Sentinel-1 data takes: the first image was generated from collects on 16 May 2015 and 18 December 2016, and the second from collects on 16 May 2015 and 15 August 2017 (note that the first collect is the same for both images). Each pixel value in Figure 8a is the ratio of the value of the corresponding pixel in the 2015–2016 correlation image to that of the corresponding pixel in the 2015–2017 correlation image.

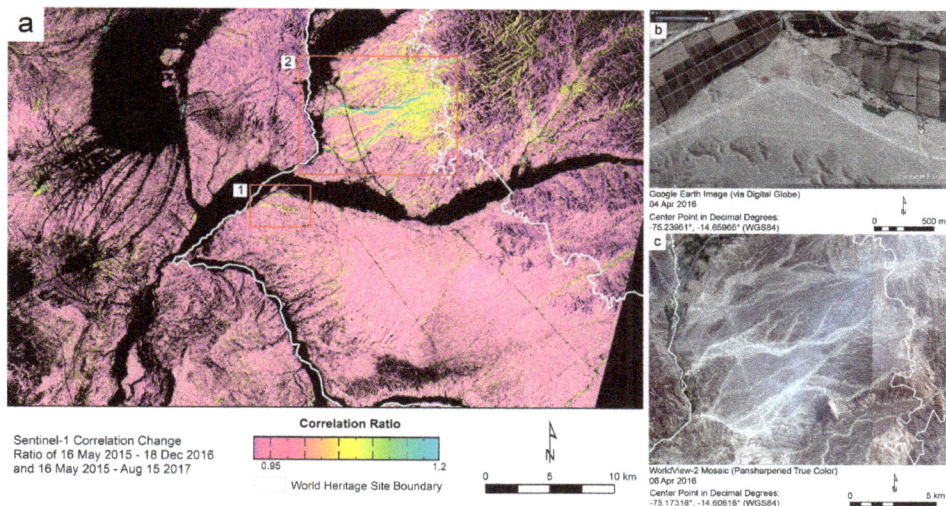

Figure 8. The correlation change image in (**a**) compares two correlation images. Red and orange areas experienced a relatively constant level of disturbance between May 2015 and August 2017. Yellow to green indicates that decorrelation during the 2016–2017 period was 5% to 20% higher than during the 2015–2016 period. Panel (**b**) is a satellite image of the area delineated and labeled 1 in (**a**); (**c**) is a satellite image of area 2. Area 1 shows two lines of atypical decorrelation: one following a road that makes a corner-like turn, and one where drivers have cut across that turn. Both the road and the shortcut are visible in (**b**). Area 2 shows evidence of flooding between December 2016 and August 2017; the apparent watercourses that this flooding appears to have followed are visible in (**c**).

The correlation change image facilitates a visual identification of areas where the occurrence of decorrelation was uneven over the entire 27-month period between May 2015 and August 2017. In effect, it highlights atypical decorrelation: areas that were similarly disturbed over both the May 2015–December 2016 period and the May 2015–August 2017 period have values close to 1, but those areas that experienced disturbance during only the latter period are shown here in brighter colors. The black areas in Figure 8a correspond to those areas that exhibited high decorrelation in both correlation images. Most of these areas are dominated by vegetation, which (as already noted) appears as high decorrelation in correlation images but does not necessarily reveal true landscape disturbance.

In interpreting Figure 8, it is helpful to note that the ENSO cycle produced torrential rainfall in the Nasca region during the first three months of 2017 [36]. The area marked by box 2 in Figure 8a includes substantial terrain, with many ridges and valleys. We suggest that the correlation change in this area (indicated by green and yellow) is the result of flooding, because the areas showing change correspond with what appear to be watercourses. Here, as in other parts of the Nasca Lines World Heritage Site, there are lines and geoglyphs that could be damaged or threatened. Correlation change images like the one in Figure 8a can be useful in assessing weather-related and other disturbances within World Heritage Sites, particularly in regions that are difficult to access.

The pink to purple areas in Figure 8a fall largely within the *pampas*, which is level and dry but crossed by several dry or intermittent streams. The apparent correlation change in the *pampas* that shows as faint, yellow smears may be due to recent alterations of the landscape, such as the construction of the Pan-American Highway, electrical lines, and roads that provide access to the electrical lines. These changes might exacerbate the damage done by heavy seasonal rainfalls, which can cause water to run through or flood the dry or intermittent streams just mentioned. This would be a cause for concern for the geoglyphs on the *pampas*.

3.2. Statistical Testing and Assessment of Disturbance

Statistical analysis of the UAVSAR decorrelation values provided further evidence of substantial disturbance to the Hummingbird geoglyph and illustrated the kind of quantitative assessment that SAR correlation images allow. The Hummingbird lies on a flat area at the northern edge of the *pampas*; a Wilcoxon rank-sum test compared the values of the correlation image pixels in the Hummingbird's immediate area (n = 8240) with a random sample of correlation pixel values (also n) in the wider surrounding area. The test's purpose was to assess whether the correlation values in and around the Hummingbird tended to be lower than those of the random sample. Figure 9 shows histograms of these two samples, and Figure 10 shows the locations of the pixels belonging to the random sample.

We chose a Wilcoxon rank-sum test for its general applicability to the ordinal comparison of two groups of observations, but also because it is nonparametric and therefore not reliant on any assumptions about the probability distributions of the groups being compared [37–39]. Preliminary analysis of the correlation values belonging to the pixels within and around the Hummingbird glyph and the pixels randomly chosen from areas of comparable topography did not indicate that the distributions of these groups were well characterized by parameterized families of probability distributions, which recommended a robust statistical approach.

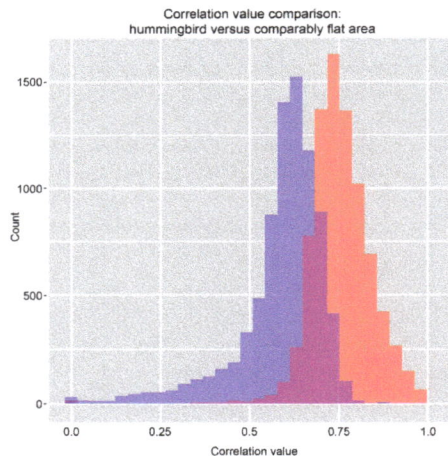

Figure 9. Histograms of the correlation values of the pixels falling within the Hummingbird and its immediate area (**blue**) and a random sample of comparable cells from nearby, flat areas without radar shadow (**red**).

Because decorrelation is generally higher on slopes, the random sample of correlation pixels was drawn only from nearby areas that are, like the Hummingbird area, flat. We used a digital elevation model (DEM) with a 30 m resolution derived from Shuttle Radar Topography Mission (SRTM) data to identify areas of the *pampas* that have a lower slope than the mean slope within the Hummingbird area, and drew pixels randomly from only these flat areas. We also identified and avoided areas where accurate SAR returns were obscured by radar shadow. Our random sample, therefore, came only from flat areas with no radar shadow. This limited the potential for bias in the test statistic that could arise from the association of slope with decorrelation, and it avoided measurement error from radar shadow.

The test suggests, at a confidence level greater than 0.999, that the correlation values within and around the Hummingbird indeed tend to be lower than those of the general area. These results indicate that, between the two UAVSAR flights, the Hummingbird and its immediate area experienced a significantly higher level of disturbance than the rest of the landscape. By statistically corroborating the clear evidence of disturbance to the Hummingbird, most notably the video taken by Greenpeace itself, these results also demonstrate the use of SAR for quantitatively identifying and testing landscape disturbance.

Center Point in DecimalDegrees: -75.15832°, -14.69540° (WGS84)

☐ Hummingbird Geoglyph Sampling Area

■ Randomly Selected Pixels

Service Layer Credits: Source: Esri, DigitalGlobe, GeoEye, i-cubed, Earthstar Geographics, CNES/Airbus DS, USDA, USGS, AEX, Getmapping, Aerogrid, IGN, IGP, swisstopo, and the GIS User Community

0   0.25   0.5   1 Kilometers

Figure 10. Satellite imagery of the northern edge of the *pampas*, with the Hummingbird near the top right corner, showing the locations of the pixels in the UAVSAR decorrelation image that were statistically compared. The 8240 pixels within and around the Hummingbird fall within the dotted rectangle; the inset shows the same area from the decorrelation image. The black dots show the locations of the 8240 pixels with which the Hummingbird pixels were compared. These pixels were randomly selected from areas without radar shadow that are, like the Hummingbird area, flat.

4. Discussion

This successful application of Sentinel-1 data presents unprecedented opportunities for continually monitoring landscape disturbance with a high temporal resolution. Together, the two Sentinel-1 satellites can acquire data over a given part of the Earth every six (not typical) to twenty-four (currently typical) days. In landscapes favorable to analysis with InSAR, such as the Nasca region, it is possible to identify areas of terrain disturbance and ascertain the disturbance intensity every time one of the satellites collects data over the area.

Though the spatial resolution of the Sentinel-1 data is not as high as that of the UAVSAR data, we found that Sentinel-1 data can be used to monitor landscape disturbance and gather information on the reason and degree of disturbance in a manner similar to that which we applied to the UAVSAR data. Because Sentinel-1 data can cover a much greater area with a far higher temporal resolution than UAVSAR or other SAR satellites that offer readily accessible data, it is of particular use in identifying and preventing damage to the cultural and natural resources of the Nasca Lines World Heritage Site.

These resources and the area's ecosystem services support sustainable tourism, agriculture, mining, and fishing industries. Without these continuing and dependable sources of income for the people living in and near the World Heritage Site, many of whom have fled the poverty of villages in the Andes, there will be a relatively high incentive for individuals to engage in commonly detrimental activities such as looting, and little incentive for the community in general to discourage such activities.

Broader Environmental Implications of This Research

The streams and rivers that flow through the *pampas* connect watersheds in the high Andes with the coastal waters of the Pacific. This coastal ecosystem is enormously productive. The river and stream networks that cross the *pampas* not only contain fertile soils, but also form habitat corridors that nurture a variety of plant and animal species that were exploited by sophisticated ancient societies, such as the Paracas and Nasca civilizations, for thousands of years. These systems rose and fell in concert with the ENSO cycle. Periods of drought or of torrential rains were catalysts for collapse, but the essentially robust environments of the stream and river courses provided the basis for renewal [27,40–44].

In many years, this region receives no measurable precipitation [45], and many of the streams and rivers flow only seasonally [44]. Flows can vary greatly from year to year, and especially from decade to decade, due to ENSO. With the loss of vegetation in the stream and river valleys, particularly of keystone species such as the *huarango* (P. limensis), the deluges that occur about once every 15 years are scouring the waterways that lead to the Rio Grande de Nasca and on to the Pacific Ocean, with effects that may be irreversible. However, as noted previously, the Rio Grande and its tributaries lie largely within the Nasca Lines World Heritage Site and thus can be considered assets to the conservation of the regional environment and its cultural and natural resources. Efforts to preserve the environment as well as the ancient glyphs, which help make the Nasca Lines World Heritage Site the second-most popular tourism destination in Peru, are likely to receive worldwide attention and generate popular understanding and support.

Disturbance of the landscape and activities that intrude on the site's ambiance (such as mining and large-scale agriculture) are incompatible with the values that render a site eligible for inscription on the World Heritage List. The Nasca Lines World Heritage Site itself provides ecosystem services in the forms of recreation and education. Further, maintaining ecosystem health in the area provides income and entrepreneurial opportunities to the local population. Environmental health here could be the basis for a strengthened eco-tourism industry that would draw from the same demographics that typically visit archaeological and historic sites. This would be a source of economic development other than mining and agriculture, which can harm the environment.

5. Conclusions

We initially used high-resolution SAR data from NASA/JPL's UAVSAR to demonstrate that InSAR can be used, in certain environments, to detect and test for subtle disturbances to a landscape that occur between SAR data takes. The disturbance to the Hummingbird geoglyph provided us with an opportunity to develop and test a means by which this can be done. Our results suggested that SAR data collected from satellite platforms could be used similarly, allowing for larger areas to be searched for landscape disturbances occurring within shorter timeframes. Our development and ground verification of correlation images from Sentinel-1 SAR data has established that this is possible.

The successful use of SAR data to monitor landscape disturbance carries with it enormous implications for the preservation of archaeological sites and the detection of human activities and natural processes that threaten cultural and natural resources. High-spatial-resolution imagery that can discern features such as the lines and glyphs of Nasca and Palpa, some of which are less than a meter or two in width, is commonly available. Before 2000, aerial photographs were the most effective tools for such tasks. Now there is Google Earth, which is free and in which many of the Nasca features are easily visible. There are also several commercial optical sensors that generate images with even higher spatial resolutions (pixels smaller than one meter), such as Quickbird, Pleiades, and WorldView-2, -3, and -4.

Yet this high-spatial-resolution spectral imagery provides what is essentially a one-off look at the landscape, the conditions frozen at the moment the imagery was obtained; the spatial resolutions of these instruments are high, but their temporal resolutions are not. Usually, these high-resolution data takes are sporadically acquired, driven by orders from customers who want to examine images acquired recently, and they can be expensive. In contrast, SAR satellites such as Sentinel-1 have short revisit cycles: Sentinel-1, as we have mentioned, acquires data at intervals of six to twenty-four days. Correlation images generated from Sentinel-1 data allow us to quantify the degree to which SAR backscatter changes between data takes, which can indicate where and how intensely the landscape has been disturbed within that time window. While the pixel sizes of the correlation images may be larger than those of commercial high-spatial-resolution images, SAR's sensitivity to structure nonetheless allows it to detect disturbance in the order of only a few meters. This provides a tool that, as this paper demonstrates, can be more useful than higher-resolution imagery. The use of free satellite SAR data, collected about every one to three weeks and indifferent to clouds or fog, provides a superior solution to many of the challenges of monitoring wide-area landscape disturbance on a continual basis.

Even within the Nasca Lines World Heritage Site, the relevance and utility of Sentinel-1 data goes far beyond the protection of the area's cultural resources. It can also be applied, using the methods described here, to monitor landscape disturbance that threatens ecosystems and crucial habitats, habitat corridors, aquifers, native vegetation, and other resources, as well as ecosystem services such as water filtration, opportunities for recreation, and biodiversity [27,28,40–44].

Acknowledgments: We would like to acknowledge the generous support of the Ministerio de Cultura del Perú and, in particular, Minister of Culture Jorge Nieto Montesinos, Minister of Culture Diana Álvarez-Calderón Gallo, Deputy Minister of Culture Luis Jaime Castillo Butters, Deputy Minister of Culture Juan Pablo de la Puente Brunke, Secretary General Mario Christofer Huapaya Nava, and Johny Isla, who is the Ministry of Culture Director of Management for the Territory of Nasca and Palpa. We also thank Pedro Gamboa Moquillaza, chief of the Servicio Nacional de Áreas Naturales Protegidas (SERNANP) del Perú; David Beresford-Jones, of the McDonald Institute of Archaeological Research, Cambridge University; Justin Moat, Research Leader of Spatial Analysis at the Royal Botanic Gardens, Kew; O.Q. Whaley, Latin America Projects Officer at the Royal Botanic Gardens, Kew; Stive Marthans, Gerente Regional de Recursos Naturales y Gestión del Medio Ambiente en Gobierno Regional de Ica; José Angel Valdivia Morón, Secretario General del Servicio Nacional Forestal y de Fauna Silvestre del Perú; and Ana María Ortiz de Zevallos Madueño, Director of the Dirección Desconcentrada de Cultura de Ica. We thank the NASA/JPL UAVSAR program and NASA's Craig Dobson for providing the data, and the NASA Space Archaeology program for its support. We thank ESA for providing Sentinel-1 data and the SNAP software. We thank Joe Mazzariello of CSRM Foundation for designing the figures and proofreading.

Author Contributions: The interpretation of the imagery was led by Douglas C. Comer. The SAR processing was led by Bruce D. Chapman. The field verification was led by Douglas C. Comer, Bruce D. Chapman, and Jacob A. Comer. The statistical analysis was led by Jacob A. Comer.

Conflicts of Interest: The authors declare no conflict of interest.

References

1. McCauley, J.F.; Schaber, G.G.; Breed, C.S.; Grolier, M.J.; Haynes, C.V.; Issawi, B.; Elachi, C.; Blom, R. Subsurface valleys and geoarchaeology of the eastern Sahara revealed by shuttle radar. *Science* **1982**, *218*, 1004–1020. [CrossRef] [PubMed]

2. Elachi, C.; Roth, L.; Schaber, G. Spaceborne Radar Subsurface Imaging in Hyperarid Regions. *IEEE Trans. Geosci. Remote Sens.* **1984**, *GE-22*, 383–388. [CrossRef]

3. Blom, R.G.; Crippen, C.; Elachi, C.; Clapp, N.; Hedges, G.R.; Zarins, J. Southern Arabian Desert Trade Routes, Frankincense, Myrrh, and the Ubar Legend. In *Remote Sensing in Archaeology*; Wiseman, J.R., El-Baz, F., Eds.; Springer: New York, NY, USA, 2007; pp. 71–87.

4. Comer, D.C.; Blom, R.G. Detection and Identification of Archaeological Sites and Features Using Synthetic Aperture Radar (SAR) Data Collected from Airborne Platforms. In *Remote Sensing in Archaeology*; Wiseman, J.R., El-Baz, F., Eds.; Springer: New York, NY, USA, 2007; pp. 71–87.

5. Comer, D.C. *Merging Aerial Synthetic Aperture Radar (SAR) and Satellite Multispectral Data to Inventory Archaeological Sites*; NCPTT: Washington, DC, USA, 2007. Available online: https://www.ncptt.nps.gov/blog/merging-aerial-synthetic-aperture-radar-sar-and-satellite-multispectral-data-to-inventory-archaeological-sites-2007-11/ (accessed on 21 September 2017).

6. Comer, D.C. Wide-area, planning level archaeological surveys using SAR and multispectral images. In Proceedings of the Geoscience and Remote Sensing Symposium, Boston, MA, USA, 7–11 July 2008; pp. 45–47.

7. Tilton, J.C.; Comer, D.C. Identifying Probable Archaeological Sites on Santa Catalina Island, California Using SAR and Ikonos Data. In *Mapping Archaeological Landscapes from Space*; Comer, D.C., Harrower, M., Eds.; Springer: New York, NY, USA, 2013; pp. 241–249.

8. Mittermayer, J.; Wollstadt, S.; Prats-Iraola, P.; Scheiber, R. The TerraSAR-X Staring Spotlight Mode Concept. *IEEE Trans. Geosci. Remote Sens.* **2014**, *52*, 3695–3706. [CrossRef]

9. Tapete, D.; Cigna, F. SAR for Landscape Archaeology. In *Sensing the Past*; Masini, N., Soldoveri, F., Eds.; Springer: New York, NY, USA, 2017; pp. 101–116.

10. Patruno, J.; Dore, N.; Crespi, M.; Potteir, E. Polarimetric Multifrequency and Multi-incidence SAR Sensors Analysis for Archaeological Purposes. *Archaeol. Prospect.* **2013**, *20*, 89–96. [CrossRef]

11. Lasaponara, R.; Masini, N.; Pecci, A.; Perciante, F.; Escot, P.E.; Rizzo, E.; Scavone, M.; Sileo, M. Qualitative evaluation of COSMO SkyMed in the detection of earthen archaeological remains: The case of Pachamacac (Peru). *J. Cult. Herit.* **2017**, *23*, 55–62. [CrossRef]

12. Amelung, F.; Galloway, D.L.; Bell, J.W.; Zebker, H.A.; Laczniak, R.J. Sensing the ups and downs of Las Vegas: InSAR reveals structural control of land subsidence and aquifer-system deformation. *Geology* **1999**, *27*, 483–486. [CrossRef]

13. Foumelis, M.; Pavlopoulos, K.; Kourkouli, P. Ground deformation monitoring in cultural heritage areas by time series SAR interferometry: The case of ancient Olympia site (Western Greece). In Proceedings of the ESA FRINGE Workshop, Frascati, Italy, 30 November–4 December 2009.

14. Marotti, L.; Prats, P.; Scheiber, R.; Wollstadt, S.; Reigber, A. Differential SAR interferometry with TerraSAR-X TOPS data: Mexico city subsidence results. In Proceedings of the EUSAR 9th European Conference on Synthetic Aperture Radar, Nuremberg, Germany, 24 April 2012; pp. 677–680.

15. Chen, F.; Guo, H.; Ma, P.; Lin, H.; Wang, C.; Ishwaran, N.; Hand, P. Radar interferometry offers new insights into threats to the Angkor site. *Sci. Adv.* **2017**, *3*. [CrossRef] [PubMed]

16. Chapman, B.D.; Blom, R.G. Synthetic Aperture Radar, Technology, Past and Future Applications to Archaeology. In *Mapping Archaeological Landscapes from Space*; Comer, D.C., Harrower, M., Eds.; Springer: New York, NY, USA, 2013; pp. 113–132.

17. Chen, F.; Lasaponara, R.; Masini, N. An overview of satellite synthetic aperture radar remote sensing in archaeology: From site detection to monitoring. *J. Cult. Herit.* **2017**, *23*, 5–11. [CrossRef]

18. Cigna, F.; Tapete, D.; Lasaponara, R.; Masini, N. Amplitude Change Detection with ENVISAT ASAR to Image the Cultural Landscape of the Nasca Region, Peru. *Archaeol. Prospect.* **2013**, *20*, 117–131. [CrossRef]

19. Tapete, D.; Cigna, F.; Masini, N.; Lasaponara, R. Prospection and Monitoring of the Archaeological Heritage of Nasca, Peru, with ENVISAT SAR. *Archaeol. Prospect.* **2013**, *20*, 133–147. [CrossRef]

20. Cigna, F.; Tapete, D. Satellite SAR Remote Sensing in Nasca. In *The Ancient Nasca World*; Lasaponara, R., Masini, N., Orefici, G., Eds.; Springer: Cham, Switzerland, 2016; pp. 529–542.

21. Lasaponara, R.; Masini, N. Following the Ancient Nasca Puquios from Space. In *Satellite Remote Sensing: A New Tool for Archaeology*; Lasaponara, R., Masini, N., Eds.; Springer: Berlin, Germany, 2012; pp. 269–290.

22. Hesse, R. Combining Structure-from-Motion with high and intermediate resolution satellite images to document threats to archaeological heritage in arid environments. *J. Cult. Herit.* **2015**, *16*, 192–201. [CrossRef]

23. Lefort, A.; Grippa, M.; Walker, N.; Stewart, L.J.; Woodhouse, I.H. Change detection across the Nasca pampa using spaceborne SAR interferometry. *Int. J. Remote Sens.* **2004**, *25*, 1799–1803. [CrossRef]

24. Ruescas, A.; Delgado, J.; Costantini, F.; Sarti, F. Change Detection by Interferometric Coherence in Nasca Lines, Peru (1997–2004). In Proceedings of the Fringe 2009 Workshop, Frascati, Italy, 30 November 2009–4 December 2009.

25. Baade, J.; Schmullius, C. High-resolution mapping of fluvial landform change in arid environments using TerraSAR-X Images. In Proceedings of the 2010 IEEE International Geoscience and Remote Sensing Symposium (IGARSS), Honolulu, HI, USA, 25–30 July 2010.

26. Chapman, B.; Comer, D.; Isla, J.A.; Silverman, H. The Measurement by Airborne Synthetic Aperture Radar (SAR) of Disturbance within the Nasca World Heritage Site. *Conserv. Manag. Archaeol. Sit.* **2015**, *17*, 270–286. [CrossRef]

27. Silverman, H. Paracas in Nazca: New Data on the Early Horizon Occupation of the Rio Grande de Nazca Drainage, Peru. *Lat. Am. Antiqu.* **2017**, *5*, 359–382. [CrossRef]
28. Orefici, G. The Paracas-Nasca Cultural Sequence. In *The Ancient Nasca World*; Lasaponara, R., Masini, N., Orefici, G., Eds.; Springer: New York, NY, USA, 2016; pp. 121–161.
29. Adelsberger, K.A.; Smith, J.R. Desert pavement development and landscape stability on the Eastern Libyan Plateau, Egypt. *Geomorphology* **2009**, *107*, 178–194. [CrossRef]
30. Bowker, M.A. Biological Soil Crust Rehabilitation in Theory and Practice: An Underexploited Opportunity. *Restor. Ecol.* **2007**, *15*, 13–23. [CrossRef]
31. Streep, A. This Man Is Greenpeace's Best Hope Outside, 13 March 2015. Available online: http://www.outsideonline.com/1959936/man-greenpeaces-best-hope (accessed on 29 September 2015).
32. Zebker, H.A.; Villasenor, J. Decorrelation in Interferometric Radar Echoes. *IEEE Trans. Geosci. Remote Sens.* **1992**, *30*, 950–959. [CrossRef]
33. ESA, Sentinel-1 Mission Objectives. Available online: https://sentinel.esa.int/web/sentinel/missions/sentinel-1/mission-objectives (accessed on 31 July 2017).
34. Strozzi, T.; Farina, P.; Corsini, A.; Ambrosi, C.; Thüring, M.; Zilger, J.; Wiesmann, A.; Wegmüller, U.; Werner, C. Survey and monitoring of landslide displacements by means of L-band satellite SAR interferometry. *Landslides* **2005**, *2*, 193–201. [CrossRef]
35. Angster, S.; Fielding, E.J.; Wesnousky, S.; Pierce, I.; Chamlagain, D.; Gautam, D.; Upreti, B.N.; Kumahara, Y.; Nakata, T. Field Reconnaissance after the 25 April 2015 M 7.8 Gorkha Earthquake. *Seismol. Res. Lett.* **2015**, *86*, 1506–1513. [CrossRef]
36. Di Liberto, T. Heavy Summer Rains Flood Peru. NOAA Climate.gov. Available online: https://www.climate.gov/news-features/event-tracker/heavy-summer-rains-flood-peru (accessed on 21 September 2017).
37. Wilcoxon, F. Individual Comparisons by Ranking Methods. *Biom. Bull.* **1945**, *1*, 80–83. [CrossRef]
38. Mann, H.B.; Whitney, D.R. On a Test of Whether One of Two Random Variables is Stochastically Larger Than the Other. *Ann. Math. Stat.* **1947**, *18*, 50–60. [CrossRef]
39. Bauer, D.F. Constructing Confidence Sets Using Rank Statistics. *J. Am. Stat. Assoc.* **1972**, *67*, 687–690. [CrossRef]
40. Beresford-Jones, D.G.; Whaley, O.Q.; Ledesma, C.; Cadwallader, L. Two millennia of changes in human ecology: Archaeobotanical and invertebrate records from the lower Ica valley, south coast Peru. *Veg. Hist. Archaeobot.* **2011**, *20*, 273. [CrossRef]
41. Beresford-Jones, D.; Arce, T.S.; Whaley, O.Q.; Chepstow-Lusty, A.J. The Role of Prosopis in Ecological and Landscape Change in the Samaca Basin, Lower Ica Valley, South Coast Peru from the Early Horizon to the Late Intermediate Period. *Lat. Am. Antiq.* **2009**, *20*, 303–332. [CrossRef]
42. Beresford-Jones, D.; Lewis, H.; Boreham, S. Linking cultural and environmental change in Peruvian prehistory: Geomorphological survey of the Samaca Basin, Lower Ica Valley, Peru. *Catena* **2009**, *78*, 234–249. [CrossRef]
43. Andrus, C.F.T.; Sandweiss, D.H.; Reitz, E.J. Climate Change and Archaeology: The Holocene History of El Niño on the Coast of Peru. In *Case Studies in Environmental Archaeology*, 2nd ed.; Reitz, E., Scarry, C.M., Scudder, S., Eds.; Springer: New York, NY, USA, 2008; pp. 143–157.
44. Whaley, O.Q.; Beresford-Jones, D.G.; Milliken, W.; Orellana, A.; Smyk, A.; Leguía, J. An ecosystem approach to restoration and sustainable management of dry forest in southern Peru. *Kew Bull.* **2010**, *65*, 613–641. [CrossRef]
45. Harris, I.; Jones, P.D.; Osborn, T.J.; Lister, D.H. Updated high-resolution grids of monthly climatic observations–the CRU TS3.10 Dataset. *Int. J. Climatol.* **2013**, 623–642. [CrossRef]

Article

Semi-Automatic Detection of Indigenous Settlement Features on Hispaniola through Remote Sensing Data

Till F. Sonnemann [1,*], Douglas C. Comer [2], Jesse L. Patsolic [3], William P. Megarry [4], Eduardo Herrera Malatesta [5] and Corinne L. Hofman [5]

[1]   Institute of Archaeology, Heritage Sciences and Art History (IADK), Otto-Friedrich-Universität Bamberg, Am Kranen 14, 96047 Bamberg, Germany

[2]   Cultural Site Research and Management (CSRM), 2113 St. Paul Street, Baltimore, MD 21218, USA; dcomer@culturalsite.com

[3]   Center for Imaging Science, Johns Hopkins University, Baltimore, MD 21218, USA; jpatso11@jhu.edu

[4]   School of Natural and Built Environment, Queen's University Belfast, Belfast BT7 1NN, UK; W.Megarry@qub.ac.uk

[5]   Faculty of Archaeology, Universiteit Leiden, Einsteinweg 2, 2333 CC Leiden, The Netherlands; e.n.herrera.malatesta@arch.leidenuniv.nl (E.H.M.);c.l.hofman@arch.leidenuniv.nl (C.L.H.)

*   Correspondence: till.sonnemann@uni-bamberg.de; Tel.: +49-951-863-3930

Received: 15 October 2017; Accepted: 29 November 2017; Published: 5 December 2017

Abstract: Satellite imagery has had limited application in the analysis of pre-colonial settlement archaeology in the Caribbean; visible evidence of wooden structures perishes quickly in tropical climates. Only slight topographic modifications remain, typically associated with middens. Nonetheless, surface scatters, as well as the soil characteristics they produce, can serve as quantifiable indicators of an archaeological site, detectable by analyzing remote sensing imagery. A variety of pre-processed, very diverse data sets went through a process of image registration, with the intention to combine multispectral bands to feed two different semi-automatic direct detection algorithms: a posterior probability, and a frequentist approach. Two 5×5 km^2 areas in the northwestern Dominican Republic with diverse environments, having sufficient imagery coverage, and a representative number of known indigenous site locations, served each for one approach. Buffers around the locations of known sites, as well as areas with no likely archaeological evidence were used as samples. The resulting maps offer quantifiable statistical outcomes of locations with similar pixel value combinations as the identified sites, indicating higher probability of archaeological evidence. These still very experimental and rather unvalidated trials, as they have not been subsequently groundtruthed, show variable potential of this method in diverse environments.

Keywords: remote sensing; direct detection; GIS mapping; Caribbean archaeology; landscape archaeology

1. Introduction

The fascination with feature identification and mapping of geometric archaeological alignments by means of remote sensing is as old as the first appearance of aerial photos [1–3]. Throughout the last centuries, it has advanced significantly, leading to new archaeological discoveries using imagery from satellites and drones [4,5]. The human eye remains an adept feature extractor and can distinguish linear or circular structures and earthworks easily from the natural soil [6]. More recently, however, automatic approaches in pattern recognition have also become common, often based on computer algorithms adopted from other disciplines [7–10], and tested for archaeological purposes to detect color [11,12], changes in topography [13,14] or different reflection patterns [15].

A different challenge is the identification of non-geometric archaeological features with more amorphous shape and structure. Without any clear geometry they pose a special problem, as the most prominent parameter for successful recognition is missing. This is the case for indigenous settlements in the Caribbean, which have been identified through assemblages of shells, ceramics bone remains, and stone tools; but not by traces of extant or sub-surface structural remains [16,17]. The irregular pattern of pre-colonial settlement vestiges has made their detection challenging for remote sensing [18]. Previous work has been dominated by traditional archaeological survey methods: the identification of surface material based on the knowledge of local scouts or landowners, and defining an approximate delineation of areas based on the surface finds on site [19,20]. The trial approach presented here, a methodological experiment exploring the use of various datasets and approaches, rather than providing any validated method or results, is nevertheless an example for a novel statistical, systematic, and therefore more objective method.

The posterior probability and the frequentist approach are two algorithms developed at Cultural Site Research and Management (CSRM) [21,22] as a Direct Detection Model (DDM), to identify the probability of sites by comparing single pixel values. Both algorithms were modified at the time of the study and ready for use at different times; this is one reason why each was applied on a different trial area. The general idea behind direct detection is that anthropogenic activities at archaeological sites, often over long periods of time, have impacted these parts of the landscape in ways that if they persist are statistically measureable in remote sensing data. The DDM has therefore two sets of input data. The first set has two parts. One is the locations of known archaeological sites. In the trial area of the northwestern Dominican Republic and Haiti, the archaeological "sites" were identified over several years by different archaeologists, mostly with the help of local guides. Each site visited was named, a number of archaeological samples taken, and georeferenced by taking one or more GPS points at the site using a handheld device. In addition the algorithm needs areas, at best also groundtruthed, with presumably no sites, which are equally important for the study. A second data set comprises a variety of remote sensing imagery. The subtle variation between already discovered areas of human activity, the sites, and areas of no human activity (non-sites) within each remote sensing band, can be used to detect difference. The difference is more likely to be detected when many different bands of available satellite or aerial data sets are combined.

The area of interest, northern Hispaniola, presents a highly diverse environment. Along the coast runs the 200 km long Cordillera Septentrional, a several hundred meter high mountain range, partly covered by temperate to tropical forest, separating the coast from the fertile floodplain of the Valle de Cibao. Large parts of the hills and the plains north of the cordillera have been cleared for pasture. The northern coast is protected by coral reefs and mangrove forests. The region has been settled through waves of immigrations, archaeologically divided into earliest lithic age period since 4000 BC [23], the archaic period from 2500 BC the later ceramic ages distinguishable by ostionoid, meillacoid and chicoid ceramics [23–26]. Shortly after the arrival of Columbus, and the foundation of the first Spanish town in the Americas at La Isabella in 1493 [27], evidence for Amerindian activity declines rapidly [28] from the archaeological record [29,30]. We can therefore postulate that most sites marked in the map are either from prehistoric or very early colonial times. Variations in topography, land use and vegetation have created a landscape that changes over few kilometers, which also affected the indigenous settlement strategy [31]. Accumulations of shells indicate Amerindian use of marine resources [32], while other sites, often on prominent location overseeing the landscape, have been identified as settlements due to their particular topographic attributes consisting of mounds and flattened areas that served as base for house construction [30,33,34].

2. Materials and Methods

Based on the availability of remote sensing datasets, and samples of already identified archaeological sites, three areas of 5×5 km^2 in different environments were initially identified for trials. All existing archaeological site datasets were merged into a single point shape file, and then split for each of the trial regions. Only the two areas in Puerto Plata, DR (1) and Montecristi, DR (2) were ultimately trialed (Figure 1). The third area in Meillac (Dep. Nord-Est), Haiti, (3) was excluded following the second round of trials. An additional 1.5×5 km^2 area (4), which had been focus of a systematic total area survey [35], was initially thought to be well suited for comparing remote sensing and ground interpretation. Unfortunately, it had to be discarded, as there were not enough known sites in the area to be implemented in the algorithms, a universal issue of sample size when doing predictive models. Most of these techniques require thousands of points in other areas to compare, particularly for the small region impossible to provide.

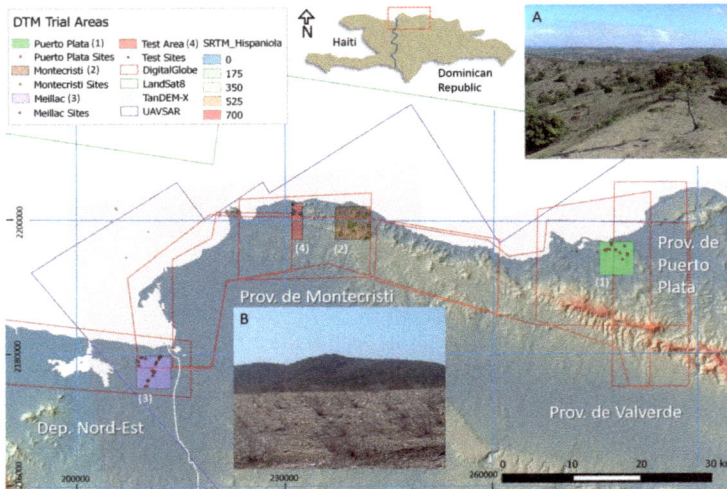

Figure 1. Initially selected trial areas in northern Hispaniola and the available remote sensing data sets superimposed on a modified DEM from JPL/NASA's Shuttle Radar Topography Mission (SRTM). The small images display the (**A**) landscape in the Puerto Plata and (**B**) the view from an archaeological site in the Montecristi region (right). (Datum: WGS84, UTM 19N).

The passive remote sensing data sets Landsat-8, Worldview-2, ASTER, and active sensors UAVSAR (Uninhabited Aerial Vehicle Synthetic Aperture Radar) as well as TanDEM-X stripmap (see Table 1 and Figure 2) were chosen based on resolution, availability, accessibility and practicality; they were either freely available, or acquired through generous data grants. Aerial imagery of northern Haiti was provided free of charge by Haiti's Centre National de l'Information Géo-Spatiale. Atmospherically corrected 15 m ASTER (Advanced Spaceborne Thermal Emission and Reflection) data was acquired through NASA/METI/AIST/Japan Space Systems and US/Japan ASTER Science Team, the low resolution however made the imagery of limited use. Additionally, several TanDEM-X data sets were acquired through a research license agreement with the DLR e.V. (Project: OTHER6189, https://tandemx-science.dlr.de/), but ultimately the uncorrected data was not utilized for the DDM. All other remote sensing data went through a series of image registration protocols to render them standard in pixel size, resolution and angle that allowed exact correlation between pixels of different data sets. To achieve this goal, all datasets were initially converted to the same georeferenced system: WGS 84, UTM 19N for the Dominican Republic (20N respectively for Haiti). Because of insufficient spatial resolution, work with Landsat-8, and ASTER was discontinued after consideration, leaving UAVSAR and Worldview-2 for further steps. The latter, multispectral data set, made available by the DigitalGlobe Foundation, covers the regions of interest in two-meter-resolution with one panchromatic and eight multispectral bands (see Table 2). The data set, with bands in the visible and near-visible range, was atmospherically corrected to reflectance values [36]. This standardized imagery removing artefacts caused by atmospheric interference. While often neglected, atmospheric correction is important and can significantly impact subsequent processing techniques like indices [37].

Table 1. Availability of initially acquired data sets for sample sites Puerto Plata, DR (1) Montecristi, DR (2) Meillac, Haiti (3) and the test site in the Montecristi province (4). The regions were picked for very light or non-existing cloud cover within the images. In bold are the data sets lastly included in the survey. MS = multispectral; PC = panchromatic.

	Dataset	Source	Bands	Resolution [m]	(1)	(2)	(3)	(4)
A	SRTM	USGS	1	30	×	×	×	×
B	LandSat-8	NASA/USGS	7 (MS) 1 (PC)	30 (MS) 15 (PC)	×	×	×	×
C	ASTER	NASA/METI/AIST	9	15	×	×	×	×
D	**UAVSAR**	**NASA/ JPL**	**6 (9)**	**5.7**	×	×	×	×
E	**WorldView-2**	**Digital Globe Foundation**	**8**	**1.85–2.07 (MS)**	×	×	×	×
F	Aerial	CNIGS (Govt. of Haiti)	3	0.7	-	-	×	-
G	TanDEM-X	DLR. e. V.	1	3	×	×	×	×

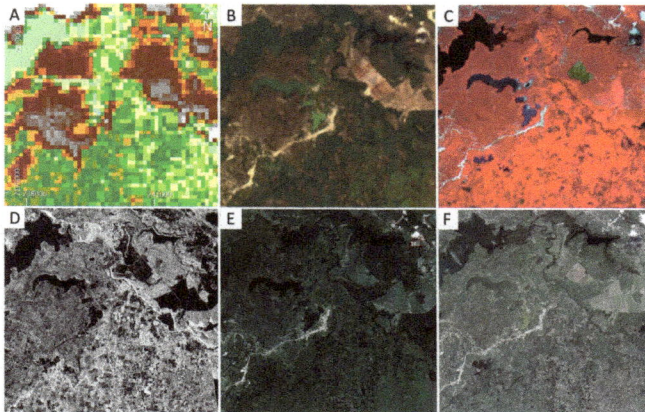

Figure 2. Overview of the initially trialed remote sensing data sets for the region Nordest Haiti. (**A**) SRTM, (**B**) LandSat, (**C**) ASTER, (**D**) UAVSAR, (**E**) Worldview-2, (**F**) Aerial. (Datum: WGS84, UTM 19N).

Table 2. Band distribution and wavelength of Worldview-2 satellite.

Band	0	1	2	3	4	5	6	7	8
Color	Pan	Coastal	Blue	Green	Yellow	Red	Red Edge	NIR1	NIR2
λ in nm	450–800	400–450	450–510	510–580	585–625	630–690	705–745	770–895	860–1040

From the original Worldview-2 data, the transformations NDVI, PCA and Tasseled Cap were applied in ArcGIS 10.3 (ESRI, Redlands, CA, USA), with the purpose to create additional bands that may improve the site identification regarding their environmental discrimination. Of these, the NDVI (Normalized Difference Vegetation Index) [38] is a unidimensional spectral index, adjusting the band information based on the principle that healthy vegetation absorbs most of the visible light and reflects most of the near-infrared light. Unhealthy or sparse vegetation reflects more visible light and less near-infrared light. The formula applied (1) used the bands red (visible) and red edge (to represent near-infrared):

$$\text{NDVI: Float ("Red Edge" } - \text{ "Red")/("Red Edge" + "Red")} \tag{1}$$

Principal Component Analysis (PCA) was applied with the intention to reduce the data dimensionality of correlated bands [39]. The method rotates the original space of features into a new space where the transformed features are pairwise orthogonal. This creates an n-dimensional space of eigenvectors, where n is the number of input dimensions (features), with the goal to orthogonalize the data set. The first principal component accounts for the maximum proportion of variance from the original dataset, the following, being orthogonal to the first one, for the next principal components, creating eventually a new coordinate system of orthogonal axes. A subset of the components is usually chosen for subsequent analysis. The method used to select these components varies by application. The first three components were included in the algorithm while the latter components were discarded as redundant. For a more detailed explanation, see [40].

Tasseled Cap Transformation or K-T (Kauth-Thomas) transform (KTT), as originally developed by [41] for LandSAT imagery, was applied on each Worldview-2 data set using bands one to eight in accordance with [42]. Tasseled cap applies predefined correction coefficients to each band and will produce eight new bands. This spectral index conversion intends to highlight changes in vegetation and soil, where the pixel values are being transferred into a new orthogonal axial system; of these the first three new bands are the most important, representing *Brightness* (red) (2), *Greenness* (green) (3) and *Wetness* or yellowness of vegetation (blue) (4). Based on the reflectance values given by [42] for each Worldview-2 component, the formulas go as such:

Brightness: Float (0.060436 × "Coastal" + 0.012147 × "Blue" + 0.125846 × "Green" + 0.313039 × "Yellow" + 0.412175 × "Red" + 0.482758 × "Red Edge" − 0.160654 × "NIR1" + 0.67351 × "NIR2") (2)

Greenness: Float (−0.140191 × "Coastal" − 0.206224 × "Blue" − 0.215854 × "Green" − 0.314441 × "Yellow" − 0.410892 × "Red" + 0.095786 × "Red Edge" + 0.600549 × "NIR1" + 0.503672 × "NIR2") (3)

Wetness and Shadow: Float (−0.270951 × "Coastal" − 0.315708 × "Blue" − 0.317263 × "Green" − 0.242544 × "Yellow" − 0.256463 × "Red" − 0.096550 × "Red Edge" − 0.742535 × "NIR1" + 0.202430 × "NIR2") (4)

NASA had captured UAVSAR (Uninhabited Aerial Vehicle Synthetic Aperture Radar) polarimetric L-band data of ~5.7 m, over the large fault zones of Hispaniola. Publicly available, this data set also covers the areas of interest. Accessed through the JPL/ASF website, the data was extracted to single band TIFF-files using [43] (downloaded product: PolSAR-polarimetric SAR-MLC). Different materials reflect radar waves with different intensities and polarizations. Among the feature differentiated are smoothness, homogeneity, as well as soil moisture, and vegetation discrimination revealed by variation in density and structure. For polarimetric SAR image representation, the Pauli color-coding is a useful application to visually differentiate the three major scattering mechanisms (1) single bounce from a plane surface backscattered towards the radar, (2) double bounce from one flat surface that is horizontal with an adjacent vertical surface, and (3) volume scattering from randomly oriented scatterers [44]. From the original seven UAVSAR bands, Pauli decomposition bands were produced through a linear combination of HH, HV, VH and VV in one three-color RGB image [45] (see Table 3).

For each region, the chosen 5×5 km^2 test areas were overlaid by a point grid of 2m-resolution, to create a matrix of 2500×2500 sampling pixels from each dataset (Figure 3). These values were then interpolated into a stack of registered raster dataset of the same 2 m resolution. After this transformation, the pixels and their attributes of each band overlapped exactly, diminishing the possibility of corner and border uncertainties. In addition, the prepared Worldview-2 data set served as base for land cover classification (Table 4), using sample regions for each attribute, made with the Image Classification toolbar of the Spatial Analyst extension in ArcGIS 10.3, to better distinguish the variety of surface coverage.

Table 3. UAVSAR bands as extracted to create Pauli decomposition bands.

Band	HHHH	HHHV	HHVV	HVHV	HVVV	VVVV		HV	HH − VV	HH + VV
Real	+	+	+	+	+	+	**Name**	Pauli3	Pauli2	Pauli1
Imaginary	−	+	+	−	+	−	**Code**	Red	Green	Blue

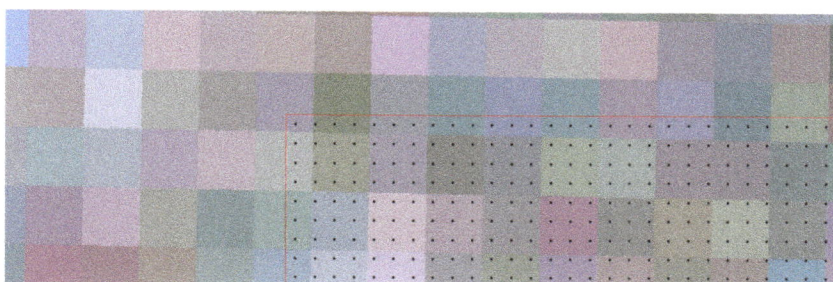

Figure 3. The image displays the necessity for point distribution and rearrangement of pixels on UAVSAR Pauli decompensated data.

Table 4. The three-band RGB combination of Worldview 2 was used to create land cover classification for each site.

	(1) Puerto Plata	(2) Montecristi	(3) Haiti	(4) Test site
1	Water	Water	Water	Water
2	Flat Surfaces	Mangrove	Mangrove	Bare Soil
3	Mangrove	Bare Earth	Structures & Roads	Forest
4	Forest	Built	Forest	Shrub
5	Eroded Land	Forest	Shrubs	
6	Pasture	Shrub	Pasture	
7	Structure	Clouds	Dump Site	

3. Results

3.1. Posterior Probability Approach

Initial tests on a modified version of the algorithm, using a posterior probability modeling [46–49] to define difference between potential areas of sites and non-sites were conducted with focus on a trial area in Puerto Plata, DR. The model records posterior probabilities generated using a linear discriminant analysis (LDA) with a Bayes plug-in classifier. LDA assumes normal distributions with similar covariance conditional on the input classes. Probabilities record the likelihood of cells belonging to a specific class and thresholds can be used to create definitive classifications. While preferable in other areas, the ordinal scale of probability values are more useful when assessing the likelihood of sites as they represent the possibility of sites being present or not. This approach had been successfully applied elsewhere and involved using known sites and alleged non-sites to build a binary classifier where each cell was assigned a posterior probability of being an archaeological site. Datasets included Woldview-2 imagery and band difference ratios similar to the NDVI, which were then reduced using PCA. The sites in the original dataset were represented by single artifact find spots around a central point which represented the site proper. Polygons were digitized around all points to delineate sites. There were 12 sites in the area with an average area of 4338 m^2. Non-Sites were generated according to the following rules:

- in surveyed areas
- A total of >100 m from known sites
- Buffers with a radius of 37 m (with an area of 4300 m^2) were generated around each site.

It was important that non-sites shared similar characteristics to actual sites, so the buffer size was chosen based on the average size of known sites in the area (4300 m^2) as calculated by material scatters. The trial results were plotted on a ROC curve (Figure 4), which demonstrates that sites were much more likely to be found in the high or higher probability areas of the posterior probability maps of the Puerto Plata trial region, and much less likely in the low probability areas.

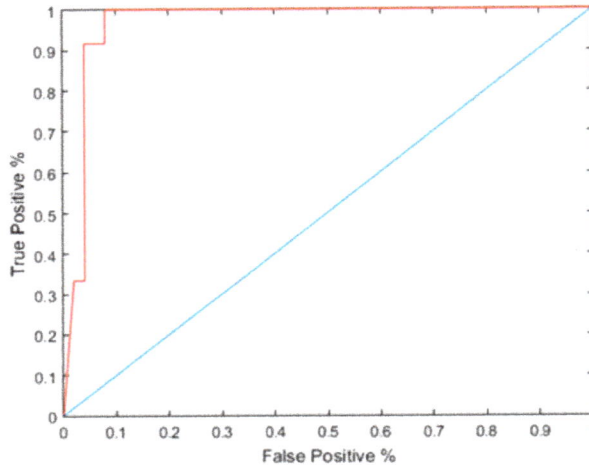

Figure 4. Receiver Operating Characteristics (ROC curve) of the posterior probability approach from Puerto Plata.

Considering the entered data sets, known sites were mostly located in the mangrove forest and on hilltops or slopes, while non-sites were distributed over the complete data set, the results showed a positive outcome, coloring similar locations in deep red (Figure 5). There are however two general caveats. Firstly, the algorithm based on a binary classification might be more effective when identifying homogenous site-types like lithic scatters. Secondly, the algorithm performed better with a larger and very accurate sample of known sites and checked known non-sites of a surveyed area. In this project, the heterogeneous nature of the sites coupled with a small number of known sites must be regarded as an impediment to this approach. When focusing on the research area, the approach may have also been hindered by the dominance of sites in the mangroves near the ocean shore, and on hilltops; the algorithm favored these areas for probable site locations. Considering the visual interpretation of the outcome by the archaeologists working in the region, and having recorded the known sites, the results generally make sense. Only based on the remote sensing imagery and without any topographic information, the highest probability of undiscovered sites is found near or on the hilltops' northern side, where additional sites were to be expected, while the region south of the range of hills is void of any probable sites. This of course remains highly speculative and is only based on landscape understanding and survey experience, and could only be confirmed by an independent accuracy assessment applied onto a different, possibly larger area, or through additional ground truthing by a total area survey of the region. Because of the clustering of a relatively limited number of sites, the Puerto Plata area (1) was seen potentially unsuitable as a trial region, and the team decided to continue with the Monte Cristi area (2), which provided more known sites, that were more broadly distributed over the area, using a different approach.

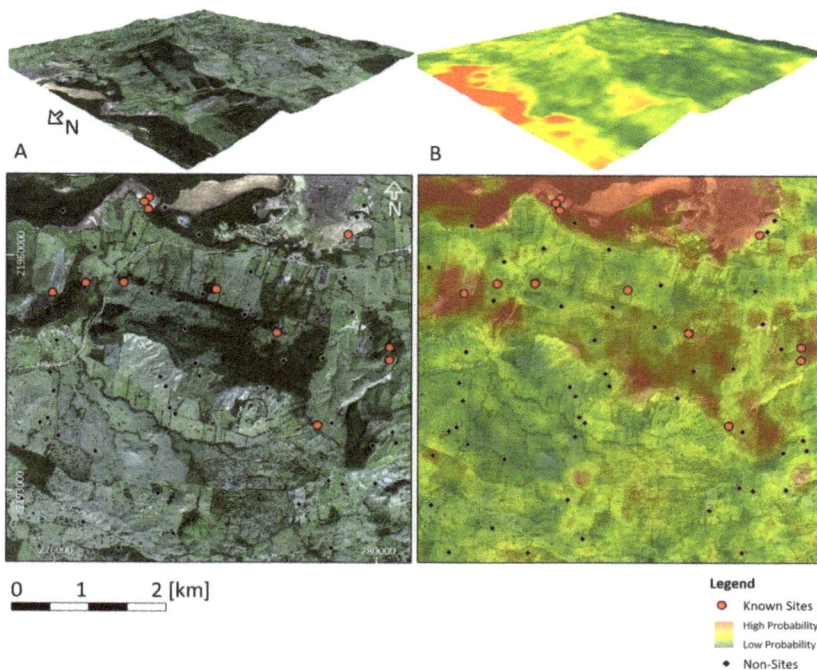

Figure 5. Posterior probability results from Puerto Plata in topographic and top down view. (**A**) The original RGB data, (**B**) overlaid by the resulting posterior probability map. (Datum: WGS84, UTM 19N).

3.2. Frequentist Protocols

A frequentist [22] approach, programmed in the statistical Software *R* (R Foundation for Statistical Computing, Vienna, Austria) [50], was applied for the Montecristi area (Figure 6). This 5×5 km^2 area is located in a hilly part of the coastal region of Montecristi, where new sites had recently been identified [31,35,51], of which 16 sites were chosen. A window of 81 pixels $\times$ 81 pixels (160×160 m^2) was set around the single pixels picked as the center of the known sites (KS) and randomly selected non-sites (NS) creating a base of information of 6561 pixels, across each band, for each point of interest. The same number of known sites and non-sites was considered. Histograms were generated for each separate band across sites and non-sites. The histograms were binned in 100 equally spaced separations (see Figure 7). A student t-test/Wilcoxon rank sum test was conducted to see if there is a significant dissimilarity between sites and non-sites.

A variety of statistical explorations were conducted. A band-difference ratio (BDR) was generated between all pairs of bands. Given two bands, their BDR is given by the pixel-wise quotient of their difference over their sum. This procedure is exactly the same as NDVI, but additionally utilizes all unique pairs of bands. The original bands (WV2) are used along with these new BDR bands in the analysis, which is performed on the WV2 bands and then the BDR bands. For each band, the 81 pixels $\times$ 81 pixels extents of KS and NS are extracted, separately. The KS pixels and NS pixels are binned respectively using the same bin widths. The KS and NS pixels in each paired bin are compared in a hypothesis test with null hypothesis that the KS pixels and NS pixels are of the same distribution. If we fail to reject the null hypothesis at significant level 0.05, then that bin is marked 0; otherwise it is marked 1. This creates a binary matrix for each of the bands. A Boolean merge is then applied, that is, these matrices are summed together.

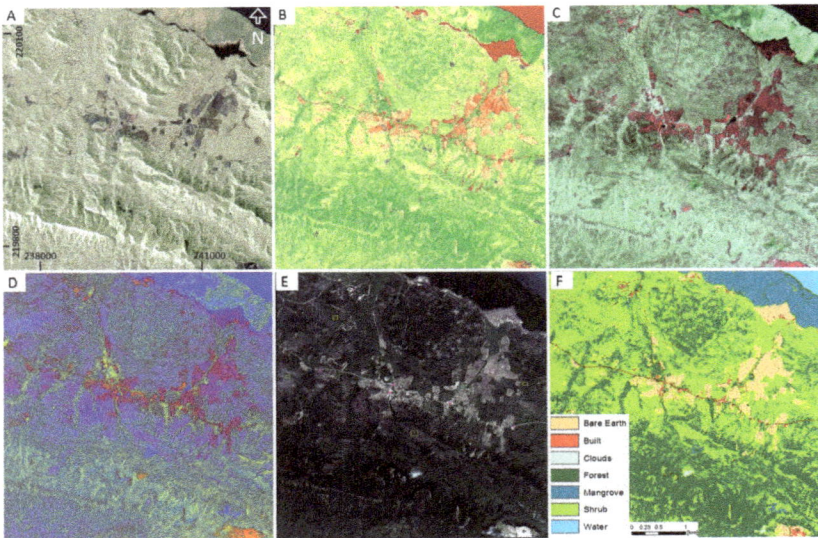

Figure 6. Various combined remote sensing data sets of the Montecristi trial area. (**A**) UAVSAR Pauli Decomposition. (**B**) Worldview-2 NDVI. (**C**) Worldview PCA. (**D**) Worldview Tasseled Cap. (**E**) Worldview RGB with rectangles defined for (**F**) land cover classification. A white cloud can be seen in the lower center of (**E**). (Datum: WGS84, UTM 19N).

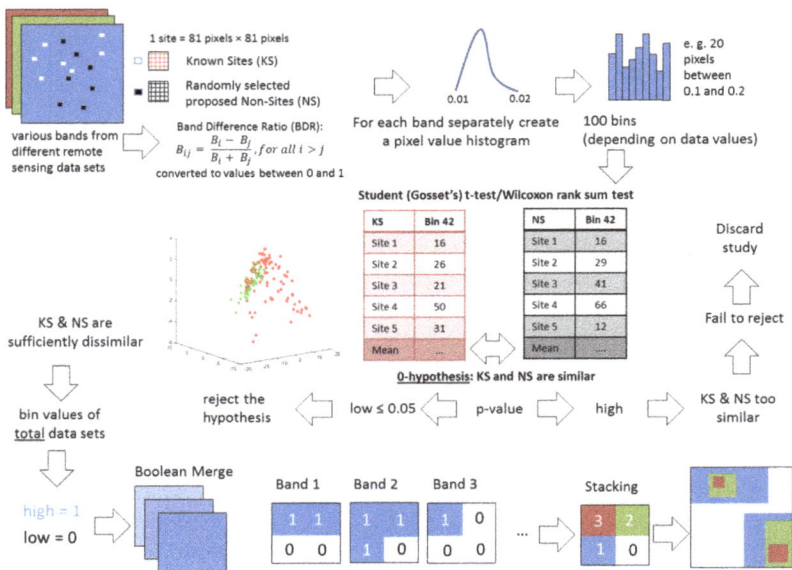

Figure 7. Sketch of the frequentist protocol algorithm. Student's (Gosset's) T-test/Wilcoxon rank sum test is applied to determine, if the distributions of site and non-site pixels in individual bins are statistically significantly different, with 0-hypothesis being they are from the same distribution.

3.3. Dominance of Bands

A variety of statistical trial calculations were applied. A band-difference ratio (BDR) was generated among every band included in the algorithm data set, to reduce the dominance of particular, as well as essentially redundant, data sets. These ratios were indices similar to the NDVI and normalized datasets. Only bands with the lowest positive response rate, a low p-value (cause for rejecting the 0-hypothesis that KS and NS were similar) were further considered for the tests. The highly diverse environment, as made visible in the land cover maps, would, one might expect, influence the success of the approach in comparison with other areas where land cover was more homogenous [52].

The frequentist protocol from [21,52] was implemented in R with different binning strategies [53,54] using Student's (Gosset's) T-Test or the Wilcoxon rank sum test (for explanations of these tests, see [55]). The bulk of the work in R was done as exploratory data analysis with mixing and matching binning strategies and hypothesis testing. The results vary strongly on different numbers and combinations, based on the variety band different ratios, and statistical tests (Figure 8).

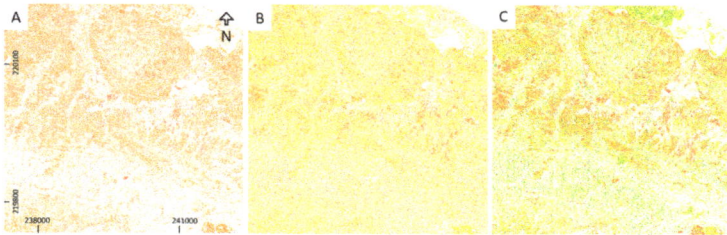

Figure 8. Results from the frequentist tests in the Montecristi area, using a total of 28 BDR bands. (**A**) Sturges rule: highest value after Boolean merge: 2, (**B**) based on Scott: highest value after Boolean merge: 4, (**C**) Sturges rule using BDR: highest combination after Boolean merge: 7. (Datum: WGS84, UTM 19N).

4. Discussion

The final image that incorporated the land cover information shows a definite response to the diverse landscape represented in the image (Figure 9). Several aspects are notable: as anticipated, without sites mapped in the mangrove area (1) it remains completely void of site activity. This, however, also appears to concern areas classified as forest, with large parts of forested areas showing no higher potential. A significant number of known sites had been identified in areas covered by forest and shrub, which, in our tests at least, do not respond well to the DDM search protocols using only the data sets available. One reason could be that more non-sites were distributed in forested areas, as large parts of the survey are covered by dense forest. Therefore, the random distribution of sites may have had an effect on the non-sites statistics, influencing the general results. In contrast, most significant high response is shown in areas with little vegetation. Here, the dimension of these sites was better defined. This expected best response rate is confirmed by the bright red colored areas surrounding these sites, showing that in these locations the algorithm shows its greatest strength. It can be expected that the reflection value of sites in bare earth areas should differ significantly from non-sites here than in forested sites, as the scatter of archaeological material is better displayed on the surface, particularly in ploughed areas, while canopy vegetation does not appear particularly affected by it.

From an archaeological interpretative view, the higher values do not necessarily represent an ancient pattern of settlement selection, but a combination of features that seem to be the trend in this particular area. It remains uncertain if the DDM corresponds to areas that follow attributes based on previous [20] and current research that served predictive models [35], a pattern that expresses tendencies, such as proximity to the sea, or other sea features such as mangrove forest, proximity to brooks, proximity to flat lands (usually less forested), and elevation less than 100 m. Modern settlements have been built near areas that combined the aforementioned features, as these also allow the development of crops. The high valued pixels of the DDM show zones in which these features have been combined, and could be a reason to have also highlighted areas of current habitation. For the northern part, the model created seems to correlate with earlier proposed indigenous activity pattern; the question remains why this may be the case. While the topography was not taken in consideration due to the low resolution available for the region, interpreted visually by archaeologists particular areas of no likely habitation coincide to areas of low probability. In the center and south, two known locations (3) are extensive sites on grassland, surrounding a former school yard. Here the high probability results appear to delineate the area of the assemblage of material. Location (4) in the southwest corner seems to pick up a small site near the recently mapped large site of El Manantial (MC-44) [18], only separated by a small gorge. The intensity at location (5) was identified as a modern dump site, while (6) represents the above mentioned small cloud.

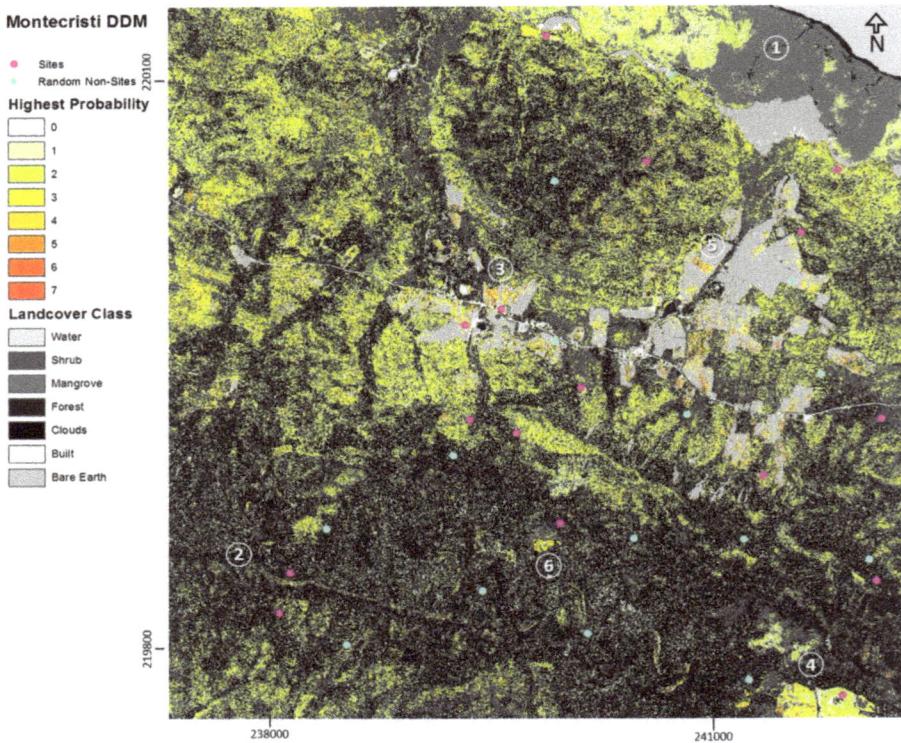

Figure 9. DDM frequentist results based on 8 UAVSAR bands, 3 Pauli Decomposition, 8 WV-2, NDVI, KTT and best BDR outcome (Datum: WGS84, UTM 19N).

5. Conclusions

Automatic detection models for archaeology, particularly the idea of predictive modeling, have been under heavy scrutiny since their appearance in archaeological research [56–58], with predominant questioning as to whether the time and effort invested served the outcome. One possibility is to leave the decision making not completely to the machines, but rather guiding them semi-automatically to a solution. More focus on the development of these types of algorithms should lead to a breakthrough in the detection of amorphous archaeological features in the future.

As it can be seen, this research is still in the experimental stage. The posterior probability approach showed some value in highlighting regions of greater archaeological potential, correlating positively with the archaeologist's idea where to find sites, but remained very vague in actual detection. Regarding the applied frequentist algorithm, it was shown in previous studies at non-forested locations that the applied algorithms was particularly useful in otherwise uniform environments to identify archaeological [21] or geological features [48]. The anticipated significant differentiation between sites and non-sites on northern Hispaniola was overshadowed by the immense environmental variation in the surveyed region. Many strong factors weigh in that made it particularly difficult for the algorithm to distinguish archaeological sites from areas with little archaeological potential. Since the algorithms were applied while being in development, due to scheduling issues within the international team, it was not possible to apply both semi-detection approaches on each of the regions. To independently compare their effectiveness in direct detection and provide an accuracy assessment, it would have been critical to apply each algorithm on the same area.

Improvements could be made, by using instead of a single point with a square of 81 pixels × 81 pixels, an average of the pixel values inside an actually determined area extent of a site, as it could have been used for the small trial area, where sites had been identified by systematic survey. This would have provided a more precise fingerprint in comparison to the non-sites. Also, picking non-sites randomly from different environments may have enhanced the probability that with very bad luck an actual not yet identified site would have been selected. A point of critique could also be the use of only two, and completely different, datasets; another might be that these data sets were used to produce synthetic bands. Also, the vegetation types or patterns in forested covered areas produce a diversity that could only be differentiated with additional data. A highly distinguishable feature of some identified sites, the topography, as identified through drone photogrammetry [34] could be an important factor to significantly improve the study, but for this the access to high resolution regional LiDAR data would be crucial. A last problem to address is validation. Both trial areas are well known and have been visited and surveyed extensively by archaeologists. Nevertheless the areas were surveyed non-systematically before, which means that most likely there are still Amerindian unknown activity areas that have not been recorded. The project, which stretched over a period of two years, left no time for subsequent validation of results by groundtruthing these potentially recognized sites, which should be a task for the future. In addition, the trial areas should be extended to other, and potentially larger areas with known sites, using the values identified in these trial areas onto a different dataset to validate if the values are identifiable as specific for an Amerindian archaeological site.

Although the study could not be completely validated in the field for logistical reasons, the results indicate the promise of semi-automatically identifying areas with non-structural archaeological potential in diverse environments: this leaves great potential for future tasks to evaluate regions for unknown and potentially threatened heritage and archaeology automatically.

Acknowledgments: The research leading to these results has received funding from the European Research Council under the European Union's Seventh Framework Programme (FP7/2007-2013)/ERC grant agreement n°319209. We would like to thank Jorge Ulloa Hung who provided the archaeological sites test data set for the Puerto Plata region. The Digital Globe Foundation provided a generous data grant of Worldview-2 imagery for the region. We thank NGA/NASA/USGS for making their SRTM DEM data available to us. The ASTER L1A data product is courtesy of the online Data Pool at the NASA Land Processes Distributed Active Archive Center (LP DAAC), USGS/Earth Resources Observation and Science (EROS) Center, Sioux Falls, South Dakota (https://lpdaac.usgs.gov/data_access).

Author Contributions: T.S. managed and oversaw all the sub elements of the study, undertook the image and data preparation in the GIS. D.C. conceived, designed the direct detection and posterior probability approach and coordinated its implementation. J.P. implemented and performed the analysis of the direct detection protocols in R. W.M. implemented the data into the posterior probability approach. E.H. contributed the archaeological site data and assisted in the archaeological interpretation. As a principal investigator of the ERC-Nexus1942, C.H. provided the funding for study, coordinated the field work in the Dominican Republic, and contributed on discussing the archaeological validity of the study.

Conflicts of Interest: The authors declare no conflict of interest.

References

1. Batut, A. *La Photographie Aérienne Par Cerf-Volant*; Gauthier-Villars: Paris, France, 1890; p. 27.
2. Capper, J.E. Photographs of stonehenge as seen from a war balloon. *Archaeol. Misc. Tracts Relat. Antiqu.* **1907**, *60*, 571. [CrossRef]
3. Millhauser, J.K.; Morehart, C.T. The ambivalence of maps: A historical perspective on sensing and representing space in mesoamerica. In *Digital Methods and Remote Sensing in Archaeology*; Forte, M., Campana, S., Eds.; Springer Books: New York, NY, USA, 2016; pp. 247–268.
4. Wiseman, J.; El-Baz, F. *Remote Sensing in Archaeology*; Springer: New York, NY, USA, 2007; pp. 1–8.
5. Comer, D.C.; Harrower, M.J. *Mapping Archaeological Landscapes from Space: In Observance of the 40th Anniversary of the World Heritage Convention*; Springer Press: New York, NY, USA, 2013; pp. 1–8.
6. Carrol, D.M.; Evans, R.; Bendelow, V.C. *Air Photo-Interpretation for Soil Mapping*; Soil Survey: Harpenden, Hertfordshire, UK, 1977; p. 5.
7. Trier, O.D.; Larsen, S.O.; Solberg, R. Automatic detection of circular structures in high-resolution satellite images of agricultural land. *Archaeol. Prospect.* **2009**, *16*, 1–15. [CrossRef]
8. Di Iorio, A.; Bridgwood, I.; Schultz Rasmussen, M.; Kamp Sorensen, M.; Carlucci, R.; Bernardini, F.; Osman, A. Automatic detection of archaeological sites using a hybrid process of remote sensing, GIS techniques and a shape detection algorithm. In Proceedings of the 3rd International Conference on Information and Communication Technologies: From Theory to Applications, Damascus, Syria, 7–11 April 2008.
9. Schuetter, J.; Goel, P.; McCorriston, J.; Park, J.; Senn, M.; Harrower, M. Autodetection of ancient Arabian tombs in high-resolution satellite imagery. *Int. J. Remote. Sens.* **2013**, *34*, 6611–6635. [CrossRef]
10. Zingman, I.; Saupe, D.; Lambers, K. Automated search for livestock enclosures of rectangular shape in remotely sensed imagery. In Proceedings of the SPIE Remote Sensing, Dresden, Germany, 17 October 2013.
11. Traviglia, A. MIVIS Hyperspectral Sensors for the Detection and GIS Supported Interpretation of Subsoil Archaeological Sites. In *Digital Discovery. Exploring New Frontiers in Human Heritage, CAA 2006, Proceedings of the 34th Conference on Computer Applications and Quantitative Methods in Archaeology, Fargo, ND, USA, 18–22 April 2006*; Clark, J.T., Hagemeister, E.M., Eds.; Archaeolingua: Budapest, Hungary, 2006.
12. Doneus, M.; Verhoeven, G.; Atzberger, C.; Wess, M.; Ruš, M. New ways to extract archaeological information from hyperspectral pixels. *J. Archaeol. Sci.* **2014**, *52*, 84–96. [CrossRef]
13. Menze, B.H.; Ur, J.A.; Sherratt, A.G. Detection of ancient settlement mounds: Archaeological survey based on the SRTM terrain model. *Photogramm. Eng. Remote Sens.* **2006**, *72*, 321–327. [CrossRef]
14. Opitz, R.S.; Cowley, D. *Interpreting Archaeological Topography: Airborne Laser Scanning, 3D Data and Ground Observation*; Oxbow Books: New York, NY, USA, 2013.
15. Stewart, C.; Lasaponara, R.; Schiavon, G. ALOS PALSAR analysis of the archaeological site of Pelusium. *Archaeol. Prospect.* **2013**, *20*, 109–116. [CrossRef]
16. Rouse, I. Pattern and process in West Indian archaeology. *World. Archaeol.* **1977**, *9*, 1–11. [CrossRef]
17. Keegan, W.F.; Hofman, C.L.; Ramos, R.R. (Eds.) *The Oxford Handbook of Caribbean Archaeology*; Oxford University Press: New York, NY, USA, 2013; p. 13.
18. Sonnemann, T.F.; Herrera Malatesta, E.; Hofman, C.L. Applying UAS photogrammetry to analyse spatial patterns of Amerindian settlement sites in the northern DR. In *Digital Methods and Remote Sensing in Archaeology*; Forte, M., Campana, S., Eds.; Springer Books: New York, NY, USA, 2016; pp. 71–87.
19. Ortega, E.; Denis, P.; Olsen Bogaert, H. Nuevos yacimientos arqueológicos en Arroyo Caña. *Bull. Mus. Hombre Dominic.* **1990**, *23*, 29–40.
20. Ulloa Hung, J. Arqueología en la Línea Noroeste de La Española Paisajes, Cerámicas e Interacciones. Ph.D. Thesis, Caribbean Research Group, Faculty of Archaeology, Leiden University, Leiden, The Netherlands, 23 April 2013.
21. Comer, D.C. Merging Aerial Synthetic Aperture Radar (SAR) and Satellite Multispectral Data to Inventory Archaeological Sites. Available online: https://www.ncptt.nps.gov/download/28370/ (accessed on 9 June 2017).
22. Comer, D.C.; Blom, R.G. Detection and identification of archaeological sites and features using Synthetic Aperture Radar (SAR) data collected from airborne platforms. In *Remote Sensing in Archaeology*; Wiseman, J., El-Baz, F., Eds.; Springer: New York, NY, USA, 2007; pp. 103–136.

23. Keegan, W.F.; Hofman, C.L. *The Caribbean before Columbus*; Oxford University Press: New York, NY, USA, 2017; pp. 40–42, 115–147.
24. Veloz Maggiolo, M.; Ortega, E.; Caba, Á. *Los Modos de Vida Meillacoides y Sus Posibles Orígenes*; Editora Taller: Santo Domingo, Dominican Republic, 1981; p. 10.
25. Sinelli, P.T. Meillacoid and the Origins of Classic Taíno Society. In *The Oxford Handbook of Caribbean Archaeology*; Keegan, W.F., Hofman, C.L., Ramos, R.R., Eds.; Oxford University Press: New York, NY, USA, 2013; pp. 221–231.
26. Ting, C.; Neyt, B.; Ulloa Hung, J.; Hofman, C.; Degryse, P. The production of pre-Colonial ceramics in northwestern Hispaniola: A technological study of Meillacoid and Chicoid ceramics from La Luperona and El Flaco, Dominican Republic. *J. Archaeol. Sci. Rep.* **2016**, *6*, 376–385. [CrossRef]
27. Deagan, K.A.; Cruxent, J.M. *Archaeology at La Isabela. America's First European Town*; Yale University Press: New Haven, CT, USA, 2002; p. 15.
28. De Las Casas, B. *Brevísima Relación de La destrucción de Las Indias*; Fundación Biblioteca Virtual Miguel de Cervantes: Alicante, Spain, 2006; p. 16.
29. Rouse, I. *The Tainos: Rise and Decline of the People Who Greeted Columbus*; Yale University Press: New Haven, CT, USA, 1993; pp. 26–48.
30. Hofman, C.L.; Ulloa Hung, J.; Herrera Malatesta, E.; Jean, J.S.; Sonnemann, T.F.; Hoogland, M.L. Indigenous Caribbean perspectives: Archaeologies and legacies of the first colonised region in the New World. *Antiquity* **2017**, in press.
31. Ulloa Hung, J.; Herrera Malatesta, E. Investigaciones arqueológicas en el norte de La Española, Entre viejos esquemas y nuevos datos. *Bull. Mus. Hombre Dominic.* **2015**, *46*, 75–107.
32. Rouse, I. Areas and periods of culture in the Greater Antilles. *Southwest. J. Anthropol.* **1951**, *7*, 248–264. [CrossRef]
33. Hofman, C.L.; Hoogland, M.L. Investigaciones arqueológicas en los sitios El Flaco (Loma de Guayacanes) y La Luperona (Unijica): Informe pre-liminar. *Bull. Mus. Hombre Dominic.* **2015**, *46*, 61–74.
34. Sonnemann, T.F.; Ulloa Hung, J.; Hofman, C.L. Mapping indigenous settlement topography in the Caribbean using drones. *Remote Sens.* **2016**, *8*, 791. [CrossRef]
35. Herrera Malatesta, E. Una Isla, Dos Mundos: Sobre la Transformación del Paisaje Indígena de Bohío A la Española. Ph.D. Thesis, Leiden University, Leiden, The Netherlands, 2017, in press.
36. Bernstein, L.S. Quick atmospheric correction code: Algorithm description and recent upgrades. *Opt. Eng* **2012**, *51*, 111719. [CrossRef]
37. Hadjimitsis, D.G.; Papadavid, G.; Agapiou, A.; Themistocleous, K.; Hadjimitsis, M.G.; Retalis, A.; Michaelides, S.; Chrysoulakis, N.; Toulios, L.; Clayton, C.R.I. Atmospheric correction for satellite remotely sensed data intended for agricultural applications: Impact on vegetation indices. *Nat. Hazards Earth Syst. Sci.* **2010**, *10*, 89–95. [CrossRef]
38. Rouse, J.W.; Haas, R.H.; Scheel, J.A.; Deering, D.W. Monitoring vegetation systems in the Great Plains with ERTS. In Proceedings of the 3rd Earth Resource Technology Satellite (ERTS) Symposium, Washington, DC, USA, 10–14 December 1973.
39. Eastman, J.R.; Filk, M. Long sequence time series evaluation using standardized principal components. *Photogramm. Eng. Remote Sens.* **1993**, *59*, 991–996.
40. Faraway, J.J. *Linear Models with R*, 2nd ed.; CRC Press: Hoboken, NJ, USA, 2015; pp. 161–171.
41. Kauth, R.J.; Thomas, G.S. The tasseled cap: A graphic description of the Spectral-Temporal Development of Agricultural Crops as Seen by Landsat. In *LARS Symposia*; Purdue University: West Lafayette, IN, USA, 1976.
42. Yarbrough, L.D.; Navulur, K.; Ravi, R. Presentation of the Kauth–Thomas transform for WorldView-2 reflectance data. *Remote Sens. Lett.* **2014**, *5*, 131–138. [CrossRef]
43. Alaska Satellite Facility. *MapReady*; NASA: Fairbanks, AK, USA, 2014.
44. Maitra, S.; Gartley, M.G.; Kerekes, J.P. Relation between degree of polarization and Pauli color coded image to characterize scattering mechanisms. In Proceedings of the SPIE Defense, Security, and Sensing, Baltimore, MD, USA, 23 April 2012.
45. *PolSARpro*, version 5; IETR (Institute of Electronics and Telecommunications of Rennes)—UMR CNRS 6164 ESA: Paris, France, 2014.

46. Chen, L.; Priebe, C.E.; Sussmann, D.L.; Comer, D.C.; Megarry, W.P.; Tilton, J.C. Enhanced Archaeological Predictive Modelling in Space Archaeology. Available online: http://arxiv.org/abs/1301.2738 (accessed on 15 January 2016).

47. Chen, L.; Comer, D.C.; Priebe, C.E.; Sussmann, D.; Tilton, J.C. Refinement of a method for identifying probable archaeological sites from remotely sensed data. In *Mapping Archaeological Landscapes from Space: In Observance of the 40th Anniversary of the World Heritage Convention*; Comer, D.C., Harrower, M.J., Eds.; Springer Press: New York, NY, USA, 2013; pp. 251–258.

48. Megarry, W.P.; Cooney, G.; Comer, D.C.; Priebe, C.E. Posterior probability modeling and image classification for archaeological site prospection: Building a survey efficacy model for identifying neolithic felsite workshops in the Shetland Islands. *Remote Sens.* **2016**, *8*, 529. [CrossRef]

49. Comer, D.C. *Institutionalizing Protocols for Wide-Area Inventory of Archaeological Sites by the Analysis of Aerial and Satellite Imagery*; Project Number 11-158; United States of America Department of Defense, Legacy Program: Washington, DC, USA, 2014.

50. R Core Team. *R Foundation for Statistical Computing*; R Core Team: Vienna, Austria, 2015.

51. Herrera Malatesta, E. Understanding ancient patterns: Predictive modeling for field research in Northern Dominican Republic. In Proceedings of the 26th Congress of the International Association of Caribbean Archaeologists, Maho Reef, Sint Maarten, Netherlands Antilles, 19–25 July 2015; pp. 88–97.

52. Tilton, J.C.; Comer, D.C. Identifying Probable Archaeological Sites on Santa Catalina Island, California Using SAR and Ikonos Data. In *Mapping Archaeological Landscapes from Space: In Observance of the 40th Anniversary of the World Heritage Convention*; Comer, D.C., Harrower, M.J., Eds.; Springer Press: New York, NY, USA, 2013.

53. Sturges, H.A. The choice of a class interval. *J. Am. Stat. Assoc.* **1926**, *21*, 65–66. [CrossRef]

54. Scott, D.W. On optimal and data-based histograms. *Biometrika* **1979**, *66*, 605–610. [CrossRef]

55. Rice, J.A. *Mathematical Statistics and Data Analysis*, 3rd ed.; Thomson Brooks/Cole Publishing: Belmont, CA, USA, 2007; pp. 435, 454.

56. Kvamme, K.L. Development and testing of quantitative models. In *Quantifying the Present and Predicting the Past: Theory, Methods, and Applications of Archaeological Predictive Modeling*; Judge, W.J., Sebastian, L., Eds.; US Department of Interior, Bureau of Land Management Service Center: Denver, CO, USA, 1988; pp. 325–428.

57. Kamermans, H.; van Leusen, M.; Verhagen, P. *Archaeological Prediction and Risk Management. Alternatives to Current Practice*; Leiden University Press: Leiden, The Netherlands, 2009; p. 7.

58. Verhagen, P.; Whitley, T.G. Integrating archaeological theory and predictive modeling: A live report from the scene. *J. Archaeol. Method Theory* **2012**, *19*, 49–100. [CrossRef]

geosciences

MDPI

Article

Landscape Pattern Detection in Archaeological Remote Sensing

Arianna Traviglia * and Andrea Torsello

Department of Environmental Sciences, Informatics and Statistics, Ca' Foscari University of Venice,
via Torino 155, 30172 Mestre (VE), Italy; atorsell@unive.it
* Correspondence: traviglia@unive.it; Tel.: +39-041-2348478

Received: 13 October 2017; Accepted: 29 November 2017; Published: 11 December 2017

Abstract: Automated detection of landscape patterns on Remote Sensing imagery has seen virtually little or no development in the archaeological domain, notwithstanding the fact that large portion of cultural landscapes worldwide are characterized by land engineering applications. The current extraordinary availability of remotely sensed images makes it now urgent to envision and develop automatic methods that can simplify their inspection and the extraction of relevant information from them, as the quantity of information is no longer manageable by traditional "human" visual interpretation. This paper expands on the development of automatic methods for the detection of target landscape features—represented by field system patterns—in very high spatial resolution images, within the framework of an archaeological project focused on the landscape engineering embedded in Roman cadasters. The targets of interest consist of a variety of similarly oriented objects of diverse nature (such as roads, drainage channels, etc.) concurring to demark the current landscape organization, which reflects the one imposed by Romans over two millennia ago. The proposed workflow exploits the textural and shape properties of real-world elements forming the field patterns using multiscale analysis of dominant oriented response filters. Trials showed that this approach provides accurate localization of target linear objects and alignments signaled by a wide range of physical entities with very different characteristics.

Keywords: remote sensing; landscape archaeology; pattern extraction; archaeological linear structures; centuriation; cadaster

---

1. Introduction

Present-day landscapes have been shaped by centuries of human action. Since their settlement, ancient communities have planned, modified, and engineered the space around them: land surveying and field system deployments were some of the first forms of large-scale "landscape engineering" performed by complex societies in antiquity. Centuriation, the iconic Roman system of land subdivision into large square plots assigned to settlers, represents one of the most advanced efforts of landscape engineering of the ancient world. Its effects can be discerned even today, with this regular territorial design model continuing to have a significant influence on present-day agrarian organization in many locations across the Mediterranean basin and surrounding regions.

Remote sensing represents one of the most effective methods for the study of landscape design and patterning and has been amply and successfully applied at any latitude to land systems prospecting. While investigation methods and technologies have been, respectively, refining and progressing over time—especially after the advent of satellite imagery on the archaeological research scene— the photo-interpretation phase, in which the nature of the objects identified on imagery are determined, have continued to be based on a visual approach and manual (digital) mapping. The current unprecedented availability of Remote Sensing data should provide considerable ground for the development of alternative and more effective procedures, and should have favored a growing interest for computerized methods that can streamline and improve detection of landscape design components on from-above imagery. Instead, automated detection of landscape patterns has seen little or no progress, nor much application, in the archaeological domain in the last decade. The current remotely sensed data "deluge" makes it urgent to envision and develop computer-driven methods that can simplify the inspection of datasets and the extraction of relevant information from them, determining a shift from traditional "human" means of detection and interpretation to automated ones.

This paper expands on the development of methods for the detection of target landscape features—constituted by field system patterns—in very high spatial resolution images, and the verification of their spatial arrangement within the framework of "Visualizing Engineered Landscapes (VEiL)", a landscape archaeology project based in Italy and focused on the centuriated landscape surrounding the Roman city of Aquileia. The targets of interest consist of a variety of similarly oriented objects (such as roads, drainage channels, etc.), disposed in parallel and orthogonal fashion concurring to demark the current Aquileian landscape organization, which reflects the one imposed by Romans from the 2nd century BC onward. The aim here is to define a workflow enabling a systematic detection of symmetrically deployed elements of the land division that concur to engineer the landscape, reducing biases derived from the subjective visual analysis of human observers. The presented approach enables the detection of similarly oriented features that are compatible with the local centurial system and automatically extracts candidate modules of the centuriation by analyzing the periodic alignment of peaks in the distribution of the oriented features. The novelty of this approach resides in that it performs computerized processes of identification directly on raster imagery (rather than e.g., on digital vector cartography) of linear elements of the landscape that align with the orientation of the local cadaster and uses them to generate all the possible centuriation module scenarios to identify the most plausible one, thus automating a process normally performed via painstaking manual calculation.

1.1. Automation in Archaeological Remote Sensing: Current State of the Research Field

Until the systematic introduction of aerial photo inspection in cadaster studies during the last third of last century, the identification of centurial systems had been largely tackled from a theoretical and speculative point of view, referring to ancient sources and epigraphic documents, or by attempting to align possible agrarian divisions with modern topographic maps, with little or no attention to the actual morphologic situation. Only the advent of airborne photography, coupled with an improved map production capacity, determined a first decisive shift in approaching centuriation scholarship [1] and laid the foundation for a new and more empirical approach to centurial grid identification proposed by the so-called Besançon Group (University of Franche-Comté, France) [2,3]. A partial dismissal of such approach in its first formulation in the last decade of last century did not stop a new generation of scholars to investigate ancient land divisions using a diachronic trans-disciplinary landscape archaeology approach based on a similar concept of archaeo-morphological analysis. Their work [4] further promoted the inclusion of geospatial datasets (including remote sensing data) and GIS analysis within the discipline. Airborne datasets continue to be fundamental in the identification of fossilized and limitedly altered centuriations in areas where post-depositional processes and phenomena have not been particularly dynamic [5,6]. Aerial images have only occasionally been substituted with satellite ones, which had the great advantage of covering broader areas; however, their use has been quite limited and confined to manual mapping of visually identified elements [7–9]. Very recently, a number of projects are successfully trialing the use of Lidar data in order to highlight grid boundaries in landscapes that have been essentially abandoned after the Roman period, with little or no occupation afterwards, and manually map them [10].

The present acceleration in the use of Remote Sensing imagery in a variety of disciplines is producing parallel rapid advancements in the application of automated approaches for detection and monitoring of phenomena and conditions (e.g., traffic management, ground conditions evaluation, ground change assessment, air pollution forecasting, etc.); nevertheless, such advancements do not appear to have invested much in the study of land design, historical and contemporary. More broadly, the whole field of automation in the domain of cultural landscapes studies is still moving the first, uncertain steps, despite several decades of development of Remote Sensing methods applied to archaeological scholarship, on one side, and while increasingly automated procedures for specific object detection are flourishing in an array of fields, on the other [11]. The reasons for lack of trust in automated methods in heritage recognition often come down to uneasiness amongst archaeological scholarship of losing control of the interpretation process [12]. Examples of implementation of automated approaches to archaeological aero-photointerpretation for determined objects are thus still scarce [13] and they broadly fall into an even smaller number of classes: template matching-based methods [14], custom algorithms [15], (GE)OBIA-based methods [16], and machine learning-based methods [15], all involving previous knowledge of the shape and characteristics of the searched object. Recent developments in Machine Learning and the appearance on the scientific scene of Convolutional Neural Networks (CNN) approaches [17,18], which gained credit for dramatically improving previous recognition performances, are generating enthusiasm for the possibilities of these high-end methods for the (semi)automatic detection of archaeological surface or sub-surface objects. Despite its potential, however, CNN has so far seen extremely limited development in the archaeological/cultural heritage field and only in applications where the shape of objects to detect (although variegated) were known and pertained to only one typology [15,19]. All these trials have in any case concentrated their attention on separated, single "entities" appearing with varying frequency within a landscape and not on systems of "related entities".

Notwithstanding limited instances, a rising awareness of the necessity of turning to automated processes to study landscape arrangement constituents as a whole is now favoring the adoption of new approaches within the discipline. VEiL project seeks to contribute to the scholarship of engineered landscapes by trialing approaches that can support the automation of tedious, time consuming tasks in landscape features extraction. This is particularly relevant for projects like this, addressing the study of patterning of diverse elements, of their repetitiveness, and of the relationships between physically connected or separated landscape components that reflect the past territorial arrangement and configuration.

1.2. The Archaeological Context of the Project

The Aquileian hinterland (Figure 1) was one of the broadest rural areas in Roman Italy to be centuriated before the first part of the 2nd century BC [20]. Aquileia is an exception in the European panorama as it was scarcely touched by 20th century urbanization, and archaeological prospections can be carried out from the immediate edge of the urban perimeter into a large surrounding area, which retains a relatively intact Roman peripheral landscape around a city of relevant size in antiquity. While patches of the Aquileian landscape still preserve the imprint of the distinctive spatial signature imposed by the Romans, with its typical repetitive patterns filtering through the noise of more modern systems [21], in other areas the centuriation elements are no longer easily identifiable or irrefutably attributable to Roman intervention.

Most of the studies undertaken have ignored the archaeological and topographical reality of the area: existing theories have drawn a variety of diverging conclusions [21] almost exclusively from modern topographic maps with noticeable "mechanicism" in reconstructions of the centurial grid obtained by juxtaposing theoretical gridirons on topographical maps. Within this volatile grid, the internal partition elements of centuriation are still mostly unknown.

A further difficulty is represented by the fact that in the Aquileian territory the Roman surveyors deployed in various moments multiple centuriations with different orientations, the chronological sequence of which is still not ultimately clarified [22], although the one that is currently more clearly and broadly preserved on the landscape pertains to the supposedly latest land division phase. This, the so-called Classic Aquileian Centuriation, is characterized by a clear and consistent orientation at about 22° W from (grid) North: initially deployed in the immediate vicinities of the newly founded

Roman colony, it was expanded likely until mid of the 1st century BC [22], using a centurial module of 20 × 20 *actus*, which was commonly used in contemporary land assignments in other areas of ancient Italy.

Figure 1. The case study area: Aquileia, located in NE Italy and its neighboring territory.

The study of the Roman cadaster in Aquileia is further complicated by the fact that the same areas have undergone later transformations in order to reclaim capacities for agricultural exploitation. Such reclamations are frequently characterized by regular plans that overlay the Roman module. It is therefore necessary to identify and expunge such posterior works in order to pinpoint the original Roman intervention. In many cases, however, these post-antiquity works have served to preserve some elements of the Roman land structuring.

1.3. Aims of the Project

Within this context, the VEiL project was envisioned and developed. The overall scope of this project is to unlock the rationale underpinning the processes of landscape engineering used by the Roman surveyors: pivotal for this is to identify and document the land system components characterized by long-term endurance in the landscape or their voids generated over the course of subsequent landscape transformations. The foundation of the research agenda is embodied by the ultimate identification of the "Classic" Aquileian cadaster module through the detection of its main axes, shape, size and extension as preserved by a variety of features characterizing the current landscape. This entails primarily the identification of the elements forming the grid, both the more evident centurial boundaries and the internal subdivisions aligning in a parallel or orthogonal fashion to the grid system. Accordingly, this project includes the reconnaissance on Remote Sensing imagery of more evident as well as subtle potential land division elements using a systematic combination of linear feature automated detection procedures based on signal processing and pattern recognition, as well as spatial analysis, field mapping and ground verification activities.

The workflow adopted to detect the grid components on imagery exploits the textural and shape properties of real-world elements forming the field patterns using multiscale analysis of dominant oriented response filters. Trials show that the proposed approach provides accurate localization of target linear objects and alignments signaled by a wide range of physical entities with very different characteristics. The spatial distribution of the extracted features can then be analyzed in order to determine candidates for starting point and module of the centuriation system.

2. Materials and Methods

When seeking to recognize historical components of current landscapes on remotely sensed imagery, the chances of identification can vary sensibly due to the nature of the elements that make up the past territorial organization and their preservation. One of the major issues faced by archaeologists when tackling the recognition of centurial systems in any sort of teledetected imagery is the identification of the land division boundaries as, in the past, they might have been formed by a variety of different natural or artificial elements, such as roads, fences/hedges, ditches, drainage and irrigation channels, tree lines and other landmarks [23,24] and they might be preserved in the modern-day territory by a likewise wide variety of artificial or natural markers. The issue is particularly intense in areas where later urban transformation has altered information embedded in the landscape and spatial signatures are obscured by the noise of more modern systems. This requires the use of the broadest possible range of geospatial data (including imagery) in order to ensure that their availability make up for shortcomings and voids rooted in each dataset and that all the landscapes elements relevant to centurial grid identification can be highlighted and properly considered.

2.1. Project Data Sources

Accordingly, the project has acquired a broad array of geospatial data ranging from base and thematic maps to remote sensing imagery, historic cartography and legacy data, all providing the necessary context to validate or expunge detected linear features.

Accurate base cartography is provided as freely accessible datasets made available by the Cartographic Division of the Regional Government of Friuli Venezia Giulia [25], the administrative Region within which Aquileia is located. The project has acquired shapefile datasets (RDN2008/TM33) at both 1:5000 and 1:25,000 scale as well as terrain height points and Lidar based DTMs (respectively 10 m and 1 m resolution) generated using airborne data collected in the past decade; in addition, previous digital and printed topographic maps (these last scanned and georeferenced by the project) have been acquired to avail a detailed sequence of modifications to the landscape as recorded in topo-maps of the last 40 years.

Much of the information relative to the landscape morphological arrangement needed for the goals of the project already appears in these basemaps. However, they are generally obtained via aero-photogrammetry: given the inherent subjectivity embedded in the image-interpretation process, the mapping of topographic elements ends up biased by the selection of the relevant mappable features made by the photogrammetric operators. As a consequence, only landscape elements deemed relevant to the mapping exercise are recorded. Thus produced, topographic maps do not, and often cannot, include information that is highly relevant to centurial grid detection, for example, cropmarks and soilmarks (visible on imagery) that can be referred to centuriation or ancient settlements; or features like tree lines made up by plants that in many cases were cyclically replanted over previous vegetation boundary markers used for many centuries to separate properties; or minor pathways and tracks that might preserve ancient ones. This makes using Remote Sensing imagery crucial in order to enable systematic recording of all the potential components of the centurial grid, irrespectively of how they have been preserved.

To prevent loss of information, the VEiL project has collated a vast remote sensing dataset, including imagery captured by various sources and sensors: these comprise (i) Black/White historical photographs; (ii) multispectral and hyperspectral imagery; (iii) RGB high-resolution aerial orthorectified images of different periods; (iv) Lidar data. For the specific goals of the work here described; data (i) to (iii) have been employed.

Historical aerial photographs (recorded starting in 1938), with a ground resolution varying from 1.5 to 3–4 m according to shooting period, represent a valuable source of information in that they provide information still embedded in the landscape in the second part of last century, but often no more visible on the ground nowadays due to later transformations (Figure 2). Soil and crop marks potentially related to the centurial grid are very visible in this type of imagery as well as are land boundaries and landscape arrangements superseded by more recent land reorganizations.

Figure 2. (a) 1938 aerial photo depicting the Aquileian landscape before post-war land rearrangements: higher frequency platting is evident; (b,c) zoom: marks referring to previous lands arrangements and roadways.

The hyperspectral data available to the project were acquired using MIVIS (Multispectral Infrared and Visible Imaging Spectrometer), a simultaneous spectral system that operates in the range of wavelengths from visible to Thermal infrared regions of the electromagnetic spectrum, with a surface spatial resolution of about 3 m × 3 m and an elevated number of channels (102) enabling a high spectral resolution (Figure 3). The images have been captured on two subsequent days during October 1998 in two acquisitions: a daytime shot (about 12 PM) and a so-called "night shot" (about 9–10 AM) that provided thermal values similar to the nocturnal ones.

The spectral characteristics of MIVIS data enables their use for monitoring vegetation health status using vegetation indexes and, therefore, provide the ability to identify alterations in the subsoil potentially related to archaeological deposits, which are reflected in the growth and biophysic parameters of the vegetation developing on top of them. Spectral characteristics can also be exploited by using soil indexes to increase the optical distinction between the wetness or the dryness of a portion of the ground, thus enabling the recognition of voids or artefacts lying immediately under the surface soil. Within the area covered by the acquisition (about 88 km^2 in extent, distributed in a polygon extending around 6.75 km in width and 12.56 km in length) over 600 multi-shaped and multi-oriented marks of potential anthropogenic origin (several of which oriented with the city plan) have been mapped using routine processing (vegetation indexes—Normalized Difference Vegetation Index (NDVI), Difference Vegetation Index (DVI), Modified Soil-adjusted Vegetation Index (MSAVI2), Generalized Difference Vegetation Index (GDVI), Green Normalized Difference Vegetation Index (GNDVI), etc. soil indexes—Principal Component Analysis PCA and Selective Principal component analysis or SPCA, Tasseled Cap) and visual detection combined with manual mapping. Besides, the MIVIS imagery displays abundance of modern landscape features aligning with the Aquileian Roman grid that, for their characteristics, can be automatically extracted with the workflow presented below (Section 2.2): the coarseness of these data enables more accurate feature extraction as the algorithm is not misled by the abundance of fine details as in the case of RGB Orthophotos. The images, however, are affected often by strong geometric distortions that alter the straightness of linear elements of the landscape, distortions that are hard to compensate for without forcing rectification procedures that dramatically alter the original aspects of the features, thus potentially misrepresenting the real geometry of landscape elements.

Figure 3. A sample of RGB (11-7-2) MIVIS imagery depicting the area N of Aquileia. Strong geometric distortions can be clearly identified along the left edge of the image.

RGB Ortho-Imagery includes coverages from 2000, 2003, 2007 and 2011 with respectively 100, 80, 70, 50 cm ground resolution, and forms the Remote Sensing dataset with the highest spatial resolution available in this project (Figure 4). This represents the most useful set of imagery for the present work as a coverage for the whole region is available for multiple years and for the increased ground resolution of the data.

Figure 4. A sample of RGB Ortho-imagery depicting the area N of Aquileia.

A custom-designed workflow has been applied on these photographic datasets in order to automatically detect similarly angled or orthogonal landscape elements having an orientation congruent with the Aquileian cadaster. As, clearly, not all correctly oriented features of the landscape have a correlation with the centurial grid, and orientation is not the only characteristic that linear features have to have in order to be considered parts of a cadaster, the extracted linear features have been filtered to remove those that do not carry a strong weight, i.e., those that are unlikely to be ancient, for example the majority of the minor drainage channels that follow the 22° W from N or 22° S from W (=248°) directions (respectively aligned with the major axes of the Centuriation, the *Cardo Maximus* and *Decumanus Maximus*) and that are likely to be simply later additions, kept parallel to a landscape dominated by Roman arrangements designed to be the most appropriate to ensure efficient water drainage. After they have been filtered, the frequency of the remaining elements oriented to the Roman grid is estimated from the images in order to compute the module with highest probability.

The goal of this and similar approaches is to examine a large set of hypotheses, to quickly exclude those with little promise and to provide reasonable candidates for further investigation based on more traditional archaeological approaches. Indeed, the automatic extraction of features can in no way provide direct confirmation or guarantee of identification, but can be used to reduce the amount of fieldwork and other investigation required to confirm elements of current landscape as ancient land division. To this effect, the proposed approach was based on hypotheses with no more than a simple range of acceptable values for the cadaster modules and a few well documented assumptions like the orientation of *cardo* and *decumanus*. Rather than being hypothesis-driven, the approach was oriented to evaluate what could be extracted, resulting in multiple hypotheses of which we deemed one more plausible than the others. It remains as a given that all of them need to be verified on the basis of a variety of other criteria and validation procedures, including fieldwork activities, legacy data re-evaluation and spatial analysis. However, thus reduced, verification can be performed on a very limited number of hypotheses vs a continuum.

2.2. Method

The workflow of feature extraction and detection implemented in © Matlab (R2016b, MathWorks Natick, MA, USA) seeks to estimate location and periodicity of dominant linear features on georeferenced images. In order to do that, the system needs to extract dominant lines aligned along the 22° W from N and 22° S from W (=68° E from N) directions, accumulate the georeferenced linear features from all images and fit the correct location and size of the centuriation on the dominant responses. For this reason, the workflow involves three processes: (i) filtering; (ii) thresholding and linear feature extraction; (iii) feature accumulation and periodicity estimation.

2.2.1. Filtering

The filtering process aims to estimate the presence of linear features by looking at the response of edge detection linear filters at various locations, orientations, and scales. To this end we independently convolved each image with a bank of Gabor filters at various orientations and scales.

Gabor filters [26] are directional wavelet filters that can detect the presence of specific frequencies at a given orientation, localized around a given location. This ability to detect directional high frequency locally periodic patterns made them the tool of choice in several image processing and applications such as character recognition [27] and fingerprint recognition [28]. These characteristics make them also perfectly suited for the present task, where drainage ditches in fields and other current land division components result in repetitive aligned linear features.

The spatial response of the filter is defined as follows:

$$g(x, y; \theta, \lambda, \sigma, \gamma, \psi) = exp\left(-\frac{x'^2 + \gamma^2 y'^2}{2\sigma^2}\right) exp\left(i\left(2\pi\frac{x'}{\lambda} + \psi\right)\right) \tag{1}$$

where

$$x' = xcos\theta y sin\theta$$
$$y' = -xsin\theta + ycos\theta$$

and θ is the filter's orientation, λ is its frequency, σ the scale of the filter, γ its aspect ratio, i.e., scale ratio along the direction versus orthogonal to the direction, and, finally, ψ is a phase offset, which will result in a phase shift in the complex response and is irrelevant for our analysis since we are interested in the absolute value of the response.

The bank of filters includes filters computed at various scales and orientations: the scales of the filters were 5, 10, 20, and 30 m, which include the fundamental scales of the sought after linear features, i.e., roads, canals, drain-lines, crop-marks, etc. The orientations were selected in a relatively narrow band around the known centuriation axes, i.e., 12°, 17°, 22°, 27°, and 32° W from North for the lines aligned with the *cardo*, and 12°, 17°, 22°, 27°, and 32° S from West for the ones aligned to the *decumanus*.

The spatial frequency was linked to the scale such that period length was equal to the scale, i.e., $\sigma/\lambda = 1$; while the aspect ratio was set to 1/3 so that the filter was fairly elongated in order to (i) increase directional accuracy; (ii) limit general high feature response; and (iii) concentrate on sizable aligned edges. Finally, the phase offset ψ was ignored (set to 0) as it does not have an effect on the absolute response value.

Given this filter bank each image was convolved with 40 filters resulting in 120 responses for RGB images (40 per channel). These responses were divided into two groups of 60 responses each: one for the filters with orientations around the *cardo* direction (22° W from N) and one for orientations around the *decumanus* direction (22° S from W). Then, we non-linearly accumulated the response across channels and scales. Here we accumulated by summing the square norm of the response at each location and direction (Figure 5). The goal of the non-linear accumulation is to perform a soft-max, i.e., obtain a continuous response that enhances the maximum values across channels and scales. The end result is 5 responses per dominant direction providing response in the location-orientation space.

Figure 5. Accumulation of the responses of the Gabor filter bank across channels and scales. Each image corresponds to responses at the same locations and different orientations of the filter: (**a**) 12° W from N; (**b**) 22° W from N; (**c**) 32° W from N.

2.2.2. Thresholding and Linear Feature Extraction

In order to extract linear feature candidates, we performed selective thresholding in the location-direction space. For each directional group, we retained locations (pixels), if the current location was a local maximum at the central direction (22° W from N for the *cardo* oriented group, 22° S from W for the *decumanus* oriented group) and its accumulated response was above an adaptive threshold. In particular, location (x,y) was retained if:

- *location maximality*

resp(x,y;θ) $\geq$ resp(x + k dx, y + k dy; θ)

with

$$
\begin{aligned}
k &= -1, -2, 1, 2 \\
dx &= -sin\theta \\
dy &= cos\theta
\end{aligned}
$$

and θ being the group's dominant direction (22° or 112° W from N)

- *directional maximality*

resp(x,y;θ) $\geq$ resp(x, y; θ + dθ)

with dθ = $-10°, -5°, 5°, 10°$

- *adaptive thresholding*

$$
resp(x, y, \theta) \geq c \max_{x,y,\theta}(resp(x,y,\theta))
$$

with the threshold parameter c being set to 0.1 in our experiments.

The result was two sets of pixel locations for each image corresponding to strong edge responses along either the *cardo* direction (22° W of N) or *decumanus* direction (22° S from W). Note that *cardo* and *decumanus* sets are independent and a pixel can appear in both sets (Figure 6).

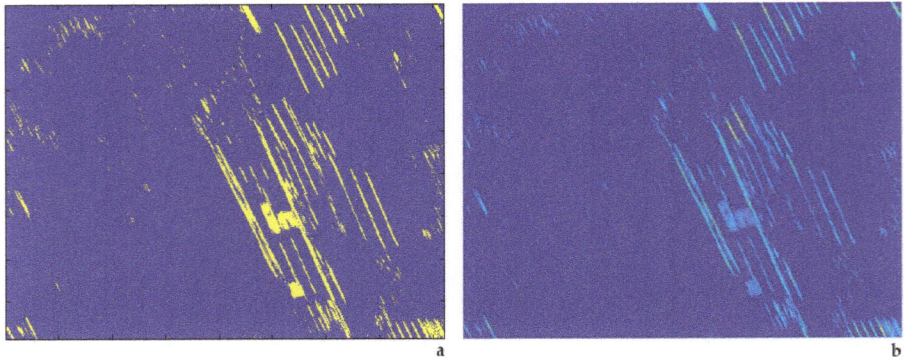

Figure 6. (**a**) Image-mask of the points at the location analyzed in Figure 5 that pass the directional thresholding process described in Section 2.2.1. (**b**) The accumulated filter-responses for those points.

2.2.3. Feature Accumulation and Periodicity Estimation

Given pixel-wise feature locations in each georeferenced image, we computed the geolocation of each point feature and accumulate those throughout all image sets, obtaining a set of georeferenced feature locations and their responses. In order to find the center and periodicity of such responses, first each linear response was down-projected and accumulated along the direction orthogonal to the group's direction, i.e., *cardo's* direction responses were accumulated along a line in the 22° S from W direction, while *decumanus* lines were accumulated along a linear space of orientation 22° N from W, thus resulting in two response histograms along the two orthogonal directions.

For ease of computation, the origin of this projected system was set at the intersection of the *via Julia Augusta* (State Route 352) and the virtual extension of the Anfora Canal (which flows toward SW, W of Aquileia) toward E, which is considered to be the most likely point of intersection of the *cardo Maximus* and *decumanus Maximus* [20].

These response histograms (Figure 7) take into account the locations of all coherently aligned responses, andare strongly influenced by the geometry of the spatial coverage, exhibiting a systematic size bias of the baseline responses visible in the histogram. However, we were not interested in the baseline responses, but rather in the peaks of the distribution, which correspond to directions with higher-than-usual aligned response. For this reason, we eliminated the baseline response by computing a centered average of the response over 50 m (smoothing with a centered average filter of size 50 m) and maintaining only the responses that were greater than 2.5 times the smoothed response.

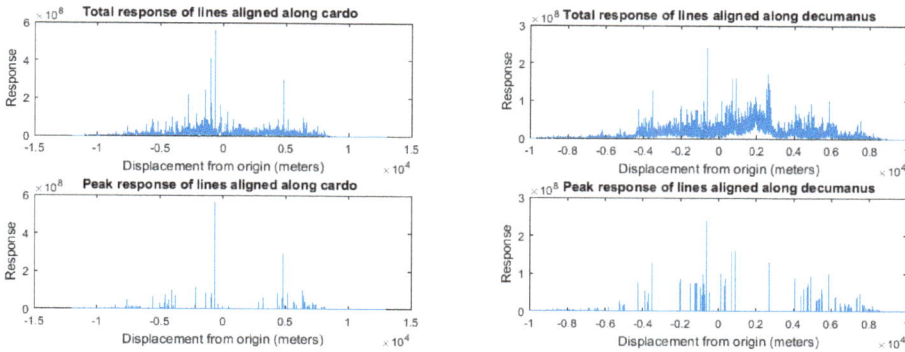

Figure 7. (**a**) Total response of the lines along the *cardo* direction accumulated along the *decumanus* direction; (**b**) Total response of the lines along the *decumanus* direction; (**c**) Peak response along the *cardo* direction; (**d**) Peak response along the *decumanus* direction.

The use of peaks provides the added advantage of making the approach more robust with respect to missing signals. Indeed, gaps due to missing alignments would simply result in fewer votes and, thus, smaller peaks, which are, however, still likely to be of much larger magnitude than the average response as a result of limited and localized lack of signal, and thus identifiable.

With the histograms of peak responses in hand, we sought to estimate the center location and periodicity of repeated aligned linear features, with the hope that these would correspond to location and module of centuriation patterns. To this end we tested centurial modules ranging from 600 m to 750 m in a 1 m increment. We operated on the two main orientations independently, and for each tested module we accumulated the response cyclically, wrapping around the accumulation with a period equal to the module.

$$wrappedHistogram(t) = \sum_{n=-\infty}^{\infty} Histogram(t + nmodulus) \qquad (2)$$

with $t = 0...module - 1$.

For each module, we retained peak value, peak location and entropy of the wrapped histogram. The Entropy is a measure of disorder or lack of information in a distribution and has maximal values when the distribution is concentrated in a few states

$$Entropy = -\sum_i p(i) \log(p(i))$$

where $p(i) = \frac{wrappedHistogram(i)}{\sum_j wrappedHistogram(j)}$.

The expectation was that the correct module would exhibit better modular alignment of (peak) directional responses, resulting in a *wrapped histogram* with a stronger and better localized peak in correspondence to the modular location of the center of the centuriation pattern and relatively little response elsewhere. Hence, the correct module would be characterized by a maximal value in the peak value and a minimal value in the entropy associated with the corresponding wrapped histogram. Once this optimal module was detected, the location offset along the orthogonal direction could be extracted from the peak location at that module value.

Note that, due to aliasing phenomena, we would have expected similar peaks to appear at any multiple of the correct modulus. However, the limits in the search range hided such effects. Indeed, to observe them we would require the maximum value of the range to be at least twice as big as the smallest modulus, while our range went from 680 to 720 m.

3. Results

Figure 8 shows the accumulated peak responses and entropies as a function of the chosen modulus. The plots on the left are for the *cardo* direction, while those on the right are for the *decumanus*. The range of *moduli* taken into consideration is between 680 and 720 m.

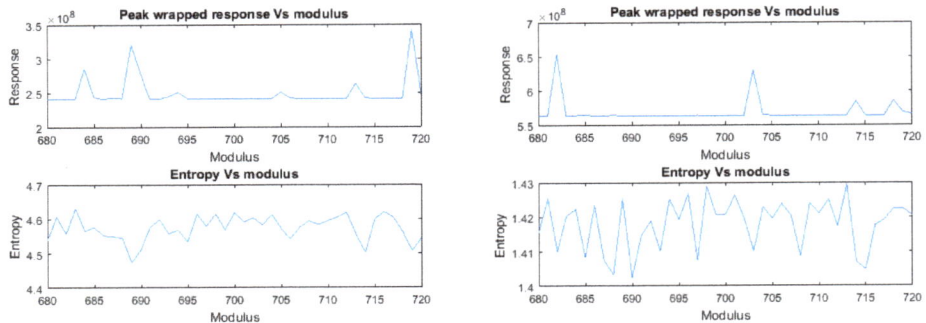

Figure 8. (a) Maximal wrapped response along the *cardo* direction as a function of the modulus; (b) Maximal wrapped response along the *decumanus* direction as a function of the modulus; (c) Entropy of the wrapped response along the *cardo* direction as a function of the module; (d) Entropy of the wrapped response along the *decumanus* direction as a function of the modulus.

The plot shows that there are a few peaks in the response in both directions. In particular, the *cardo* presents the strongest peak at 719 m, while the *decumanus* presents a maximum at 683 m. In both cases, the peaks correspond to minima in entropy, so, taken in isolation, they would seem to be good candidates. However, when taking into consideration both directions, these measures are not consistently good candidates for a centurial module. In fact, the closest peak in the orthogonal directions (684 m in the *cardo* direction and 718 in the *decumanus* direction) do not correspond to local minima in the entropy, suggesting that these peaks are not the result of a square grid, but rather something else, presumably Moiré patterns due to periodicity on the orthogonal direction, probably due to modern landscape linear components. Moreover, these values (718–719 m, 683–684 m) are not consistent with a centurial module and therefore can be disregarded.

The only peak presents in both direction that is consistently a local minimum for the entropy is the peak around 704 m (705 m in the *cardo* direction and 703 m in the *decumanus* direction).

4. Discussion

The presented periodicity of the feature accumulation suggests a value of 704–705 m for the grid module, which cannot exclude a 706 m result due to the resolution, a value that is fully compatible with a centurial cell. The result is quite relevant as other metrological analyses recently undertaken [29] have highlighted that the 20 × 20 *actus* module of Aquileia measured likely 704–705 m × 704–705 m rather than conventional 710 m. The trend is known in other early centuriations deployed in Central Italy in the initial period of the Roman colonial expansion and centuriation deployment [30] and the phenomenon connected to the variations in size of the Roman foot (*pes*). This represents a primary outcome of the current work, as previous scholarship related to the Aquileian cadaster has derived hypothetical grid models (of varying number of *actus*) based on the assumption that the centurial module would have been generated as a multiple of an *actus* with a standard size of 35.51 m; the outcomes of this trial show instead it could be comprised between 35.25 m and 35.3 m, thus determining a notable variation of the grid size on the long distances. A slightly smaller than "standard" *centuria* would also explain current unaccounted for metrological errors and mismatches verifiable on the long distance (from the centurial *umbilicus* of Aquileia) projecting a virtual grid with a standard module of 710 m: many already recognized and validated *limites*, like the ones of Sedegliano for example [21], would not be aligned to the grid-lines in this case. Redefining the measures of the "Classic" Aquileian centurion's module has implications also in the definition of its relationship with the other centuriations that have been detected in the Friulian plain [21], establishing a relationship of antecedence with those characterized by a longer (and likely later) module, thus challenging current theories [21,22] that indicate the "Classical" centuriation as succeeding other deployments.

A generated grid of 706 m × 706 m, overlaid over the remote sensing imagery (Figure 9) and aligned along the *via Julia Augusta* and the Anfora Canal with origin at their intersection, shows the overlap of some of the major elements of the cadaster with the grid-lines. Several centuriation constituents line up rather accurately with the proposed virtual grid, especially in the area W of Aquileia and the main *cardo*. An interesting feature is that several major roads that connect to the city are aligned to the cadaster oblique to the square grid with varying, but regular ratios, like the sector of the via Gemina (exiting Aquileia at NE, linking it to ancient Emona, present Ljubljana) nearer to the city fits the grid with a ratio of 1:1.

Figure 9. Estimated grid overlaid over an Ortho Image of Aquileia.

Notwithstanding these positive empirical outcomes, the workflow is still showing some flaws that are being investigated, like the presence of very strong peaks in the wrapped histogram referring to repetitiveness of elements at distances that do not have meaning within the "space dimension" of a Roman cadaster, and that are likely generated by later landscape elements aligned with the Roman system of which they mimicked orientation and configuration. In future work, it will be likely necessary that the linear objects, once extracted, undergo through individual optical analysis to determine their congruence to the centurial grid based on a number of other variables and information (especially of archaeological nature) that can validate their historicity, rather than simply filtering them out based on a number of physical parameters. This will reduce the number of features that add complexity to the calculation.

5. Conclusions

In studying ancient regularly designed landscapes characterized by repetitive patterns, where highlighting the components of the system in order to determine repetitiveness is crucial to the goal of the research, the use of automated detection and extraction of cadaster element methods enables not only to reduce manual mapping time, but also to improve the quality and number of correct identifications. The provisional results of this work bring attention to the need of customizable tools and workflow that can streamline processes that would otherwise be alienating and time consuming. Moreover, when mapping visually identified features is undertaken via manual tracing, the possibility of introducing unwanted errors is very likely, as even a misalignment equal to 1° of orientation can generate relevant errors on long distance. Automation of processes appear therefore a desirable solution.

Clearly, as in any sort of automation, the proposed workflow has to be customized based on the goal and research questions of the project, and the design of a process aiming to extract specific objects has to start from the landscape characteristics (including the morphology) and the nature, forms, model and patterning of the system components. This experience shows that land division elements, once extracted, are likely to demand visual analysis on a one to one basis to determine their congruence to the field system based on a number of variables and criteria before bringing them into more complex statistical analysis. This process will likely be automated in a near future using approaches from Artificial intelligence and Machine Learning, but the application of these methods to the archaeological domain, as we have seen, is still in its infancy. Thus, at the moment, such vetting process will still have to be based on a combination of automated filtering and visual assessment. The results of this automated workflow are in any case just the first step of a much more complex process were theoretic models have to be substantiated and validated by spatial analysis, field mapping and ground verification activities and corroborated by archaeological information and investigation.

Acknowledgments: Arianna Traviglia's work was funded by the European Union's Horizon 2020 research and innovation program under the Marie Skłodowska-Curie grant agreement No 656337.

Author Contributions: A. Traviglia and A. Torsello conceived and designed the research; A. Torsello performed the experiments; A. Traviglia and A. Torsello analyzed the data; A. Traviglia and A. Torsello wrote the paper. The founding sponsor had no role in the design of the study; in the collection, analyses, or interpretation of data; in the writing of the manuscript, and in the decision to publish the results.

Conflicts of Interest: The authors declare no conflict of interest.

References

1. Alfieri, N.; Schmiedt, G. *Atlante Aerofotografico Delle Sedi Umane in Italia: La Centuriazione Romana, Parte Terza*; Istituto Geografico Militare: Florence, Italy, 1989. (In Italian)

2. Clavel-Lévêque, M.; Conso, D.; Favory, F.; Guillaumin, J.-Y.; Robin, Ph. *Corpus Agrimensorum Romanorum. I. Siculus Flaccus, Les Conditions des Terres*; Jovene: Napoli, Italy, 1993. (In Italian)

3. Chouquer, G.; Clavel-Lévêque, M.; Favory, F.; Vallat, J.-P. *Structures Agraires en Italie Centro-méridionale. Cadastres et Paysages Ruraux*; École française de Rome: Rome, Italy, 1987. (In Italian)

4. Palet, J.M.; Orengo, H. The Roman centuriated landscape: Conception, genesis and development as inferred from the Ager Tarraconensis case. *Am. J. Archaeol.* **2011**, *115*, 383–402. [CrossRef]

5. Guaitoli, M. (Ed.) *Lo sguardo di Icaro: Le Collezioni dell'Aerofototeca Nazionale per la Conoscenza del Territorio*; Campisano Editore: Rome, Italy, 2003. (In Italian)

6. Ceraudo, G.; Ferrari, V. Fonti tradizionali e nuove metodologie d'indagine per la ricostruzione della centuriazione attribuita all'ager Aecanus nel Tavoliere di Puglia. *Agri Centuriati* **2009**, *6*, 125–141. (In Italian)

7. Marcolongo, B.; Mascellani, A. Immagini da satellite e loro elaborazioni applicate alla individuazione del reticolato romano nella pianura veneta. *Archeol. Ven.* **1978**, *1*, 131–146. (In Italian)

8. Baggio, P. Il contributo delle immagini da satellite per l'analisi dell'area aquileiese. In *Canale Anfora. Realta' e Prospettive tra Storia, Archeologia e Ambiente. Aquileia e Terzo di Aquileia 29 Aprile 2000*; Special issue of Quaderni Aquileiesi 6–7; Editreg: San Francisco, CA, USA, 2000; pp. 42–45.

9. Masini, N.; Lasaponara, R. Satellite-based recognition of landscape archaeological features related to ancient human transformation. *J. Geophys. Eng.* **2006**, *3*, 230–235. [CrossRef]

10. Mlekuz, D. Watching Stones Grow: Aerial Archaeology of the Karst. 2017. Available online: https://www.academia.edu/31650010/Sledovi_rimske_zemlji%C5%A1ke_razdelitve_na_Krasu_Traces_of_roman_land_division_on_the_Karst (accessed on 20 June 2017).

11. Szeliski, R. *Computer Vision: Algorithms and Applications*; Springer: London, UK, 2011.

12. Traviglia, A.; Cowley, D.; Lambers, K. Finding common ground: Human and computer vision in archaeological prospection. *AARGnews-Newsl. Aer. Archaeol. Res. Group* **2016**, *53*, 11–24.

13. Lambers, K.; Traviglia, A. Automated detection in remote sensing archaeology: A reading list. *AARGnews-Newsl. Aer. Archaeol. Res. Group* **2016**, *53*, 25–29.

14. Trier, O.D.; Larsen, S.O.; Solberg, R. Automatic detection of circular structures in high-resolution satellite images of agricultural land. *Archaeol. Prospect.* **2009**, *16*, 1–15. [CrossRef]

15. Zingman, I.; Saupe, D.; Penatti, O.A.B.; Lambers, K. Detection of fragmented rectangular enclosures in very high resolution remote sensing images. *IEEE Trans. Geosci. Remote Sens.* **2016**, *54*, 4580–4593. [CrossRef]

16. De Guio, A.; Magnini, L.; Bettineschi, C. *GeOBIA Approaches to Remote Sensing of Fossil Landscapes: Two Case Studies from Northern Italy*; Traviglia, A., Ed.; AUP: Amsterdam, The Netherland, 2015; pp. 45–53.

17. Krizhevsky, A.; Sutskever, I.; Hinton, G.E. ImageNet Classification with Deep Convolutional Neural Networks. In *Advances in Neural Information Processing Systems 25*; Pereira, F., Burges, C.J.C., Bottou, L., Weinberger, K.Q., Eds.; Curran Associates: Red Hook, NY, USA, 2012; pp. 1097–1105.

18. Ciresan, D.; Meier, U.; Masci, J.; Gambardella, L.M.; Schmidhuber, J. Flexible, High Performance Convolutional Neural Networks for Image Classification. In Proceedings of the International Joint Conference on Artificial Intelligence (IJCAI), Barcelona, Spain, 16–22 July 2011; pp. 1237–1242.

19. Trier, Ø.D.; Zortea, M.; Tonning, C. Automatic detection of mound structures in airborne laser scanning data. *J. Archaeol. Sci. Rep.* **2015**, *2*, 69–79. [CrossRef]

20. Chiabà, M. Dalla fondazione all'età tretarchica. In *Moenibus et Portu Celeberrima*; Ghedini, F., Bueno, M., Novello, M., Eds.; Istituto Poligrafico dello Stato: Rome, Italy, 2009; pp. 7–22. (In Italian)

21. Prenc, F. *Le Pianificazioni Agrarie di età Romana Nella Pianura Aquileiese*; Editreg: Trieste, Italy, 2002. (In Italian)

22. Bianchetti, A. La Centuriazione. In *Terra di Castellieri Archeologia e territorio nel Medio Friuli*; Bianchetti, A., Ed.; Cre@ttiva: Tolmezzo, Italy, 2004; pp. 103–140. (In Italian)

23. Chartrain, A. Da Lattara a Montpellier: Una prima archeologia del territorio centuriato. Età romana, età del Ferro. *Agri Centuriati* **2009**, *6*, 143–158. (In Italian)

24. Dellong, E. Apprendre la cadastration antique à partir de la cartographie des structures archéologiques: L'exemple de la cité de Narbonne et de son proche territoire. *Agri centuriati* **2011**, *7*, 94–112. (In Italian)

25. L'Infrastruttura Regionale dei Dati Ambientali e Territoriali (IRDAT). Available online: http://irdat.regione.fvg.it/ (accessed on 20 June 2017).

26. Feichtinger, H.; Strohme, T. *Advances in Gabor Analysis*; Birkhäuser: Boston, MA, USA, 2003.

27. Peeta, P.B.; Ramakrishnan, A.G. Word Level Multi-script Identification. *Pattern Recognit. Lett.* **2008**, *29*, 1218–1229.

28. Jain, A.K.; Prabhakar, S.; Hong, L.; Pankanti, S. Filterbank-based fingerprint matching. *IEEE Trans. Image Process.* **2000**, *9*, 846–859. [CrossRef] [PubMed]
29. Peterson, J.; University of East Anglia, Norwich, UK. Personal communication, 2017.
30. Camerieri, P.; Manconi, D. La Pertica della colonia latina di Spoletium nel quadro dei nuovi studi sulle centuriazioni della valle umbra. *Agri centuriati* **2010**, *7*, 263–273. (In Italian)

geosciences

MDPI

Review

Quantitative Examination of Piezoelectric/Seismoelectric Anomalies from Near-Surface Targets

Lev Eppelbaum

School of Earth Sciences, Faculty of Exact Sciences, Tel Aviv University, Ramat Aviv, Tel Aviv 6997801, Israel; levap@post.tau.ac.il; Tel.: +972-3-6405086

Received: 21 August 2017; Accepted: 13 September 2017; Published: 19 September 2017

Abstract: The piezoelectric and seismo-electrokinetic phenomena are manifested by electrical and electromagnetic processes that occur in rocks under the influence of elastic oscillations triggered by shots or mechanical impacts. Differences in piezoelectric properties between the studied targets and host media determine the possibilities of the piezoelectric/seismoelectric method application. Over a long time, an interpretation of obtained data is carried out by the use of methods developed in seismic prospecting. Examination of nature of piezoelectric/seismoelectric anomalies observed in subsurface indicates that these may be related (mainly) to electric potential field. In this paper, it is shown that quantitative analysis of piezoelectric/seismoelectric anomalies may be performed by the advanced and reliable methodologies developed in magnetic prospecting. Some examples from mining geophysics (Russia) and ancient metallurgical site (Israel) confirm applicability of the suggested approach.

Keywords: piezoelectric/seismoelectric anomalies; subsurface geophysics; archaeology; quantitative analysis; interpretation methodology

1. Introduction

The piezoelectric and seismo-electrokinetic phenomena are manifested by electrical and electromagnetic processes that occur in rocks under the influence of elastic oscillations triggered by shots or mechanical impacts (hits) ([1–32]).

Because the manifestation patterns of the above phenomena are different in different rocks, these phenomena can be used as a basis for geophysical exploration techniques. In this paper, it is assumed that the studied piezoelectric and seismoelectric anomalies cannot be separated from one another, since the anomalous targets with the contrast piezoelectric properties as a rule occur in sedimentary host deposits where the seismoelectric effects take a place.

The piezoelectric method is an example of a successful application of piezoelectric/seismo-electrokinetic phenomena in exploration geophysics. It has been successfully applied in mineral exploration and environmental features research in Russia, USA, Canada, USA, Australia and other countries. The greatest contribution to the piezoelectric method application in subsurface geophysics (since the mid-1950s) was made by Naum Neishtadt (1929–2016) (Figure 1).

Figure 1. Prof. Naum Neishtadt, one of the founders of the piezoelectric method of geophysical prospecting (1927–2016).

Interpretation of seismoelectric/piezoelectric anomalies is often carried out using procedures similar to those employed in seismic prospecting. In the paper, it is shown that quantitative interpretation of expressed anomalies observed in near-surface may be performed using the methods developed for potential geophysical fields.

2. A Brief Background

This method is based on the piezoelectric activity of rocks, ores, and minerals. It enables direct exploration for pegmatite, apatite-nepheline, essentially sphalerite, and ore-quartz deposits of gold, tin, tungsten, molybdenum, zinc, crystal, and other raw materials. This method also enables differentiation of rocks such as bauxites, kimberlites, etc., from the host rocks, by their electrokinetic properties [14].

Classification of some rocks, ores, and minerals by their piezoactivity is given in Table 1. These objects (targets) transform wave elastic oscillations into electromagnetic ones. It should be take into account that, sometimes, anomalous bodies may be detected not by positive, but by negative anomalies, if low-piezoactive body occurs in the higher piezoactive medium.

The piezoelectric method is an example of successful application of piezoelectric/seismo-electrokinetic phenomena in exploration and environmental geophysics and designed for delineation of targets differing from the host media by piezoelectric properties [14,30]. This method can be employed in surface, downhole, and underground modes.

Experimental investigations enabled obtaining the main equation of the piezoelectric effect [30]:

$$\varepsilon \frac{\partial \mathbf{E}}{\partial t} + \sigma \mathbf{E} = \text{curl} \mathbf{H}$$

where ε is the dielectric constant, and σ is the electric conductivity of rock. It explains why the piezoelectric effect may be registered by both observations of intensities of electric field E and magnetic field H.

Recent testing of piezeoelectric effects of archaeological samples composed from the fired clay have shown values of $(3.0–4.0) \times 10^{-14}$ C/N (Coulomb/Newton).

An observation scheme for ground surveys using the piezoelectric method is presented in Figure 2. The conventional piezoelectric measurements are conducted using electrodes, while the geophones play a subsidiary role for monitoring intensity of the elastic oscillation generation and behavior of the initial seismic field.

Table 1. Classification of some rocks, ores, and minerals by their piezoactivity d, 10^{-14} Coulomb/ Newton (after [14,30], with modifications).

Piezoactivity Group	Rock/Ore/Mineral	$D_{min}-D_{max}$	D_{aver}
I	Quartz-tourmaline-cassiterite ore	0.8–28.0	15.7
	Antimonite-quartz ore	0.2–1.35	0.6
	Apatite-nepheline ore	0–5.0	0.9
	Galenite-sphalerite ore	0.2–7.7	3.3
	Ijolite	0.1–8	1.3
II	Melteigite	0.2–5.0	1.6
	Pegmatite	0.1–4.8	1.3
	Skarn with galenite-sphalerite mineralization	0.1–3.0	0.6
	Sphalerite-galenite ore	0.3–7.7	3.8
	Turjaite	0.9–4.8	2.2
	Urtite	0.1–32.5	3.4
	Juvite	0.2–5.4	1.8
III	Aleurolite silificated	0–0.5	0.2
	Aplite	0–1.7	0.6
	Breccia aleurolite-quartz	0.1–0.4	0.2
	Gneiss	0–1.4	0.3
	Granite	0–1.6	0.4
	Granodiorite	0–0.2	0.1
	Quartzite	0–3.3	0.6
	Pegmatite ceramic	0–1.0	0.1
	Sandstone silificated and tourmalinised	0.1–1.4	0.5
	Feldspars	0–0.4	0.15
	Porphyrite	0–0.3	0.1
	Ristschorrite	0.3–0.9	0.5
	Schist argillaceous	0–0.6	0.1
	Hornfels	0–0.4	0.2
	Skarn sphaleritic-garnet	0–1	0.3
	Skarn pyroxene-garnet	0–0.2	0.1
IV	Aleurolite, amphibolites, andesite, gabbro, greisens, diabase, sandstone	0–0.1	0.05
	Argillite, beresite, dacite, diorite-porphyrite, felsite-liparite, limestone, tuff, fenite	0	0

I (highly active): piezo-activity of samples is greater than 5.0×10^{-14} C/N; II (moderately active): piezo-activity of samples is $(0.5–5.0) \times 10^{-14}$ C/N; III (weakly active): piezo-activity of samples is less than 0.5×10^{-14} C/N; IV (non-active): piezo-activity of samples are near zero.

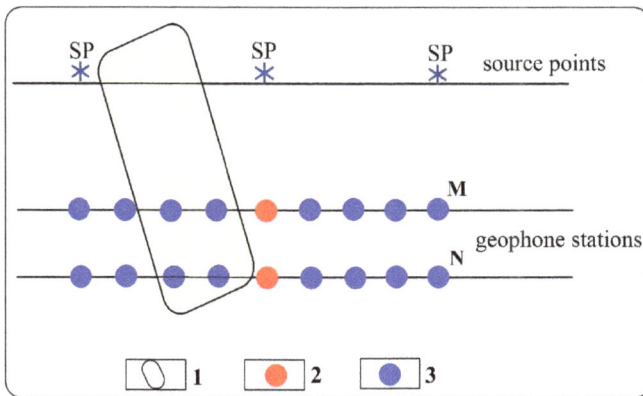

Figure 2. Piezoelectric measurement array (after [14] with some modifications): (1) projection of the piezoactive body to the earth's surface; (2) geophone; and (3) electrode.

3. Can Piezoelectric and Seismoelectric Effects Be Related to Potential Fields?

The seismo-electrokinetic (electrokinetic) phenomenon E in rocks is triggered by the potential gradient due to the displacement of liquid phase relative to the solid "skeleton" of the rock under the elastic wave influence. Essential similarity of this phenomenon to potential produced by water flow in rocks proposes a similar origin for these phenomena [33].

Firstly, Ivanov [34] discovered the seismoelectric effect E in the rocks of sedimentary complex. He proposed a filtration hypothesis consisting of that in the sedimentary rocks occurs a double diffusion layer at the boundary between hard particles and liquid. A propagation of elastic waves causes a relative distortion of electrolyte and hard phase; thus, equilibrium condition in the double electric layer is distorted. This effect generates appearance of so-called filtration potential U, flow (streaming) potential, an instantaneous value of which is determined over a part of the length of the elastic wave by the equation of Helmholtz–Smoluchowsky [35]:

$$U = \frac{1}{4\pi} \int_{\rho_1}^{\rho_2} \frac{k\varepsilon\zeta}{\eta\sigma} d\rho, \tag{1}$$

where ρ_1 and ρ_2 are the instantaneous pressure values in the section under consideration; k is the coefficient calculating the dynamics of the elastic wave distribution, ε is the dielectric constant, ζ is the potential of the double electric layer, η is the solution viscosity, and σ is the conductivity.

Frenkel [36] offered a mathematical description of the seismo-electric phenomenon E, which was based on the Ivanov's [34] hypothesis regarding the electro-filtration nature of this phenomenon. Frenkel described compression and stretching of rock particles and pores, as well as of the pore moisture, under the influence of the elastic wave propagation. Moreover, he proposed an equation that described the propagation of elastic waves in moisturized soil. For calculating electro-kinetic potential, this equation uses the Helmholtz–Smoluchowski equation that describes the intensity of longitudinal electric field for stationary flow of liquid through the pores of the solid "skeleton":

$$E = i \frac{4\varepsilon\zeta\chi\omega^2}{\pi\sigma r^2\mu} f\rho_2 \left(\frac{K_2\beta}{\rho_2\beta'\omega_0^2} - 1 \right) u, \tag{2}$$

where μ is the coefficient of viscosity of the medium, i is the complex-valued electric field intensity, $\beta = \frac{1}{f(1+\alpha)}$, α is the parameter that describes the mechanical properties of the medium, $\beta' = 1 + (\beta - 1)\frac{K_2}{K_0}$, K_0 is the coefficient of compressibility of the solid phase, $\alpha = \frac{K_1}{f}$ is the coefficient of permeability, r is the pore radius, f is the porosity, K_1 is the coefficient of permeability of the soil, ρ_2 is the true specific gravity of the pore moisture, K_2 is the coefficient of compressibility of the liquid phase, ω_0 is the propagation velocity of the longitudinal elastic wave, and u is the displacement.

Equation (2) shows that the electric field intensity is proportional to porosity and is independent of the pore radius, because the α coefficient is proportional to the square of this radius. With the assumed values of displacement, the electric field intensity is proportional to the square of frequency of the elastic oscillations. In the above formula, the author assumes that the period of time required to produce the electro-kinetic potential gradient is negligibly short compared to the oscillation period $2\pi/\omega$. This is why the value of E at any given moment practically coincides with the value corresponding to the instantaneous value of the relative velocity.

To confirm Ivanov's [34] suggestion on the nature of this effect in rocks, Volarovich and Parkhomenko [1] put experiments to reproduce this phenomenon on the artificially moistened rock samples under the laboratory conditions. They found that before the artificial moistening, the dolomite sample did not show any elastic oscillations when electrified; after the moistening, the appearance of an electric potential was observed on its faces. At the same time, the sign of the charge did not depend on which side of the sample the charges were taken from, but determined by the gradient of the pressure drop.

Parkhomenko [5] stated that further study of the phenomenon *E* will advance the current perception of electrokinetic phenomena, and that it may lead to the development of a new geophysical exploration technique for determination of porosity properties of rocks. The author emphasized the importance of studying the behavior of the phenomenon *E* in various sedimentary rocks, its dependencies on the medium saturation, chemical composition of the pore moisture, and the values of ζ-potential, apparent conductivity and polarizability.

Parkhomenko [6] established that the magnitude of the phenomenon *E* is a function of several variables, the most important of which are the medium saturation, concentration of salts in the liquid phase, electrochemical properties of the solid phase, texture of the rock, and the frequency of the applied seismic field. Specific surface of the electric double layer was found to be the key factor.

Butler [37] applied the Laplace equation to solve some problems of seismoelectric effects. Haines et al.'s [26] constructions in seismoelectric imaging are based on the potential electric quasi-static current dipole. Jardani et al. [38] several times underlined the role of electrostatic potential in seismoelectric imaging. Mahardika and Revil [39] noted a necessity employment of electrostatic potential for calculation of seismoelectric response generated at the boundary of two mediums.

Antonova [40] applied the Laplace equation for calculation of finite piezoelectric body with open electric boundaries. Jandaghian and Jafari [41] assumed that the electric potential field in the piezoelectric layer is satisfied to the Maxwell static electricity equation. Jouniaux and Zyserman [32] gave a description of the electric potential within the electric double layer by seismo-electric and electro-seismic measurements.

All the aforementioned facts testify that the seismoelectric and piezoelectric anomalies observed in subsurface can be considered as anomalies of quasi-potential field.

Absence of reliable procedures for solving the direct and inverse problems of piezoelectric anomalies (PEA) drastically hampers further progression of the method. Therefore, it was suggested to adapt the tomography procedure, widely used in the seismic prospecting, to the PEA modeling. Diffraction of seismic waves has been computed for models of circular cylinder, thin inclined bed and thick bed [42]. As a result, spatial-time distribution of the electromagnetic field caused by the seismic wave has been found. The computations have shown that effectiveness and reliability of PEA analysis may be critically enhanced by considering total electro- and magnetograms as differentiated from the conventional approaches. Distribution of the electromagnetic field obtained by solving the direct problem was the basis for an inverse problem, i.e., revealing depth of a body occurrence, its location in a space as well as determining physical properties. At the same time, this method has not received a wide practical application taking into account complexity of real geological media.

4. Short Description of the Interpretation Methodology Developed in Magnetic Prospecting

Careful analysis of piezoelectric/seismoelectric anomalies shows (see Section 3) the possibility of application for quantitative analysis of these effects in advanced methodologies developed for magnetic prospecting in complex physical-geological conditions: rugged terrain relief, oblique polarization and complex media [43–48]. Employment of these methodologies (improved modifications of tangents, characteristic points and areal methods) for obtaining quantitative characteristics of ore bodies, environmental features and archaeological targets (models of horizontal circular cylinder, sphere, thin bed, thick bed and thin horizontal plate were utilized) may have significant importance [49].

According to analogy with magnetic field, such parameter as "piezoelectric moment" (*PM*) can be calculated. The formulas for calculation of *PM* for the models of thin bed, horizontal circular cylinder (HCC) and thick bed are presented below.

(1) Thin bed:

$$A_e = 0.5A_T \cdot h, \tag{3a}$$

where A_e is the piezoelectric moment, A_T is the total intensity of the piezoelectric (seismoelectric) anomaly, and h is the depth of the upper edge of a thin bed.

(2) HCC:

$$A_e = A_T h_c^2 / k_m, \text{ where } k_m = \left(3\sqrt{3}/2\right)\cos\left(30^0 - \theta/3\right). \tag{3b}$$

where h_c is the depth to the center of the HCC, and parameter θ indicates some generalized parameter (its determination is given in detail in [43,50].

(3) Thick bed:

$$A_e = \frac{A_T}{2k'_m}, \tag{3c}$$

where k'_m is determined from special relationships [48].

If anomalies are observed on an inclined profile, then the obtained parameters characterize a certain fictitious body. The transition from fictitious body parameters to those of the real body is performed using the following expressions (the subscript "*r*" stands for a parameter of the real body) [43]:

$$\left\{ \begin{array}{l} h_r = h + x\tan\omega_0, \\ x_r = -h\tan\omega_0 + x \end{array} \right\}, \tag{4}$$

where h is the depth of the upper edge (center of HCC) occurrence, x_0 is the location of the source's projection to plan relative to the extremum having the greatest magnitude, and ω_0 is the angle of the terrain relief inclination ($\omega_0 > 0$ when the inclination is toward the positive direction of the x-axis).

5. Application of the Proposed Methodology: Field Cases

5.1. Employment of the Interpretation Methodology in Ore Geophysics

5.1.1. Gold-Bearing Quartz Deposit Ustnerinskoe (Eastern Yakutia, Russia)

Surface measurements at a gold-bearing quartz deposit Ustnerinskoe (Eastern Yakutia, Russia) showed wide anomaly with intensity of about 1700 microVolt (Figure 3). As it follows in Table 1 (Group I), quartz is one of the most piezoactive minerals. This anomaly is produced by integral effect from several quartz-mica zones. A form of this anomaly indicates that it can be examined as thick bed (or intermediate model between thick bed and thin plate). Results of interpretation position of angle points and center of the upper edge of the anomalous target (see Figure 3) coincide well with geological data. *PM* calculated for this target (coefficient k'_m was obtained from [43]) consists of $\approx$300 µV.

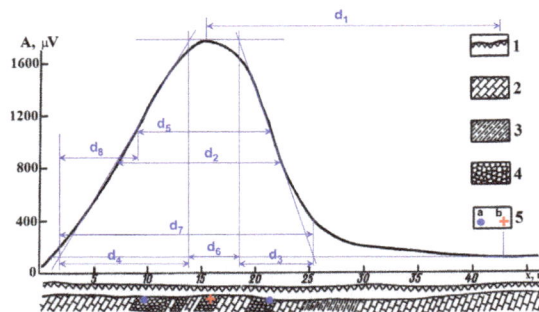

Figure 3. Quantitative analysis of piezoelectric measurements at a gold-bearing quartz deposit Ustnerinskoe (Yakutia region, Russia) (initial geological-geophysical data from [14]): (1) deluvium; (2) limestone; (3) quartz-mica shales; (4) veined quartz; and (5) obtained parameters for the model of thick bed: (a) angle points; and (b) center of the anomalous body.

5.1.2. Gold Quartz Deposit (Central Yakutia, Russia)

A gold-quartz deposit of central Yakutia occurs in conditions of very rugged relief (Figure 4). It should be noted that upper part of the crystal-quartz body is influenced by different weathering processes and its piezoelectric properties were eliminated. For transfer from fictitious to real coordinates of the anomalous object, Equation (4) was applied. Results of interpretation (here a model of thin bed was used) are in line with the available geological data.

The following terms are taken from the plot:

d_1 = distance between the maximum and minimum of the anomaly;

d_2 = distance between the left and right branches at the level of semiamplitude;

d_3 = difference in abscissae of the points of intersection of an inclined tangent with horizontal tangents on one branch;

d_4 = the same on the other branch (d_3 is selected from the plot branch with conjugated extremums, $d_3 \leq d_4$), and the x-axis is oriented in this direction);

d_5 = distance between the middle point of the left and right tangents;

d_6 = distance between d_3 and d_4;

$d_7 = d_3 + d_4 + d_6$;

and

d_8 = distance between the ending of parameter d_4 and beginning of parameter d_5.

Calculated *PM* factor is $1/2 \cdot 600(\mu V) \cdot 7(m) = 2100 \ \mu V \cdot m$.

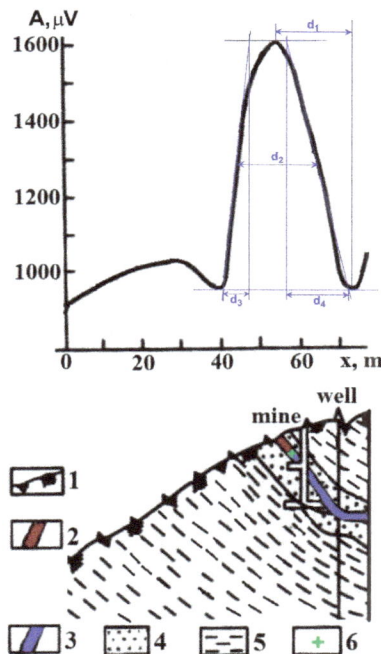

Figure 4. Quantitative examination of piezoelectric anomaly observed at one of gold-quartz deposits of Yakutia region (Russia) (initial geological-geophysical data from [30]: (1) soil-vegetation layer; (2) oxidized upper part of quartz vein; (3) quartz vein; (4) sandstone; (5) siltstone; and (6) determined position of the center of upper edge of anomalous body.

5.1.3. Crystal-Quartz Deposit Pilengichey (Subpolar Ural, Russia)

Piezoelectric profile across the central zone of the crystal-quartz deposit Pilengichey (Subpolar Ural) displays two clear anomalies with intensity about 1000 microVolt (Figure 5). Anomaly "A" and "B" were interpreted uding the HCC and thin bed models, respectively. The obtained results indicate that, if position of the upper edge center of thin bed received a good fit to the geological data, HCC position occurs above the geological body. This not large discrepancy may be explained by some host media inhomogeneities. *PM* for anomaly A is $\approx$1300 μV$\cdot$m^2, and for anomaly B—720 μV$\cdot$m.

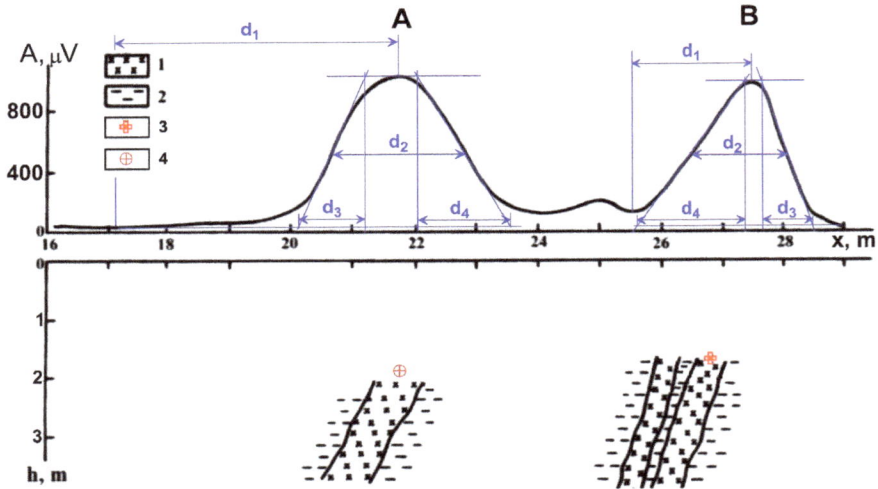

Figure 5. Quantitative analysis of piezoelectric anomaly in the crystal-quartz deposit Pilengichey of the Subpolar Ural (Russia) (initial geological-geophysical data from [30]. (1) ore-quartz zone; (2) host rocks, siltstone; results of quantitative examination ((3) and (4)): (3) position of the center of HCC inscribed to the upper part of the anomalous body; and (4) position of the center of upper edge of a thin bed.

5.2. Case Study at the Archaeological Site Tel Kara Hadid (Southern Israel)

Field piezoelectric observations were conducted at an ancient archaeological site Tel Kara Hadid with gold-quartz mineralization in southern Israel, within the Precambrian terrain at the northern extension of the Arabian-Nubian Shield [14]. The area of the archaeological site is located eight kilometers north of the town of Eilat, in an area of strong industrial noise. Ancient river alluvial terraces (extremely heterogeneous at a local scale, varying from boulders to silt) cover the quartz veins and complicate their identification. Piezoelectric measurements conducted over a quartz vein covered by surface sediments (approximately of 0.4 m thickness) produced a sharp (500 microVolt) piezoelectric anomaly (Figure 6). Values recorded over the host rocks (clays and shales of basic composition) were close to zero. The observed piezoelectric anomaly was firstly quantitatively interpreted by the use of methodologies developed in magnetic prospecting for the model of thick and intermediate bodies [48]. Calculated *PM* here is $\approx$95 μV.

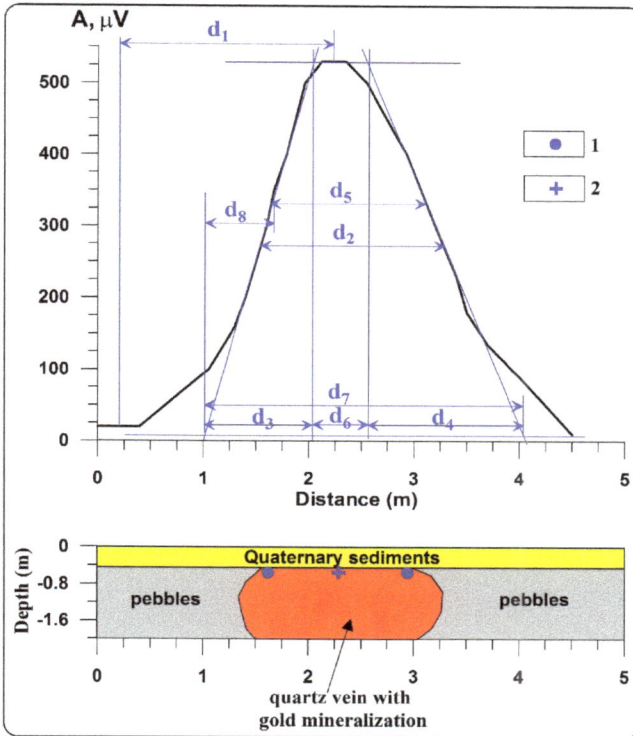

Figure 6. Quantitative analysis of piezoelectric anomaly from gold-containing quartz vein (Tel Karra Hadid, southern Israel) (initial geological-geophysical data after [14]). Results of interpretation: (1) location of angle points of anomalous target; and (2) position of the center of the upper edge of anomalous target.

6. Discussion and Conclusions

The presented physical characteristics of piezoactive rocks, ores and minerals give a wide spectrum of targets for searching of which seismo/piezoelectric method may be employed. Examined peculiarities of seismo/piezoelectric signals propagation in near-surface enable accepting these observations as quasi-potential ones.

For the first time, it was shown in detail that seismoelectric/piezoelectric anomalies in the near-surface geophysics may be analyzed using effective and reliable methods (improved version of characteristic points, tangents and areal) earlier developed in magnetic prospecting. It was proposed to employ such parameter as "piezoelectric moment" for classification of seismoelectric/piezoelectric anomalies. Effectiveness of this methodology was tested on several ore targets (Russia) and archaeological object (Israel). Obviously, further employment of seismoelectric/piezoelectric modifications in archaeology may include (besides quartz bodies), first, any clay (fired clay) targets (clay was widely used in antiquity as a building material and as a matter for construction of various domestic targets). Boulytchov [19] has shown applicability of seismoelectric/piezoelectric method for delineation of underground caves (ancient caves account for at least 5–7% of archaeological targets).

At the same time, it must be underlined that proposed technology does not contradict to the conventional methodologies of piezoelectric/seismoelectric data analysis and could be applied as independent interpretation "method".

For effective integration of piezo/seismoelectric interpretation results with other geophysical methods, some apparatus developed in theory of information [51] and wavelet theory [52] can be effectively applied.

Acknowledgments: The author would like to thank two anonymous reviewers, who thoroughly reviewed the manuscript, and their critical comments and valuable suggestions were helpful in preparing this paper.

Conflicts of Interest: The authors declare no conflict of interest.

References

1. Volarovich, M.P.; Parkhomenko, E.I. Piezoelectric effect of rocks. *Acad. Sci. USSR Geophys.* **1955**, 215–222.
2. Neishdadt, N.M.; Osipov, L.N. On using of seismoelectric effects of the second type observed by pegmatites searching. *Trans. VITR (All-Union Inst. Tech. Prospect. Methods)* **1958**, *11*, 63–71. (In Russian)
3. Neishdadt, N.M.; Osipov, L.N. Piezoelectric method. In *Borehole and Mine Geophysics*; Nedra Publisher: Moscow, Russia, 1959; pp. 153–168, 251–256, and 371–372. (In Russian)
4. Neishtadt, N.M. Searching pegmatites using seismo-electric effect of the second kind. *Sov. Geol.* **1961**, *1*, 121–127. (In Russian)
5. Parkhomenko, E.I. *Electrification Phenomena in Rocks*; Plenum Press: New York, NY, USA, 1971.
6. Parkhomenko, E.I. Main peculiarities of seismoelectric effect of sedimentary rocks and ways of its using in geophysics. In *Physical Properties of Rocks and Minerals under High Pressure and Temperature*; Nauka Publisher: Moscow, Russia, 1977; pp. 201–208. (In Russian)
7. Kondrashev, S.N. *The Piezoelectric Method of Exploration*; Nedra, Moscow, Engl. Transl.; University of British Columbia: Vancouver, BC, Canada, 1980.
8. Sobolev, G.A.; Demin, V.M.; Narod, B.B.; White, P. Tests of piezoelectric and pulsed-radio methods for quartz vein and base-metal sulfides prospecting at Giant Yellowknife Mine, N.W.T., and Sullivan Mine, Kimberley, Canada. *Geophysics* **1984**, *49*, 2178–2185. [CrossRef]
9. Neishdadt, N.M.; Mazanova, Z.V.; Suvorov, N.D. The application of piezoelectric method for searching ore-quartz deposits in Yakutia. In *Seismic Methods of Studying Complicated Media in Ore Regions*; NPO Rudgeofizika: Leningrad, Russia, 1986; pp. 109–116. (In Russian)
10. Maxwell, M.; Russel, R.D.; Kepic, A.W.; Butler, K.E. Electromagnetic responses from seismically excited targets: Non-Piezoelectric Phenomena. *Explor. Geophys.* **1992**, *23*, 201–208. [CrossRef]
11. Neishtadt, N.M.; Mazanova, Z.V.; Suvorov, V.D.; Popov, A. Technology of the piezoelectric method application in ore-quartz deposits using the Ametist-type station. In Proceedings of the Transaction of SEG-EAGE Moscow Geophysical Conference and Exhibition, Moscow, Russia, 16–19 August 1993; pp. 76–77.
12. Butler, K.E.; Russell, R.D.; Kepic, A.W.; Maxwell, M. Mapping of a Stratigraphic Boundary by its Seismoelectric Response. In Proceedings of the SAGEEP 1994 Conference, Englefield, OH, USA, 27 March 1994; pp. 689–699.
13. Kepic, A.W.; Maxwell, M.; Russell, R.D. Field trials of a seismoelectric method for detecting massive sulfides. *Geophysics* **1995**, *60*, 365–373. [CrossRef]
14. Neishtadt, N.; Eppelbaum, L.; Levitski, A. Application of seismo-electric phenomena in exploration geophysics: Review of Russian and Israeli experience. *Geophysics* **2006**, *71*, B41–B53. [CrossRef]
15. Haartsen, M.W.; Pride, S.R. Electroseismic waves from point sources in layered media. *J. Geophys. Res.* **1997**, *102*, 24745–24769. [CrossRef]
16. Mikhailov, O.V.; Haarsten, M.W.; Toksoz, N. Electroseismic investigation of the shallow subsurface: Field measurements and numerical modeling. *Geophysics* **1997**, *62*, 97–105. [CrossRef]
17. Sasaoka, H.; Yamanaka, S.; Ikea, M. Measurements of electric potential variation by piezoelectricity of granite. *Geophys. Res. Lett.* **1998**, *25*, 2225–2228. [CrossRef]
18. Beamish, D. Characteristics of near surface electrokinetic coupling. *Geophys. J. Int.* **1999**, *137*, 231–242. [CrossRef]
19. Boulytchov, A. Seismic-electric effect method on guided and reflected waves. *Phys. Chem. Earth Part A Solid Earth Geod.* **2000**, *25*, 333–336. [CrossRef]

20. Neishtadt, N.M. Application of piezoelectric method in ore deposits. In Proceedings of the Transaction of the 15th Conference of Israel Mineral Science and Engineering Association, Jerusalem, Israel, 12–13 April 2000; pp. 74–78.

21. Zhu, Z.; Haartsen, M.W.; Toksöz, M.N. Experimental studies of seismoelectric conversions in fluid-saturated porous media. *J. Geophys. Res. Solid Earth* **2000**, *105*, 28055–28064. [CrossRef]

22. Gershenzon, N.; Bambakidis, G. Modeling of seismo-electromagnetic phenomena. *Russ. J. Earth Sci.* **2001**, *3*, 247–275. [CrossRef]

23. Tiesseyre, K.P. Anomalous piezoelectric effects found in the laboratory and reconstructed y numerical simulation. *Ann. Geophys.* **2002**, *45*, 273–278.

24. Butler, K.E.; Russell, R.D. Cancellation of multiple harmonic noise series in geophysical records. *Geophysics* **2003**, *68*, 1083–1090. [CrossRef]

25. Pride, S.R.; Garambois, S. Electroseismic wave theory of Frenkel and more recent developments. *J. Eng. Mech.* **2005**, *131*, 898–907. [CrossRef]

26. Haines, S.S.; Pride, S.R.; Klemperer, S.L.; Biodi, B. Seismoelectric imaging of shallow targets. *Geophysics* **2007**, *72*, G9–G20. [CrossRef]

27. Dupuis, J.C.; Butler, K.E.; Kepic, A.W.; Harris, B.D. Anatomy of a seismoelectric conversion: Measurements and conceptual modeling in boreholes penetrating a sandy aquifer. *J. Geophysl Res.* **2009**, *114*, B10306. [CrossRef]

28. Glover, P.W.J.; Jackson, M.D. Borehole electrokinetics. *Lead. Edge* **2010**, *29*, 724–728. [CrossRef]

29. Schakel, M.D.; Smeulders, D.M.J.; Slob, E.C.; Heller, H.K.J. Seismoelectric interface response: Experimental results and forward model. *Geophysics* **2011**, *76*, N29–N36. [CrossRef]

30. Neishtadt, N.M.; Eppelbaum, L.V. Perspectives of application of piezoelectric and seismoelectric methods in applied geophysics. *Russ. Geophys. J.* **2012**, *51*, 63–80. (In Russian)

31. Gershenzon, N.I.; Bambakidis, G.; Ternovskiy, I. Coseismic electromagnetic field due to the electrokinetic effect. *Geophysics* **2014**, *79*, E217–E229. [CrossRef]

32. Jouniaux, L.; Zyserman, F. A review on electrokinetically induced seismo-electrics, electro-seismics, and seismo-magnetics for Earth sciences. *Solid Earth* **2016**, *7*, 249–284. [CrossRef]

33. Fridrichsberg, D.A. *Course of Colloidal Chemistry*; Chemistry Publisher: S.-Petersburg, Russia, 1995. (In Russian)

34. Ivanov, A.G. The electroseismic effect of the second kind. *Izv. Acad. Sci. USSR (Trans. Sov. Acad. Sci.)* **1940**, *5*, 699–727, (In Russian, transl. to English).

35. Probstein, R.F. *Physiochemical Hydrodynamics: An Introduction*, 2nd ed.; Wiley & Sons: New York, NY, USA, 1994.

36. Frenkel, Y.I. On the theory of seismic and seismoelectric phenomena in a moist soil. *Izv. Acad. Sci. USSR* **1944**, 133–150, (In Russian, transl. to English). [CrossRef]

37. Butler, K.E. Seismoelectric Effects of Electrokinetic Origin. Ph.D. Thesis, The University of British Columbia, Vancouver, BC, Canada, 1996.

38. Jardani, A.; Revil, A.; Slob, E.; Söllner, W. Stochastic joint inversion of 2D seismic and seismoelectric signals in linear poroelastic materials: A numerical investigation. *Geophysics* **2010**, *75*, N19–N31. [CrossRef]

39. Mahardika, H.; Revil, A. Seismoelectric conversion generated from water-oil boundary in unsaturated porous media. In Proceedings of the Transactions of SEG Meeting, Houston, TX, USA, 22–27 September 2013; pp. 1852–1857.

40. Antonova, E. Finite Elements for Electrically Unbounded Piezoelectric Vibrations. Ph.D. Thesis, McGill University, Montreal, QC, Canada, 2000.

41. Jandaghian, A.A.; Jafari, A.A. Investigating the Effect of Piezoelectric layers on Circular Plates under Forced Vibration. *Int. J. Adv. Des. Manuf. Technol.* **2012**, *5*, 1–9.

42. Alperovich, L.S.; Neishtadt, N.M.; Berkovitch, A.L.; Eppelbaum, L.V. Tomography approach and interpretation of the piezoelectric data. In Proceedings of the Transactions of the IX General Assembly of the European Geophysical Society, Strasbourg, France; 1997. 59/4P02. p. 546. Available online: https://www.researchgate.net/profile/Lev_Eppelbaum/publication/240527113_Tomography_approach_and_interpretation_of_the_piezoelectric_data/links/0deec531d5667bc6ab000000.pdf (accessed on 15 September 2017).

43. Eppelbaum, L.V.; Itkis, S.E.; Khesin, B.E. Optimization of Magnetic Investigations in the Archaeological Sites in Israel. In *Filtering, Modeling and Interpretation of Geophysical Fields at Archaeological Objects*; Special Issue of *Prospezioni Archeologiche*; 2000; pp. 65–92. Available online: https://www.researchgate.net/publication/

250613019_Optimization_of_magnetic_investigations_in_the_archaeological_sites_in_Israel (accessed on 15 September 2017).

44. Eppelbaum, L.V.; Khesin, B.E.; Itkis, S.E. Prompt magnetic investigations of archaeological remains in areas of infrastructure development: Israeli experience. *Archaeol. Prospect.* **2001**, *8*, 163–185. [CrossRef]

45. Eppelbaum, L.V.; Khesin, B.E.; Itkis, S.E. Archaeological geophysics in arid environments: Examples from Israel. *J. Arid Environ.* **2010**, *74*, 849–860. [CrossRef]

46. Eppelbaum, L.V. Archaeological geophysics in Israel: Past, Present and Future. *Adv. Geosci.* **2010**, *24*, 45–68. [CrossRef]

47. Eppelbaum, L.V. Study of magnetic anomalies over archaeological targets in urban conditions. *Phys. Chem. Earth* **2011**, *36*, 1318–1330. [CrossRef]

48. Eppelbaum, L.V. Quantitative interpretation of magnetic anomalies from thick bed, horizontal plate and intermediate models under complex physical-geological environments in archaeological prospection. *Archaeol. Prospect.* **2015**, *23*, 255–268. [CrossRef]

49. Eppelbaum, L.V. Quantitative Analysis of Piezoelectric and Seismoelectric Anomalies in Subsurface Geophysics. In Proceedings of the Transactions of the 13th EUG Meeting Geophysical Research Abstracts, Vienna, Austria, 23–28 April 2017; Volume 19.

50. Eppelbaum, L.V.; Mishne, A.R. Unmanned Airborne Magnetic and VLF investigations: Effective Geophysical Methodology of the Near Future. *Positioning* **2011**, *2*, 112–133. [CrossRef]

51. Eppelbaum, L.V. Geophysical observations at archaeological sites: Estimating informational content. *Archaeol. Prospect.* **2014**, *21*, 25–38. [CrossRef]

52. Eppelbaum, L.V.; Alperovich, L.; Zheludev, V.; Pechersky, A. Application of informational and wavelet approaches for integrated processing of geophysical data in complex environments. In Proceedings of the 2011 SAGEEP Conference, Charleston, SC, USA, 10–14 April 2011; pp. 24–60.

geosciences

MDPI

Article

Geoarchaeological Core Prospection as a Tool to Validate Archaeological Interpretation Based on Geophysical Data at the Roman Settlement of Auritz/Burguete and Aurizberri/Espinal (Navarre) †

Ekhine Garcia-Garcia [1,2,3,*], James Andrews [4], Eneko Iriarte [1,5], Roger Sala [3], Arantza Aranburu [1,2], Julian Hill [4] and Juantxo Agirre-Mauleon [1]

[1] Aranzadi Society of Sciences, Donostia-San Sebastian 20014, Basque Country; arantza.aranburu@ehu.eus (A.A.); zuzendaritza@aranzadi.eus (J.A.-M.)
[2] Department of Mineralogy and Petrology, University of the Basque Country (EHU/UPV), Leioa 48940, Basque Country
[3] SOT Archaeological Prospection, Barcelona 08198, Spain; roger_sala_bar@yahoo.es
[4] MOLA (Museum of London Archaeology), Mortimer Wheeler House, London N1 7ED, UK; jandrews@mola.org.uk (J.A.); jhill@mola.org.uk (J.H.)
[5] Geography, History and Communication department, University of Burgos, Burgos 09001, Spain; eiriarte@ubu.es
* Correspondence: ekhinegarcia@yahoo.com
† This paper is an extended version of our paper published in Garcia-Garcia, E.; Agirre-Mauleon, J.; Andrews, J.; Aranburu, A.; Arrazola, H.; Etxegoien, J.; Fuldáin, J.; Hill, J.; Iriarte, E.; Legorburu, M.; et al. Geoarchaeological core prospection investigation to improve the archaeological interpretation of geophysical data: Case study of a Roman settlement at Auritz (Navarre). *Archaeol. Polona* **2015**, *53*, 88–91.

Received: 21 August 2017; Accepted: 3 October 2017; Published: 13 October 2017

Abstract: Geophysical survey methods are broadly used to delimit and characterize archaeological sites, but the archaeological interpretation of geophysical data remains one of the challenges. Indeed, many scenarios can generate a similar geophysical response, and often interpretations can not be validated without access to the subsoil. In large geophysical surveys many anomalies are detected and validation through archaeological trenches can not be afforded. This paper analyses the validity of geoarchaeological core survey to check the archaeological interpretations based on geophysical results. The Roman site located at Auritz/Burguete and Aurizberri/Espinal (Navarre), provides a great case of study as many investigations have been carried out. After the gradiometer survey performed in 2013 a sediment core survey was designed. 132 cores were drilled using a hand-held coring machine and the sediments were analysed in situ. Site delimitation and archaeological interpretations based on magnetic data could be improved or corrected. In this regard, the core survey proved to be an useful methodology as many anomalies could be checked within reasonable time and resources. However, further geophysical investigations trough GPR revealed unexpected remains in areas where no archaeological deposits were identified through coring. Excavations showed poor conservation level in some of those areas, leading to thin archaeological deposits hard to identify at the cores. The sediment core survey, therefore, was proved to be inconclusive to delimit the archaeological site.

Keywords: Geoarchaeology; Geophysics; Archaeological Interpretation; Roman; Navarre

1. Introduction

Geophysical prospection has proved to be a useful tool in helping archaeologists improving their knowledge of archaeological sites. In last decades, their use has notably grown and specialized literature has been developped (e.g., [1–5]).

Thanks to the improvement in both the sensitivity of the sensors and the spacial resolution of the acquired data, high fidelity images of the subsurface can be now produced. Moreover, the development of multichannel geophysical systems allows large areas to be surveyed in a moderately short time (e.g., [6–9]).

The archaeological interpretation of geophysical data consists in offering plausible explanations for anomalies detected by geophysical systems [10], and is one of the challenges of the archaeological geophysics. Indeed, many scenarios can generate a similar geophysical response and, in most circumstances, it is the expert knowledge of the surveyor, alongside insight into the site's geophysical and archaeological characteristics, which guides further interpretation. However, any interpretation process is limited and the possibility of scenarios with erroneous or inaccurate interpretations should be considered.

In most cases, the validation through excavation would be the best option in order to check the interpretations. Nevertheless, it is not always possible, particularly in commercial projects, or only a small fraction of the surveyed area can be excavated. Often the least invasive approach is preferred, particularly in non-threatened sites. The sediment core survey is less invasive and less time-consuming than archaeological trenches. Therefore, more anomalies can be checked with this method, which is one of the benefits of this methodology. However, the amount of recovered information is highly dependent on the drilled location and, due to the limited view of sediments, interpretation problems can arise [11]. Therefore, the context is more limited than the archaeological trenches and it could not be sufficient for answering some questions.

This paper analyzes the validity of geoarchaeological survey to check the archaeological interpretations based on geophysical results, as an extension of a previous contribution [12].

1.1. The Background of the Research

The sediment core survey is a broadly used technique to characterize the stratigraphical sequence in geological investigation. After the 1970s, it has also been extensively used in archaeological research [11]. A typical application is related to the reconstruction of the past landscape and its relationship to archaeological record (e.g., [13,14]). In some projects, sediment coring is used as a cost-effective archaeological evaluation tool to delineate archaeological sites, analyse sediment or soil sequences and built geoarchaeological site formation models, which can be used to support subsequent excavation work or further research (e.g., [11,15,16]).

Some studies combine the use of geoarchaeological and geophysical investigations, but relative few studies have described the use of core sampling to target geophysical anomalies and validate the archaeological interpretation based on geophysical data (e.g., [17,18]). In many cases, the sediment core sampling is used, together with archaeological trenches, to access to soil samples that are characterized to have complementary data or to analyse the interactions between soil physical properties and geophysical techniques (e.g., [19–21]). Indeed, the understanding of this interactions helps improving further interpretations. However, in some low budget projects, or in commercial projects where time is a constraint, there is no access to soil characterization techniques. This paper provides an example from a Roman site, where a sediment core sampling was used to target geophysical anomalies and validate the archaeological interpretation.

The site was discovered during a walk over survey in 2008, and the archaeological trenches excavated in 2012 revealed substantial Roman masonry building foundations [22]. Since then, geophysical survey methods have been used to delimit and characterize the settlement.

The site is situated south of the town of Auritz/Burguete, on a high plateau close to the crest line of the Pyrenees (Figure 1a). The Pyrenees are characterized by complex structural geology and form a natural boundary between the Iberian Peninsula and the continent [23,24]. A number of mountain streams rise in the area to the north of the site and flow south to combine as the River Urrobi. The site is situated on the banks of this River, on a spacious plain or terrace surrounded by higher mountain peaks. Stepped, terrace topography is clear on site with the elevation dropping downwards towards and parallel to the Urrobi river in a north-south alignment along the length of the site. There is not a detailed geomorphological study of the Urrobi basin but it is known that the geological context is formed by quaternary fluvial terrace deposits of clay, silt and gravel [25].

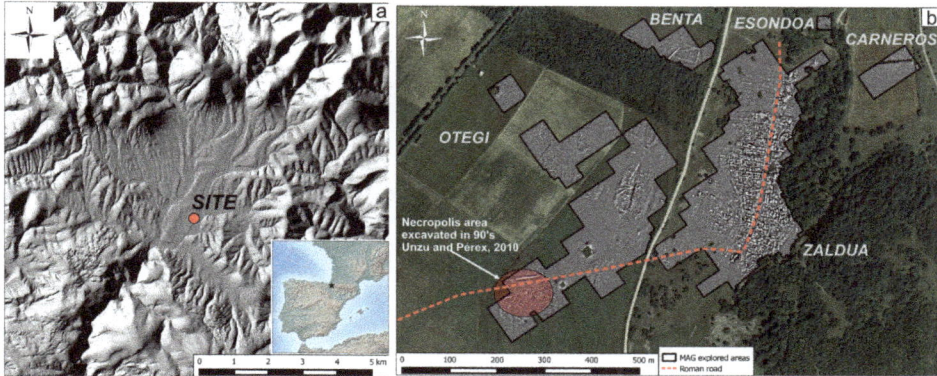

Figure 1. (**a**) Geographical location of the studied archaeological site on an elevation map created from the LiDAR data provided by the © Instituto Geográfico Nacional de España (5 m cell). (**b**) Magnetic response map in the overall explored area (−8 nT black, +8 nT white). The Roman road and the necropolis area excavated are indicated. Modified from [26] and reproduced by permission of ARANZADI Society of Science.

A magnetic survey performed in 2013 covered ca. 18 ha and revealed the main layout of the settlement [26]. Results revealed differences in the magnetic response that had not been well explained and became, together with the archaeological description of the site, one of the main concerns of the investigation team (Figure 1b).

2. Research Objectives

Following the magnetic survey, a sediment core sampling strategy was designed to refine interpretation of the prospection data. The main goal was to better characterize the archaeological deposits of the site, and to correlate the distribution and characteristics of the archaeological deposits with the magnetic response. In addition, some of the core sampling was undertaken with the specific goal of evaluating the geophysical interpretations [27]. This paper will focus in this second goal.

Besides this, the validity of cores to identify occupation areas will be discussed on this paper.

3. Applied Methodology

The strategy is based on different techniques applied at different scale, starting from a broad context and focusing on the interesting areas for more detailed investigations (Table 1).

The methodology and results of the magnetic survey can be found in [26].

Cores were drilled with a mechanical petrol-driven 2-stroke power auger (Van Walt window corer equipped with a Cobra TT engine). This is a hand-held coring machine, which drives 1 m long, open-sided steel gouges into the ground (Figure 2A). The sediment-filled gouge is then recovered from the ground using the two-person jacking device (Figure 2B). Two gouges (of 100 mm and 60 mm diameter) were fitted with 1 m long steel extension rods to reach a maximum of 3 m depth. In general, the wider auger was used, reserving the second one to areas where the penetration was harder.

Table 1. Applied survey techniques.

Technique	Extension	Resolution	Device
Magnetometer Survey	18 ha	$0.25m \times 0.5$ m	Bartington Grad 601-dual
Sediment core survey	132 cores (18 ha)	not regular	Van Walt window corer
GPR survey	3.3 ha	0.2×0.025 m	IDS Hi-Mod (200 MHz–600 MHz)
Excavation	186.5 m^2	-	-

Figure 2. (**A**) Photograph of the power auger (Cobra TT engine) used for drilling; (**B**) Photograph of ARANZADI and MOLA team removing the gouge manually.

Once the gouges were retrieved, the deposits were cleaned and recorded on-site from the open-sided auger sample. Sediments were photographed and described according to standard sedimentary criteria [28,29]. Preliminary interpretations of the depositional conditions represented by the soils and sediments within each core were made, and the deposits or contexts grouped into broad stratigraphic units.

GPR prospecting was performed using an IDS Hi-Mod instrument with a single multi-frequency antenna (200 MHz and 600 MHz). Measurements were taken every 2.5 cm along parallel traverses, generally 0.2 m apart, and in zigzag mode. The time window was set at 90 ns for the 200 MHz antenna and at 60 ns for the 600 MHz antenna. Data were processed using GPR_SLICE software to generate horizontal maps [30].

Archaeological excavation was carried out in accordance with MOLA recording techniques as set out in the Museum of London Archaeological Site Manual [31]. Selected features and layers were additionally recorded with the total station theodolite.

4. The Layout of the Surveys

4.1. The Core Survey

Based on the magnetic survey, the area of investigation was divided into four zones. In a first campaign 105 locations were drilled across the site, but the cores were not equally distributed.

The main focus of the survey was on Zaldua area, which contains the most of the settlement. Here, the survey comprised cores acquired on nine east–west transects (Figure 3a). The transects were positioned to create a grid of points covering the entire zone but also to include a number of specific points to resolve anomalies apparent in the geophysical survey.

In 2014 survey of the Otegi zone was less intensive and cannot be considered representative of the whole area. In 2016, some more cores were drilled in order to complete information about unanswered questions (Figure 3b).

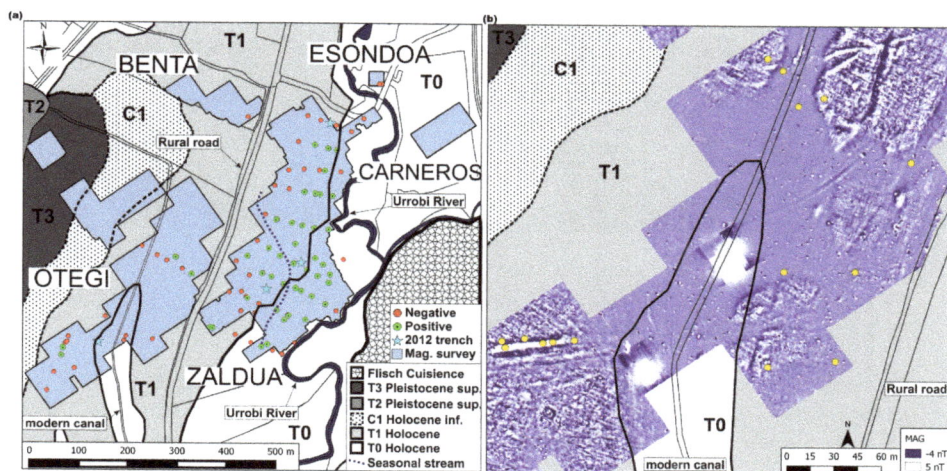

Figure 3. (**a**) Location of the cores drilled in 2014 over the geological plan and areas of the magnetic survey (T: *Terrace*; C: *Colluvium*). The cores showing archaeological deposits are highlighted as positives, and the ones where no archaeological remains were identifies as negatives. (**b**) Location of the cores drilled in Otegi area in 2016, over the magnetic response map.

4.2. Complementary Geophysical Surveys and Archaeological Excavations

After the first campaign of sediment core survey was finish in 2014, new investigations were carried out on the site. In particular, some areas with low magnetic contrast or without clearly identified anomalies were analysed using GPR as the main system [32,33] (Figure 4a). In addition, four verification trenches and an area of 165 m^2 were excavated in 2015 and 2016 respectively (Figure 4b).

5. Results and Discussion

5.1. Comparison with Geophysical Anomalies

In areas where well defined magnetic anomalies were targeted, the cores generally confirmed the interpretation from the geophysical survey. Furthermore, the extracted sequences allowed these interpretations to be refined and permitted a better understanding of the origin of the anomalies. However, some discordance was found and new areas of interest were identified.

5.1.1. Target: Buildings

The magnetic survey had provided a great amount of information regarding building location and characteristics. It could be established that the buildings are mainly organized along the road and oriented perpendicularly to it, leading to an irregular urban layout. In general, the magnetic contrast allowed a good description of the main walls and the inner distribution of the buildings. Some areas, however, had shown a lack of magnetic contrast or poor definition of anomalies, even where the existence of archaeological deposits was known [26]. The cores were mainly drilled in three different areas, expecting to obtain complementary information to understand the differences on magnetic contrast.

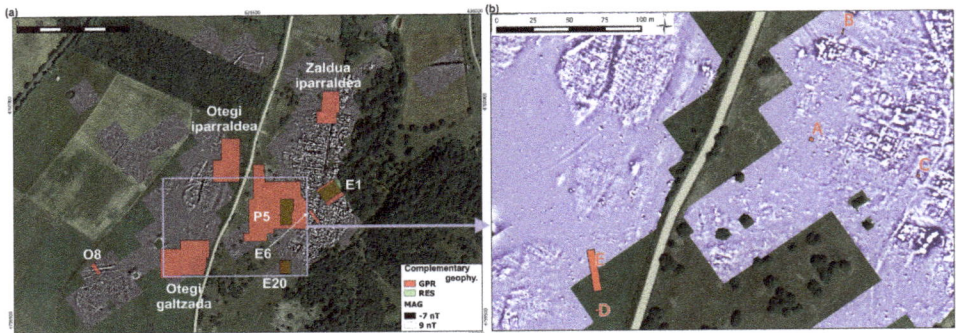

Figure 4. (**a**) Location of areas where complementary surveys (GPR and electric survey) have been performed, over the magnetic results (−7 nT black, 9 nT white); (**b**) Location of the archaeological trenches excavated in 2015 and 2016 over the magnetic results (−7 nT blue, 9 nT white)

The cores BH1, BH2 and BH3 had been located at the south-west part of the Zaldua area. Archaeological remains (walls and pavements) had been observed during installation of a water channel. The magnetic contrast was, however, weak and the anomalies were not clearly defined (Figure 5a). One of considered possibilities was that the remains were deeper in comparison with other areas. The sedimentary sequence of BH1 was a natural sequence consisting of turf and top soil layers over fluvial terrace silt and gravels (Table 2). The other two encountered solid stone obstructions at 0.22 m and 0.37 m respectively. As BH1 demonstrated bedrock is situated below 2.0 m bgl, the obstructions are likely to be walls or stone/rubble from a collapsed wall. Because the remains are located just below the organic layer, the lack of magnetic contrast in not related to an increase of the deepness. The sequence, however, has no evidence of charcoal or pottery flecks. Therefore, it can be conjectured that only the basement levels have been conserved, which would explain the lack of the magnetic contrast.

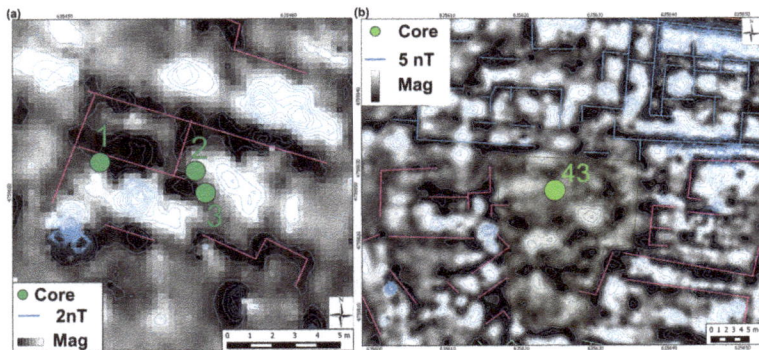

Figure 5. (**a**) Location of the BH1, BH2 and BH3 over the magnetic survey results, where some linear anomalies attributed to a building are identified. The area is highlighted in the general context of Figure 6. (**b**) Location of the BH43 over the magnetic survey results, where an area without remarkable magnetic anomalies and surrounded by buildings have been attributed to a possible open area. The area is highlighted in the general context of Figure 6.

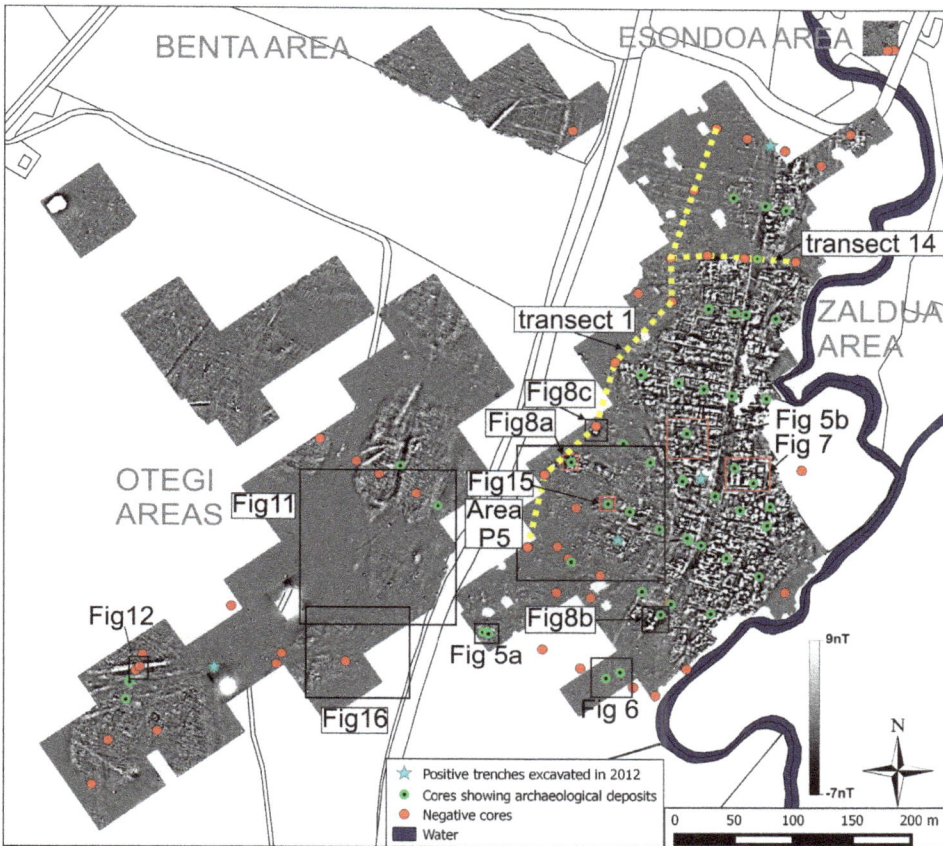

Figure 6. Core location over the magnetic response map. The areas highlighted in other figures are indicated. The transects 1 and 14 mentioned further are indicated in yellow dashed lines.

Table 2. Description of cores BH1, BH2 and BH3. Occ: *Occasional*; Mod: *Moderate*; Med: *Medium*.

Core	Thickness	Lithology	Description	Interpretation
BH01	0–0.5 m	Turf		Topsoil
	0.05–0.3 m	Silt, clayey	mod light brown friable sandy silt with roots	Topsoil
	0.3–0.78 m	Silt, sandy	as above but fewer roots	Archaeology
	0.78–1 m	Clay, silty	firm orange brown silt clay with manganese and iron staining (flecks) occ. small clasts	Fluvial silts
	1.22–2 m	Silt, clayey	as above 0.78 m (increasingly firm with depth). No archaeology Increase in clay occ. small/med clasts.	Fluvial silts
BH02	0–0.05 m	Turf	mod light brown friable sandy silt with roots	Topsoil
	0.05–0.22 m	Silt, sandy		Obstruction
BH03	0–0.05 m	Turf	mod light brown friable sandy silt with roots	Topsoil
	0.05–0.37 m	Silt, sandy		Obstruction

The cores BH6 and BH7 had been drilled at the south-east part of the Zaldua area. Analysis of aerial images revealed the existence of a building, and the existence of the walls was proven simply by removing the superficial turf layer. This building was not detected in the magnetic survey, but was clearly distinguished in GPR and electric surveys (Figure 7).

Figure 7. (**a**) Pictures of the sediment cores extracted on locations BH6 and BH7; (**b**) Magnetic results of the area where the location of the cores can be seen; (**c**) GPR results over the same area, where the building is clearly defined (time slice at 0.37–0.54 m bgl, v = 8 cm/ns, high reflection in dark). The area is highlighted in the general context of Figure 6.

The BH7 targeted the south of the building. A clean and very stiff layer of silt/clay was observed between 0.66 m and 0.87 m bgl, attributed to a clay floor. The deposits above this layer didn't show neither charcoal or pottery flecks, but had been considered preliminarily archaeological based on disturbance evidences (i.e., unsorted gravels). The BH6 was drilled outside of the inferred building. Results show a thick archaeological deposit between 0.22 and 0.95 m consisting on differentiate layers where frequent charcoal and building material flecks had been observed. In this case, the archaeological layer found in the inner part of the building consisted in a clean levelling layer without evidence of magnetically enhanced deposits. However, the thick archaeological layer outside the building suggests that the walls could show a negative magnetic contrast.

Finally, the cores BH41 and BH42 were drilled on the main part of the settlement, were the magnetic contrast is strong (Figure 8). A series of floor layers were observed in BH41 (0.64–1.42 m bgl). In BH42, a possible collapsed mortared wall or mortar-rich construction debris interrupt a sequence of archaeological dumps (0.52–1.1 m bgl). The existence of thicker archaeological layers in those cores is in agreement with the stronger magnetic contrast detected on this area.

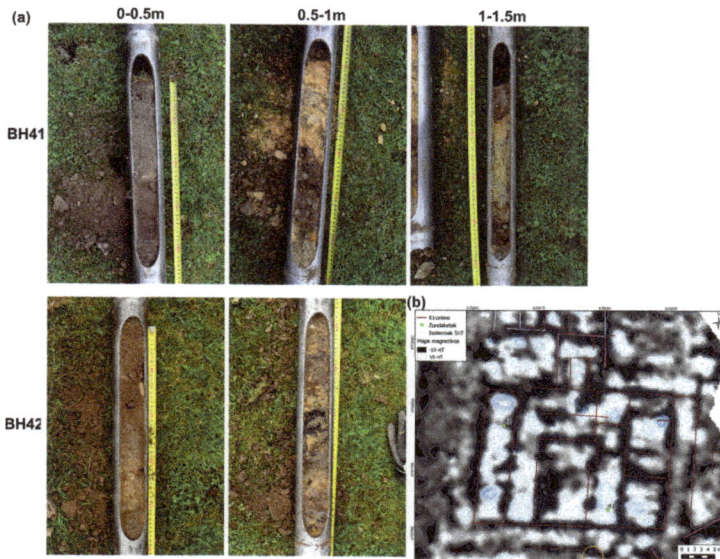

Figure 8. (**a**) Pictures of the sediment cores extracted on locations BH41 and BH42. (**b**) Magnetic results of the area (±10 nT) where the location of the cores can be seen. The area is highlighted in the general context of Figure 6.

5.1.2. Target: Combustion Signatures

Magnetic data had provided several combustion evidences on the surveyed area. Some of them were attributed to kilns or massive fires, whereas others seem to be related to hearths. Some of those location had been included into the core survey in order to validate the interpretations based on magnetic data.

The core BH23 was drilled on the north limit of the P5 area, where magnetic data showed evidences of combustion processes but without any coherent shape of anomalies (Figure 9A). It contains archaeological deposits between 0.3 m and 0.9 m consisting in a silt matrix where many heat affected clay, charcoal lumps and red sandstone clasts can be observed (Figure 10). The absence of well defined stratification agrees to the characteristics of magnetic data, and it can be conjectured that a conflagration happened in this location.

Core 12 targeted an anomaly interpreted as a kiln (Figure 9B). A series of archaeological layers containing fragments of slag were observed over a floor of firm clay with charcoal flecks. The absence of any heat scorching on the clay surface suggests that the core did not locate a kiln. However the abundance of slag in the above layers suggests metalworking was taking place nearby. Therefore, observations support the interpretation of a building associated with fire industry.

Core 33 targeted an intense magnetic anomaly isolated from the main occupation area, attributed to the traces of combustion related activity (Figure 9C). The core showed a burnt clay surface immediately below the superficial layer (Figure 10). The anomaly was then attributed to a recent fire for vegetation burning instead to an archaeological feature, confirmed by local people.

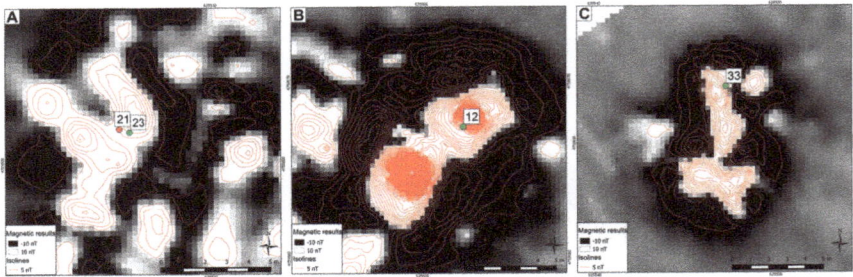

Figure 9. (**A**) Location of the sediment core 23 over the magnetic response map; (**B**) Location of the sediment core 12 over the magnetic response map; (**C**) Location of the sediment core 33 over the magnetic response map. The magnetic contrast is the same in all the figures (−10 nT black, 10 nT white). The areas are highlighted in the general context of Figure 6.

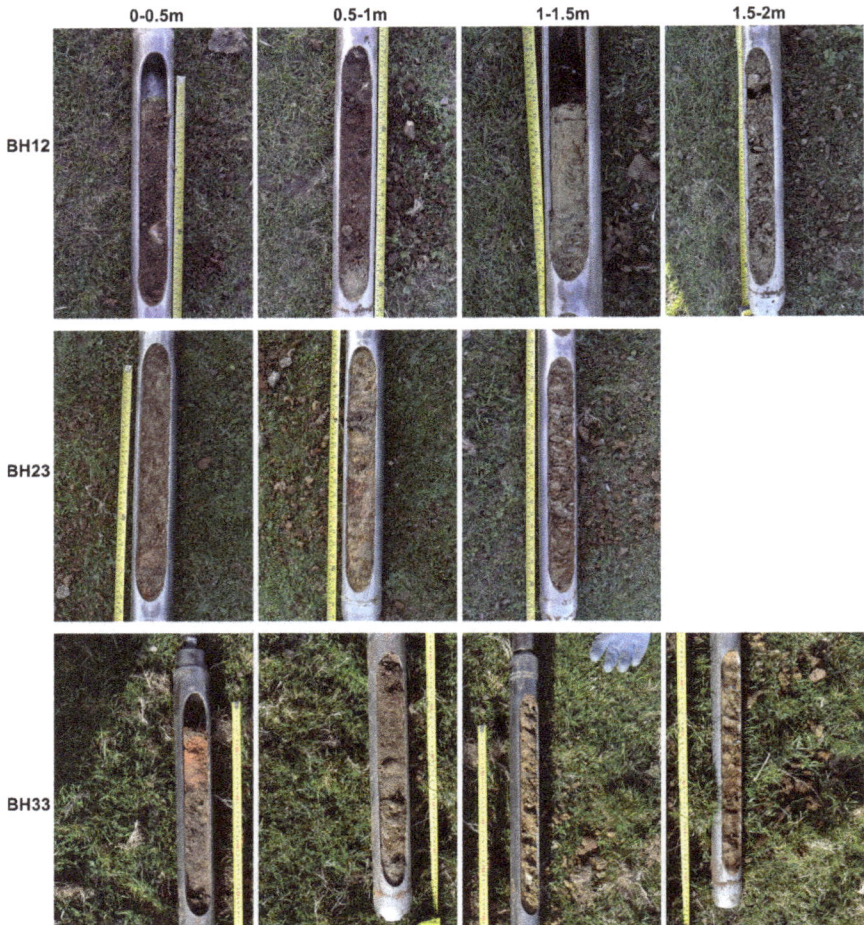

Figure 10. Pictures of the sediments extracted in the cores BH12, BH23 and BH33.

5.1.3. Target: Open Areas

The magnetic data layout revealed some areas without relevant anomalies surrounded by anomalies attributed to buildings. Those areas where preliminarily attributed to open areas that could be integrated into building complexes.

In the west part of Zaldua, magnetic data suggested the existence of a rectangular area surrounded by coherently oriented buildings (P5). Standing in a small hill, this area was preliminarily interpreted as a possible singular area of the site [26,32]. Four cores were drilled within this area: BH17, BH19, BH24 and BH25 (Figure 11a). In BH17 archaeological layers were encountered and comprised 2 possible gravel surfaces with levelling or dumped layers beneath (0.24–0.33 m bgl and 0.42–0.53 m bgl). The same sequence was observed 15 metres north-west in BH24 (0.25–0.39 m bgl and 0.5–0.64 m bgl). The levels and thickness of the gravel layers suggest than this area had been surfaced twice at least in the south-east corner of P5. In contrast, the sedimentary sequence observed in BH25 and BH19 consisted of turf and topsoil layers over fluvial terrace silts and gravels, and had no evidence of gravel floors. Given the contrasting results, the core survey cannot corroborate the interpretation of P5 as an archaeological unit.

Further investigations focused in this area. The GPR surveys show complementary information to that obtained in the magnetic survey [32]. The south-eastern limit of the P5 rectangle appears clearly on the time-slice sequence, allowing a more detailed description than the one obtained from magnetic data. Furthermore, new buildings could be described, improving notably the architectural information obtained from the magnetic survey. As it can be seen in Figure 11b, the cores where the gravel levels were found have been drilled in areas where the GPR energy is reflected, whereas the others are located in areas without reflection. The non homogeneous stratigraphy of that areas was then confirmed, but as the coherent orientation of buildings is corroborated by the GPR results, the unity of the area is not discarded.

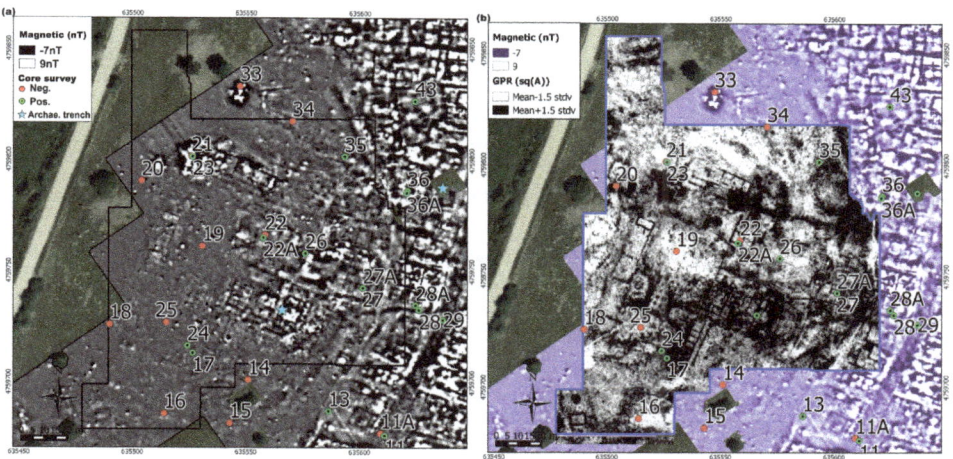

Figure 11. P5 area. (**a**) Location of cores over the magnetic data; (**b**) Location of the cores over the GPR data. Time slice at 39–56 m bgl (v = 8.34 cm/ns). The area is highlighted in the general context of Figure 6.

Core 43 was drilled through another potential open space (Figure 5b). In contrast to P5 this much smaller area is surrounded by buildings. The sedimentary sequence is also very different to that of P5, no evidence was found of buildings or floor layers so the core supports the interpretation of an open space (Table 3). The use of the area also differs, since a massive and homogeneous layer of mixed dumped material (0.5–1.3 m bgl) may suggest this area was used for the disposal of rubbish.

Table 3. Description of core BH43. Mod: *Moderate*; Freq.: *Frequent*; CBM: *Common Building Material*.

Core	Thickness	Lithology	Description	Interpretation
BH43	0–0.06 m	Turf		Topsoil
	0.06–0.3 m	Silt, clayey	Mod compact light brown friable sandy silt with roots	Topsoil
	0.3–0.5 m	Clay, silty	Mod compact mid brown silty sandy clay with small clasts; further subsoil	Topsoil
	0.5–1.13 m	Silt, clayey	Mod compact/friable grey brown clay silt with sand with freq flecks charcoal and CBM	Archaeology
	1.13–1.30 m	Silt, sandy	Mod compact yellow sandy silt with some clay; orange-black patches at 1.16–1.20	Fluvial Silts
	1.30–1.50 m	Gravel	Mod compact mid yellow brown gravel	Fluvial Gravel
	1.50–1.60 m	Silt, clayey	Grey brown silt with some sand and clay	Fluvial Silts
	1.60–2 m	Gravel	Orange brown clast supported gravel, poorly sorted gravel of mixed lithology in a yellow silty matrix	Fluvial Gravel

5.1.4. Target: The Road

The road is detected as a linear and weakly magnetic zone which is easily identifiable when is surrounded by other anomalies, but difficult to discern in other areas, as for example at the north of the main occupation area in Zaldua. Borehole 62 in transect 14 was drilled through the missing road section (Figure 6). The road is visible as a compact gravel horizon (0.16–0.57 m bgl). The continuity of the road in that point was confirmed.

5.1.5. Magnetic Disturbance in Otegi Area

In Otegi area the geophysical survey revealed variations in background magnetic disturbance, that difficult the identification of archaeological remains [26]. This disturbance had been preliminarily attributed to differences in soil properties, not related to archaeological occupation.

A series of cores were drilled in 2014 in order to clarify the origin of this magnetic disturbance, and no archaeological evidence was observed [27]. It had been appointed that the disturbance could be related to differences in soil horizon composition and its thickness.

Finally, additional cores were drilled in 2016, in order to clarify the origin of this magnetic disturbance (Figure 12). The magnetically stable areas coincide with areas where sedimentary deposits were identified, whereas in disturbed areas the deposits are related to the erosion of the bedrock and alterite type thin soil formation. Based on this data, the difference on magnetic response can be attributed to the activity of sedimentary processes related to the nowadays canalised water string. In that campaign soil samples have been collected to be analysed in laboratory in order to better understand the mineralogical composition that leads to the magnetic disturbance. This research is still ongoing and the result will be published in the future.

5.1.6. Geophysical Anomaly Parallel to the Road in Otegi Area

In the north of the necropolis a positively contrasted magnetic anomaly was detected running parallel to the road and located next to it. The GPR survey results show a reflective horizon consistent with magnetic data at approximately 80 cm below ground level (bgl), but without homogeneous intensity or coherent shape, as might be expected for a construction element (Figure 13).

The sedimentary sequence in BH89 targeting this anomaly, comprised natural sequences of fluvial terrace silts above fluvial terrace gravels at 0.80 m bgl (Table 4). The core survey cannot explain the magnetic anomaly. The GPR reflection, however, could be produced by the fluvial terrace gravels.

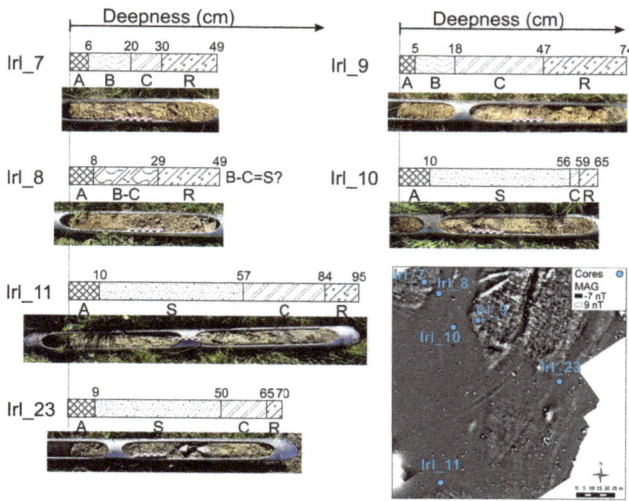

Figure 12. Detail of the location and results of cores drilled in 2016 to clarify the origin of magnetic disturbance in Otegi area. The area is highlighted in the general context of Figure 6. A horizon: Topsoil; surficial organic-rich mineral layer; B horizon: Subsoil; contains lixiviated and precipitated iron oxides and illuviated clays; C horizon: Parent material; barely edaphized but fragmented rock; R horizon: Bed rock; partially weathered bedrock, marls and limestone, at the base of the soil profile; S: Sedimentary layers.

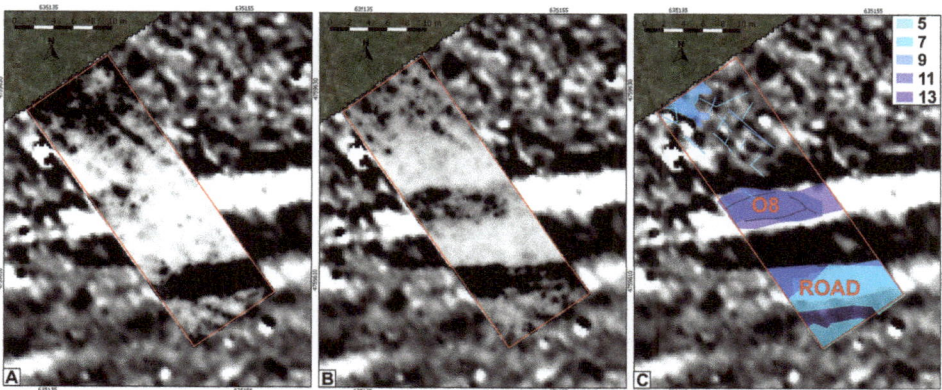

Figure 13. Detail of the geophysical anomaly parallel to the road in Otegi area (The area is highlighted in the general context of Figure 6). GPR results over magnetic response map (−5 nT black, 6 nT white). (**a**)Time slice at 0.36–0.55 m bgl; (**b**) Time slice at 0.61–0.80 m bgl. (**c**)Interpretation scheme of GPR results.

Table 4. Description of core BH80.

Core	Thickness	Lithology	Description	Interpretation
BH80	0–0.05 m	Turf		Topsoil
	0.05–0.3 m	Silt, clayey	Topsoil	Topsoil
	0.3–0.8 m	Silt	Very compact mid orange brown silt with some clay	Fluvial silts
	0.8–1 m	Gravel, silty	Large clasts shale gravel/bedrock	Fluvial gravels

5.2. Presence and Absence of Archaeological Deposits

Between the 105 drilled cores, 73 showed archaeological deposits including the road and shallow obstructions, which had been attributed to walls. The majority of the cores reached undisturbed alluvial terrace deposits (silts or gravels) at between 1 and 2 m below modern ground level [27]. The excavations performed in 2012, however, had revealed thicker archaeological deposits, near to 2 m thick, especially on the main area of Zaldua. That means that the trenches had been placed in areas where stratigraphy was especially thick and that the volume of archaeology across the site is lower than estimated until the core survey. An interpolated raster of the archaeological deposit thickness is shown for the Zaldua area in Figure 14a.

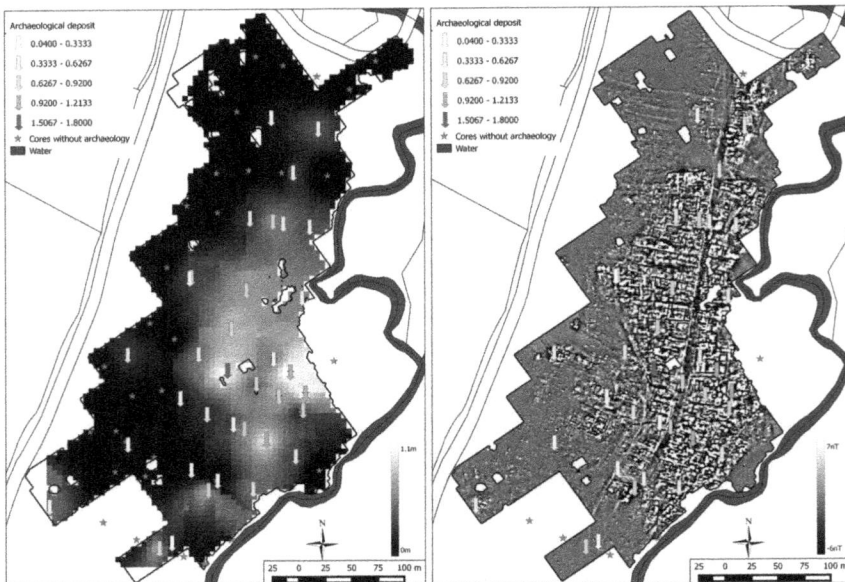

Figure 14. (**a**) Cores of Zaldua area over the interpolated raster for the thickness of the archaeological deposit. (**b**) Cores over the magnetic gradiometer results. Cores where no archaeological deposits were identified are signalled in red. The arrow's colour reflects the thickness of the identified archaeological deposits.

Comparing with magnetic results, cores that showed natural sequences with no archaeology are mainly those at the edge of the surveyed area, outside the main focus of settlement area. Specifically, the northern transect (transec 14, Figure 6) confirmed an absence of archaeological material and strengthened the interpretation of this area as the northern edge of the main settlement [26].

Similarly, the western transect (transect 1, Figure 6) did not detect any archaeological deposits apart from the BH23 located on the north-western limit of a possible square P5. Then, the western limit of the occupation seems to be coincident with that deduced from the magnetic survey.

The core survey across Otegi zone was less intensive and archaeological deposits were recorded in only five of the 19 holes drilled. The archaeological origin of the magnetically altered areas was, therefore, discounted. The occupation seems to be restricted to the necropolis area and to the surroundings of the road.

Therefore, the core survey across the site mirrors the presence and absences seen in the magnetic survey. However, when the GPR survey was performed, results revealed archaeological remains in areas were the core survey did not identify any archaeological evidence. Those areas are coincident with the areas where the magnetic contrast is not enough to detect any significant anomaly.

Figure 15 shows the location of the cores BH22 and BH22A over magnetic and GPR results. The first encountered an obstruction and, because of that, the second one was drilled. In this one no archaeological deposits were identified. The GPR results, however, clearly show a reflective rectangular anomaly attributed to the remains of a building.

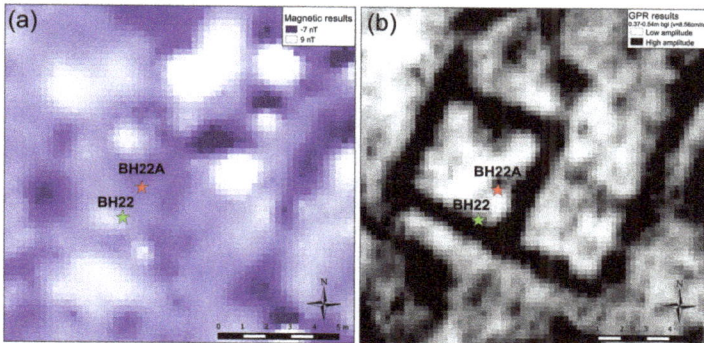

Figure 15. (a) Cores BH22 (Obstruction) and BH22A (negative) over the magnetic results; (b) GPR results where showing that the core BH22 encountered a wall whereas the BH22A was located within a building. The area is highlighted in the general context of Figure 6.

Similarly, Figure 16 shows the location of a group of cores drilled in 2016 in Otegi area. Core Ark24 revealed a gravel deposit at 0.35 m bgl, corresponding to the reflective area detected in GPR results. Even that it is clear from the GPR that they have been drilled in the interior of buildings, no evidence of archaeological deposits were encountered in cores Ark25 and Ark26. Excavations performed in trench E revealed thin archaeological layers where only the building basements were preserved (Figure 17). In that condition, the identification of archaeological sediments in the core is almost impossible, as they are the same materials of the geological sequence.

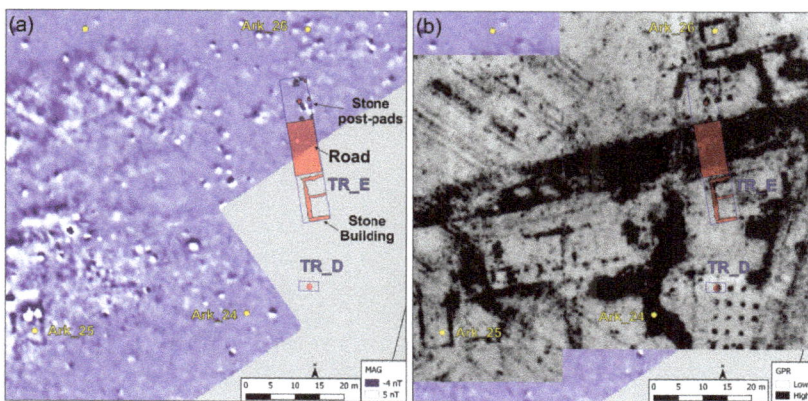

Figure 16. Detail of cores drilled in Otegi area in 2016. The trenches D and E (TR_D and TR_E, respectively), together with the significant encountered archaeological features are also indicated. The area is highlighted in the general context of Figure 6. (a) Over the magnetic results; (b) Over GPR results (time-slice at 0.38–59 m bgl, v = 8.1 cm/ns).

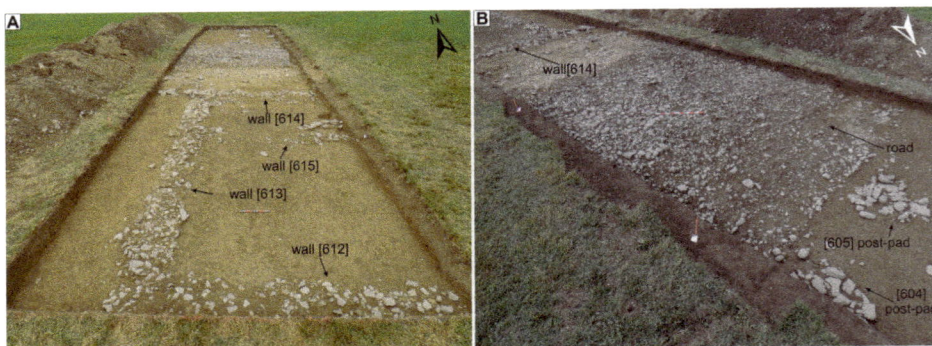

Figure 17. Photographs of the archaeological remains encountered at the trench E. (**A**) Stone basement of a rectangular building; (**B**) The road and some of the post-pads. Photographer: Antonietta Lerz.

6. Conclusions

The sediment core survey approach to complement geophysical data has proven to be useful to improve the archaeological interpretation of this Roman site. Indeed, many anomalies could be targeted with limited cost in time and resources. The cores provide feedback on observations and assumptions based on geophysical data, which are hard to validate without access to the subsoil. Even the limited amount of extracted sediments, interpretations could be mainly validated and complemented. Some interpretations were discovered to be erroneous, and further investigations could be designed to resolve those anomalies.

However, the research highlights the necessity of using complementary analysis in order to have a better understanding of the geophysical properties of the involved soils. This was demonstrated in particular by the magnetic disturbance at Otegi area. It could be correlated to the shallowness of the bedrock, but the specific mechanism that produces the disturbance in not still understood. Soil analyses, such as the measurement of the magnetic and electric properties and micromorphological studies, could help to characterize better the extracted soils and determine the origin of such disturbance.

While the sediment core survey is sometimes proposed as a tool to delimit the extent of the archaeological sites (e.g., [11,34]), in this study it revealed to be less efficient. Results mainly mirror the presence and absences seen in the magnetic survey but, in some areas classified as non archaeological, indisputable evidence of remains had been found in further GPR investigations. Archaeological excavations conducted in 2015 and 2016 revealed that in those areas only the basement of the building remains, without evidence of any other archaeological deposit. The lack of magnetic contrast and the non identification of archaeological deposits in the cores can be therefore understood. This study proves that the identification of archaeological areas can be difficult through this method.

In summary, this study demonstrates that sediment coring is a powerful technique to target geophysical anomalies and provide feedback on archaeological interpretation in a reasonable time and cost. While the support of further analysis would help to better characterize the interactions of the environmental features and processes with the geophysical techniques, many of the targeted anomalies could be resolved by a simple observation. In addition, the extracted sediments can be sampled to further geoarchaeological investigation at inconclusive areas. Therefore, a combination between geophysical survey and sediment coring enables a more accurate archaeological interpretation and provides a robust basis for further research.

It is interesting to remark that the example outlined here is based on a Roman site of Navarre, a chronology in which often clear traces of human activity lie into the ground. Although the principles are applicable within any other chronologies and regions, depending on the site's characteristics, the variations producing the targeted geophysical anomalies can be indiscernible by simple observation. In those cases, the validity of sediment coring to obtain a quick feedback would be compromised.

Acknowledgments: The different interventions of this research project were made possible by the volunteer work of a great number of people from the Aranzadi Science Society and the Museum of London Archaeology. In particular, the authors would like to thank all those who contributed to core surveys: Euken Alonso, Haizea Arrazola, Mikel Enparantza, Joxe Etxegoien, Juan J. Fuldáin, Enrique Lekuona, Juan Mari Mtz. Txoperena, Mary Nicholls, Peter Rauxloh, Begoña Yubero and Rafa Zubiria. The municipalities of affected areas and Eusko Kultur Fundazioa, together with some particulars, gave economical support. Finally, the authors wish to thank the Arrobi Society, which owns part of the lands explored, and to the municipality of Erro Valley, which owns the other part, for giving permission to conduct all the interventions related to the research.

Author Contributions: E. Garcia-Garcia participated in all the stages of the research after the discovery of the archaeological site in 2012 as part of her PhD research, and wrote the paper; J. Andrews participated in the core survey performed in 2014 and prepared the evaluation report; E. Iriarte participated in both core survey campaigns in 2014 and 2016, and contributed to the analysis of the results; R. Sala participated in the geophysical campaigns performed at this archaeological site; A. Aranburu supervised the research as promoter of the PhD research of E. Garcia-Garcia. J. Hill and J. Agirre-Mauleon supervised the research as directors of the archaeological project of the site.

Conflicts of Interest: The authors declare no conflict of interest. The founding sponsors had no role in the design of the study; in the collection, analyses, or interpretation of data; in the writing of the manuscript, and in the decision to publish the results.

Abbreviations

The following abbreviations are used in this manuscript:

MDPI	Multidisciplinary Digital Publishing Institute
DOAJ	Directory of open access journals
ca.	circa
bgl	Below Ground Level
GPR	Ground Penetrating Radar
MHz	Mega Hertz (10^6 Hz)
ns	nano seconds (10^{-9} s)
MOLA	Museum of London Archaeology

References

1. Scollar, I.; Tabbagh, A.; Hesse, A.; Herzog, I. *Archaeological Prospecting and Remote Sensing*; Topics in Remote Sensing; Cambridge University Press: Cambridge, UK, 1990.
2. Clark, A. *Seeing Beneath the Soil: Prospecting Methods in Archaeology*, 2nd ed.; Batsford: London, UK, 1996.
3. Gater, J.; Gaffney, C. *Revealing the Buried Past: Geophysics for Archaeologists*; Tempus: Stroud, UK, 2003.
4. Sala, R.; Tamba, R.; Garcia-Garcia, E. Application of Geophysical Methods to Cultural Heritage. *Elements* **2016**, *12*, 19–25.
5. Schmidt, A.; Linford, P.; Linford, N.; David, A.; Gaffney, C.; Sarris, A.; Fassbinder, J. *EAC Guidelines for the Use of Geophysics in Archaeology: Questions to Ask and Points to Consider*; Europae Archaeologiae Consilium: Namur, Belgium, 2016.
6. Trinks, I.; Johansson, B.; Gustafsson, J.; Emilsson, J.; Friborg, J.; Gustafsson, C.; Nissen, J.; Hinterleitner, A. Efficient, large-scale archaeological prospection using a true three-dimensional ground-penetrating Radar Array system. *Archaeol. Prospect.* **2010**, *17*, 175–186.
7. Bossuet, G.; Thivet, M.; Trillaud, S.; Marmet, E.; Laplaige, C.; Dabas, M.; Hullin, G.; Favard, A.; Combe, L.; Barres, E.; et al. City Map of Ancient Epomanduodurum (Mandeure-Mathay, Franche-Comté, Eastern France): Contribution of Geophysical Prospecting Techniques. *Archaeol. Prospect.* **2012**, *19*, 261–280.
8. Gaffney, C.; Gaffney, V.; Neubauer, W.; Baldwin, E.; Chapman, H.; Garwood, P.; Moulden, H.; Sparrow, T.; Bates, R.; Löcker, K.; et al. The Stonehenge Hidden Landscapes Project. *Archaeol. Prospect.* **2012**, *19*, 147–155.
9. Garcia-Garcia, E.; de Prado, G.; Principal, J.E. *Working with buried remains at Ullastret (Catalonia). Proceedings of the 1st MAC International Workshop of Archaeological Geophysics*; Monografies d'Ullastret, Museu d'Arqueologia de Catalunya-Ullastret: Girona, Catalonia, 2016; Volume 3.

10. Sala, R.; Garcia-Garcia, E.; Tamba, R. Archaeological Geophysics—From Basics to New Perspectives. In *Archaeology, New Approaches in Theory and Techniques*; InTech: Rijeka, Croatia, 2012; pp. 133–166. Available online: https://www.intechopen.com/books/archaeology-new-approaches-in-theory-and-techniques/archaeological-geophysics-from-basics-to-new-perspectives (accessed on 10 October 2017).

11. Canti, M.G.; Meddens, F.M. Mechanical Coring as an Aid to Archaeological Projects. *J. Field Archaeol.* **1998**, *25*, 97–105, doi:10.1179/jfa.1998.25.1.97.

12. Garcia-Garcia, E.; Agirre-Mauleon, J.; Andrews, J.; Aranburu, A.; Arrazola, H.; Etxegoien, J.; Fuldáin, J.; Hill, J.; Iriarte, E.; Legorburu, M.; et al. Geoarchaeological core prospection investigation to improve the archaeological interpretation of geophysical data: Case study of a Roman settlement at Auritz (Navarre). *Archaeol. Polona* **2015**, *53*, 88–91.

13. Carey, C.; Howard, A.J.; Jackson, R.; Brown, A. Using geoarchaeological deposit modelling as a framework for archaeological evaluation and mitigation in alluvial environments. *J. Archaeol. Sci.: Rep.* **2017**, *11*, 658–673.

14. Revelles, J.; Cho, S.; Iriarte, E.; Burjachs, F.; van Geel, B.; Palomo, A.; Piqué, R.; Peña-Chocarro, L.; Terradas, X. Mid-Holocene vegetation history and Neolithic land-use in the Lake Banyoles area (Girona, Spain). *Palaeogeogr. Palaeoclimatol. Palaeoecol.* **2015**, *435*, 70–85.

15. Horlings, R.L. Archaeological Microsampling by Means of Sediment Coring at Submerged Sites. *Geoarchaeology* **2013**, *28*, 308–315.

16. Neal, C. The Potential of Integrated Urban Deposit Modelling as a Cultural Heritage Planning Tool. *Plan. Pract. Res.* **2014**, *29*, 256–267, doi:10.1080/02697459.2014.929839.

17. Chapman, H.; Adcock, J.; Gater, J. An approach to mapping buried prehistoric palaeosols of the Atlantic seaboard in Northwest Europe using GPR, geoarchaeology and GIS and the implications for heritage management. *J. Archaeol. Sci.* **2009**, *36*, 2308–2313.

18. Ellwood, B.B.; Harrold, F.B.; Marks, A.E. Site Identification and Correlation Using Geoarchaeological Methods at the Cabeço do Porto Marinho (CPM) Locality, Rio Maior, Portugal. *J. Archaeol. Sci.* **1994**, *21*, 779–784.

19. Cuenca-Garcia, C.; Sarris, A.; Makarona, C.; Hafez, I. Integrated Geophysical and In-situ Soil Geochemical Survey at Dromolaxia-Vizakia (Hala Sultan Tekke, Cyprus). *Archaeol. Polona* **2015**, *53*, 72–75.

20. Nowaczinski, E.; Schukraft, G.; Hecht, S.; Rassmann, K.; Bubenzer, O.; Eitel, B. A Multimethodological Approach for the Investigation of Archaeological Ditches -Exemplified by the Early Bronze Age Settlement of Fidvár Near Vráble (Slovakia). *Archaeol. Prospect.* **2012**, *19*, 281–295.

21. Schneidhofer, P.; Nau, E.; Leigh McGraw, J.; Tonning, C.; Draganits, E.; Gustavsen, L.; Trinks, I.; Filzwieser, R.; Aldrian, L.; Gansum, T.; et al. Geoarchaeological evaluation of ground penetrating radar and magnetometry surveys at the Iron Age burial mound Rom in Norway. *Archaeol. Prospect.* **2017**. Available online: http://onlinelibrary.wiley.com/doi/10.1002/arp.1579/full (accessed on 10 October 2017).

22. Agirre-Mauleon, J.; Txoperena, J.M.; Puldain, J. *Proyecto de Prospección Arqueológica en Los Términos Municipales de Luzaide-Valcarlos, Orreaga-Roncesvalles, Auritz-Burguete, Erroibar-Valle de Erro y Artzibar-Valle de Arce*; Unpublished Technical Report, 2012.

23. Barnolas, A.; Pujalte, V. La Cordillera Pirenaica: Definición, límites y división. In *Geología de España*; Sociedad Geológica de España-Instituto Geológico y Minero de España: Madrid, Spain, 2004; pp. 231–338.

24. Gibbons, W.; Moreno, T. (Eds.) *The Geology of Spain*; Geology of Series, Geological Society of London: London, UK, 2002.

25. Geological Map of Navarre. Government of Navarre 1997. Electronic Resource. Available online: http://geologia.navarra.es/ (accessed on 9 October 2017).

26. Garcia-Garcia, E.; Mtz. Txoperena, J.M.; Sala, R.; Aranburu, A.; Agirre-Mauleon, J. Magnetometer Survey at the Newly-discovered Roman City of Auritz/Burguete (Navarre). Results and Preliminary Archaeological Interpretation. *Archaeol. Prospect.* **2016**, *23*, 243–256.

27. Andrews, J.; Hill, J.; Nicholls, M.; Rauxloh, P. *Iturissa Roman Town (Navarra). Geo–archaeological Evaluation Report*; Unpublished Technical Report, 2014.

28. Jones, A.; Tucker, M.N.; Hart, J. The description and analysis of Quaternary stratigraphic field sections. In *Technical Guide, Number 7*; Quaternary Research Association: London, UK, 1999.

29. Tucker, M. *Sedimentary Rocks in the Field*; John Wiley and Sons: Chichester, UK, 1982.

30. Goodman, D.; Nishimura, Y.; Rogers, J.D. GPR time slices in archaeological prospection. *Archaeol. Prospect.* **1995**, *2*, 85–89.

31. Museum of London Archaeology Service. *Archaeological Site Manual*, 3rd ed.; Museum of London: London, UK, 1994.

32. Garcia-Garcia, E.; Agirre-Mauleon, J.; Aranburu, A.; Arrazola, H.; Hill, J.; Etxegoien, J.; Mtz. Txoperena, J.M.; Rauxloh, P.; Zubiria, R. The Roman settlement at Auritz (Navarre): Preliminary results of a multi-system approach to asses the functionality of a singular area. *Archaeol. Polona* **2015**, *53*, 92–94.

33. Garcia-Garcia, E. Geofisika Tekniken Karakterizazioa Euskal Herriko Antzinateko Aztarnategi Arkeologikoetan. Ph.D. Thesis, Euskal Herriko Unibertsitatea (EHU/UPV), Leioa, Basque Country, 2017.

34. Rapp, G.R.; Hill, C.L. *Geoarchaeology: The Earth-Science Approach to Archaeological Interpretation*, 2nd ed.; Yale University: New Haven, CT, USA, 2006.

geosciences

MDPI

Article

Comparison Study to the Use of Geophysical Methods at Archaeological Sites Observed by Various Remote Sensing Techniques in the Czech Republic

Roman Křivánek

Institute of Archaeology of the Czech Academy of Sciences, Prague, v.v.i., Department of Archaeology of Landscape and Archaeobiology, Letenská 4, 118 01 Prague 1, Czech Republic; krivanek@arup.cas.cz; Tel.: +420-2-5701-4033

Received: 4 August 2017; Accepted: 1 September 2017; Published: 7 September 2017

Abstract: A combination of geophysical methods could be very a useful and a practical way of verifying the origin and precise localisation of archaeological situations identified by different remote sensing techniques. The results of different methods (and scales) of monitoring these fully non-destructive methods provide distinct data and often complement each other. The presented examples of combinations of these methods/techniques in this study (aerial survey, LIDAR-ALS and surface magnetometer or resistivity survey) could provide information on some specifics and may also be limitations in surveying different archaeological terrains, types of archaeological situations and activities. The archaeological site in this contribution is considered to be a material of this study. In case of Neolithic ditch enclosure near Kolín were compared aerial prospection data, magnetometer survey and aerial photo-documentation of excavated site. In the case of hillforts near Levousy we compared LIDAR data with aerial photography and large-scale magnetometer survey. In the case of the medieval castle Liběhrad we compared LIDAR data with geoelectric resistivity measurement. In case of a burial mound cemetery we combined LIDAR data with magnetometer survey. In the case of the production area near Rynartice we combined LIDAR data with magnetometer and resistivity measurements and result of archaeological excavation. Fortunately for successful combination of geophysical and remote sensing results, their conditions and factors for efficient use in archaeology are not the same. On the other hand, the quality and state of many prehistoric, early medieval, medieval and also modern archaeological sites is rapidly changing over time and both groups of techniques represent important support for their comprehensive and precise documentation and protection.

Keywords: archaeological prospection; remote sensing; data integration; condition assessment

1. Introduction

Remote sensing techniques represent the quickest and fully non-destructive source of information on the scale of whole archaeological sites and/or particular segments of cultural (archaeological) landscapes. Their main advantages are new extensive identification of sites and quick and long-time possibility of passive/active collecting of various data. For archaeological heritage is also very important the progressive documentation and qualitative monitoring of condition changes (including various hazard risks) of different archaeological situation or landscapes. Aerial survey and later LIDAR have been for a few decades the most intensively applied RS methods with many published results [1–4]. Current increasing importance of wider RS techniques in archaeology confirms more about this issue oriented publications [5–11]. Results of RS data were also in some conferences compared with results of ground non-invasive prospection methods [12,13]. From the beginning of their application in archaeology geophysical methods in the sphere of archaeological prospection have had different goals, scale or accuracy based on the surface monitoring of changes in physical properties of subsurface layers [14,15]. From the point of view of the methodology of measurement (use of terrestrial sensors), geophysical methods are sometimes included in a broader group of various remote sensing techniques. However, from a practical point of view of surveyed responses of subsurface physical changes, geophysical prospection in archaeology can offer 2D (or also 3D) information about archaeological sites and especially about individual archaeological features, structures and layers [16,17]. Therefore from an archaeological point of view, the specific application of the geophysical method in archaeology is more often ranked among archaeological prospection methods, methods of non-destructive (or non-invasive) archaeology, methods of field archaeology or near-surface geophysical methods. All of these views are based on the use of geophysical methods in archaeology and are justified.

In the Czech Republic, various remote sensing techniques have been applied in archaeology up until now in varying intensities and scales of use. Aerial photography and LIDAR (respectively ALS —airborne laser scanning) have been the most intensively used results of RS by archaeologists for a longer period of time [18–23]. In recent years, some interdisciplinary projects and works have extended these data to include multispectral data and drone surveys results [24]. Comparison of potential of chosen satellite images (from satellite systems IKONOS, QuickBird and CORONA purchased through a Czech company ArcData to Dept. of Archaeology at the University of West Bohemia in Pilsen) with aerial archaeological survey photographs has been done in four selected regions [25]. The subsequent cooperation of archaeologists specialised in aerial archaeology and geophysicists working in archaeology also had a rather long tradition with many projects, outputs and compared results. The examples presented in this paper should demonstrate different ways of subsequent application of remote sensing techniques and surface geophysical methods, their interdependence and differences in use for archaeology.

2. Materials and Methods

As study material of this contribution, we use surface and subsurface relics of selected archaeological situations and sites. Five chosen examples compare results from different types of archaeological sites (enclosure, fortified area, medieval settlement, funeral and production area), but there are also examples from sites with other characteristic human activities (for example mining area, communication, military area or polyfunctional site).

As study methods, we use the complexity of applying various remote sensing techniques and geophysical methods and, above all, comparison of their results. Suitable methods for the geophysical survey of any archaeological situation are closely related to the current field conditions of measurement. For years, magnetometer surveys have been the main and most intensively applied geophysical method in Bohemian archaeology. Two types of magnetometers have been used over the past two decades. Since 1998, a gradient variant of the caesium vapour magnetometer Smartmag SM-4g (Scintrex Ltd., Concord, ON, Canada) was performed, with an approximate network of 1.0×0.25 m (Kolín, Rynartice), with some details in a 0.5×0.2 m density (Údraž). Since 2010, a five-channel Magneto-Arch magnetometer system with FMG-650B (Sensys GmbH, Bad Saarow, Germany) fluxgate gradiometers has been used with a data density of 0.5×0.2 m (Levousy) or 0.25×0.1 m. This is the most powerful method for the identification of various sunken features or burned situations on arable fields, meadows or pastures, but on a much smaller scale we can also use it also in some forested terrains. Magnetic susceptibility measurement in open archaeological situations (in situ) is also used in archaeological excavations. Geoelectric resistivity measurement as a second geophysical method also has a long tradition of use in Bohemian archaeology. Since the beginning of the 21st century, the RM-15 instrument (Geoscan Research, Bradford, UK) has been used for geoelectrical resistivity surveys primarily in a Wenner or Schlumberger configuration and a common data density of 1×1 m (Liběhrad), some details in 0.5×0.5 m (Rynartice). This method was applied in various agricultural and also forested terrains for the identification of various remains of features with an expected stone construction and their destructions. The results of these geophysical methods are presented in this paper. However, in some other specific cases we combine their results with other geophysical methods: electromagnetic or radar measurement (detection of stone structures or unfilled areas beneath pavements, tiles or floors) or thermometry (unfilled areas inside sacred architecture or buildings). Geophysical data were in paper compared with aerial photographs (from the archive of the Institute of Archaeology, CAS, Prague, v.v.i.) or Digital Terrain Model of the 5th generation (DMR 5G provided as fully public service from the Czech Office for Surveying, Mapping and Cadastre [26]. In this study, were used oblique aerial photographs of M. Gojda, but also in partial cases some orto-photographs from fully public web services (at present time five different years of public aerial photography are on public web server [27]). LIDAR-ALS data were also public and in case of filtered Digital Terrain Model of the 5th generation (DMR 5G) they represents image of the natural or man-modelled terrain with total mean error of the height 0.18 m in an open terrain or 0.3 m in a forested terrain (previous DMR 4G had total mean error 0.4 m in an open terrain and 1 m in a forested terrain which was insufficient for archaeological purposes).

3. Results

3.1. Ditch Enclosures

Neolithic roundels, Eneolithic interrupted ditch enclosures, Bronze Age oval ditch enclosures and La Tène quadrangular enclosures are common types of different ditch enclosures or enclosed systems found in the Czech landscape. Most of them have been identified on agricultural land thanks to characteristic crop marks from aerial photographs, and only a few of these enclosures can be distinguished in forested areas from LIDAR data. Long-term or deep ploughing of fields did not allow the preservation of sufficiently distinguished anthropogenic relief formations (low chance for identification in LIDAR). However, possibilities of aerial prospection are also highly variable depending on soil pedology and the geology of ploughed fields. For example, Neolithic roundels are very often localised on sloped (eroded) terrains with varying layer of loess. For aerial prospection, these conditions are not optimal for the identification of sunken features such as ditch enclosures, but are very suitable for successful geophysical (magnetometer) prospection [28]. In the case of the Neolithic roundel (1) near Kolín (Kolín district), preliminary magnetometer survey helped precisely before the start of the rescue archaeological excavation of the Kolín bypass road. Remains of the Neolithic roundel with a diameter of over 200 m were identified on three fields around the junction of present roads (Figure 1b). The results were locally disturbed by different electric power lines and also subsurface cable and metallic gas pipe lines (high magnetic violet-blue lines in Figure 1b). The roundel probably originally had four entrances, and subsequent archaeological excavations in the corridor of the bypass confirmed three circular ditches and a fourth unfinished ditch [29,30]. The Neolithic roundel was never identified on loess from aerial photographs (see Figure 1a), and the aerial documentation of uncovered parts of the enclosure was carried out during rescue excavations (see Figure 1c–e). Additional magnetic susceptibility measurements in situ on vertical profiles also documented changes in the fills of open ditches-more about fills in [31]. Another measurement of magnetic susceptibility observed the superposition of the end of the outer unfinished ditch of the roundel with a Neolithic long house (different phases of Neolithic settlement and activities at the site). Result of magnetometer survey in this case helped (without any previous successful RS data) to manage efficient rescue archaeological excavation of endangered parts of the one of the largest bohemian Neolithic roundels.

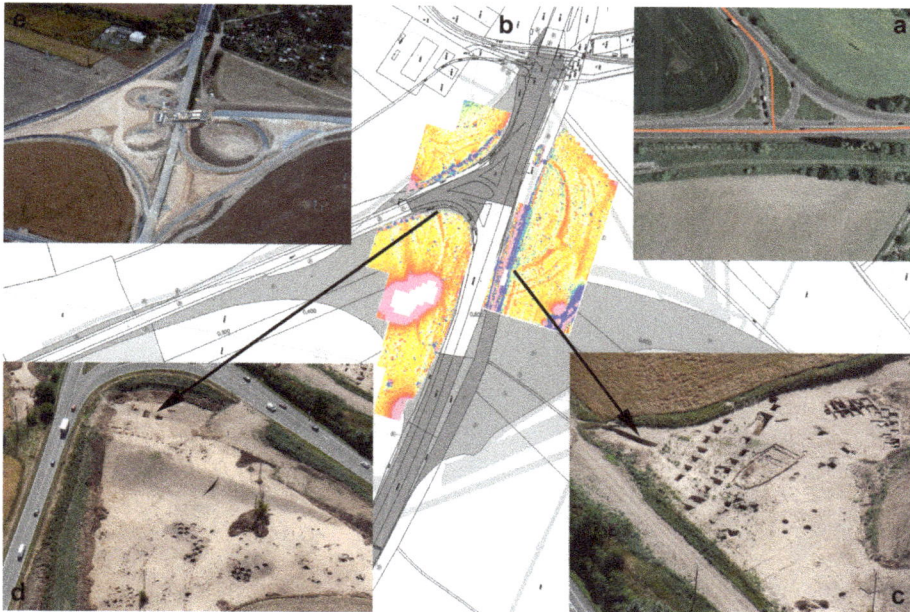

Figure 1. Kolín, district Kolín–Neolithic roundel. The combination of results of (**a**) aerial photography before excavation, (**b**) results of magnetometer survey in place of planned corridor of bypass road and (**c**–**e**) aerial photodocumentation of uncovered remains of enclosure and other settlement (source: archive of the Institute of Archaeology, CAS, Prague, v.v.i.–photo: M. Gojda, surveyed area: approx. 2.15 ha, geophysical survey: Křivánek 2008).

3.2. Fortified Settlement Areas

Prehistoric and early medieval hillforts were the most common types of fortified and often internally structured settlements in Bohemia (until the 13th century AD). Their positions in the landscape were targeted at strategic places, on promontories, hilltops, near important communications, rivers, sources of raw materials, etc. Aerial photography is very intensively used for the overall documentation of hillforts situated in agricultural areas, whereas LIDAR is often more helpful in forested areas. Preserved remains of fortification systems can be identified on the surface from identified linear elevations or depression, while some subsurface remains of destroyed fortifications can be observed from vegetation (linear crop marks) or soil changes. The application of geophysical prospection then offers much more detailed information about the real extent, structure, entrances, internal settlement and other activities at hillforts [32,33]. Magnetometer measurements are often focused on the verification of large-scale unexcavated areas of sites or on the verification of particular new situations from aerial photographs or field artefact collections. In some specific areas (such as gates, parts of ramparts, paths, places with an expected stone construction), these data then add particular geoelectric resistivity measurements. The final interpretation of geophysical data then depends on more regionally and/or locally specific limits (variability of geology, pedology, level of soil erosion or modern-recent disturbances of measurements, preservation of the original terrain, etc.), also including the often repeated use of the same site in different periods. In the case of the Late Bronze Age and early medieval hillfort near Levousy (Litoměřice district), a magnetometer survey of all accessible terrains of fields includes sunken or locally burned features with highly varied dating of these activities. LIDAR images of the fortified site show how the strategic claystone promontory above the Ohře River was chosen in different periods (Figure 2a). Archaeological excavation of inner ramparts by trenches confirmed the first settlement in the Neolithic and Eneolithic periods, while the first fortification of part of the site is expected in the Late Bronze Age; settlement continuation in the Iron Age has been confirmed. The Slavic hillfort (9th and 10th century AD) was enlarged to an area of over 12 ha of the fortified site [34]. However, the promontory also played an important military function during the Austro-Prussian War in the second half of the 19th century. Results of a magnetometer survey then include magnetic anomalies (sunken features) from various prehistoric periods, remains of the ploughed-out perimeter rampart, remains of internal divisions (baileys) of the early medieval hillfort, but also later lines of military polygons (the remains of two "reduta" fortifications) from the 19th century and also local magnetic disturbances from modern agricultural and orchard activities at the site (see Figure 2c). Military remains and different types of settlement were also confirmed by aerial photographs (straight lines in Figure 2b). Additional local resistivity survey confirmed the ploughed-out stone construction of destroyed parts of the perimeter rampart. Result of a large-scale magnetometer survey in this case could be (with results of aerial surveys) used as a new level of information for evidence and protection of immovable archaeological monument (hillfort with repeated prehistoric and early medieval settlement and modern military use).

3.3. Abandoned Medieval Sites

The Bohemian countryside today includes many ruins of abandoned medieval strongholds, villages and small castles. These settlement areas and places of lower regional nobility were later abandoned, destroyed, forgotten or deeply changed by the modern landscape and land-use changes. Their remains in agricultural areas are in a very poor condition and often without any possibility to identify regular shape and extent of origin medieval structures. The situation seems to be somewhat better in forested areas, where better surface and subsurface remains with stone components can be found. Details in LIDAR data is also used for the preliminary identification of these remains in often forested and not easily accessible areas (Figure 3a,b). Their precision is highly variable and is connected with the present state of the original terrain of sites, with qualitative changes in the landscape and also with recent timber harvesting and new afforestation. Systematic surface geodetic and archaeological documentation is then important for more precise evidence of these sites. Geophysical survey could also help in some cases of preserved clear situations and landscapes. This was also the case of the ruins of small Liběhrad Castle near Libčice nad Vltavou (Praha-západ district) abandoned in the 16th century; a medieval village known from written sources from the end of 10th century with later Zbraslav Monastery property during the 13th/14th century [35]. Modern landscape changes meant that geodetic-archaeological surface documentation and application of geophysical methods were very limited. The northern part of the original castle was partly destroyed by the construction of the railway, and the state of the remodelled and forested landscape within the electro-magnetic disturbances of the electrified railway line did not facilitate the efficient use of magnetic or electromagnetic methods. Newerthless, the results of a geoelectric resistivity survey of the accessible inner part of castle identified some subsurface remains of internal castle settlement structures (Figure 3c,d). High resistivity anomalies confirmed remains of the collapsed perimeter wall on the rampart and other stone destructions (blue areas in Figure 3d). Low resistivity anomalies indicated the possible destruction of another structured settlement (sunken features, platforms) inside (yellow areas in Figure 3d). A detailed geodetic survey of the surface combined with geophysical measurement showed an efficient way for the detailed documentation of evidence and the protection of a highly remodelled archaeological site. Result of limited resistivity survey (combined with LIDAR-ALS data new geodetic plan of site) contributed to more detailed non-destructive information about partly damaged medieval castle.

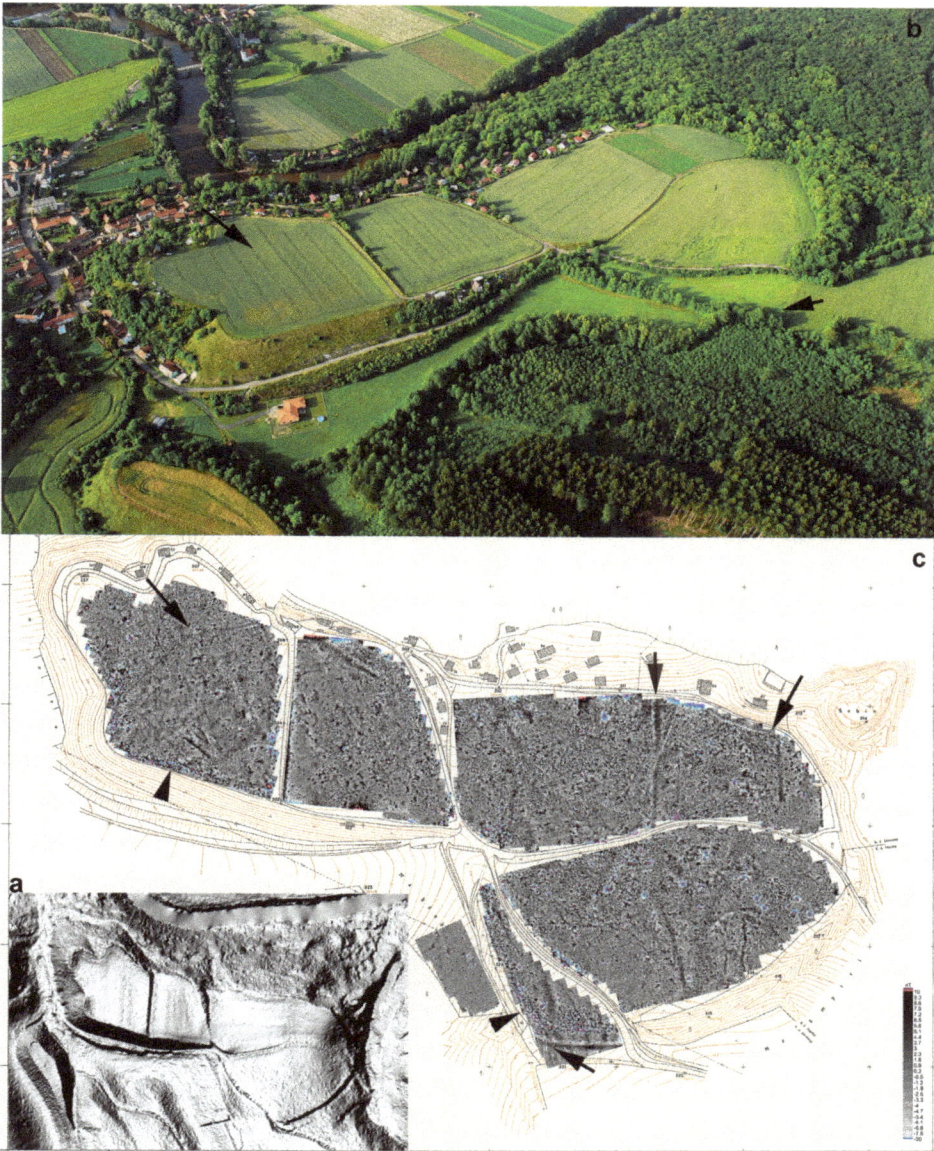

Figure 2. Levousy, district Litoměřice–prehistoric and early medieval hillfort. The combination of results of (**a**) LIDAR–DMR 5G in shaded relief (left bottom corner), (**b**) aerial photography and (**c**) large-scale magnetometer prospection of site including linear remains of ploughed out rampart (short triangular arrows), ditch fortifications and modern military activities (longer arrows) from 19th century (source: archive of the Institute of Archaeology, CAS, Prague, v.v.i.–photo: M. Gojda and www.geoportal.cuzk.cz-Copyright © ČÚZK, surveyed area: approx. 9.8 ha, geophysical survey: Křivánek 2015).

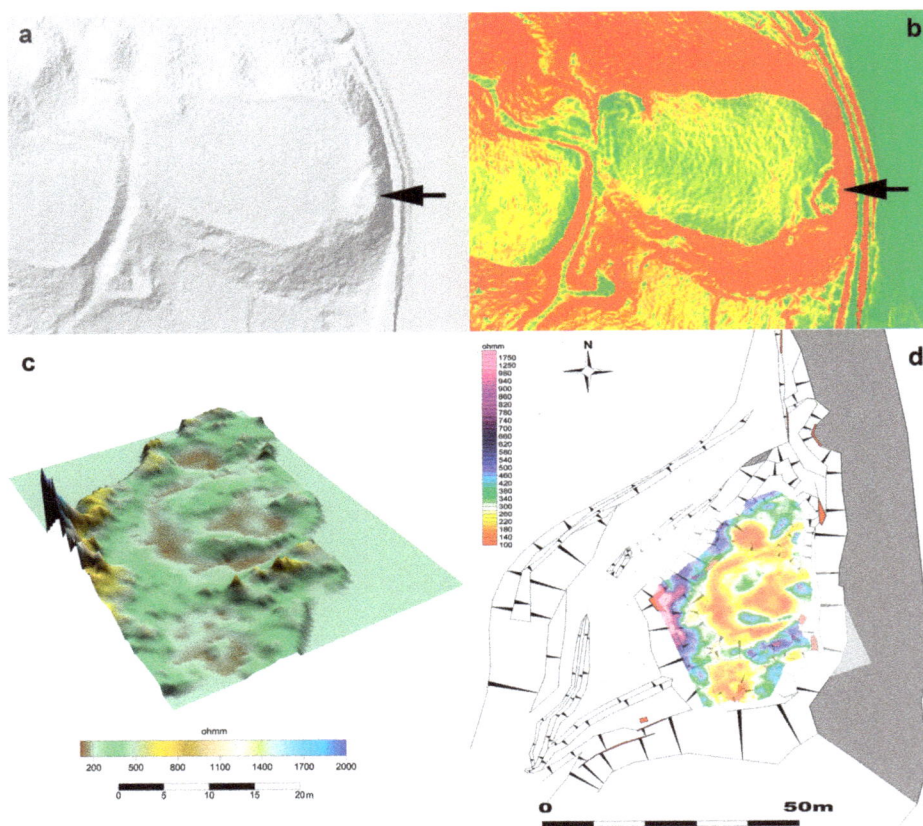

Figure 3. Libčice n. V., district Praha-západ–medieval Liběhrad Castle. The combination of results of (**a**) LIDAR–DMR 5G in shaded relief and (**b**) inclination of slopes (arrows points place of castle), (**c**) 3D-presentation of resistivity measurement results and (**d**) geodetic plan of remains of abandoned castle with resistivity data (source: [35] (Figures 15 and 22) and www.geoportal.cuzk.cz-Copyright © ČÚZK, surveyed area: approx. 1100 m², geophysical survey: Křivánek 2006–2007).

3.4. Funeral Sites

Prehistoric or early medieval burials and barrow cemeteries are a very typical type of archaeological sites, especially in forested parts of south and west Bohemia. Most of these remains with funeral activity have been looted, and some barrow cemeteries were also excavated by archaeologists very early (beginning in the 19th century), some on agricultural land were ploughed-out, and others in forests were damaged or remodelled by repeated afforestation. However, due to characteristic elevations, many of them were (or their remains) inside forests can still be safely distinguished by LIDAR data, and it is possible to separate groups of burial mounds, different shapes of burial cemeteries and/or the different sizes of individual burial mounds. As a second step, we can then apply more surface archaeological methods, including geophysical techniques. Magnetometer measurements are often combined with geoelectric resistivity measurements, but sometimes also with electromagnetic and even GPR measurement. The efficiency and choice of suitable geophysical methods in forested areas depends greatly on the type and material of burial mounds, the type of landscape, geology or the funeral activity. In the case of the Bronze Age barrow cemetery near Údraž (Písek district), a magnetometer survey was aimed at verifying potential outer flat graves and observing the state of the subsurface preservation of a chosen group of three funeral features (Figure 4b). The result confirmed the anticipated coincidence of three elevations, from LIDAR (Figure 4a) or surface geodetic documentation [36,37], with magnetic anomalies of barrows (the slightly magnetic layers of burial mounds) and a few another anomalies in the flat terrain, possible sunken feature outside of barrows [38–40] (red small areas out of barrows in Figure 4b). Results from the magnetometer most likely showed the not fully oval shape of burial mounds. The varied character of magnetic anomalies over individual barrows could indicate the different state of subsurface preservation of barrows and burials inside. In the case of two barrows, interruptions are visible inside features (probably remains of old excavations or illegal looting with a metal detector); in the case of the third barrow, it seems to be the central place without any disturbed or relocated material. Result of a partial magnetometer survey (together with LIDAR-ALS data and without successful aerial prospection-forest) helped to verify the state and extent of partially excavated prehistoric burial mound cemetery.

3.5. Production Areas

Abandoned medieval or modern archaeological sites with specific production activity (glass-works, charcoal pens, pitch-production) are typical for the more mountainous border regions around Bohemia. In the case of the Czech-Saxon Switzerland National Park in north Bohemia, the remains of more pitch-production centres were identified during field surveys in forested areas [41]. Due to the typical quite high waste heaps from this production, we could identify also some of these elevation details from LIDAR (Figure 5a). However, due to highly variable and rugged terrain with typical sandstone formations and valleys, identification is quite complicated and limited by the character of the local terrain. More suitable geophysical methods were used to distinguish between smaller production features such as pitch-furnaces and the scope of waste materials. The combination of detailed field surface survey with magnetometer prospection seemed to be best way to identifying the remains of abandoned medieval or Early Modern production sites in intensively forested regions. In the case of the medieval pitch-production centre near Rynartice (Děčín district), a magnetometer survey helped distinguish the most highly magnetic parts of the terrain (Figure 5b) between two elevations formed by massive layers of burned ashes (violet-blue high magnetic anomalies between magnetic red areas in Figure 5c). This result provided a better and much more efficient verification of the site by archaeological excavation [37,39]. Subsequent archaeological trenches [42] uncovered the remains of a furnace from the 14th/15th century for pitch distillation (Figure 5e,f). The original source of the highly magnetic anomaly was shown to be a combination of heavily burned clay materials and also partly the stone construction of the furnace with neovolcanic material [43]. The results of the excavation with the uncovered pitch-furnace were also supplemented with another geoelectric resistivity measurement combined with magnetic data (Figure 5d). A survey of the wider area confirmed the presence of more pitch-production activities—the wider extent of production near a water source with a partly preserved path. Result of combined magnetometer and resistivity survey in this case could (with result of archaeological excavation, partial result of LIDAR-ALS data and without possible results of aerial survey) illustrate efficient combination of methods in the study of similar forested and separate medieval/modern production areas.

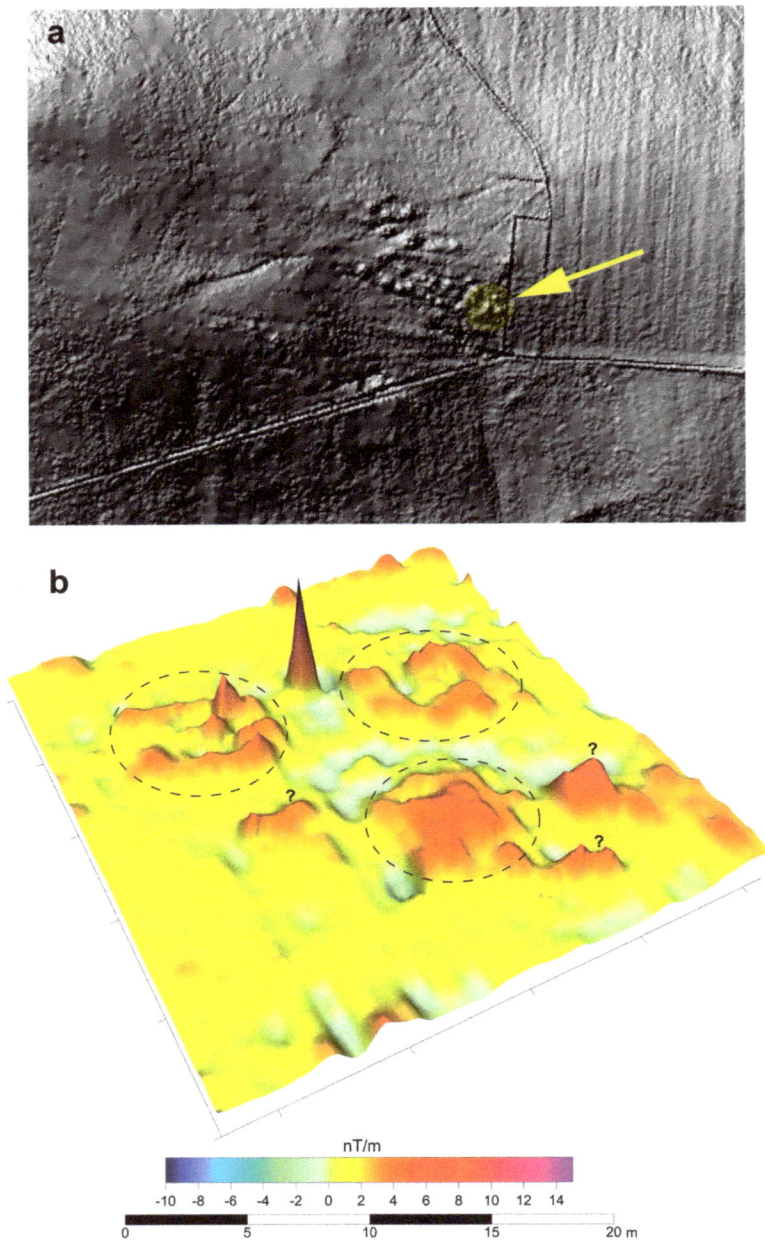

Figure 4. Údraž, district Písek–Bronze Age burial mound cemetery. The combination of (**a**) LIDAR–DMR 5G documentation of the whole site (yellow arrow points group of verified barrows) and (**b**) detail of 3D-presentation of magnetometer measurement result over three barrows (source: [38] (Figure 2) and www.geoportal.cuzk.cz-Copyright © ČÚZK, surveyed area: 575 m², geophysical survey: Křivánek 2006).

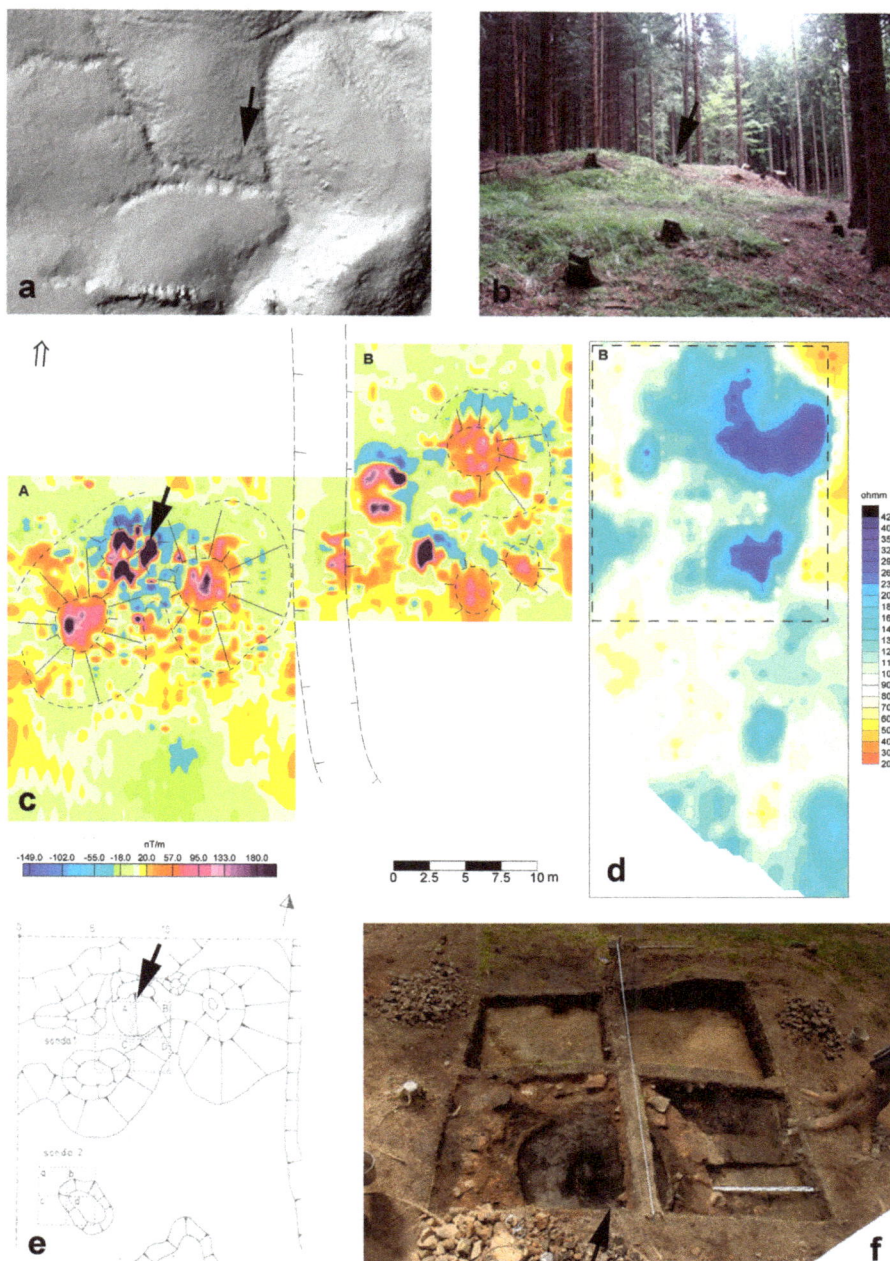

Figure 5. Rynartice, district Děčín–medieval pitch production area. The combination of results of
(**a**) LIDAR–DMR 5G, (**b**) photodocumentation of site before excavation, (**c**) result of magnetometer
survey of areas A and B, (**d**) results of resistivity surveys of area B, (**e**) plan with archaeological trenches
and (**f**) final uncovered medieval pitch-furnace (arrow points place of pitch-furnace; source: [40,42]
and www.geoportal.cuzk.cz-Copyright © ČÚZK, surveyed area: 0.15 + 0.1 ha, geophysical survey:
Křivánek 2004–2005).

4. Discussion

In this paper, five examples from various types of archaeological sites document different possibilities of the chosen remote sensing techniques and applied geophysical methods. Their information is irreplaceable and their subsequent cooperation is very often necessary and leads to the more efficient study of archaeological sites. However, from the view of specific and different principles of collecting data we can see also some specific advantages and disadvantages of particular methods and techniques. The quality and level of results of remote sensing techniques is very often dependent on suitable time and the conditions of application of airborne or space-borne sensors. For example, quality of data from passive satellite systems is limited by the type of low (CORONA), high (LANDSAT) or very high (IKONOS) spatial resolution. Quality of active ALS-LIDAR is for example very connected with density of data collection but also with subsequent processing, classification, filtration or interpolation of origin data. Active aerial archaeology is then for example very dependent on actual conditions of monitored areas (existence of crop/soil/snow/shadow-marks) and height above the terrain. However, success of all of passive and active remote sensing methods is closely related to other field conditions like terrain surface variability or level of impact of land-use changes. Another very limiting factor of many RS techniques, except the LIDAR-ALS, is vegetation and mainly intensity of afforestation of terrain. A big advantage of all RS techniques is then the possibility of use of changing data in longer horizon of collecting data. These sets of repeated data could illustrate how fast and important are modern changes of archaeological sites and terrains. Their most efficient use is on the scale of whole archaeological sites or the cultural (archaeological) landscape, where we can observe surface changes of individual sites in time as spatial contexts of different archaeological sites and wider areas. Collected data from RS techniques play an important role in protection and recording of immovable archaeological heritage, but also in other archaeological disciplines (like for example regional history, settlement dynamics, networks or environmental archaeology). Disadvantages of successful application of geophysical method in archaeology are very often connected with the actual state of preservation of physically distinguishable archaeological situations beneath the soil. The results of geophysical measurements are always dependent on the state of the archaeological site and field conditions, the quantity and intensity of land-use changes, landscape changes, pedology and geology of studied area. In very variable conditions of archaeological sites we cannot always use the same geophysical methods for identification of similar remains of subsurface situations or activities. For example somewhere we can find clear results from aerial photographs with less readable results from magnetometer measurements (sand-gravel sediments and terraces), somewhere else we can see clear results from magnetometer surveys without clearly visible features from aerial prospection (loess or clay sediments). Regional and local pedological and geological conditions play a very important role in archaeo-geophysical prospection. Possibilities of geophysical prospection by various methods are then also very dependent on vegetation and/or local disturbances for different methods. Their most efficient use is on the scale of particular archaeological sites or their parts (only some newly extensively and intensively applied geophysical techniques offer monitoring of some parts of the archaeological landscape). The main advance of geophysical methods is in the possibility of identifying or distinguishing the subsurface remains of smaller individual features and archaeological situations. Various geophysical methods (in the field, but sometimes also in the laboratory) can be applied at different scales of new archaeo-geophysical information. Some geophysical measurements we can use before archaeological verification of the site, others during archaeological investigation and another also after finished particular destructive excavation of site. However, from the perspective of archaeological sites and landscape protection, we can see a very close interdependence between these methods. Neverthless, an effective combination of these methods cannot be done without the support of further information and data (e.g., from the fields of historical cartography, geography, geospatial data, geochemistry, etc.).

5. Conclusions

Valid for all non-destructive geophysical methods and remote sensing techniques is that agricultural activities such as the ploughing of fields, afforestation, mining or building activities play a very important role in the real preservation of surface and subsurface archaeological layers. Together with subsequent erosion or irreversible landscape changes they play also a very important role in the real preservation of surface and subsurface archaeological layers, preservation and possibilities for monitoring archaeological sites. From all of combined data it is clear that quality (and in some regions also quantity) and state of archaeological sites is rapidly, and somewhere also dramatically, changing over short time. Archaeology together within cultural heritage management will need more complex and precise documentation of endangered sites and landscapes.

Acknowledgments: All sources of funding of the study should be disclosed. Please clearly indicate grants that you have received in support of your research work. Research work has been supported mainly during research project of the Institute of Archaeology in Prague AV0Z80020508; some results were produced also during the project of Regional cooperation "Non-destructive geophysical surveys of significant and endangered archaeological sites in the Ústí Region, Czech Republic" R300021421 (CAS, Křivánek 2014–2016). Clearly state if you received funds for covering the costs to publish in open access.

Conflicts of Interest: The author declares no conflict of interest. The founding sponsors had no role in the design of the study; in the collection, analyses, or interpretation of data; in the writing of the manuscript, and in the decision to publish the results.

References

1. Rackowski, W. *Archeologia Lotnicza—Metoda Wobec Teorii (Aerial Archaeology—Method in the Face of Theory)*; Wydawnictwo Naukowe UAM: Warszawa, Poland, 2002; ISBN 83-232-1194-9. (In Polish)
2. Doneus, M. *Die Hinterlassene Landschaft—Prospektion und Interpretation in der Landschaftsarchäologie*; Mitteilungen der Prähistorischen Kommission der Österreichischen Akademie der Wissenschaften; Verlag der Österreichischen Akademie der Wissenschaften: Wien, Austria, 2013; ISBN 978-3-7001-7197-3. (In German)
3. Bewley, R.H.; Crutchley, C.A.; Shell, C.A. New light on an Ancient Landscape: Lidar Survey in the Stonehenge World Heritage Site. *Antiquity* **2005**, *79*, 636–647. [CrossRef]
4. Devereux, B.J. The Potential of Airborne Lidar for Detection of Archaeological Features under Woodland Canopies. *Antiquity* **2005**, *79*, 648–660.
5. *Remote Sensing in Archaeology: An Explicitly North American Perspective*; Johnson, J.K., Ed.; University of Alabama Press: Tuscaloosa, AL, USA, 2006; ISBN 978-0-8173-5343-8.
6. *Remote Sensing in Archaeology*; Wieseman, J.R., El-Baz, F., Eds.; Springer: New York, NY, USA, 2007; ISBN 978-038-7444-4536.
7. *Remote Sensing for Archaeology and Cultural Heritage Management*; Lasaponara, R., Masini, N., Eds.; Aracne: Rome, Italy, 2008; ISBN 978-88-548-2030-2.
8. Parcak, S.H. *Satellite Remote Sensing for Archaeology*; Routledge: London, UK, 2009; ISBN 978-0415448789.
9. *Remote Sensing for Archaeological Heritage Management*; EAC Occasional Paper No. 5; Cowley, D., Ed.; Archaeolingua: Budapest, Hungary, 2011; ISBN 978-963-9911-20-8.
10. Cower, D.C.; Harrower, M.J. *Mapping Archaeological Landscapes from Space*; Springer: New York, NY, USA, 2013; ISBN 978-1-4614-60732.
11. *Sensing the Past. Contribution from the ArcLand Conference on Remote Sensing for Archaeology*; Posluschny, A., Ed.; Habelt-Verlag for ArcLand: Bonn, Germany, 2015; ISBN 978-3774939640.
12. *Interpreting Archaeological Topography: Airborne Laser Scanning, 3D Data and Ground Observation*; Occasional Publication of AARG No. 5; Opitz, R.S., Cowley, D., Eds.; Oxbow Books: Oxford, UK, 2013; ISBN 978-1842175163.
13. *A Sense of the Past: Studies in Current Archaeological Applications of Remote Sensing and Non-invasive Prospection Methods*; Kamermans, H., Gojda, M., Eds.; BAR. International Series; Archaeopress: Oxford, UK, 2014; ISBN 978-1-4073-1216-3.
14. *Archäologische Prospektion: Luftbildarchäologie und Geophysik*; Becker, H., Ed.; Arbeitshefte des Bayerischen Landesamtes für Denkmalpflege; Bayerisches Landesamt für Denkmalpflege: München, Germany, 1996; ISBN 3-87490-541-1. (In German)

15. Gater, J.; Gaffney, C. *Revealing the Buried Past: Geophysics for Archaeologists*; Tempus Publishing: Stroud, UK, 2003; ISBN 978-0-7524-2556-6.

16. Witten, A.J. *Handbook of Geophysics and Archaeology*; Equinox Publishing: London-Oakville, UK, 2006; ISBN 1-904768-59-8.

17. *Mittelneolithische Kreisgrabenanlagen in Niederösterreich: Geophysikalisch-archäologische Prospektion-ein interdisziplinäres Forschungsprojekt. Mitteilungen der Prähistorischen Kommission der Österreichischen Akademie der Wissenschaften*; Neubauer, W., Melichar, P., Eds.; Verlag der Österreichischen Akademie der Wissenschaften: Wien, Austria, 2010; ISBN 978-3-7001-6684-9. (In German)

18. *Ancient Landscape, Settlement Dynamics and Non-Destructive Archeology—Czech Research Project 1997-2002*, 1st ed.; Gojda, M., Ed.; Academia: Praha, Czech Republic, 2004; ISBN 978-80-200-1215-X.

19. Gojda, M. *Archeologie a Dálkový Průzkum: Historie, Metody, Prameny (Archaeology and Remote Sensing: History, Methods, Data)*; Academia: Praha, Czech Republic, 2017; ISBN 978-80-200-2644-6. (In Czech)

20. Křivánek, R. 2.3. Geophysical Prospection. New Perspectives for Settlement Studies in Bohemia. In *Ancient Landscape, Settlement Dynamics and Non-Destructive Archeology—Czech Research Project 1997-2002*, 1st ed.; Gojda, M., Ed.; Academia: Prague, Czech Republic, 2004; pp. 39–71, ISBN 978-80-200-2644-6.

21. Křivánek, R. Geofyzikální měření při ověřování výsledků leteckých průzkumů v severozápadních Čechách. In *Archeologické Výzkumy v Severozápadních Čechách v Letech 2003–2007 (Sborník k životnímu jubileu Zdeňka Smrže)*; Ústav Archeologické Památkové Péče Severozápadních Čech v Mostě: Most, Czech Republic, 2008; pp. 385–397, ISBN 978-80-86531-05-2.

22. Gojda, M.; John, J. *Archeologie a Letecké Laserové Skenování Krajiny (Archaeology and Airborne Laser Scanning of the Landscape)*; Západočeská universita v Plzni: Plzeň, Czech Republic, 2013; ISBN 978-80-261-0194-9. (In Czech)

23. Křivánek, R.; Gojda, M. Půdní a Geologické Podmínky při Leteckém a Geofyzikálním Průzkumu–Soil and Geological Conditions in Aerial and Geophysical Surveys. *Živá Archeol. (re)Konstr. Exp. Archeol.* **2010**, *11*, 92–95. (extended variant of paper on http://www.zivaarcheologie.cz/CasopisZivaArcheologieCisla.html (accessed on 9 August 2017)) (In Czech)

24. Chytráček, M.; Chvojka, O.; Egg, M.; John, J.; Kozáková, R.; Křivánek, R.; Kyselý, R.; Michálek, J.; Stránská, P. A Disturbed Late Hallstatt Period Princely Grave with a two-wheeled Chariot and Bronze Vessels in Sedlina Forest near Rovná in South Bohemia: A Preliminary Report. In *Archaeological Sites in Forests—Strategies for their Protrection, Inhalte–Projekte–Dokumentationen, Sriftenreihe des Bayerisches Landesamt fur Denkmalpflege*; Volk Verlag: Munchen, Germany, 2017; pp. 83–90, ISBN 978-3-86222-226-1.

25. Gojda, M.; John, J. Dálkový Archeologický Průzkum Starého Sídelního Území Čech: Konfrontace Výsledků Letecké Prospekce a Analýzy Družicových Dat. *Archeol. Rozhl.* **2009**, *61*, 467–492. (In Czech)

26. Český Úřad Zeměměřický a Katastrální (ČÚZK) (Czech Office for Surveying, Mapping and Cadastre). Available online: www.geoportal.cuzk.cz (accessed on 9 August 2017).

27. Public Web Server with Base Maps and Current Air Maps–Copyright © 1996–2017 Seznam.cz, a.s. Available online: https://mapy.cz (accessed on 2 July 2017).

28. Křivánek, R. Geophysical Surveys of Abandoned Quadrangular Enclosures ("Viereckschanzen") from La Tène Period in Bohemia. *ISAP News (The newsletter of the International Society for Archaeological Prospection)*. Issue 36. August 2013, pp. 5–7. Available online: http://www.archprospection.org/isapnews/isapnews-36 (accessed on 1 August 2013).

29. Šumberová, R.; Malyková, D.; Vepřeková-Žegklitzová, J.; Pecinovská, M. Sídlení Aglomerace v Prostoru Dnešního Kolína. Záchranný Výzkum v Trase Obchvatu Města (Settlement Aglomerations in Today's Kolín Area, Central Bohemia. Rescue Excavations along the Town by-pass Road). *Archeol. Rozhl.* **2010**, *62*, 661–679. (In Czech)

30. *Cesta Napříč Časem a Krajinou: Katalog k Výstavě Nálezů ze Záchranného Archeologického Výzkumu v Trase Obchvatu Kolína 2008-2010*; Šumberová, R., Ed.; Archeologický ústav AV ČR: Praha, Czech Republic, 2012; ISBN 978-80-87365-52-6. (In Czech)

31. Lisá, L.; Bajer, A.; Válek, D.; Květina, P.; Šumberová, R. Micromorphological Evidence of Neolithic Rondel-like Ditch Infillings: Case Studies from Těšetice-Kyjovice and Kolín, Czech Republic. *Interdiscip. Archaeol. Nat. Sci. Archaeol.* **2010**, *4*, 135–146.

32. Křivánek, R. Changes of Structure and Extent of Early Medieval Strongholds in Central Bohemia from Geophysical Surveys of Sites. Archaeological prospection. In Proceedings of the 10th International Conference

on Archaeological Prospection, Vienna, Austria, 29 May–2 June 2013; Neubauer, W., Trinks, I., Salisbury, R.B., Einwögerer, C.H., Eds.; Austrian Academy of Sciences Press: Wien, Austria, 2013; pp. 281–284.

33. Křivánek, R. Hillfort Investigations in the Czech Republic. In *Field Archaeology from Around the World. Ideas and Approaches. Springer Briefs in Archaeology*, 1st ed.; Carver, M., Gaydarska, B., Eds.; Springer International Publishing: Cham, Switzerland, 2015; pp. 157–161, ISBN 978-3-319-09818-0.

34. Zápotocký, M. Raně Středověké Sídelní Komory na Dolní Ohři. *Archeol. Rozhl.* **1992**, *44*, 185–215.

35. Hložek, J.; Křivánek, R. Komplexní Nedestruktivní Výzkum Hradu Liběhradu. *Castellologica Bohemika* **2009**, *11*, 297–312. (In Czech)

36. Krištuf, P.; Rytíř, L. Mohylová Pohřebiště na Okrese Písek. Metoda Zaměřování Mohylových Pohřebišť—Burial Mound Cemeteries in the Písek Region. Method of Visual Survey of the Burial Mound Cemeteries. In *Opomíjená Archeologie 2005-2006–Neglected Archaeology 2005–2006*; Krištuf, P., Šmejda, L., Eds.; University of west Bohemia: Plzeň, Czech Republic, 2007; pp. 133–140, ISBN 978-80-254-1062-2. (In Czech)

37. Chvojka, O.; Krištuf, P.; Rytíř, L. *Mohylová Pohřebiště na Okrese Písek: Barrow Cemeteries in the Písek District. Archeologické Výzkumy v Jižních Čechách. Supplementum*, 1st ed.; Jihočeské Museum: České Budějovice, Czech Republic, 2009; ISBN 978-80-87311-02-8. (In Czech)

38. Křivánek, R. Possibilities and Limitations of Surveys by Caesium Magnetometers in Forested Terrains of Archaeological Sites. In *Študijné Zvesti Archeologického Ústavu SAV 41–2007, Archaeological Prospection–Topics and Abstracts, Proceeding of the 7th International Conference on Archaeological Prospection, Nitra, Slovakia, 11–15 September 2007*; Kuzma, I., Ed.; Archeologický Ústav SAV: Nitra, Slowakia, 2007; pp. 202–204, ISBN 978-80-89315-00-0.

39. Křivánek, R. Využití Geofyzikálních Metod ARÚ Praha při Průzkumech Archeologických Lokalit v Jižních Čechách. *Archeol. Výzkumy Již. Čechách* **2007**, *20*, 435–451. (In Czech)

40. Křivánek, R. Příklady Využití Magnetometrických Metod při Průzkumech Zalesněných Archeologických Lokalit. *Služ. Archeol.* **2008**, *20*, 70–77. (In Czech)

41. Lissek, P. Povrchový Průzkum Dehtářských Pracovišť v Českém Švýcarsku. *Archeol. Tech.* **2005**, *16*, 72–78. (In Czech)

42. Lissek, P. *Archeologický Výzkum Areálu Dehtářského Pracoviště "Pod Purkarticemi" na k. ú. Rynartice, okr. Děčín v Roce 2005*; Investorská zpráva pro Správu Národního parku české Švýcarsko (Report for the Investor of Archaeological Excavation); Ústav archeologické památkové péče severozápadních Čech v Mostě: Most, Czech Republic, 2015.

43. Křivánek, R. Detailní Měření Magnetické Susceptibility v Odkrytých Archeologických Situacích (Detailed Measurement of Magnetic Susceptibility in an Open Archaeological Situation). *Archeol. Rozhl.* **2008**, *60*, 695–724. (In Czech)

MDPI

Article

Fusion of Satellite Multispectral Images Based on Ground-Penetrating Radar (GPR) Data for the Investigation of Buried Concealed Archaeological Remains

Athos Agapiou [1],*, Vasiliki Lysandrou [1], Apostolos Sarris [2], Nikos Papadopoulos [2] and Diofantos G. Hadjimitsis [1]

[1] Remote Sensing and Geo-Environment Laboratory, Eratosthenes Research Centre, Department of Civil Engineering and Geomatics, Cyprus University of Technology, Saripolou 2-8, 3603 Limassol, Cyprus; vasiliki.lysandrou@cut.ac.cy (V.L); d.hadjimitsis@cut.ac.cy (D.G.H.)

[2] Laboratory of Geophysical-Satellite Remote Sensing & Archaeoenvironment, Foundation for Research & Technology-Hellas (F.O.R.T.H.), Nik. Foka 130, Rethymno, 74100 Crete, Greece; asaris@ret.forthnet.gr (A.S.); nikos@ims.forth.gr (N.P.)

* Correspondence: athos.agapiou@cut.ac.cy; Tel.: +357-25-002471

Academic Editors: Deodato Tapete and Jesus Martinez-Frias
Received: 28 April 2017; Accepted: 2 June 2017; Published: 6 June 2017

Abstract: The paper investigates the superficial layers of an archaeological landscape based on the integration of various remote sensing techniques. It is well known in the literature that shallow depths may be rich in archeological remains, which generate different signal responses depending on the applied technique. In this study three main technologies are examined, namely ground-penetrating radar (GPR), ground spectroscopy, and multispectral satellite imagery. The study aims to propose a methodology to enhance optical remote sensing satellite images, intended for archaeological research, based on the integration of ground based and satellite datasets. For this task, a regression model between the ground spectroradiometer and GPR is established which is then projected to a high resolution sub-meter optical image. The overall methodology consists of nine steps. Beyond the acquirement of the in-situ measurements and their calibration (Steps 1–3), various regression models are examined for more than 70 different vegetation indices (Steps 4–5). The specific data analysis indicated that the red-edge position (REP) hyperspectral index was the most appropriate for developing a local fusion model between ground spectroscopy data and GPR datasets (Step 6), providing comparable results with the in situ GPR measurements (Step 7). Other vegetation indices, such as the normalized difference vegetation index (NDVI), have also been examined, providing significant correlation between the two datasets ($R = 0.50$). The model is then projected to a high-resolution image over the area of interest (Step 8). The proposed methodology was evaluated with a series of field data collected from the Vésztő-Mágor Tell in the eastern part of Hungary. The results were compared with in situ magnetic gradiometry measurements, indicating common interpretation results. The results were also compatible with the preliminary archaeological investigations of the area (Step 9). The overall outcomes document that fusion models between various types of remote sensing datasets frequently used to support archaeological research can further expand the current capabilities and applications for the detection of buried archaeological remains.

Keywords: enhancement; fusion; ground spectroscopy; ground-penetrating radar (GPR); GeoEye; geophysics; remote sensing archaeology

1. Introduction

The scientific field of remote sensing has exhibited great potential over the last years for archaeological investigations, even in challenging environments [1–7]. Optical and radar satellite datasets have been used to reconstruct the archaeo-environment [8,9], to map recent land use and land cover changes in areas with archaeological interest [10–12], to detect traces linked with archaeological remains [13,14] or even to investigate spatial patterns between archaeological sites [15].

The integration of various remote sensing data has attracted the interest of scholars. For instance, in [16,17], the potential use of multispectral satellite images and archive aerial imageries with ground geophysical prospection was demonstrated over the same archaeological area. In addition, Morehart [18] has exploited various forms of landscape information like historic records and maps together with archive aerial images, and medium- and high-resolution satellite multispectral datasets, in order to study and identify buried features. Others have applied a multi-disciplinary study in the area of the Vésztő-Mágor tell, Hungary, using archive aerial images, high resolution multispectral datasets, ground spectroscopy and geophysics to reconstruct pre-historic landscapes [16].

Despite the wide use of various such non-invasive remote sensing techniques, these are frequently employed separately. Indeed, in most cases the different processed and geocoded remote sensing data were inserted in a Geographical Information System (GIS), followed by simple superimposition of the diverse ortho-rectified datasets for visual and/or semi-automatic multilayer analysis (e.g., [8,9,19,20]). Some studies have also presented interesting results on data integration and fusion by employing quantitative means for multiple data sources [21–25]. Some other similar cases studies related to the exploitation of multi-sensor data for archaeological research can be found in [24,25].

The synthesis and interpretation of the overall results from the different remote sensing sensors help researchers to identify "hot spots" which can be considered as potential archaeological targets. Table 1 summarizes some of the most basic advantages and limitations of the three remote sensing methods used here: (a) satellite datasets; (b) ground-penetrating radar (GPR) and (c) ground spectroscopy. The table provides basic attributes of these methods, such as spatial resolution and spectral range; spatial extent; potential extraction of 3D underground information; soil penetration, and the capability to use archive datasets. It is therefore understood that none of these technologies can be considered as "ideal" for archaeological research. For instance, high-resolution multispectral images can see beyond the visible part of the spectrum and therefore detect vegetation anomalies but at the same time their spatial resolution (e.g., 0.5 m in the panchromatic band) can be problematic for identification of small-scale localized buried features. In contrast, ground spectroscopy can be used to collect "accurate" spectral signatures (ground truth) for small targets with limited spatial extent. The GPR method is the only one of the three methods that penetrates the upper layers of ground surface and produces a 3D reconstruction of the underground targets with increased spatial resolution.

Table 1. General characteristics of satellite remote sensing; ground-penetrating radar (GPR) and ground spectroscopy for archaeological research.

Characteristics	Satellite Remote Sensing (for Optical High Resolution)	Ground-Penetrating Radar	Ground Spectroscopy
Spatial resolution	Medium–High	High–Very High (adjustable)	High–Very High (adjustable)
Spectral range	Visible-Near Infrared (Multispectral)	Microwave	Vis-NIR (Hyperspectral)
Spatial extent	Several km^2	Several Hectares	Several m^2
3D visualization	No	Yes	No
Soil penetration	No	Yes	No
Archive data	Yes	No	No
Type of information	Raster	Point	Raster

2. Aims

Buried archaeological remains can have a different response to the backscattering signal captured by the various remote sensing sensors, which are used either on top of the ground surface or located in a distance (Figure 1). For example, both satellite images and ground spectral signatures have the capability to record the vegetation status over archaeological landscapes and therefore sense any vegetation-stressed conditions. This vegetation stress formed due to the presence of buried archaeological remains (known in the archaeological research as "crop marks phenomenon"), is detectable in the near infrared part of the electromagnetic spectrum (i.e., passive sensors). This indirect approach allows researchers to "examine" the upper layers of the archaeological context (up to 0.50 m below the surface) without any real penetration as occurs with active sensors like GPR.

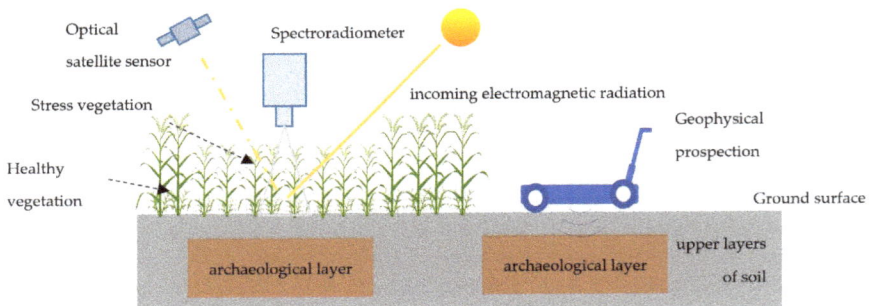

Figure 1. Detection of buried remains either using an indirect method (left, passive sensors) or a direct method (right, active sensors). The first category can include both satellite and ground spectroradiometer measurements (i.e., reflectance values in several bands) over cultivated areas where vegetation stress is used as a proxy for the presence of buried archaeological remains in the upper layers of soil. In contrast, direct methods such as GPR can penetrate soil and record the backscattering signal.

Since each remote sensing technique uses different technology, their results can provide useful but still partial information regarding the concealed archaeological remains. Thus, the integration of various technologies is considered as a key step towards the holistic understanding of the landscape. However, the fusion and integration of data gathered from different sensors is quite challenging. Although some fusion techniques already exist, like pan-sharpening between multi-spectral and panchromatic bands of the same or different satellite sensors [26,27], or as more recently shown, fusion between archive greyscale and recent multispectral satellite images [28], fusion methodologies of more diverse datasets need to be developed and tested for archaeological research. Studies have shown that it is feasible to scale-up ground spectroradiometric data (i.e., ground spectral signatures) to existing [29] or even forthcoming satellite sensors [30].

This study aims to go a step forward by investigating the integration of ground spectroradiometric and GPR data, based on a fusion regression model and then projected to a high resolution optical image. The result is an enhanced optical satellite image capable of improving the initial raw satellite information.

3. Methodology

The enhancement of the final satellite image was based exploring three different datasets: (a) a high-resolution GeoEye image with spatial resolution of 50 cm; (b) ground hyperspectral signatures collected from the GER-1500 spectroradiometer (Spectra Vista Corporation, New York, NY, USA) with a spectral range between 350–1050 nm (Visible-Near Infrared spectrum) and (c) GPR measurements with a 250 MHz antenna. Both GER-1500 and GPR measurements were collected moving along parallel transects 0.5 m apart with 5 cm sampling along the lines. The ground spectroradiometer and GPR data have been used to develop the local fusion regression model.

All the datasets have undergone the initial pre-processing steps: geometric and radiometric corrections have been applied to the GeoEye image, while during the collection of the ground spectral signatures, sun irradiance was measured using a calibrated spectralon panel. The latter allows us to standardize the results and minimize the sun illumination noise. More details regarding the different datasets and their result can be found in [16,31]. The methodology applied consists of nine (9) steps, described in Figure 2.

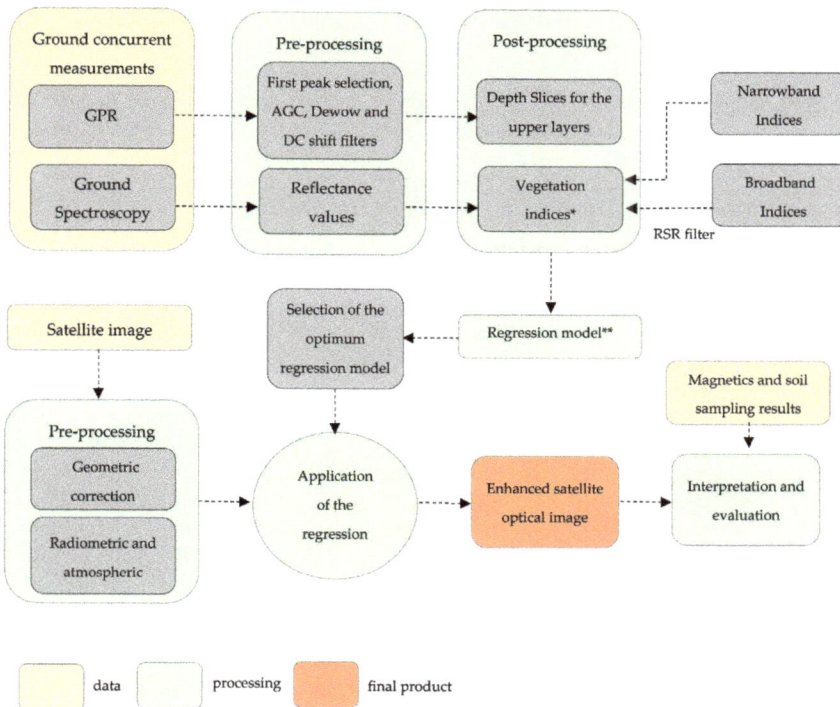

Figure 2. Overall proposed methodology for the enhancement of the optical remote sensing image based on the application of the fusion regression model between the ground data gathered from GPR and ground spectroscopy (* see Supplementary Table S1; ** see Table 2). RSR: relative spectral response; AGC: Automatic Gain Control; DC: Direct Current.

Step 1 is with regard to the collection of ground remote sensing data over an archaeological area of interest. GPR and spectroradiometric measurements have been collected concurrently to minimize any background noise from climatic changes such as rainfall, humidity etc. These datasets were then processed separately using standard techniques which are described in more detail in [16] (Step 2). Several filters have been applied for the GPR measurements, while hyperspectral reflectance values with a 1 nm interval have been calculated based on the incoming and backscattering radiance values. The following step (Step 3) employs the calculation of the broadband reflectance values of the selected satellite sensor (i.e., GeoEye-1) using its relative spectral response (RSR) filter (see more in [16]). The RSR filter is used as the band-average relative spectral radiance response of the specific sensor. As is known, each satellite sensor has its own RSR filter which describes the capability of the sensor to capture the electromagnetic radiation. The calculation of vegetation indices, including both the broadband (i.e., based on multispectral bands—Step 3) and narrowband (i.e., based on hyperspectral bands—Step 2) reflectance values is performed in Step 4. Overall, 71 vegetation indices have been estimated upon the equations shown in the Supplementary Table S1 (see also more details in [32]).

The next step (Step 5) considers the regression analysis between these 71 vegetation indices and the GPR measurements for the upper layers of soil (i.e., 0.00–0.60 m). In detail, linear; exponential; Fourier; Gaussian; polynomial (second–third order); power; rational and sum of sin regression models have been evaluated. In the various regression models the GPR measurements were considered as the independent X variable, while the vegetation indices (from the spectroradiometer) were considered as the dependent Y variable. Based on the regression analysis (Pearson correlation coefficient—R) the optimum fusion model was selected (Step 6). Before the application of the selected model to the satellite image (Step 8), an evaluation of the proposed model was performed over the common area of interest using the ground data (see Step 7). Finally, the overall results are evaluated (Step 9). It should be mentioned that the proposed enhancement technique to the GeoEye-1 image or in any other satellite image is easily applicable to any GIS environment using raster calculators.

4. Case Study: The Vésztő-Mágor Tell

The Vésztő-Mágor Tell is located in the southeastern Great Hungarian Plain, at a latitude of 46°56′24.36″ and longitude of 21°12′32.67″ (Figure 3). The tell covers about 4.25 hectares and rises to a height of about 9 m above mean sea level. A monastery on top of the tell was the main focus of the archaeological excavations that were initiated in 1968. Excavations revealed cultural layers dated from the Late Middle Neolithic period continuing until the Late Neolithic period (ca. 5000–4600 B.C). After the Late Neolithic (Tisza Culture) the site was abandoned to be reoccupied in the Early Copper Age (Tiszapolgár Culture). Habitation of the settlement continued until the Early/Middle Bronze Age (Gyulavarsánd Culture) with a small interruption during the end of the Middle Copper Age [33–36]. In the 11th century AD a church and then a monastery of the Csolt-clan were established on the Vésztő-Mágor Tell. The monastery continued to operate until the end of the 14th century, while the two towers of the church were standing until 1798. During the early 19th century, the monastery, which was constructed at the southern part of the tell was completely destroyed by the operations carried out during the construction works of a wine cellar [33,35,37]. Since the 1980s the site has been one of the Hungarian National Parks.

Figure 3. Area covered from the different techniques: Grey and blue polygons indicate the areas covered from the geophysical surveys (magnetic and GPRsurvey, respectively), the red polygon indicates the area covered from the ground spectroradiometric measurements while the whole area is covered by the GeoEye image (no polygon used) (Latitude: 46°56′24.36″; Longitude: 21°12′32.67″, WGS84).

The area was intensively investigated in the previous years as part of the Körös Regional Archaeological Project [38] with various non-invasive techniques aiming to reconstruct the organization of Neolithic tell-based settlements of Vésztő-Mágor and Szeghalom-Kovácshalom. These techniques included intensive, gridded and surface collections as well as magnetometry, ground-penetrating radar, electrical resistance tomography, hyperspectral spectroradiometry, and soil chemistry. Based on the stylistic attributes of the recovered ceramics, it appears that the tell at Vésztő-Mágor was inhabited during the Neolithic period, possibly for several centuries [38].

The area covered by ground spectroscopy is indicated in Figure 3 with red dashed polygon, while geophysical surveys from GPR and magnetic methods are shown with blue and grey polygons, respectively. Measurements taken from ground spectroradiometer and GPR were used to set up the model (see Steps 1 and 5).

5. Results

5.1. Results from GPR and Spectroradiometer

Details regarding the overall results of the in-situ investigation around the Vésztő-Mágor tell can be found in [16]. Sometimes similar interpretations were noticeable, for instance the NDVI—as calculated from the ground spectroradiometer—with the first GPR depth slice of 0.00–0.20 m below ground surface (Figure 4). Indeed, a linear anomaly at around 80 m in the Y-axis (indicated with red arrow in Figure 4) was highlighted by both techniques. This feature is supposed to be part of a ditch which surrounded the tell (see more in [38]). In the GPR depth slice of Figure 4 (right) positive waveform values indicate that the polarity of the signal is same as the incident wave, or negative if the polarity of the reflected wave is opposite to the incident wave.

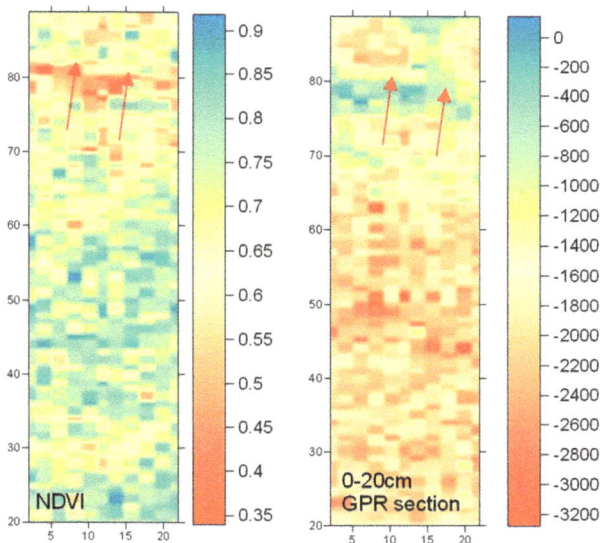

Figure 4. The normalized difference vegetation index (NDVI) as calculated from the ground spectroradiometer over the red area of Figure 3 (left) and the first depth slice (i.e., 0.00–0.20 m below ground surface) from the GPR measurements. A linear anomaly, indicated with red arrows, at around 80 m of the Y-axis was detected by both methodologies.

GPR sensor could penetrate up to a depth of 2 m below ground surface, providing 10 depth slices, each with a thickness of 20 cm. The GPR slices with increasing depth were created after the application of the first peak selection; AGC (Automatic Gain Control), Dewow and DC (Direct Current) shift filters were used to amplify the average signal amplitude and remove the initial DC bias. A velocity of 0.1 m/ns for the transmission of the electromagnetic waves (250 MHz antennas) was assumed based on the hyperbola fitting. GPR interpretation can be very complicated since each slice should be treated separately, but at the same time the overall interpretation is linked with the archaeological context of the site. As shown in Figure 5, scatterplots between the first depth slice of the GPR (0–20 cm below the ground surface) and the last depth slices beyond the 1 m depth, are completely uncorrelated. Pearson correlation coefficient (R) starting from top of the surface until a depth of 2 m is shown in Table 2. The R value was selected to be shown (instead of R square) since the aim was to identify any existing correlation between the two datasets. The upper layers of soil, which are considered as the "optimum depth" for indirect methods such as satellite images and ground spectroscopy, are indicated in the first row (i.e., 0.00–0.60 m) of Figure 5.

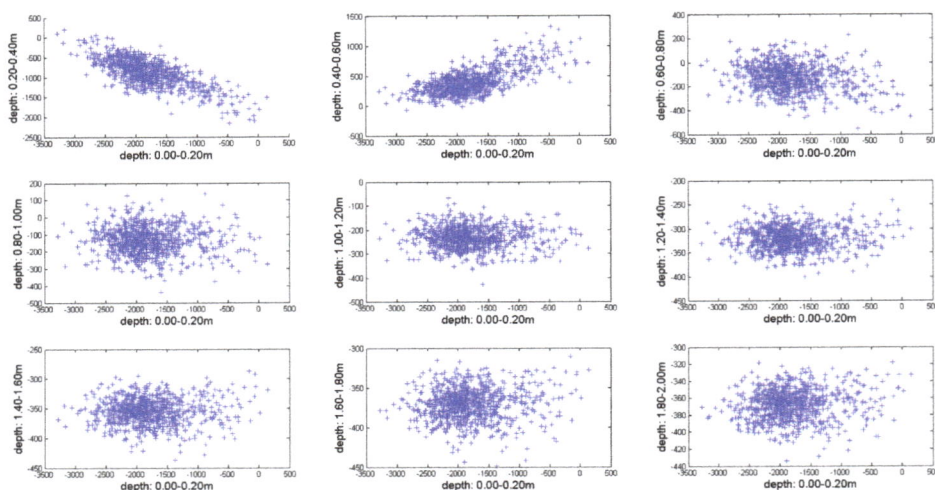

Figure 5. Scatterplots of the first slice of the GPR results (i.e., depth 0.00–0.20 m below ground surface) against the rest of the depth slices of thickness 0.20 m until 2.00 m below ground surface. The Pearson correlation coefficient starting from the top of the surface to a 2 m depth was estimated as −0.74 for the slice 0.20–0.40 m; 0.67 for the slice 0.40–0.60 m; −0.22 for the slice 0.60–0.80 m; −0.07 for the slice 0.80–1.00 m; −0.04 for the slice 1.00–1.20 m; −0.01 for the slice 1.20–1.40 m; 0.03 for the slice 1.40–1.60 m; 0.06 for the slice 1.60–1.80 m and 0.10 for the slice 1.80–2.00 m. The upper layers of soil (i.e. 0.00–0.80 m) are shown in the first row.

Table 2. The Pearson correlation coefficient (R) starting from top of the surface down to a 2 m depth for GPR results of the first slice compared to the rest of the slices.

Slice (in m Below Surface)	R Value
0.20–0.40	−0.74
0.40–0.60	0.67
0.60–0.80	−0.22
0.80–1.00	−0.07
1.00–1.20	−0.04
1.20–1.40	−0.01
1.40–1.60	0.03
1.60–1.80	0.06
1.80–2.00	0.10

Similarly, Figure 6 presents the scatterplot of the NDVI index against the GPR measurements for various depth slices. It is noticeable that the variance of the data in the slices representing depths larger than 0.80 m below ground surface is increasing. The *R* value from this analysis is shown in Table 3. The variance observed in Figure 6 is linked to the fact that both NDVI and GPR results are influenced by different factors: on one hand NDVI indicates vegetation status on top of the surface, while on the other hand GPR results depend on the electromagnetic properties of the underground targets and the surrounding soil. Figure 6 also demonstrates the difficulty in achieving a high correlation regression model between different remote sensing datasets even though the spectroradiometer and GPR measurements have been acquired concurrently over the same area and in the same conditions.

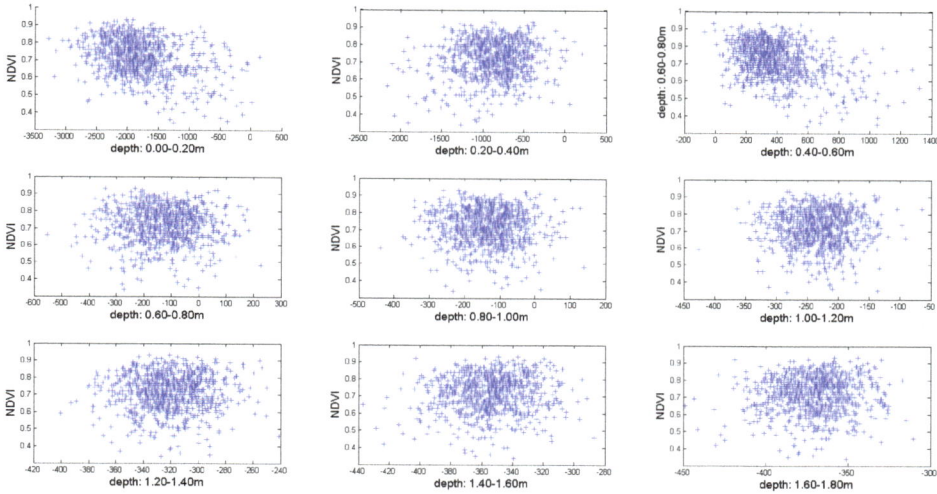

Figure 6. Scatterplots of the NDVI index against the 10 depth slices with a width 0.20 to 2 m below ground surface. The Pearson correlation coefficient of the NDVI compared to the GPR measurements starting from top of the surface to a depth of 1.80 meters was estimated as −0.35 for the slice 0.00–0.20 m; 0.12 for the slice 20–0.40 m; −0.39 for the slice 0.40–0.60 m; −0.01 for the slice 0.60–0.80 m; −0.03 for the slice 0.80–1.00 m; 0.01 for the slice 1.00–1.20 m; 0.05 for the slice 1.20–1.40 m; 0.08 for the slice 1.40–1.60 m and 0.05 for the slice 1.60–1.80 m. The upper layers of soil (i.e. 0.00–0.80 m) are shown in the first row.

Table 3. The Pearson correlation coefficient (*R*) starting from top of the surface until a 2 m depth between GPR results and the NDVI.

Slice (in m Below Surface)	*R* Value
0.00–0.20	−0.35
0.20–0.40	0.12
0.40–0.60	−0.39
0.60–0.80	−0.01
0.80–1.00	−0.03
1.00–1.20	0.01
1.20–1.40	0.05
1.40–1.60	0.08
1.60–1.80	0.05

5.2. Setting up the Local Regression Model for the Enchacement of the Optical Data

In order to find an "optimum" fusion model between the spectroradiometric and the GPR measurements, the first three upper layers corresponding to a depth up to 0.60 m have been used. The GPR measurements and the various vegetation indices mentioned in the Supplementary Table S1 (including also the Vis—NIR bands) were correlated with the following mathematical models shown in Table 4:

Table 4. General mathematical equations used for the regression model between the spectroradiometric measurements and the different GPR amplitude value slices.

No.	General Equations	Name
1.	$f(x) = a \times x + b$	Linear
2.	$f(x) = a \times \exp(b \times x)$	Exponential (one term)
3.	$f(x) = a \times \exp(b \times x) + c \times \exp(d \times x)$	Exponential (two terms)
4.	$f(x) = a0 + a1 \times \cos(x \times w) + b1 \times \sin(x \times w)$	Fourier (one term)
5.	$f(x) = a0 + a1 \times \cos(x \times w) + b1 \times \sin(x \times w) + a2 \times \cos(2 \times x \times w) + b2 \times \sin(2 \times x \times w)$	Fourier (two terms)
6.	$f(x) = a1 \times \exp(-((x - b1)/c1)^2$	Gaussian (one term)
7	$f(x) = a1 \times \exp(-((x - b1)/c1)^2) + a2 \times \exp(-((x - b2)/c2)^2)$	Gaussian (two terms)
8.	$f(x) = p1 \times x^2 + p2 \times x + p3$	Polynomial (second order)
9.	$f(x) = p1 \times x^3 + p2 \times x^2 + p3 \times x + p4$	Polynomial (third order)
10.	$f(x) = a \times x^b$	Power (one term)
11.	$f(x) = a \times x^b + c$	Power (two terms)
12.	$f(x) = (p1 \times x + p2)/(x + q1)$	Rational (first degree)
13.	$f(x) = (p1 \times x^2 + p2 \times x + p3)/(x + q1)$	Rational (second degree)
14.	$f(x) = a1 \times \sin(b1 \times x + c1)$	Sum of Sin (one term)
15.	$f(x) = a1 \times \sin(b1 \times x + c1) + a2 \times \sin(b2 \times x + c2)$	Sum of Sin (two terms)

Where $f(x)$ refers to the pseudo GPR amplitude value per slice and x is the vegetation index/band (see Supplementary Table S1).

The Pearson correlation coefficient (R) was estimated for each combination (i.e., 10 slices of GPR $\times$ 71 vegetation indices and four spectral bands). The final analysis indicated that the linear model (Model 1 of Table 4), despite its simplicity, performed superior results compared to the rest of the mathematical models mentioned in Table 4. Some of them showed very low performance (e.g., power models, numbers 10–11 of Table 4) and therefore could not assist towards the aims of this study.

Table 5 displays the Pearson correlation coefficient (R) results of the linear regression model (Model 1 of Table 5) for each vegetation index (see Table S1) and depth slice (up to 0.80 m). Below this depth, the coefficient was found to be extremely low ($R \rightarrow 0.00$). Within the first three depths of GPR slices (i.e., depth range 0.00–0.60 m, "optimum depth" for the passive remote sensing techniques) a higher Pearson correlation coefficient (R) (up to almost $R = 0.5$) is recorded, which indicates that a regression model between the optical images and geophysical measurements can be established. Among the various vegetation indices examined, it seems that the red-edge position (REP) index (Equation (37), of Table S1) is the most suitable for use, at least for this linear correlation and for the specific depth between the two datasets. However, the REP (hyperspectral) index cannot be applied to multispectral and high spatial resolution sensors such as the GeoEye. On the contrary, the broadband NDVI index, which has also provided some promising results for a couple of surface depth slices, is more applicable to these sensors.

Table 5. The results of the Pearson correlation coefficient (R) with range of -1 to $+1$ for each vegetation index and for each depth slice (up to 0.80 m) for the linear regression model. Higher correlation (either negative or positive) is highlighted in the table.

Depth/Index	BAND1	BAND2	BAND3	BAND4	EVI	Green NDVI	NDVI	SR	MSR	MTVI2
0.000 to 0.200 m	0.23	0.15	0.22	−0.33	−0.32	−0.41	−0.37	−0.29	0.28	0.32
0.200 to 0.400 m	−0.12	−0.11	−0.07	0.11	0.09	0.22	0.15	0.10	−0.10	−0.10
0.400 to 0.600 m	0.26	0.17	0.27	−0.33	−0.38	−0.44	−0.41	−0.33	0.32	0.37
0.600 to 0.800 m	0.04	0.05	0.03	0.05	0.00	−0.01	0.00	−0.02	0.02	0.00

Depth/Index	RDVI	IRG	PVI	RVI	TSAVI	MSAVI	GEMI	ARVI	SARVI	OSAVI
0.000 to 0.200 m	−0.38	0.24	−0.37	0.38	−0.37	−0.38	−0.32	−0.35	0.40	−0.37
0.200 to 0.400 m	0.14	−0.01	0.12	−0.16	0.15	0.16	0.09	0.13	−0.20	0.15
0.400 to 0.600 m	−0.41	0.31	−0.39	0.42	−0.41	−0.42	−0.34	−0.40	0.43	−0.41
0.600 to 0.800 m	0.02	0.00	0.03	0.00	0.00	0.00	0.03	−0.01	−0.01	0.00

Depth/Index	DVI	SRxNDVI	CARI	GI	GVI	MCARI	MCARI2	mNDVI	SR705	mNDVI2
0.000 to 0.200 m	−0.36	0.00	0.07	−0.19	0.21	−0.14	−0.34	−0.30	−0.36	−0.41
0.200 to 0.400 m	0.12	0.00	−0.14	−0.02	0.01	−0.09	0.09	0.06	0.16	0.20
0.400 to 0.600 m	−0.38	0.02	0.06	−0.27	0.29	−0.20	−0.36	−0.36	−0.39	−0.45
0.600 to 0.800 m	0.04	0.01	0.05	−0.01	0.01	0.02	0.04	−0.01	−0.01	0.00

Depth/Index	MSAVI	mSR	mSR2	MSR	MTCI	mTVI	NDVI	NDVI2	OSAVI	RDVI
0.000 to 0.200 m	−0.36	−0.14	−0.37	−0.30	−0.45	−0.34	−0.35	−0.40	−0.35	−0.37
0.200 to 0.400 m	0.14	0.02	0.17	0.09	0.29	0.09	0.13	0.19	0.13	0.12
0.400 to 0.600 m	−0.40	−0.19	−0.40	−0.35	−0.45	−0.36	−0.40	−0.44	−0.40	−0.40
0.600 to 0.800 m	0.00	−0.02	0.00	−0.01	0.00	0.04	0.00	0.00	0.00	0.02

Depth/Index	REP	SIPI	SIPI2	SIPI3	SPVI	SR	SR1	SR2	SR3	SR4
0.000 to 0.200 m	−0.49	0.38	0.30	0.30	−0.36	−0.26	−0.34	−0.29	−0.37	0.23
0.200 to 0.400 m	0.38	−0.16	−0.08	−0.08	0.12	0.07	0.14	0.09	0.18	0.00
0.400 to 0.600 m	−0.47	0.42	0.36	0.36	−0.38	−0.32	−0.38	−0.34	−0.40	0.30
0.600 to 0.800 m	−0.01	0.00	0.00	0.00	0.04	−0.02	−0.01	−0.01	−0.01	0.00

Depth/Index	TCARI	TSAVI	TVI	VOG	VOG2	ARI	ARI2	BGI	BRI	CRI
0.000 to 0.200 m	−0.10	−0.35	−0.33	−0.40	0.39	0.17	−0.06	0.32	−0.06	−0.24
0.200 to 0.400 m	−0.11	0.12	0.08	0.20	−0.22	0.02	0.16	−0.12	−0.11	0.11
0.400 to 0.600 m	−0.14	−0.39	−0.36	−0.42	0.40	0.22	0.00	0.36	−0.14	−0.29
0.600 to 0.800 m	0.05	0.00	0.04	−0.01	0.01	−0.05	−0.03	0.00	−0.01	−0.05

Depth/Index	RGI	CI	LIC	NPCI	NPQI	PRI	PRI2	PSRI	SR5	SR6
0.000 to 0.200 m	0.26	−0.02	−0.06	0.06	0.27	−0.24	0.29	0.26	−0.30	−0.24
0.200 to 0.400 m	−0.03	−0.14	−0.10	0.14	−0.25	0.05	−0.08	−0.03	0.13	0.34
0.400 to 0.600 m	0.33	−0.09	−0.14	0.15	0.26	−0.28	0.34	0.33	−0.33	−0.16
0.600 to 0.800 m	0.00	−0.01	−0.02	0.02	−0.03	−0.01	0.01	0.00	−0.02	0.01

Depth/Index	SPRI	VS	MVSR	fWBI	SG	max	min
0.000 to 0.200 m	−0.24	−0.34	−0.33	−0.37	0.16	0.40	−0.49
0.200 to 0.400 m	0.34	0.13	0.12	0.09	−0.11	0.38	−0.25
0.400 to 0.600 m	−0.16	−0.38	−0.38	−0.38	0.18	0.43	−0.47
0.600 to 0.800 m	0.01	0.00	0.00	0.07	0.05	0.07	−0.05

Relative strong positive correlation
Relative low correlation
Relative strong negative correlation

5.3. Comparison of the Fusion Model from the Spectroradiometric Measurements with the GPR Results

Based on the previous findings a linear enhancement was applied to the spectroradiometric data using the REP and NDVI results. The first three columns of Figure 7 demonstrate the original GPR depth slices for the upper layers of soil (range from 0.0 to 0.60 m depth) which have been collected over the area indicated with red polygon at Figure 3, while the next six columns of Figure 7 show the enhanced images derived from the linear regression models after the REP and NDVI indices. All the results of Figure 7 share the same range as the original GPR measurements for direct visual comparison. The interpolation between the point samples of the GPR measurements and enhanced spectroradiometric results was performed using the kriging method.

Enhanced images derived from the REP index tend to give generally very comparative interpretation results with the actual GPR section for the upper layers of soil. Enhanced images below this "optimum" depth (i.e., >0.60 m below ground surface) cannot be considered reliable as shown earlier from the correlation coefficient and therefore these results are not demonstrated here.

The linear anomaly indicated in the first upper layers of the GPR section at around 80 m in the Y-axis (see dashed parallel lines in the first image of Figure 7) is also visible in the enhanced vegetation image generated from the REP index. According to the GPR data, this anomaly seems to consist of two parallel features, about 2 m apart, running in a west to east direction [16]. This feature has been hypothesized to belong to the north boundary of the old monastery's yard which acted possibly as Neolithic fortifications ditches [38].

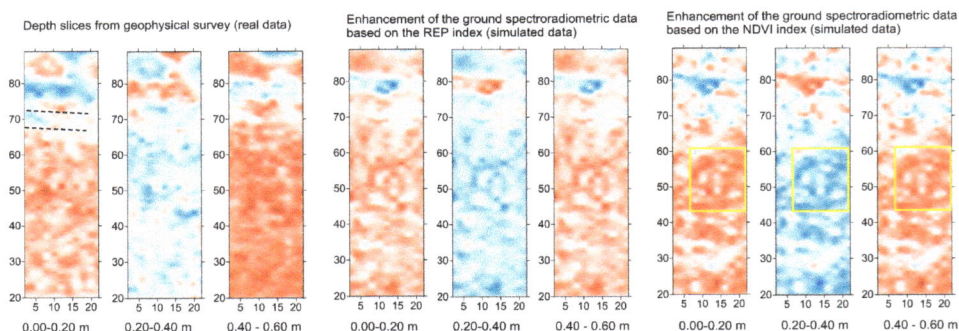

Figure 7. First three columns from the left: GPR real data for the upper layer of soil (depth from 0.0–0.6 m); next three columns: enhanced results generated from the red-edge position (REP) regression model; the last three columns: enhanced results generated from the NDVI regression model. The two parallel linear features are indicated in the first column while some circular false positive marks are indicated within the yellow sqaure in the NDVI regression results (Latitude: 46°56'24.36"; Longitude: 21°12'32.67", WGS84).

Similar interpretation results between the original GPR data and the enhanced NDVI sections are also observed for the first three depth slices (i.e., 0.0–0.60 m depth). Again, the results are very poor below this "optimum" depth. Some circular "false positive" marks were also observed at around 50 m in the Y-axis of the enhanced NDVI images for all slices (see Figure 7, yellow squares). Overall, the results and especially those originated by the employment of the REP index, are found to be very promising in enabling the enhancement of the optical remote sensing data.

5.4. Application to the Vésztő-Mágor Tell Case Study

The final steps (Steps 8 and 9 of the overall methodology) refer to the application of the regression model to a satellite imagery. Due to current limitations (i.e., spatial resolution) of existing hyperspectral satellite sensors—such as the Earth Observing (EO)-1 Hyperion sensor with 30 meters of spatial resolution—the projection of any fusion model generated from hyperspectral vegetation index—like the REP index—is still restricted. In contrast, fusion models generated from broadband vegetation indices—such as the NDVI index—can be applied to high resolution multispectral images. A GeoEye image was selected since the spatial resolution of the sensor (0.50 m) is similar to the ground transects applied for both GPR and spectroradiometric measurements. The Red Green Blue (RGB) composite of the GeoEye image is shown in Figure 8a, while the overall results from the GPR and magnetic prospection are shown in Figure 8b,c, respectively. Figure 8d is an enhanced NVDI image after the linear regression model was applied to the GeoEye image (shown in Figure 8a). Some linear features have been already identified from the GPR prospection in the northern part of the site, while other anomalies have been also detected around the old monastery's yard (see more about the geophysical results in references [16,38]). Magnetic survey, GPR and soil coring suggest that the Vésztő-Mágor tell may have been surrounded by a system of three large ditches and palisades, with a possible entrance in the northwest. Preliminary coring into the linear anomalies also suggests that some of the ditches may exceed 4 m in depth. Structural similarities with other Neolithic fortifications led archaeologists to suspect that part of the ditch and palisade system at the Vésztő-Mágor Tell was established during this period. The number of geophysical anomalies located outside the ditch and palisade system drops dramatically, suggesting that the Late Neolithic occupations were restricted to the tell itself (i.e., no evidence for a surrounding settlement around the tell), and a substantial horizontal settlement was not established in the surrounding area [38].

Figure 8. (**a**) GeoEye image shown in the RGB composite; (**b**) GPR anomalies results; (**c**) magnetic prospection results and (**d**) enhanced NDVI image for the upper layer between 0.20–0.40 m below the surface, overlaid with GPR and magnetic results (Latitude: 46°56′24.36″; Longitude: 21°12′32.67″, WGS84). Two areas (Area A and Area B) have been selected for further interpretation (see Figure 12).

A comparison between the enhanced NDVI images and the magnetic prospection results is provided in Figure 9. The magnetic data after despiking and equalization of dynamic range lay in the range of ±15 nT/m. Figure 9a,b shows the overall results from the magnetic prospection and the interpretation of the magnetic anomalies, respectively (see more in [16]). The NDVI index derived from the GeoEye image is presented in Figure 9c. The next figure (Figure 9d) presents the enhanced NDVI image for the upper layer between 0.00 and 0.20 m below ground surface, derived from the linear regression model. The latest image can provide additional information compared to the initial NDVI image. Linear features on the north of the Tell (part of the large ditches and palisades acting as fortification of the Tell?) are clearly identified in the enhanced image (Figure 9d) as well as in the magnetic results (Figure 9b). The eastern part of the same feature and other circular features on the southern part are also noticeable. It should be mentioned that this region was explored in past by excavations. All these features, indicated with arrows in Figure 9d, are aligned with the magnetometer results and the rest of the non-invasive techniques applied in the area around the Tell. Of course, magnetic results provide more details and visible results compared to the enhanced NDVI image, since the correlation results and the regression model could explain only a portion of the in-situ geophysical measurements. It should be mentioned that similar outcomes (features) have been also spotted in the following slices (i.e., 0.20–0.60 m) thus permitting all this information to be used for extracting the 3D profile of the hidden features.

It is important to stress the advantages of the proposed methodology in relation to the standard interpretation of the satellite images. As seen in the RGB GeoEye and the NDVI images (Figures 8a and 9c respectively), the interpretation results of these images are limited, while only a small part can be detected through interpretation process (e.g., part of the lower circular feature south to the tell in Figure 9c). In contrast, the enhanced NDVI image overcomes these interpretation limitations, providing additional information for the area around the Vésztő-Mágor Tell.

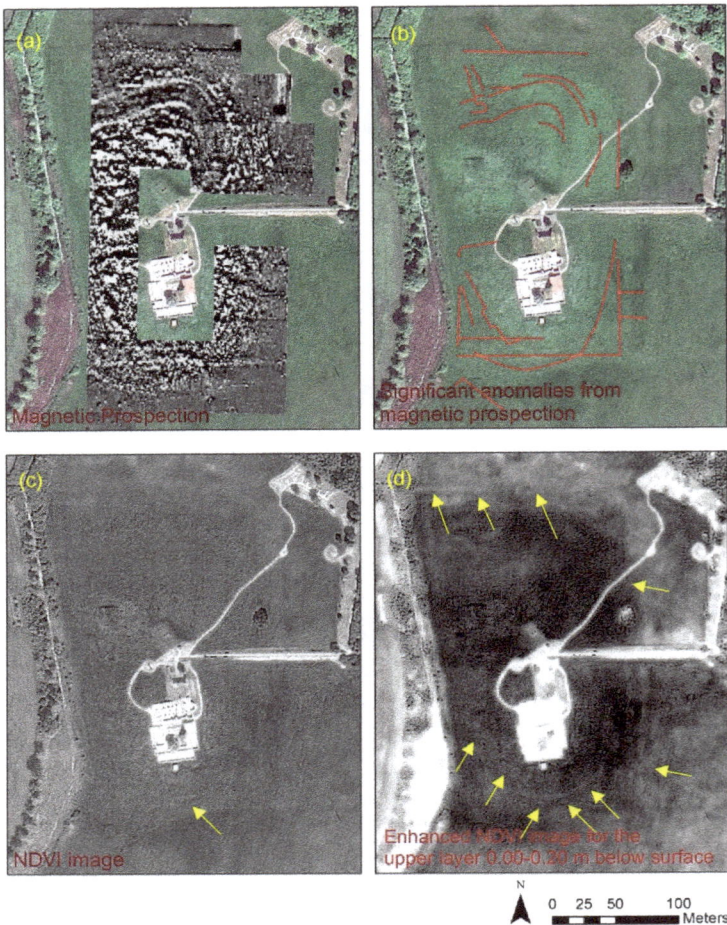

Figure 9. (a) Magnetic prospection results around the Vésztő-Mágor Tell (see more in [14,36]); (b) interpretation results of the magnetic prospection (see more in [16,38]); (c) raw NDVI image and (d) enhanced NDVI image for the upper layers at 0.00–0.20 m below the surface. Linear and circular features in correspondence to the 8b interpretation are indicated with yellow arrows (Latitude: 46°56′24.36″; Longitude: 21°12′32.67″, WGS84).

In order to further improve the interpretation of the enhanced NDVI images, pseudo color composites and spatial filters have been also applied. Figure 10, shows the pseudo RGB color composite of the enhanced NDVI image for three different slices (i.e., 0.00–0.60 m below surface, with a 20 cm depth interval). The pseudo RGB color composite could enhance further the results derived from the individual NDVI enhanced image. As revealed, several linear and circular features found near the Tell are detectible and in line with the magnetometer interpretation results (see Figure 9b). The southern and eastern fortification ditches (?), with an either linear or circular ground plan (shown with yellow arrows in Figure 9), are outlined due to their different contrast from the surrounding area. Other anomalies with lower contrast values on the northern part of the Tell can also be noticed, especially a feature which is linked to a path or wall going to the north and then running to the west, nowadays invisible. Based on the geophysical prospection results these are signs of the surrounding ditches of the Tell which become visible from the reflective signals of the GPR and in a relative good correlation to the magnetic results. Figure 11, presents the same pseudo RGB color composite as in Figure 10, superimposed to a Sobel spatial filter image. The latest filter has been applied for the most top upper layer of soil (i.e., 0.00–0.20 m). A Sobel filter can improve the enhancement of various edges, detected in the images and therefore can be used to advance the interpretation practice. The gain of this approach, in addition to enhancing the above-mentioned anomalies, is that it allows for capture of the linear anomaly on the north of the Tell which was not detected before (i.e., in Figure 10). According to the GPR data this anomaly seems to consist of two parallel features, about 2 m apart, running in the direction west to east.

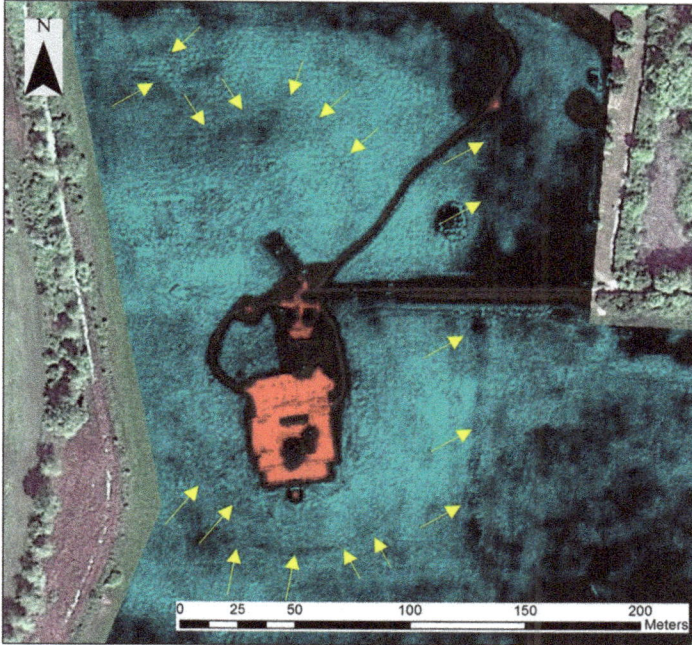

Figure 10. Pseudo RGB composite of the enhanced NDVI images for three different slices of the "optimum" depth (i.e., 0.00–0.60 m below the surface). Linear and circular features in correspondence to the 8b interpretation are indicated with yellow arrows (Latitude: 46°56′24.36″; Longitude: 21°12′32.67″, WGS84).

The Sobel spatial filter was also found to be very useful in enhancing other linear features around the Vésztő-Mágor tell. Figure 12 demonstrates this application with specific focus on two areas (see Area A and Area B in Figure 8). The Sobel filter could minimize the background noise, and enhance some of the most prominent edges of the image. Slices below the "optimum" depth (i.e., >0.60 m depth) bear much noise (see Figure 12, slice 0.60–0.80 m) and therefore are difficult to be interpreted. In contrast, the upper layers of soil (i.e., 0.00–0.60 m) have revealed some linear and circular features. While a good contrast can be observed in the most top layer of soil (i.e., 0.00–0.20 m) of Figure 12, this is decreasing as we move lower in depth (i.e., similar findings with the model 1 of Table 4). However, for the next two slices (i.e., 0.20–0.60 m) some linear features are still visible, as indicated with the arrows of Figure 12. It should be mentioned that these features go beyond the area of investigation of the geophysical prospection, providing therefore some additional "hot spot" areas for future investigations.

Figure 11. Pseudo RGB composite of the enhanced NDVI images for three different slices of the "optimum" depth (i.e., 0.00–0.60 m below surface) superimposed to a Sobel spatial filter applied to the most top upper layer of soil. Linear and circular features, in correspondence to the 8b interpretation, are indicated with yellow arrows (Latitude: 46°56′24.36″; Longitude: 21°12′32.67″, WGS84).

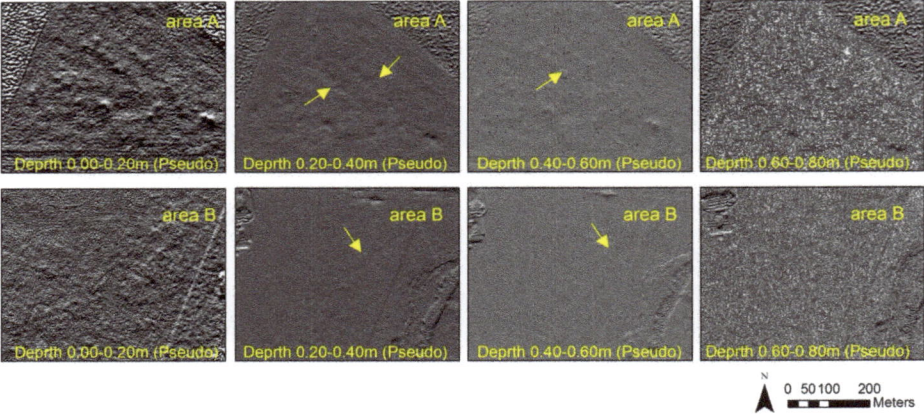

Figure 12. Enhanced images using the NDVI index for the slices between 0.00 and 0.80 m below the ground surface after the application of the Sobel filter. Top: Focus on the northern area (Area A) for each depth. Bottom: Focus on the southern area (Area B) for each depth. Area A and Area B are indicated in Figure 8 (Latitude: 46°56′24.36″; Longitude: 21°12′32.67″, WGS84).

6. Discussion

The proposed methodology aims to enhance optical satellite data intended for archaeological research based on the integration of different remote sensing datasets. The proposed regression approach provided acceptable but not completely satisfactory results. Evidently, there are many considerations regarding the regression model and some of the most important ones are discussed here. Essentially, when the regression model is established between the ground data and then projected to the satellite image (or part of this image) it is normally considered that the areas have similar context (i.e., same soil properties; same archaeological findings etc.). However, this is only an assumption, which is not always valid. The proposed enhanced methodology can provide more reliable results in cases were the archaeological background knowledge permits us to assume a relatively "homogenous" area. In addition, as the results of Figure 7 indicated, someone should proceed with caution when interoperating the data as the enhancement processing might provide false positive alarms or by-products of the processing (e.g., the linear anomaly observed in Figure 7 at around 50 m from NDVI model). Moreover, it is also expected that reflectance values (either from satellite image or calculated from the spectroradiometer) have good correlation with the geophysical prospection measurements. This is only valid if either crop or soil marks can be detected from the reflectance values for the "optimum" depth. Therefore, a correlation threshold between the optical and active data (i.e., geophysical measurements) can be set by the researchers to avoid any potential errors.

Additionally, the correlations between GPR and the vegetation indices from the ground data seem to vary with each depth but nonetheless each regression model in this "optimum" depth is a linear function of the input optical data. Therefore, as a linear enhancement, the generated enhanced optical image can only improve the initial optical image and its information. Of course, this is a usual limitation for any other enhancement technique including also the vegetation indices. Moreover, GPR variations that are not already in the optical image, and therefore cannot be explained by the regression model, cannot be mapped in the enhanced optical image.

Concerning the collection of ground data, both spectroradiometer and GPR sensors have been collected simultaneously to eliminate the influence (i.e., noise) from other external factors (mainly climatic changes i.e., rainfall). Though this limits the applicability of the proposed technique, future research is expected to be carried by the authors in proposing an "image-based" regression model. It is expected that noise which will be recorded within these regression models will be removed by setting gains and offset values.

Most important is the fact that such enhanced methodology should be a primary step prior to a larger archaeolandscape analysis, and not the final "model". As for every fusion, this technique will only work well in the case where both techniques produce satisfactory results. The purpose of enhanced techniques is primarily to identify "hot spot" areas where ground truth data (i.e., geophysical prospection) can take place. However, regression models—such as those discussed here—allow scientists to exploit the current capabilities of optical remote sensing to detect buried relics. Given the fact that such approach is quite new, further developments can additionally improve the final outcomes and provide more reliable results.

7. Conclusions

This paper discusses an approach applicable to any existing high-resolution satellite sensor in order to enhance image interpretation for the detection of hidden archaeological remains. The overall methodology consists of nine steps, with the most significant ones being those determining the optimum "regression" model based on the ground data (Steps 5 and 6).

The linear regression model developed in this study for the "optimum depth", was found to be the most encouraging (with Pearson coefficient up to almost 50%), providing promising results especially when elaborating the various hyperspectral indices (see Figure 7, REP index). Even though such high resolution hyperspectral datasets are not currently available, the use of airborne/ unmanned aerial vehicles (UAVs) equipped with small hyperspectral sensors could be a reliable alternative solution.

The proposed approach can be also applied in areas that have been partially examined through geophysical prospection in the past. In this case, these (archive) results can be further exploited for developing regression models and for calibration purposes in order to examine a wider area of interest. As in this case study presented here, archaeological excavations that will be carried out in the area can be used as a "calibration factor" to improve and enhance further the results.

Despite its simplicity, the linear regression model was found to be the most appropriate for the integration between the two datasets, which can be used for the enhancement of the satellite image. The application of the regression models in several satellite images acquired during different periods and therefore in different phenological stages of the crops can improve the interpretation for the detection of buried archaeological remains.

The final outcomes have demonstrated that further enhancement of optical remote sensing data is feasible, contributing in the prospection of archaeological targets. Nevertheless, additional experiments are expected to be carried out under different environmental conditions (e.g., climate and soil characteristics) and employing different satellite datasets. This will allow us to maximize its possible use as a supporting tool to archaeological investigation using also more advanced mathematical models such as machine learning and neural networks.

Supplementary Materials: The following are available online at www.mdpi.com/2076-3263/7/2/40/s1, Table S1: Vegetation indices applied in the study (see more details in [32]).

Acknowledgments: The present communication is under the "ATHENA" project H2020-TWINN2015 of European Commission. This project has received funding from the European Union's Horizon 2020 research and innovation programme under grant agreement No. 691936. The fieldwork campaign was supported by USA-NSF (U.S.-Hungarian-Greek Collaborative International Research Experience for Students on Origins and Development of Prehistoric European Villages) and the Wenner-Gren Foundation (International Collaborative Research Grant, "Early Village Social Dynamics: Prehistoric Settlement Nucleation on the Great Hungarian Plain"). William A. Parkinson, Richard W. Yerkes, Attila Gyucha and Paul R. Duffy provided their expertise and support in the fieldwork activities.

Author Contributions: Athos Agapiou conceived and performed the regression experiments; Vasiliki Lysandrou contributed to the interpretation of the regression model and its projection to the satellite image. Apostolos Sarris and Nikos Papadopoulos described the study area and performed the geophysical prospection. All authors have contributed to the interpretation of the results.

Conflicts of Interest: The authors declare no conflict of interest.

References

1. Lasaponara, R.; Masini, N. Satellite remote sensing in archaeology: Past, present and future perspectives. *J. Archaeol. Sci.* **2011**, *38*, 1995–2002. [CrossRef]
2. Giardino, J.M. A history of NASA remote sensing contributions to archaeology. *J. Archaeol. Sci.* **2011**, *38*, 2003–2009. [CrossRef]
3. Agapiou, A.; Lysandrou, V. Remote sensing archaeology: Tracking and mapping evolution in European scientific literature from 1999 to 2015. *J. Archaeol. Sci. Rep.* **2015**, *4*, 192–200. [CrossRef]
4. Keay, J.S.; Parcak, H.S.; Strutt, D.K. High resolution space and ground-based remote sensing and implications for landscape archaeology: The case from Portus, Italy. *J. Archaeol. Sci.* **2014**, *52*, 277–292. [CrossRef]
5. Reinhold, S.; Belinskiy, A.; Korobov, D. Caucasia top-down: Remote sensing data for survey in a high altitude mountain landscape. *Quat. Int.* **2016**, *402*, 46–60. [CrossRef]
6. Barone, M.P.; Desibio, L. A remote sensing approach to understanding the archaeological potential: the case study of some Roman evidence in Umbria (Italy). *Int. J. Archaeol.* **2015**, *3*, 37–44. [CrossRef]
7. Barone, M.P.; Desibio, L. Landscape archaeology of southern Umbria (Italy) using the GPR technique. In Proceedings of the 2015 8th International Workshop on Advanced Ground Penetrating Radar (IWAGPR), Florence, Italy, 7–10 July 2015.
8. Alexakis, D.; Sarris, A.; Astaras, T.; Albanakis, K. Detection of neolithic settlements in thessaly (Greece) through multispectral and hyperspectral satellite imagery. *Sensors* **2009**, *9*, 1167–1187. [CrossRef] [PubMed]

9.   Alexakis, A.; Sarris, A.; Astaras, T.; Albanakis, K. Integrated GIS, remote sensing and geomorphologic approaches for the reconstruction of the landscape habitation of Thessaly during the Neolithic period. *J. Archaeol. Sci.* **2011**, *38*, 89–100. [CrossRef]

10.   Banerjee, R.; Srivastava, K.P. Reconstruction of contested landscape: Detecting land cover transformation hosting cultural heritage sites from Central India using remote sensing. *Land Use Policy* **2013**, *34*, 193–203. [CrossRef]

11.   Agapiou, A.; Alexakis, D.D.; Lysandrou, V.; Sarris, A.; Cuca, B.; Themistocleous, K.; Hadjimitsis, D.G. Impact of Urban Sprawl to archaeological research: the case study of Paphos area in Cyprus. *J. Cult. Herit.* **2015**, *16*, 671–680. [CrossRef]

12.   Cerra, D.; Plank, S.; Lysandrou, V.; Tian, J. Cultural Heritage Sites in Danger—Towards Automatic Damage Detection from Space. *Remote Sens.* **2016**, *8*, 781. [CrossRef]

13.   Chen, F.; Lasaponara, R.; Masini, N. An Overview of Satellite Synthetic Aperture Radar Remote Sensing in Archaeology: From Site Detection to Monitoring. *J. Cult. Herit.* **2017**, *23*, 5–11. [CrossRef]

14.   Stek, D.T. Drones over Mediterranean landscapes. The potential of small UAV's (drones) for site detection and heritage management in archaeological survey projects: A case study from Le Pianelle in the Tappino Valley, Molise (Italy). *J. Cult. Herit.* **2016**, *22*, 1066–1071. [CrossRef]

15.   Siart, C.; Eitel, B.; Panagiotopoulos, D. Investigation of past archaeological landscapes using remote sensing and GIS: a multi-method case study from Mount Ida, Crete. *J. Archaeol. Sci.* **2008**, *35*, 2918–2926. [CrossRef]

16.   Sarris, A.; Papadopoulos, N.; Agapiou, A.; Salvi, M.C.; Hadjimitsis, D.G.; Parkinson, A.; Yerkes, R.W.; Gyucha, A.; Duffy, R.P. Integration of geophysical surveys, ground hyperspectral measurements, aerial and satellite imagery for archaeological prospection of prehistoric sites: The case study of Vésztő-Mágor Tell, Hungary. *J. Archaeol. Sci.* **2013**, *40*, 1454–1470. [CrossRef]

17.   Malfitana, D.; Leucci, G.; Fragalà, G.; Masini, N.; Scardozzi, G.; Cacciaguerra, G.; Santagati, C.; Shehi, E. The potential of integrated GPR survey and aerial photographic analysis of historic urban areas: A case study and digital reconstruction of a Late Roman villa in Durrës (Albania). *J. Archaeol. Sci. Rep.* **2015**, *4*, 276–284. [CrossRef]

18.   Morehart, C.T. Mapping ancient chinampa landscapes in the Basin of Mexico: A remote sensing and GIS approach. *J. Archaeol. Sci.* **2012**, *39*, 2541–2551. [CrossRef]

19.   Traviglia, A.; Cottica, D. Remote sensing applications and archaeological research in the Northern Lagoon of Venice: the case of the lost settlement of Constanciacus. *J. Archaeol. Sci.* **2011**, *38*, 2040–2050. [CrossRef]

20.   Gallo, D.; Ciminale, M.; Becker, H.; Masini, N. Remote sensing techniques for reconstructing a vast Neolithic settlement in Southern Italy. *J. Archaeol. Sci.* **2009**, *36*, 43–50. [CrossRef]

21.   Keay, S.; Earl, G.; Hay, S.; Kay, S.; Ogden, J.; Strutt, K.D. The role of integrated survey methods in the assessment of archaeological landscapes: the case of Portus. *Archaeol. Prospect.* **2009**, *15*, 154–166. [CrossRef]

22.   Ciminale, M.; Gallo, D.; Lasaponara, R.; Masini, N. A multiscale approach for reconstructing archaeological landscapes: Applications in northern Apulia (Italy). *Archaeol. Prospect.* **2009**, *16*, 143–153. [CrossRef]

23.   Piro, S.; Mauriello, P.; Cammarano, F. Quantitative integration of geophysical methods for archaeological prospection. *Archaeol. Prospect.* **2000**, *7*, 203–213. [CrossRef]

24.   Kvamme, K.L. Integrating Multidimensional Geophysical Data. *Archaeol. Prospect.* **2006**, *13*, 57–72. [CrossRef]

25.   Kvamme, K.; Ernenwein, E.; Hargrave, M.; Sever, T.; Harmon, D.; Limp, F. *New Approaches to the Use and Integration of Multi-Sensor Remote Sensing for Historic Resource.* Identification and Evaluation, Report of SERDP Project, SI-1263, University of Arkansas, Center for Advanced Spatial Technologies. 10 November 2006. Available online: http://s3.amazonaws.com/academia.edu.documents/10387950/si-1263-part1.pdf? AWSAccessKeyId=AKIAIWOWYYGZ2Y53UL3A&Expires=1496724950&Signature=BKPfKdlCCLxJ9Q% 2BoU30JtrM3MP8%3D&response-content-disposition=inline%3B%20filename%3DNew_Approaches_to_ the_Use_and_Integratio.pdf (accessed on 6 June 2017).

26.   Aiazzi, B.; Baronti, S.; Lotti, F.; Selva, M. A comparison between global and context-adaptive pansharpening of multispectral images. *IEEE Geosci. Remote Sens. Lett.* **2009**, *6*, 302–306. [CrossRef]

27.   Garzelli, A. Pansharpening of Multispectral Images Based on Nonlocal Parameter Optimization. *IEEE Trans. Geosci. Remote Sens.* **2015**, *53*, 2096–2107. [CrossRef]

28.   Agapiou, A.; Alexakis, D.D.; Sarris, A.; Hadjimitsis, D.G. Colour to grayscale pixels: Re-seeing grayscale archived aerial photographs and declassified satellite CORONA images based on image fusion techniques. *Archaeol. Prospect.* **2016**, *23*, 231–241. [CrossRef]

29. Schaepman, E.M.; Ustin, L.S.; Plaza, J.A.; Painter, H.T.; Verrelst, J.; Liang, S. Earth system science related imaging spectroscopy—An assessment. *Remote Sens. Environ.* **2009**, *113*, S123–S137. [CrossRef]

30. Agapiou, A.; Alexakis, D.D.; Sarris, A.; Hadjimitsis, D.G. Evaluating the potentials of Sentinel-2 for archaeological perspective. *Remote Sens.* **2014**, *6*, 2176–2194. [CrossRef]

31. Agapiou, A.; Sarris, A.; Papadopoulos, N.; Alexakis, D.D.; Hadjimitsis, D.G. 3D pseudo GPR sections based on NDVI values: Fusion of optical and active remote sensing techniques at the Vészto-Mágor tell, Hungary. In *Archaeological Research in the Digital Age, Proceedings of the 1st Conference on Computer Applications and Quantitative Methods in Archaeology Greek Chapter (CAA-GR), Rethymno Crete, Greece, 6–8 March 2014*; Papadopoulos, C., Paliou, E., Chrysanthi, A., Kotoula, E., Sarris, A., Eds.; Institute for Mediterranean Studies-Foundation of Research and Technology (IMS-Forth): Rethymno, Greece, 2015.

32. Agapiou, A.; Hadjimitsis, D.G.; Alexakis, D.D. Evaluation of broadband and narrowband vegetation indices for the identification of archaeological crop marks. *Remote Sens.* **2012**, *4*, 3892–3919. [CrossRef]

33. Hegedűs, K. Vésztő-Mágori-domb. In *Magyarország Régészeti Topográfiája VI. Békés Megye Régészeti Topográfiája: A Szeghalmi Járás 1982 IV/1*; Ecsedy, I., Kovács, L., Maráz, B., Torma, I., Eds.; Akadémiai Kiadó: Budapest, Hungary, 1982; pp. 184–185. (In Hungarian)

34. Hegedűs, K.; Makkay, J. Vésztő-Mágor: A Settlement of the Tisza Culture. In *The Late Neolithic of the Tisza Region: A Survey of Recent Excavations and their Findings*; Tálas, L., Raczky, P, Eds.; Szolnok County Museums: Budapest-Szolnok, Hungary, 1987; pp. 85–104.

35. Makkay, J. Vésztő–Mágor. Ásatás a szülőföldön. Békés Megyei Múzeumok Igazgatósága, Békéscsaba. 2004. Available online: http://mek.oszk.hu/07600/07616/07616.pdf (accessed on 6 June 2017).

36. Parkinson, W.A. Tribal boundaries: Stylistic variability and social boundary maintenance during the transition to the Copper Age on the Great Hungarian Plain. *J. Anthropol. Archaeol.* **2006**, *25*, 33–58. [CrossRef]

37. Juhász, I. A Csolt nemzetség monostora. In *A középkori Dél-Alföld és Szer*; Kollár, T., Ed.; pp. 281–304. Available online: http://opac.regesta-imperii.de/lang_en/anzeige.php?sammelwerk=A+k%C3%B6z%C3%A9pkori+D%C3%A9l-Alf%C3%B6ld+%C3%A9s+Szer (accessed on 6 June 2017).

38. Gyucha, A.; Yerkes, W.R.; Parkinson, A.W.; Sarris, A.; Papadopoulos, N.; Duffy, R.P.; Salisbury, B.R. Settlement Nucleation in the Neolithic: A Preliminary Report of the Körös Regional Archaeological Project's Investigations at Szeghalom-Kovácshalom and Vésztő-Mágor. In *Neolithic and Copper Age between the Carpathians and the Aegean Sea: Chronologies and Technologies from the 6th to the 4th Millennium BCE. International Workshop Budapest 2012*; Hansen, S., Raczky, P., Anders, A., Reingruber, A., Eds.; Habelt-Verlag: Bonn, Germany, 2015; pp. 129–142.

geosciences

MDPI

Article

A Manifold Approach for the Investigation of Early and Middle Neolithic Settlements in Thessaly, Greece

Tuna Kalayci [1,*], François-Xavier Simon [1,2] and Apostolos Sarris [1]

[1] Laboratory of Geophysical—Satellite Remote Sensing & Archaeo-Environment,
Foundation of Research & Technology, Hellas (F.O.R.T.H.), Melissinou & Nik. Foka 130, 74100 Rethymno,
Crete, Greece; francois-xavier.simon@inrap.fr (F.-X.S.); asaris@ims.forth.gr (A.S.)

[2] INRAP/Institut National de Recherches Archéologiques Préventives, Direction Scientifique et Technique,
75104 Paris, France

* Correspondence: tuna@ims.forth.gr; Tel.: +30-28310-25-146 (ext. 56627)

Received: 28 July 2017; Accepted: 31 August 2017; Published: 6 September 2017

Abstract: The IGEAN (Innovative Geophysical Approaches for the study of Early Agricultural villages of Neolithic) Thessaly project focused on Early and Neolithic settlements in Thessaly, Central Greece. The aim of the project was to highlight in an extensive way differences in settlement layouts while investigating commonalities as a way to understand Neolithic use of space. To accomplish this, a suite of geophysical prospection techniques (geomagnetic, electromagnetic induction, and Ground Penetrating Radar (GPR)), aerial platforms (historic aerial imagery and Remotely Piloted Aerial Systems (RPAS)) as well as very high resolution spaceborne sensors were integrated to acquire comprehensive pictures of settlements. Results of the IGEAN project provide archaeological information on the dynamic character of enclosures, the structure of architectural features and open spaces within sites as an indication of economic or communal spaces. At the same time, they demonstrated the importance of employing a suite of different geophysical techniques to reveal different aspects of the hindered prehistoric settlements that could not be highlighted with a single geophysical approach.

Keywords: Neolithic Thessaly; geophysical prospection; integrated approaches; enclosures; settlement patterns

---

1. Introduction

The Neolithic period in Europe is widely considered as an important period because migrant hunters and gatherers gave way to permanent agrarian societies preoccupied with the cultivation of food-crops and animal husbandry for sustenance. Neolithic dwellers conceptualized their surroundings for the first time in durable architectural forms as a community in flat extended settlements, but most notably on large habitation mounds—or combining the above two.

The southern Balkan Peninsula served as a gateway for the Neolithization of Europe [1]. In particular, early farming communities appeared on an extensive scale in Thessaly, Greece and provided the seeds for a new cultural landscape. The early chronology and high density of Neolithic tell-settlements (locally called *magoules*) make Thessaly a key area for understanding the pathways in which the Neolithic emerged and spread towards the continent.

In consideration of the significance of the region, the IGEAN (Innovative Geophysical Approaches for the study of Early Agricultural villages of Neolithic Thessaly) (http://igean.ims.forth.gr) [2] project investigated the intra-site patterns of Neolithic settlements using a suite of multi-scalar geospatial methodologies. Furthermore, the project surpasses conventional remote sensing studies, deviating from the recent multi-sensor practice and adopting a manifold approach [3]. As a result of this strategy, Thessaly now provides a wide range of instrumental datasets and makes it possible to pose key questions related to: potential geographical factors behind the establishment of early farming communities, the spatial development of agriculturalist villages, the variations in the use of space and land-use patterns, the internal organization of structures, the differences in types of settlements, and the functions of settlement enclosures.

The IGEAN Project focused on numerous prehistoric sites in Thessaly. Due to the nature of geospatial practice, it is rarely possible to determine exact periods of habitation. In order to have better chronological control over archaeological information inferred from geospatial data, this paper solely focuses on five settlements with only Early Neolithic and Middle Neolithic habitation phases. Among these sites, Almyros 2, Karatsantagli, Perdika 2, and PerivleptoKastraki 2 are dated to both periods, whereas Kamara is dated only to the Middle Neolithic.

2. Background

Thessaly is a lowland area located in Central Greece (Figure 1). It isa closed geographic region with mountainous borders on three sides. Karditsa Basin in the west and Larisa Basin in the east cover most of Thessaly [4]. These major basins are elongated in northwest–southeast and are divided by low ridges. Sometime during the Plio-Quaternary, lakes filled in the submerged basins, but the land was eventually formed again with uplifting. Despite its dynamic tectonic evolution, Thessaly is a relatively smooth-relief area. Specifically, in Larissa Basin, the maximum altitude difference does not exceed 50 m [5].

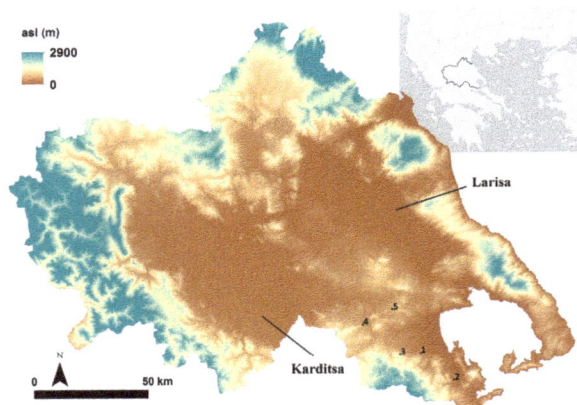

Figure 1. Thessaly is located in Central Greece. Larisa Basin in the east and Karditsa Basin in the west forms the majority of region; these basins are separated by low hills. Sites under investigation are: (1) Almyros 2; (2) Kamara; (3) Karatsantagli; (4) Perdika 2; and (5) Perivlepto/Kastraki2.

The Pineios River is the bearer of the active hydrographic network of Thessaly. It runs about 200 km and empties into the Aegean Sea while forming a delta [6].The Pineios River and its tributaries have been important features in shaping the landscape and the human activities since the prehistoric times; alluviation as well as periods of pedogenesis must have heavily affected the locations of prehistoric habitations [7]. In consideration of episodic floods for every 25 to 50 years [6], one can further argue that the Pineios River had greatly influenced agricultural production since the onset of early farming communities.

Thessaly contains remarkable evidence of continuous habitation in all phases of the Neolithic period, mainly in the form of *magoules* of which almost 350 have been identified to date. These habitation mounds have attracted the interest of scientific inquiries since the beginning of the twentieth century. The major source of information is due to the early excavations carried out at Sesklo and

Dimini [8]. Wace and Thompson [9] added further scholarly contributions and their work is still widely referenced today. This fortunate and early initiation of scientific work, however, also created a "Thessalocentrism", transforming Thessaly into a homogenous cultural core and currently shaping the archaeology of neighboring regions [10].

During the 1980s (and onwards), the research in Neolithic Thessaly was geared towards the identification of new sites, understanding environmental contexts, and constructing settlement patterns. Halstead [11] suggested that Neolithic settlements were established within a wide range of micro-environments, rather than on exclusively arable lands. Gallis [12] produced the first catalogue of the *magoules* and provided information about their physical characteristics, extent, and chronological framework. The analysis carried out by Van Andel and Runnels [13] proposed that Neolithic communities were built on floodplains in order to take advantage of spring floodwater farming. Perlès [1] examined the distribution of Early Neolithic sites and argued that they were not concentrated near water sources, but instead were clustered in territories, resulting in variations in site densities as well as distances between sites.

A rejuvenated approach to Neolithic Thessaly has been emerging around the discussions on formation of agencies [14], detecting complexities in social organizations [15], and negotiating identities [16]. These studies make use of existing rich datasets thanks to the long research history of the Neolithic Thessaly while providing in-depth discussions on Neolithic praxis. Recent advancements in computational technologies tacked with theoretical paradigms also provide means for an understanding of the everyday life in Neolithic Thessaly [17].

3. Aims

Within a new scientific framework of investigating Neolithic cultures, the GeoSatReSeArch Lab of the Foundation of Research & Technology, Hellas (FORTH) initiated from 2013 to 2015 a multi-year geophysical and remote sensing campaign to study the physical landscape and living dynamics of Neolithic settlements within the coastal hinterlands of eastern Thessaly. The IGEAN project is the first systematic and extensive geophysical survey of Neolithic *magoules* in Thessaly. A suite of innovative multi-technique (manifold approach), multi-sensor geophysical instruments and geospatial technologies (magnetometry, ground penetrating radar (GPR), electromagnetics, Remotely Piloted Aerial Systems (RPAS), and historic aerial imagery as well as very high resolution satellite data) has revealed massive information related to the spatial characteristics of Neolithic settlements. It is anticipated that the results of this study will assist to achieve further archaeological inferences. The focus of this study is two-fold. Methodologically, it shows the importance of following manifold geospatial approach to draw a complete or close-to-complete picture of past habitations. The claim is that only using various sensors/approaches the actual settlement biographies can be achieved. Deploying a single type of sensor/techniques (ground-based, airborne, or space-borne) may not only provide deficient information, but it may also lead into erroneous inferences about the past use of space.

From an archaeological point of view, the study aims to reveal significant spatial variations between Neolithic settlements while emphasizing peculiar commonalities in the Neolithic use of space. This holistic approach is an important step towards a better understanding of everyday life in Thessalian farming communities. These differences and shared traits can only derive through an integrated geoinformatics methodology that has advantages over traditional excavations.

4. Methodology

4.1. Aerial Photography and Satellite Remote Sensing

For the broader exploration of prehistoric settlements and the documentation of their environmental contexts, the IGEAN Project employed satellite and aerial remote sensing. Very high-resolution (VHR) data from GeoEye-1 and WorldView-2 platforms covered the target sites.

The spatial resolution of these sensors (panchromatic: ~0.5 m and multispectral: ~2.0 m) enables the detection of small details on the ground. Following the orthorectification process, pansharpeningfurther increased the spatial resolution of multispectral bands for a more detailed representation of sites and their immediate surroundings.

In addition, 26 historical aerial photographs were acquired from the Hellenic Military Geographical Service (GYS). Historic imagery was georeferenced using available higher accuracy spatial data. Furthermore, the IGEAN Project also used the orthophotography archive of the National Cadastre and Mapping Agency (Ktimatologio). These images covered the time spam between 2007 and 2009.

A Droidworx CX4 RPAS (Remotely Piloted Aerial System), also referred as UAV (Unmanned Aerial Vehicle), was used for the acquisition of visible light imagery of Neolithic settlements. Moreover, thanks to the capability of collecting data from the near-infrared portion of the electromagnetic spectrum, the system provided information on archaeological proxy data. Finally, high-resolution high-accuracy digital elevation models (DEMs) of settlements created from the RPAS assisted in making archaeological interpretations and broader inferences.

4.2. Geophysical Prospection

The manifold geophysical research of the IGEAN consists of single and multi-sensor geomagnetic prospection, Ground Penetrating Radar (GPR) surveys, and multi frequency Electromagnetic Induction (EMI) investigations. Details of instrument sampling parameters and approximate depth of investigations are given in Table 1. The integrated approach drastically increases the efficiency of detection of subsurface archaeological features, and thus, provides a more complete picture of past occupation [18].

Table 1. The instrumentation of the IGEAN project and their parameter values.

Instrument	Transect (Dx in m)	Sampling (Dy in m)	~Sensing Depth (m)
Bartington G601	0.5	0.125	1.0–2.0
SENSYS	0.5	0.05	1.0–2.0
Noggin	0.5	0.025	0.5–3.0
Geophex GEM-2	1.0	0.25	1.0–2.0

Sensorik & Systemtechnologie (SENSYS) MX Compact Survey System was used for the geomagnetic data collection. The SENSYS MX is a multi-channel measurement system equipped with FGM600 magnetometers. The selected sensor configuration was an 8-sensor setup with 50 cm separation in order to optimize between prospection coverage and investigation resolution. A Javad Triumph 1 Global Navigation Satellite System (GNSS) receiver was used for the positioning of the multi-channel system. A custom data processing methodology was developed for the analysis of magnetic data [19].

The rugged Mediterranean topography does not always make it possible to conduct a geomagnetic survey with a wide array system. Tall and sturdy bushes, orchards, vineyards, and rock outcrops create obstacles so that it is sometimes necessary to employ single sensor gradiometers. Bartington G601 gradiometers replaced the SENSYS MX array when such challenging areas were encountered.

For the GPR data collection, a Noggin system with a 250 MHz antenna was used. For the prospection, grids were designated to the areas of interest and data were collected in parallel transects. Data were recorded along each line of every grid with a constant step of 0.025 m, along parallel transect 0.5 m apart. The transects that derived from GPR were processed using EKKO View Deluxe, GFP Edit4, EKKO Mapper4, and EKKO Project 2 by Sensors and Software.

A broadband multi-frequency Geophex GEM-2 electromagnetic sensor was operated at five different frequencies (5010, 13,370, 22,530, 31,290, and 40,050 Hz). The instrument is carried at a height of approximately 0.3 m above the ground surface. The acquisition frequency was set to 1 Hz.

A custom-built Python script allowed the transformation of the EM data into magnetic susceptibility and electromagnetic conductivity based on vertical electric sounding (VES) measurements for instrument calibration [18].

5. Results

5.1. Almyros 2 (22.736° E, 39.168° N)

The WorldView-2 image (acquisition date: 4 June 2012) reveals four major soil marks related to the site (Figure 2). M0 is located to the east of the proposed location of the settlement. It may be due to past human activities. The grey colored patch is surrounded by a darker rectilinear feature in the west and the south. Due to the sharp turn it makes, this feature is less likely to be related to a Neolithic ditch, but later occupational phases cannot be ignored. M1 represents the extent of Almyros 2. The brighter core of M1 overlaps with the core of the settlement and the circular shape of the soil mark coincides with the extent of the site. At a first glance, M2 seems to be unrelated to M1. However, geomagnetic survey (see below) suggests M2 was also bounded by a settlement enclosure, and thus, must have been of anthropogenic origin (suggesting two neighboring tell sites?). To the south of Almyros 2, M3 has also the potential to represent past human activity. On the other hand, the geomagnetic survey results over M3 indicate the opposite, and, thus, M3 must be a natural soil formation. These two opposing cases between M2 and M3 marks suggest the need for the cautious interpretation of aerial and satellite imagery of archaeological features.

Figure 2. The true color WorldView-2 imagery (acquisition date: 4 June 2012) revealing soil marks related to the settlement.

5.1.1. Geomagnetic Survey

Geomagnetic data reveal a wide enclosure (A1) running around the settlement (Figure 3). A secondary and thinner enclosure is located closer to the core of the settlement supporting the outer one in the eastern section. In consideration of their shapes and positive magnetic signatures, they were most likely to be ditches. A2 is another similar feature encircling the core of the settlement. Even though not complete in its shape, the feature is much thinner and better defined than other documented enclosures. In addition, taking into account well-defined openings/gates in A2, it is highly possible that the feature was a (mudbrick) wall-enclosure rather than a ditch.

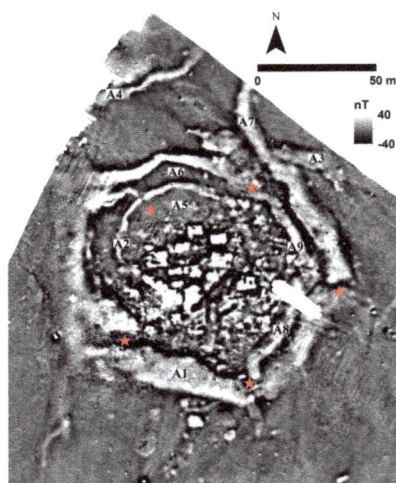

Figure 3. Geomagnetic prospection results from Almyros 2. Red stars indicate possible entrances through the ditch system.

Of additional interest is the enclosure formed by (A3) and (A7) where its half-circular shape suggests a smaller neighboring site. Even though the geomagnetic data coverage is not complete, satellite data provide evidence for its extent and size as a soil-mark. Due to the nature of the geophysical data, it is not possible to comment on the contemporaneity of habitation or use. However, considering the relatively large size of A3 with respect to the small-scale site, which it is bounded, one may suggest this "settlement" had some significant meaning for the Neolithic dwellers of Almyros 2. A similar possible enclosure (A4) is located to the north of Almyros 2. Unlike A3, it lacks context and useful geophysical information for further analysis. Thus, the feature is excluded from the final interpretation.

Geomagnetic data also reveal information on a series of buildings at the core of the settlement which are primarily aligned in East–West direction. Their signatures suggest these features were mainly made of stone walls and possibly clay floors. Even though it seems like they suffered a burning event the buildings remained intact in most of the cases. This is also to say that their destruction must have been minimal and maybe so in control. The wall-enclosure (A2) further complicates the settlement pattern. In consideration of its circular shape and spanning area, A2 divides the distribution of the buildings into two spatial clusters. When investigated with this division in mind, it becomes clear that in their interior buildings are better defined and their shape and orientation are more consistent. It is harder to delineate buildings outside of A2, suggesting different architectural structures or a different treatment during and after use.

Finally, an empty zone (A5) to the north of the core of the settlement attracts attention. The zone is magnetically quiet suggesting a decision for keeping the area free of human occupation; or a "cleaning" prior to the destruction/abandonment of the settlement. A similar observation can be made for the arching space between the outer ditch (A6) and the wall-enclosure (A2).

5.1.2. Electromagnetic Induction Survey

The magnetic susceptibility map confirms the circular shape of the settlement (Figure 4a). The ditch system and the wall-enclosure are also clearly defined in the data. Overall, the susceptibility increases towards the core and south of the settlement, most probably of the anthropogenic activities in the particular sections of it. The empty zone also traced in the geomagnetic data (A5, A6) further contributes to this separation.

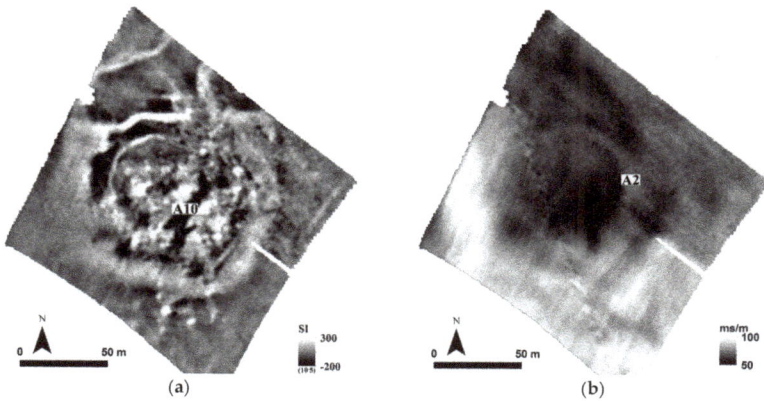

Figure 4. Magnetic susceptibility (**a**); and electromagnetic conductivity (**b**) results from Almyros 2.

Investigated in relation to geomagnetic data (Figure 5), susceptibility map may be used to distinguish architectural, if not cultural phases in Almyros 2. Although the settlement phasing cannot be proposed in an absolute certainty, there is reasonable doubt thanks to different penetration depths of instruments; magnetic susceptibility (due to in-phase component of the electromagnetic induction) measures features shallower than geomagnetic measurements. Under the assumption that shallower features belong to a later phase of occupation the spatial shifts of archaeological features (as one progresses from one depth to another) indicate a change in the layout of settlement pattern.

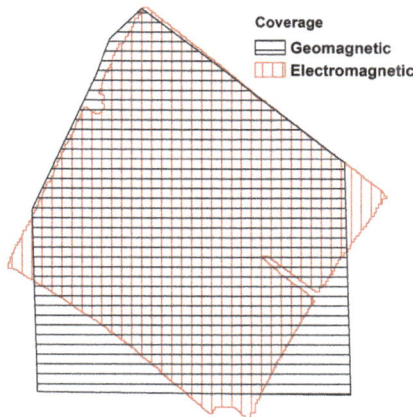

Figure 5. The overlay for the prospection coverage by different methods at Almyros 2.

In the geomagnetic data, the end point of feature A7 (the earlier phase) joins with the end point of A3, closing the space for the secondary site. In magnetic susceptibility map the terminal of A7 (the later phase) was shifted and was bent towards the Almyros 2, suggesting that the secondary site was linked to the main settlement at a later period.

The second phasing information is related to the eastern portion of the ditch system. In the geomagnetic data (the earlier phase), there is clear evidence for a double ditch system; the main outer ditch and the secondary inner ditch. In the magnetic susceptibility map (the later phase), the secondary inner ditch (A8 and A9) mostly disappears and only its highly magnetic portions remain visible.

Specific areas with low magnetic susceptibility values require elaboration. The first is the smaller site bounded by the ditch (A3 and A7). Susceptibility values are in contrast with the main settlement, suggesting low level of human presence; at least for the area covered by the EMI survey. The second is the arching space (A6). Susceptibility values reveal that the space contained almost no anthropogenic activities. The only exception for this is the break with relatively higher susceptibility values due to the gate in the wall-enclosure A2. Next, the linear feature (A10) running on northeast–southwest direction seems to be a later intrusion to the site, as it cuts through a series of geomagnetic features. If A10 indeed depicts an event, which cleared the marks of past anthropogenic activities, abovementioned archaeological areas with low susceptibility values indicate excessive clearing during the habitation of the site or just prior to its abandonment. Finally, minimal human activity alongside the ditch-system is evident through documented lower susceptibility values. Particularly, it is significant to encounter these areas at critical gateways of the ditch.

Electrical conductivity map shows a clear distinction between the northeast (resistive) and southwest (conductive) surroundings of the settlement (Figure 4b). As expected, the site itself is part of the high resistivity section, related to the compaction of the soil, due to intensive human activities of the past. It is possible that anthropogenic soils "spilled" towards northeast; or there was another locus of habitation in this area, as already suggested by the geomagnetic data.

The conductivity data further expose the circular shape of the Neolithic settlement. Its shape is particularly well defined in the north and east thanks to higher conductivity area between the wall-enclosure (A2) and the main ditch system. This layout further supports the reconstruction of the circular wall-enclosure to the south. Finally, resistivity values further increase towards the south where structural remains appear to be denser in contrast with the north of settlement.

5.2. Kamara (22.880° E, 39.078° N)

The most informative high-altitude view of Kamara comes from a 1971 aerial photograph (Figure 6). Recent intrusions, including a mid-size construction on the site, hinder a complete evaluation. This transformation of the landscape emphasizes the importance of historic imagery, especially for the areas witnessing industrial scale agriculture and urbanization. The remnants of the main settlement are visible after the soil mark M4. Feature M5 further suggests that Kamara was also extended towards the south. The geophysical prospection could not cover this southern portion due to modern obstacles, and, thus, the aerial photography remains in use for determining the possible extent of the site.

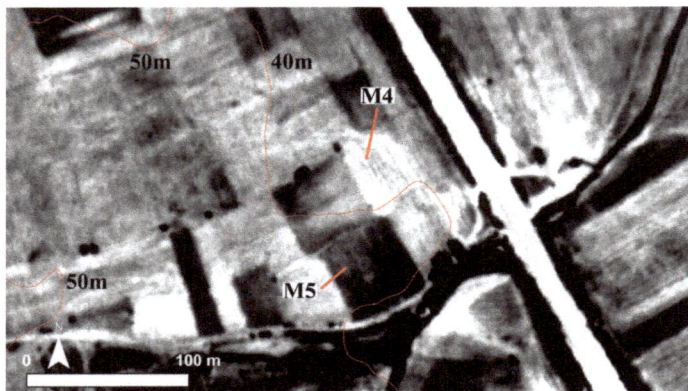

Figure 6. The 1971 historic aerial imagery from Kamara with modern elevation contours. The data predate modern intrusions to the site revealing two distinct soil marks.

Geomagnetic Survey

Geomagnetic data from Kamara reveal the location and sizes of architectural features related to potentially Neolithic houses (Figure 7). They are not well preserved and there is indication of collapse. Nevertheless, their geomagnetic signatures suggest they were rather rectangular in shape and were mainly built by clay rich material. Considering its compact layout and relatively heterogeneous magnetic distribution in comparison to the rest of the settlement, B1 might have been the core of the site.

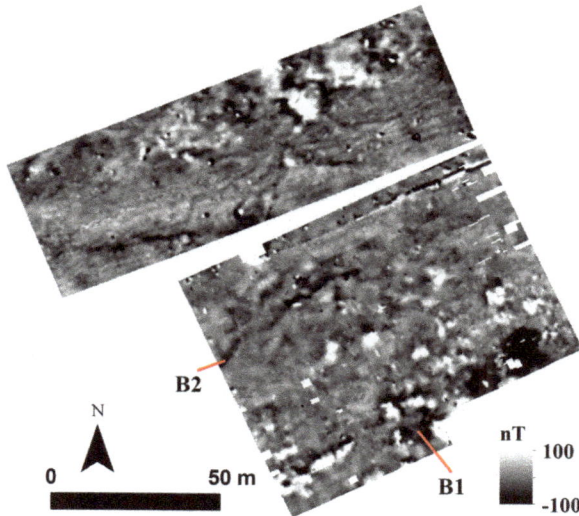

Figure 7. Geomagnetic prospection results from Kamara.

Other rectangular magnetic features are detected in the north. They are separated from the rest of the Neolithic settlement with a distance of ~80 m. These northerly features do not bear a similar soil mark in historic aerial photography and their sizes are significantly larger than the architectural features of Kamara. Therefore, it can be suggested that they were most probably dated to a later period than Middle Neolithic Period.

There is no indication of a settlement enclosure in the form of a ditch or a circuit wall. If the soil mark on the historic aerial imagery is an accurate indicator for the size of the settlement, the enclosure should have been encountered immediately to the north of the soil mark. A weak candidate for an enclosure is B2. However, its magnetic signature (negative magnetic gradient) is contra to the well-established signatures (positive magnetic gradient) of other surveyed Neolithic settlements. A wall-enclosure may be ruled out due to the irregular shape of the feature. Therefore, B2 must have been a formation with natural origins.

5.3. Karatsantagli (22.701° E, 39.161° N)

Historic aerial photography from 1945 shows two features (possibly enclosures) which are most probably related with the site (Figure 8). M5 (north) is a semi-arc bounding the settlement in the north. A high resolution Digital Terrain Model (DTM) acquired from the RPAS additionally supports the claim that M5 (north) was built in part of the site as the elevation contour of the settlement succinctly traces this feature (Figure 9). Furthermore, the extent of the site to the east and to the south as documented by the geomagnetic prospection (see below) merges with this feature, providing a better representation of the limits of the site.

North–south running anomaly M5 (south) must delineate the limit of the site in the west. The feature slightly cuts the elevation counters, and thus, raises a doubt for its prehistoric origins. Nevertheless, the feature appears to be the continuation of the ditch-enclosure (C2) as documented in the geomagnetic dataset. Evaluated together, geophysics data, historic aerial photography, and RPAS imagery offer the most complete picture.

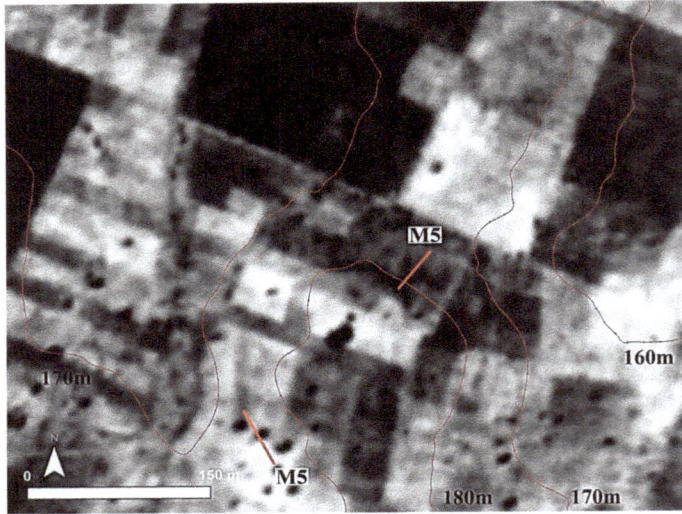

Figure 8. Historic aerial photograph dated to 1945 reveals two potential ditches, which are no longer visible in modern imagery. The modern elevation contours range between 160 m and 180 m within the area.

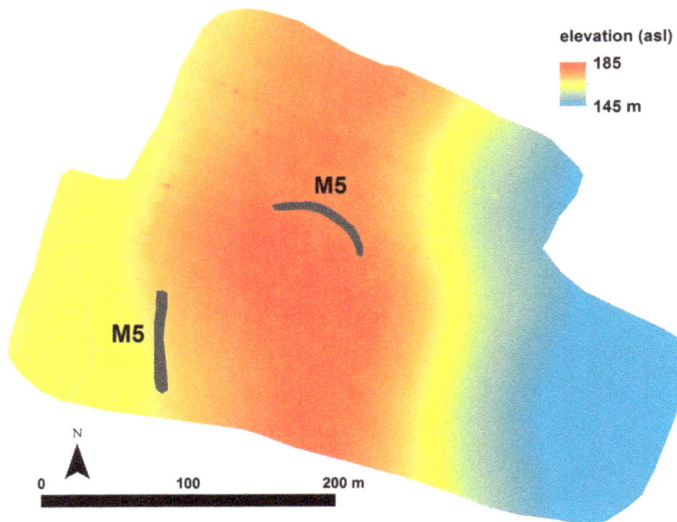

Figure 9. A Digital Terrain Model is processed from RPAS imagery. The details of the model provide further proof for the location and extent of settlement enclosures.

5.3.1. Geomagnetic Survey

Geomagnetic data show a clear separation between anthropogenic and non-anthropogenic areas (Figure 10). This is most evident in the southern and eastern side of the settlement where there is a drastic increase in the homogeneity of the readings. In the west, a wide (~3 m) and highly magnetic feature (C1) is probably a ditch-enclosure and tightly bounds the settlement. Further to the west of C1, the remnants of a curvilinear feature (C2) with a larger span might have been another enclosure, but poorly preserved. Magnetically homogenous area in between C1 and C2 suggests minimum anthropogenic activity. Neither of these ditch features are detected in the south and in the east, but rather the settlement terminates abruptly without any specific architectural marker.

Within the habitation limits of the site, there are architectural features with visible internal partitions. Their magnetic signatures indicate that they had stone wall foundations and probably mudbrick walls and/or clay floors. A clear evidence for the former comes from C3 where the walls have apparent negative magnetic gradient within the clay rich soil, as suggested by the high magnetic susceptibility values of the background. Feature C4 provides evidence for burning; high magnetic values after burning with respect to the background indicates the architectural use of magnetically rich material, i.e., clay.

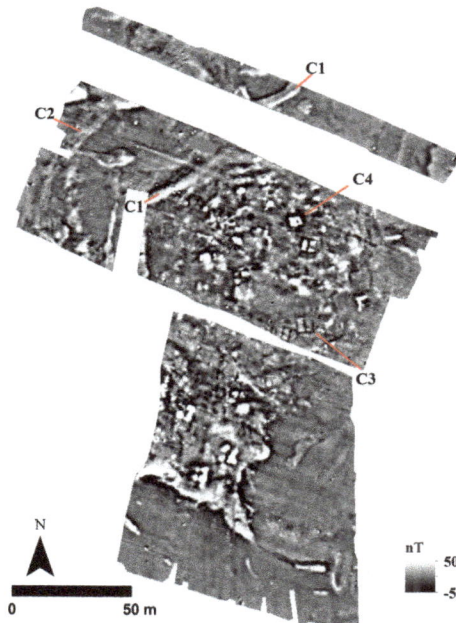

Figure 10. Geomagnetic prospection results from Karatsantagli.

5.3.2. Electromagnetic Induction Survey

Magnetic susceptibility values are high and exhibit a close correlation with magnetic results (Figure 11a). The proposed ditch-enclosure (C1) has high susceptibility values, further supporting the hypothesis for its use. Further to the west of the C1, high susceptibility values highlight the core of the settlement. In the eastern edge of the survey area, values are lower than the average susceptibility readings, and thus, further marking the end of habitation. Another low susceptibility area is detected abutting the ditch-enclosure C1. Like in the case of Almyros 2, the area might be indicating minimal anthropogenic activity around the opening of the ditch.

Due to its lower spatial resolution, the EMI survey does not provide architectural details of domestic dwellings. However, it further helps to discriminate buildings which faced pyro activities and their variable intensities. This is the most visible for the burnt structure, C4.

Electrical conductivity dataareanother evidence for the separation between the site (lower-conductivity) and off-site (C5) (higher conductivity) (Figure 11b). A sizeable area at the center of the site (C6) has the lowest conductivity values as a hint towards built environment. Geomagnetic data, however, do not reveal any structures for the same area (Figure 10; also see Figure 12 for an overlay of extent). Therefore, it is probable that the area was devoid of buildings, but had a very compact surface, suggesting C6 was devoted to specialized activities.

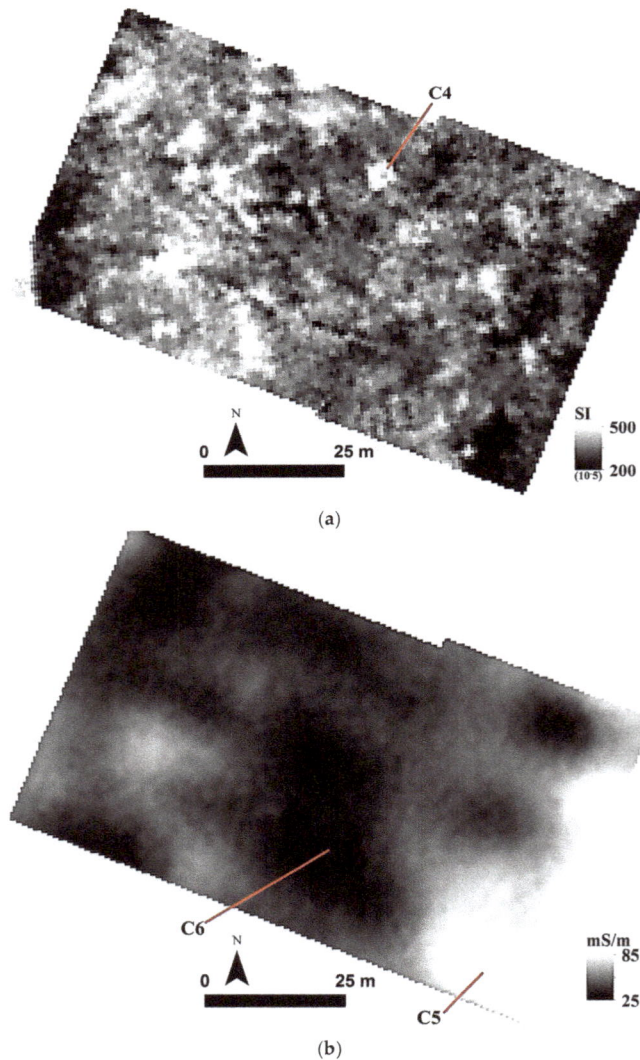

(a)

(b)

Figure 11. Magnetic susceptibility (**a**); and electromagnetic conductivity (**b**) results from Karatsantagli.

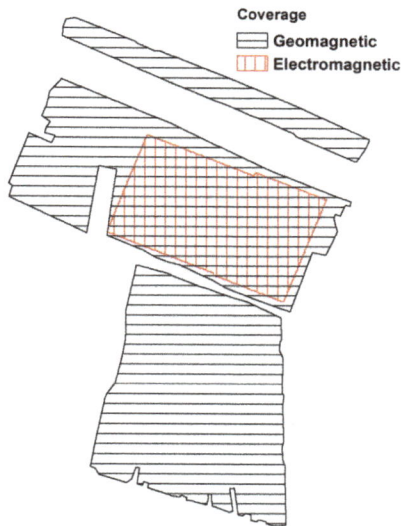

Figure 12. The overlay for the prospection coverage by different methods at Karatsantagli.

5.4. Perdika 2 (22.613° E, 39.245° N)

GeoEye-1 imagery (acquisition date: 3 June 2010) provides the most useful information when viewed in false color (bands 4-2-1) (Figure 13). Due to limited vegetation coverage on site, the use of near-infrared part of the electromagnetic spectrum is not optimal. Nevertheless, this band combination is still useful for emphasizing the soil marks. For instance, northwest–southeast running M6 is identified as a ditch-enclosure and the shape and the extent of Perdika 2 is visible in the south.

Figure 13. False color (Bands 4-2-1) GeoEye-1 imagery (acquisition date: 3 June 2010).

5.4.1. Geomagnetic Survey

Geomagnetic data reveal a complex site layout (Figure 14). Multiple settlement cores and enclosures indicate various architectural and even cultural phases. It is almost impossible to provide secure chronologies of features due the nature of geophysical prospection. On the other hand, superimposition of magnetic signatures can be tentatively used to define the occupational history of Perdika 2.

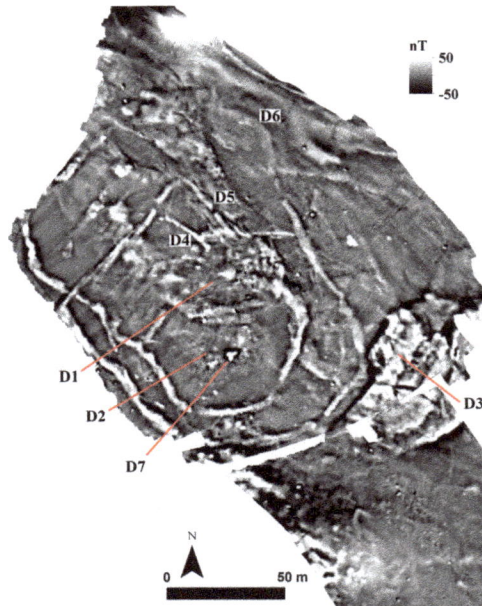

Figure 14. Geomagnetic prospection results from Perdika 2.

Survey results suggest three settlement cores. Cluster D1 appears to be the oldest core that most probably had expanded towards the east acquiring its ellipsoidal shape. The expansion of ditch-enclosures, first D4, then D5, and finally D6 supports this hypothesis. As such, we can say that Perdika 2 was initially a smaller settlement and eventually expanded, in both its core and the area bounded by its ditch-system.

Cluster D2 must have been a later core since its primary enclosing ditch (D6) cuts the core D1. The secondary ditch makes use of northern section of D1 and expands the settlement area. If the proposed reconstruction and superimposition of D2 system is correct, the outermost ditch of D2 must have made use of existing southern ditch of D1. This also suggests that the previous layout of the site was known to its new inhabitants, further suggesting little to no hiatus in between the phases.

The third and the youngest core (D3) is the largest and bears no spatial relationship with the first two. It was built on top of the outermost ditch of D2 and thus can be dated to a later period than the largest extent of D2. There are no immediately visible architectural elements in D3, but the heterogeneity and the strength of geomagnetic data suggest that the area was the host for concentrated anthropogenic, and probably domestic activities.

Geomagnetic data provide limited information on architectural environment and this is valid for all three proposed phases. Even though it is possible to map areas of occupation (e.g., settlement cores) and to detect targets as potential structures (e.g., D7), there is no visible settlement pattern within the bounds of ditch-enclosures.

5.4.2. Electromagnetic Induction Survey

Magnetic susceptibility map confirms the multifaceted ditch system as well as the location and extent of settlement cores (D1, D2, and D3) (Figure 15). Both of these feature sets are characterized by high susceptibility values. There are areas of low susceptibility values within the bounds of the settlement, suggesting relatively constricted anthropogenic use. There is no evidence for architectural features other than a possible two roomed building (D8) within core D1.

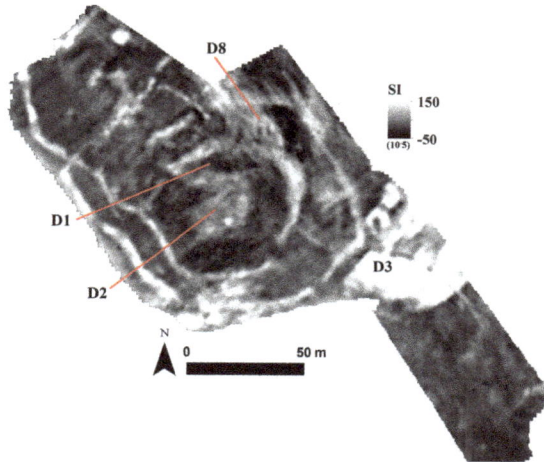

Figure 15. Magnetic susceptibility survey results from Perdika 2.

Electrical conductivity data show pockets of high conductivity (low resistivity) areas (Figure 16). These pockets do not form a major pattern. Nevertheless, the cluster in the north (D9) mimics the orientation of outermost enclosure of D2, albeit with a slight rotation. The major high conductivity area is the core D3 itself.

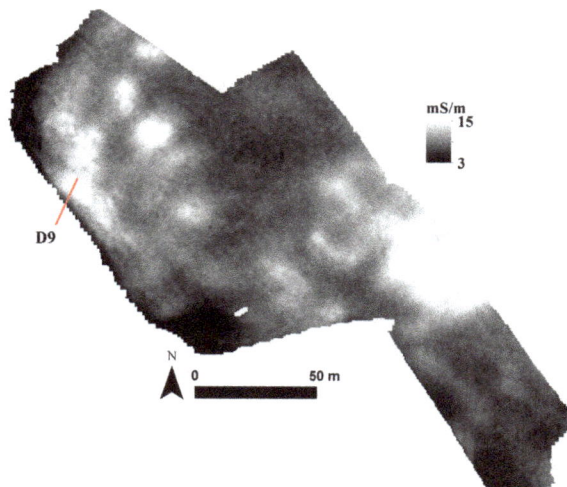

Figure 16. Electrical conductivity data from Perdika 2.

5.4.3. Ground Penetrating Radar Survey

GPR signals document a structure with multiple rooms aligned in east–west direction (D10) and located at ~70–80 cm below the surface (Figure 17). The particular structure is located at the central section of the site, which was the locus of the GPR survey (Figure 18). The fact that this multi-room building is vaguely visible in magnetic data suggests intact, non-burnt architecture with substantial stone wall foundations. To the south of D10, there are indications of another structure (D11) with little to no preservation. This structure appears to be smaller than D10, but is still within the bounds of settlement core D2. There is no substantial evidence for built environments outside of the D2. This is particularly important for relative dating of the D1 with respect to D2. GPR signals do not pick up building material even at deeper levels within D1, supporting the hypothesis that D1 is dated to an earlier period than D2 and its associated ditch enclosures.

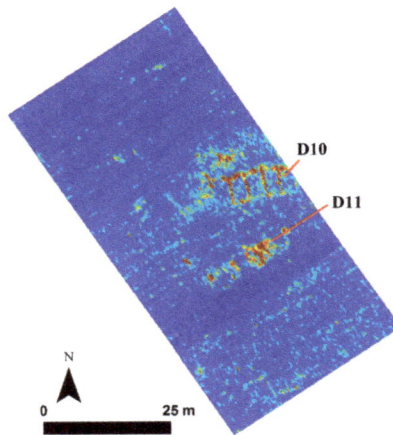

Figure 17. GPR survey results from Perdika 2 at 70–80 cm depth. Reddish colorsindicate high reflectance anomalies.

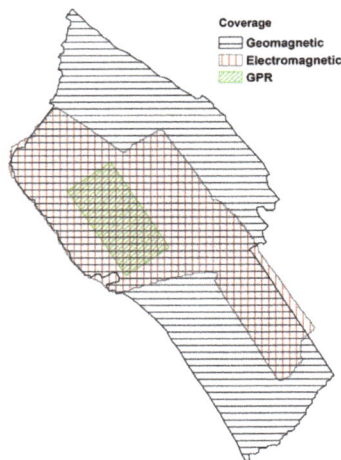

Figure 18. The overlay for the prospection coverage by different methods at Perdika 2.

The enclosures are completely absent in the GPR data. The lack of high reflections in the form of building materials further supports the hypothesis that these enclosures were ditches rather than circuit walls. Therefore, GPR data compliment other geophysical prospection methodologies for the determination of building materials making these enclosures (Figure 18).

5.5. Perivlepto/Kastraki 2 (22.707° E, 39.296° N)

Droidworx RPAS imagery reveals two circular soil marks. Feature M7 represents the Perivlepto/Kastraki 2 (Figure 19). The southern circular core and two arching marks to the northeast match with the results of geophysical prospection (see below). The jagged edges in the north reveals the impact of modern plowing and raises concerns for the preservation of material culture. Feature M8 is an unexpected extension to the site. The separation further suggests a dual occupation at Kastraki 2, M7 representing the main habitation.

Figure 19. Droidworx RPAS imagery (acquisition date: 23 April 2015) reveals two soil marks.

5.5.1. Geomagnetic Survey

Traces of enclosures are visible surrounding the core of the settlement (Figure 20). Their higher magnetic values and shapes suggest these enclosures were probably ditches. Immediately, to the north of the core, the half-circular features E1 and E2 bound the settlement. It is not certain if these ditches were initially complete enclosures, but filled/destroyed later or if they were purposefully built as such. There is a similar setting in the southern portion of the settlement. Ditches bound the site, but are only visible in parts (E8 and E9). In consideration of their spanning angles, enclosures in the north (E1, E2) and in the south (E8, E9) were not built to meet. Therefore, it is possible to argue that Kastraki 2 never had a complete ditch running around the site, but they were opportunistically located in order to serve for specific uses.

Southern portions of E1 and E2 are magnetically empty. In comparison to the rest of the data, these two areas (E3 and E4) must have contained very limited number of features and must have hosted a different kind of human use of space, but not habitation. The prospection further reveals highly burnt structures to the south of the settlement (E5). They are aligned on east–west axis except for the one located further north from this cluster (E6). Their magnetic signatures suggest intensive burning and mask any visible internal partition.

To the north of the settlement, and 50 m away from the northernmost, there exist indications of habitation in the form of free standing buildings (E7). Moreover, in consideration of the documented soil mark on satellite imagery, a secondary site to the Kastraki 2 is also a possibility. It is speculative to suggest their contemporaneity. However, if their habitations had overlapped sometime during the Neolithic, this layout of spilled-over buildings outside of the bounds of Kastraki 2 remains as a curious case.

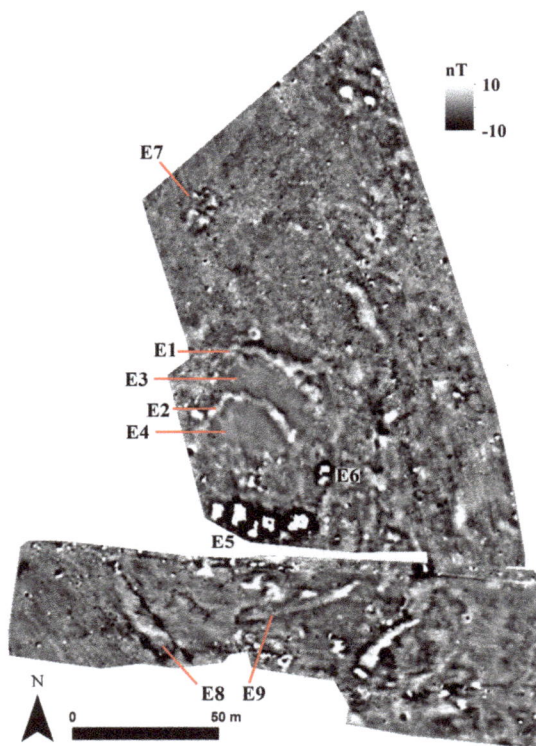

Figure 20. Geomagnetic prospection results from Perivlepto/Kastraki 2.

5.5.2. Electromagnetic Induction Survey

Magnetic susceptibility confirms locations and extent of ditch-enclosures (E1 and E2) in the north (Figure 21a). Furthermore, E3 and E4 have lower magnetic susceptibilities indicating minimal cultural transformation of soils. Only three of the structures in cluster E5 show up as high magnetic susceptibility localities. This might be explained by differential post-deposition of structures or initial variations in terrace levels.

The conductivity map reveals an area with high resistivity values, initially suggesting the concentration of architectural elements (Figure 21b). Contrary to the expectations, however, structures are located at the fringes of this area. Furthermore, areas E3 and E4 show low level of human activity and no architectural features as revealed in magnetic susceptibility and geomagnetic datasets. These contrasting observations can be resolved by suggesting that these areas were compacted surfaces for particular use(s) and were devoid of major built features.

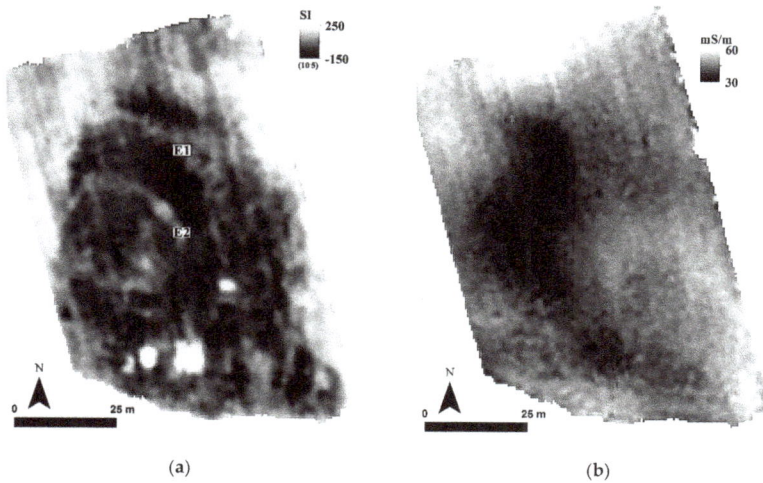

Figure 21. Magnetic susceptibility (**a**); and electromagnetic conductivity (**b**) results from Perivlepto/Kastraki 2.

5.5.3. Ground Penetrating Radar Survey

GPR successfully maps a series of structures (Figure 22). The survey provides further details of the east–west aligned building cluster (E5). To the north of this cluster, a group of scattered buildings is visible. Geomagnetic and magnetic susceptibility data (Figures 20, 21a and 23) do not suggest a firing event for this second group, suggesting a different treatment of architectural features in close vicinities. Within the bounds of the site, GPR supports the hypothesis of significant surface compaction for the areas of E1 and E2. High reflectance zones overlap with high resistivity and low magnetic susceptibility values. The soil mark on RPAS imagery (M7) succinctly describes this compaction as well.

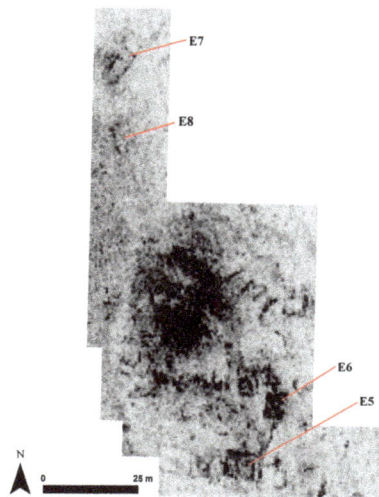

Figure 22. GPR survey results from Perdika 2 at 90–100 cm depth. Darker colors indicate high reflectance anomalies.

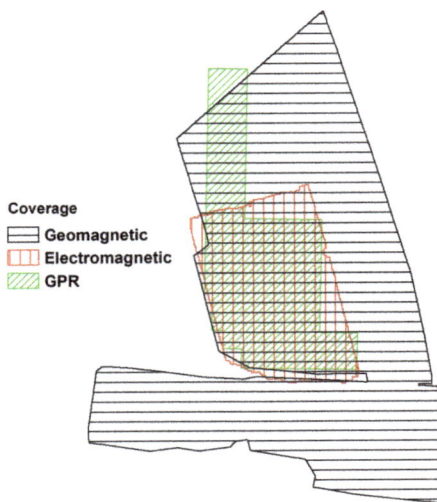

Figure 23. The overlay for the prospection coverage by different methods at Perivlepto/Kastraki 2.

GPR survey adds E8 to the building inventory. In consideration of the soil mark on satellite imagery, the E7–E8 group further increases the possibility of a secondary site in the north. In this secondary site, the buildings surround the soil mark as in the case for Kastraki 2. Therefore, one may claim a similar use of space.

6. Discussion

6.1. Methodological Advancements

The IGEAN Project emphasized the importance of integrated geospatial methodologies for the solution of archaeological problems. Through the manifold approach, the limitation of one sensor is compensated by another. Therefore, a more complete picture of past use of space can be drawn. This case became especially evident for Almyros 2; geomagnetic prospection and electromagnetic induction surveys were helpful in exploring the changes in settlement pattern in two consequent architectural (or cultural) phases. Despite the power of this approach, it would be naïve to suggest geospatial technologies will eventually substitute the norm of archaeological surveys and excavations.

Site enclosures, locations of houses, preservation conditions, pyro-activities, deliberately avoided areas within sites and many other spatial and attribute data were gathered thanks to the workhorse of archaeological geophysics, i.e., the magnetometer. Furthermore, multi-sensor (multi+) geomagnetic prospection provided large area coverage due to instrument's high data collection efficiency. It is now possible to explore geomagnetic landscapes and locate archaeological features in much wider contexts [20]. On the other hand, one can further argue that it is only possible through the manifold approach (multiple methods) to truly investigate differences and commonalities in inter-site settlement patterns.

Electromagnetic induction proved to be useful both for providing information on variations in use-of-space within and between settlements and for complementing the results of geomagnetic prospection. In particular, the evaluation of magnetic susceptibility in relation to geomagnetism made it possible to observe architectural transitions from one phase to another. Electrical conductivity data reveal possibly compacted surfaces within sites and differentiate between ditches and fortification walls. The study of architectural plans further supported the existence of these open spaces, hinting at a peculiar use of space in Early and Middle Neolithic settlements in Thessaly.

GPR survey exposed features that were not detected by other geophysical instruments, refining the shape and extent of buildings that were documented with other methods, and provided further proof for land-use practices. Specifically, the GPR, in tandem to magnetic survey, was instrumental in discriminating between types of building materials, mudbrick or stone foundations. Finally, the methodology offered means to understand the structure of ditch-enclosures.

Despite their strengths, geophysical sensors also have intrinsic limitations. These limitations are further pronounced in Neolithic contexts where the detection of prehistoric anthropogenic activities requires high resolution and highly sensitive data collection. Especially, electromagnetic induction surveys tend to have low spatial resolutions, hindering a finer delineation of Neolithic use of space. Furthermore, the major raw material of the Neolithic period, i.e., the clay, presents opportunities and obstacles for the collection and interpretation of geophysical data. Magnetic features located closer to the surface tend to mask anomalies of a lower magnetization or deeper strata [21]. This magnetic superimposition creates a bias for the interpretation of geomagnetic maps and over emphasizes the use of materials with mudbrick content; a Neolithic fact one can enthusiastically accept. The same setting, however, works against the favor of GPRs where high clay content tend to attenuate GPR signals. Overall, these Neolithic limitations are minimized using manifold methodology.

Integration of aerial and space-borne systems to the ground arsenal is now almost a standard procedure. RPAS is becoming an ordinary tool in archaeological projects and very high resolution satellite imagery is more available due to an increase in the number of space platforms. Furthermore, online data repositories keep providing invaluable spatial datasets. The IGEAN project capitalized over these advancements and complimented the results of geophysical prospection while contextualizing settlements in their larger environmental settings.

6.2. Archaeological Advancements

The IGEAN project produced comparable Early and Middle Neolithic settlement layouts. These plans are invaluable for revealing variations and commonalities between settlements in great clarity (Figure 24). Discrimination between ditch-enclosures and circuit walls, investigating the reasons behind "incomplete" ditch features, observing the dynamics of enclosure evolution, resolving burnt and unburnt buildings, delineating highly compacted surfaces within the settlements, locating the areas with minimum anthropogenic activities all contribute for a better understanding of everyday life in Neolithic Thessaly.

It appears that settlement agglomeration resulted in a clear separation of inside from the outside world. The demarcation is the most evident when there is an enclosure around a Neolithic settlement as the results of the IGEAN project suggest. The interpretation of enclosures which amounts to a growing presence in recent archaeological data, directly affects the image of the relations between communities (e.g., warfare [22]) or within the community (e.g., neighbors [23]). Second, it is now clear that ditch-enclosures were not static features despite their monolithic sizes, but they were part of the settlement dynamics. In this regard, they were not the determinants of the Neolithic inside/outside binary opposition, but were signifiers of it.

The separation between burnt and unburnt buildings is a complex one. At Almyros 2 and Kamara, all documented structures were burnt, accidentally or deliberately. In addition, in consideration of the data from Karatsantagli and Kastraki 2, deliberate burning might have been the case, as suggested by Stevanović for Neolithic domestic architecture [24]. Under the assumption of contemporaneity, the IGEAN Project reveals selective burning of the houses within the same settlement.

Finally, the IGEAN Project is instructive for details of use of space. It seems like Neolithic inhabitants provided extra care to minimize activities around enclosure entrances. This must have been necessary for the ease of movement between the site and the off-site. This is most evident at Almyros 2. Similarly, open air surfaces within settlements might be hinting towards a specific economic activity or indicating the bounds of ceremonial spaces or communal gathering places.

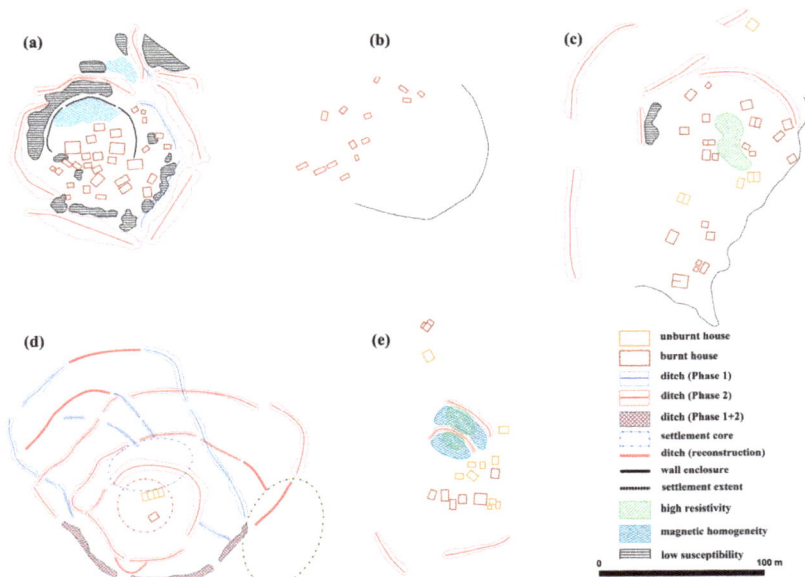

Figure 24. Proposed settlement layout reconstructions from: (**a**) Almyros 2; (**b**) Kamara; (**c**) Karatsantagli; (**d**) Perdika 2; and (**e**) Perivlepto/Kastraki 2. These vector data can also be viewed with the help of an online GIS served at http://igean.ims.forth.gr.

7. Conclusions

This study explores Early and Middle Neolithic settlements in Thessaly, Greece using a manifold geospatial approach. Integration of satellite, aerial, and ground remote sensing methodologies fully exploit individual sensor capabilities to hypothesize "close-to-complete" settlement layouts. Therefore, it is now possible to highlight variations and commonalities between these settlements and provide a grounded speculation of a "Neolithic way of use of space".

First, settlement enclosures show a greater variation than previously anticipated. There is convincing evidence for ditches thatare not fully encircling settlements. This suggests that these features were not always necessarily built for warfare. Second observation is related to the openings in these enclosures. Successive cleaning of doorways appears to be a societal habit of Neolithic dwellers. Third, clean and compacted Neolithic surfaces inside settlements point towards a specialized use of space. It is not possible to further comment on these features without archaeological excavations, but the lack of anthropogenic markers speculatively suggests exclusionary/prohibitionary use. Finally, discriminatory house burning may be indicating social differentiation.

Acknowledgments: The project was implemented under the "ARISTEIA" action of the "Operational Programme Educational and Lifelong Learning" and is co-funded by the European Social Fund (ESF) as well as National Resources.

Author Contributions: Authors equally contributed to the paper.

Conflicts of Interest: The authors declare no conflict of interest. The funding sponsors had no role in the design of the study; in the collection, analyses, or interpretation of data; in the writing of the manuscript, and in the decision to publish the results.

References

1. Perlès, C. *The Early Neolithic in Greece: The First Farming Communities in Europe*; Cambridge University Press: Cambridge, UK, 2001.
2. Sarris, A.; Kalayci, T.; Simon, F.-X.; Donati, J.C.; Cuenca Garcia, C.; Manataki, M.; Cantoro, G.; Karampatsou, G.; Kalogiropoulou, E.; Argyriou, N.; et al. Cultural Variations of the Neolithic Landscape of Thessaly. *Archaeol. Prospect.* **2015**, *53*, 355–359.
3. Sarris, A. Multi+ or Manifold Geophysical Prospection? In *Archaeology in the Digital Era*; Earl, G., Sly, T., Chrysanthi, A., Murrieta-Flores, P., Papadopoulos, C., Romanowska, I., Wheatley, D., Eds.; University of Southampton: Southampton, UK, 2012; pp. 761–770.
4. Wijnen, M.-H.J.M.N. The Early Neolithic I Settlement at Sesklo: An Early Farming Community in Thessaly, Greece. In *Analecta Praehistorica Leidensia*; Leiden University: Leiden, The Netherlands, 1981; Volume 14, pp. 1–146.
5. Caputo, R.; Bravard, J.-P.; Helly, B. The Pliocene-Quaternary Tecto-sedimentary Evolution of the Larissa Plain (Eastern Thessaly, Greece). *Geodin. Acta* **1994**, *7*, 219–231. [CrossRef]
6. Migiros, G.; Bathrellos, G.D.; Skilodimou, H.D.; Karamousalis, T. Pinios (Peneus) River (Central Greece): Hydrological—Geomorphological Elements and Changes during the Quaternary. *Cent. Eur. J. Geosci.* **2011**, *3*, 215–228. [CrossRef]
7. Demitrack, A. The Late Quaternary Geologic History of the Larissa Plain Thessaly, Greece: Tectonic, Climatic, and Human Impact on the Landscape. Ph.D. Thesis, Stanford University, Stanford, CA, USA, 1986.
8. Tsountas, C. *Ai Proistorikai Akropoleis Diminioukai Sesklou*; ΕνΑθηναιςΑρχαιολογικήΕταιρεία: Athens, Greece, 1908.
9. Wace, A.J.B.; Thompson, M.S. *Prehistoric Thessaly: Being Some Account of Recent Excavations and Explorations in North-Eastern Greece from Lake Kopais to the Borders of Macedonia*; Cambridge University Press: Cambridge, UK, 1912; Volume 89, p. 294.
10. Andreou, S.; Fotiadis, M.; Kotsakis, K. Review of Aegean Prehistory V: The Neolithic and Bronze Age of Northern Greece. *Am. J. Archaeol.* **1996**, *100*, 537–597. [CrossRef]
11. Halstead, P. *Strategies for Survival: An Ecological Approach to Social and Economic Change in the Early Farming Communities of Thessaly, N. Greece*; University of Cambridge: Cambridge, UK, 1984.
12. Gallis, K.I. *Atlas Proisrorikon Oikosmon Tis Anatolikis Thessalikis Pediadas*; Society of Historical Research of Thessaly: Larisa, Greece, 1992.
13. Van Andel, T.H.; Runnels, C.N. The Earliest Farmers in Europe. *Antiquity* **1995**, *69*, 481–500. [CrossRef]
14. Nanoglou, S. Subjectivity and Material Culture in Thessaly, Greece: The Case of Neolithic Anthropomorphic Imagery. *Camb. Archaeol. J.* **2005**, *15*, 141–156. [CrossRef]
15. Souvatzi, S. Social Complexity Is Not the Same as Hierarchy. In *Socialising Complexity: Structure, Interaction and Power in Archaeological Discourse*; Kohring, S., Wynne-Jones, S., Eds.; Oxbow: Oxford, UK, 2007; pp. 37–59.
16. Pentedeka, A. Negotiating Identities and Exchanging Values. In *Balkan Dialogues: Negotiating Identity Between Prehistory and the Present*; Gori, M., Ivanova, M., Eds.; Routledge Studies in Archaeology; Taylor & Francis: Didcot, UK, 2017; pp. 131–155.
17. Papadopoulos, C.; Hamilakis, Y.; Kyparissi-Apostolika, N. Light in a Neolithic Dwelling: Building 1 at Koutroulou Magoula (Greece). *Antiquity* **2015**, *89*, 1034–1050. [CrossRef]
18. Simon, F.-X.; Kalayci, T.; Donati, J.C.; Garcia, C.C.; Manataki, M.; Sarris, A. How Efficient Is an Integrative Approach in Archaeological Geophysics? Comparative Case Studies from Neolithic Settlements in Thessaly (central Greece). *Near Surf. Geophys.* **2015**, *13*, 633–643.
19. Kalayci, T.; Sarris, A. Multi-Sensor Geomagnetic Prospection: A Case Study from Neolithic Thessaly, Greece. *Remote Sens.* **2016**, *8*, 966. [CrossRef]
20. Kvamme, K.L. Geophysical Surveys as Landscape Archaeology. *Am. Antiq.* **2003**, *68*, 435–457. [CrossRef]
21. Yerkes, R.W.; Sarris, A.; Frolking, T.; Parkinson, W.A.; Gyucha, A.; Hardy, M.; Catanoso, L. Geophysical and Geochemical Investigations at two Early Copper Age Settlements in the Körös River Valley, Southeastern Hungary. *Geoarchaeology* **2007**, *22*, 845–871. [CrossRef]
22. Runnels, C.N.; White, C.; Payne, C.; Wolff, N.P.; Rifkind, N.V.; LeBlanc, S.A. Warfare in Neolithic Thessaly: A Case Study. *Hesperia* **2009**, *78*, 165–194. [CrossRef]

23. Halstead, P. Neighbours from hell? The Household in Neolithic Greece. In *Neolithic Society in Greece*; Halstead, P., Ed.; Sheffield Academic Press: Sheffield, UK, 1999; pp. 77–95.

24. Stevanović, M. The Age of Clay: The Social Dynamics of House Destruction. *J. Anthropol. Archaeol.* **1997**, *16*, 334–395. [CrossRef]

geosciences

MDPI

Article

Geometric Analysis on Stone Façades with Terrestrial Laser Scanner Technology

Juan Corso [1], Josep Roca [2] and Felipe Buill [3,*

[1]   Laboratorio de Modelización Virtual de la Ciudad (LMVC), Universitat Politècnica de Catalunya, 08034 Barcelona, Spain; juan.corso@upc.edu
[2]   Centre of Land Policy and Valuations, Universitat Politècnica de Catalunya, 08034 Barcelona, Spain; josep.roca@upc.edu
[3]   Division of Geotechnical Engineering and Geosciences, Universitat Politècnica de Catalunya, 08034 Barcelona, Spain
*   Correspondence: felipe.buill@upc.edu; Tel.: +34-934011933

Received: 11 July 2017; Accepted: 30 September 2017; Published: 10 October 2017

Abstract: This article presents a methodology to process information from a Terrestrial Laser Scanner (TLS) from three dimensions (3D) to two dimensions (2D), and to two dimensions with a color value (2.5D), as a tool to document and analyze heritage buildings. Principally focused on the loss of material in stone, this study aims at creating an evaluation method for loss control, taking into account the state of conservation of a building in terms of restoration, from studying the pathologies, to their identification and delimitation. A case study on the Cathedral of the Seu Vella de Lleida was completed, examining the details of the stone surfaces. This cathedral was affected by military use, periods of abandonment, and periodic restorations.

Keywords: Terrestrial Laser Scanning; orthoimage; heritage; remote sensing; preservation; archaeology

1. Introduction

Historic buildings in urban areas often show accelerated deterioration. In historic stone buildings, many different types of damage can occur including: loss of stone material, discoloration, deposits, detachment, fissures, deformation, and damage from previous intervention [1]. Detection of material degradation in historic buildings with traditional methods, such as manual mapping or simple eye examination by an expert [2], are considered time-consuming [3] and laborious procedures, so TLS (Terrestrial Laser Scanner) technology and image processing methodologies are being developed, allowing for the detection of pathologies [4,5], their evolution [6], identification of deformations [7], changes in material [8], and stone façade documentation [9–11].

Understanding this need, this article proposes a methodology for the geometric analysis of stone façades, based on TLS surveys, focusing on the information generated to preserve the architectural heritage [12,13] and its diffusion, as a means to study and analyze [14,15] historical contexts [16,17]. The methodology developed in this article proposes the conversion of three-dimensional (3D) data to two dimensions (2D) [18,19], and two dimensions plus a color value (2.5D) as depth databases (Figure 1).

With precise control of data transformations, the conversion of 2.5D to 3D can be achieved, which is a necessity for the patrimonial documentation [20–24], to the point of generating programs for the work of 3D images [25–28]. This analysis is possible thanks to the use of a geographic information system (GIS) on complex buildings [29,30]. With an emphasis on 3D GIS applied to restoration [31–34].

The goal of applying this process is to perform an analysis that emphasizes the geometric detail of the patrimonial façades, composed mainly of stone. This information would be useful for the diagnosis of stone alterations. As a complementary process, the metric images were used to generate planes in vectorial format using a semi-automatic process [35–38], given the quality of definition and thickness of the lines generated.

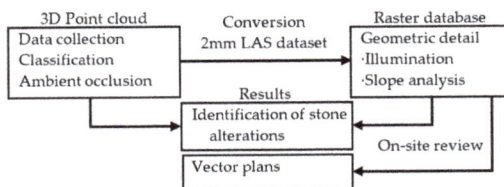

Figure 1. Methodology and structure of the article.

The case study is on the Cathedral of the Seu Vella de Lleida, which is located on the hill known as Turó de Lleida that dominates the city in the Segrià region. The cathedral is designed in a transitional style between Romanesque and Gothic. The Seu Vella de Lleida has Romanesque forms and the monumentality of the Gothic era; it has Gothic cross-vaults. This patrimonial building has suffered damage from several military contests as well as from periods of abandonment [39]. The military uses damaged the building, the most serious of which was caused by civil war bombings—primarily in the terraces and decks—ruining decorative elements like the pinnacles, gables, gargoyles, and the sills of the roof terrace.

During the second half of the 19th century, with the Catalan movement of the Renaissance, the Cathedral was revalued, and around 12 June 1918 it was declared a national monument [40]. In 1948, the army ceded it and restoration began in 1949. During this restoration, the architects Francisco Pons Sorolla and Alexander Ferran made modifications and even replaced some elements. These elements include the traceries and mullions of the Gothic windows, which were reconstructed copying the originals with molds and cement.

There are two types of stone in the study area: the Lleida stone and the Vinaixa stone. They both have intergranular porosity, which is more visible in the Vinaixa stone. A major part of the walls and the base of the cloister of the church are composed of Lleida stone; it is a sandstone rich in carbonates and the texture allows classification like a calcarenite. The Vinaixa stone is found in all the elements that are more or less worked, including arches, capitals, frontons, and traceries. Almost the entirety of the upper part of the cloister is also built with Vinaixa stone [39].

The case study involved surveying the stone with the laser scanner technology, and focused mainly on the gallery of the cloister, which is the green zone shown in Figure 2, which was also a viewpoint of the city of Lleida in the 13th and 14th centuries [39]. This methodology was also applied to other areas of the cathedral with similar stone type characteristics, including to an area of medieval wall, the perimeter of the building, and the door of the Apostles.

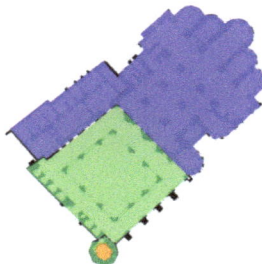

Figure 2. Construction phases: the blue represents the 13th century; the green represents the 14th century; and orange is the 15th century [39].

2. Survey with Terrestrial Laser Scanner Technology

For data collection, a phase difference scanner was used, the Focus^{3D} 120 (FARO®, Stuttgart, Germany), which allows a wide 360° H·305° V capture range, with an accuracy of 0.6 mm to 10 m, and 0.95 mm to 25 m. With a systematic error of ≤±2 mm at 25 m, it also captures the information of reflected intensity of the laser and photographic color, with RGB color information assigned to the points, through an integrated camera of 70 megapixels.

Positions were chosen on each arch of the scanned corridor, in the courtyard of the Seu Vella. The scanning resolution was 1/8, with 4× quality in relation to the scene program, which was used to operate the scanner. The position was obtained by an average of 38 million measures. In total, seven positions were chosen for the gallery of the cloister. In the other zones of the cathedral, 17 positions were selected.

3. Information Stored in the 3D Point Cloud

With the aim of documenting the surface geometry of the scanned elements, a set of high resolutions measures, of at least one point every 2 mm [3], was required. The information generated was stored as scalar values as a 3D cloud of points, which would later be converted into 2D raster databases. The resolution used and the ability store data in 3D clouds of points and 2D raster maintains the following analyses in the database: curvature analysis of the surfaces, direction of surfaces, classification of surfaces, and controlled lighting on surfaces, as shown in Figure 3.

Figure 3. 3D point cloud. CANUPO classification and illumination by ambient occlusion [11]. Classification: floor is shown in gray, the walls are blank, the ornamental elements are shown in blue, and the linear elements in red.

3.1. Classification of the 3D Point Cloud

The technique used for this classification, CANUPO, is a set of applications designed for the classification of natural environments, including rivers, coastal environments, cliffs, etc. This process is defined as:

> ... a multi-scale measure of the point cloud dimensionality around each point. The dimensionality characterizes the local 3D organization of the point cloud within spheres centered on the measured points and varies from being 1D (points set along a line), 2D (points forming a plane) to the full 3D volume. By varying the diameter of the sphere, we can thus monitor how the local cloud geometry behaves across scales [41].

Classification was completed by dividing walls as planes, and the other surfaces, on scales between 5 cm and 30 cm, as shown on the right side of Figure 3. Linear and ornamental elements are also classified, on scales between 2 cm and 10 cm, with red denoting linear elements, dark blue defining ornamental elements, and light blue for curved surfaces.

This classification not only included constructive elements; CANUPO classifies everything in the architectural heritage, which requires the interpretation of a specialist. An example of these possible interpretations is the description of isolated elements that make up the building, like the tombs of nobles. In the classification, we identified changes in surfaces, such as edges, planes, or volumes. The identifier of the point cloud was introduced as data, and a number for each point was assigned. The return field was used as a second classification, which can hide elements that are not taken into account in subsequent steps of conversion to raster images—such as the ceiling, furniture, or the floor itself—so that clear information on walls of the façade remained.

3.2. Details of Surfaces with Lighting Processes

For geometric analysis, apply lightning processes to the point cloud is useful. These techniques allow the identification of areas that are poorly visible, including concavities of the surface in rough areas, alterations of the material, or simply holes in the surface.

For this kind of analyses, we used the Screen Space Ambient Occlusion [42], with the Plugin qSSAO in Cloudcompare. This is a shading 3D method which adds realism to the models, through the attenuation of the light due to reflection and occlusions that are produced by the geometry of the model (Figure 4).

The calculation of the surface occlusions starts from a virtual lighting process. This lighting obscures the areas in which light rays bounce between nearby surfaces; the more concave the surface the more impacts it receives, and the greater value it acquires. If the surface is more exposed to the exterior, the fewer impacts it receives. If the surface is perpendicular to the sky, it will not darken, keeping its base color.

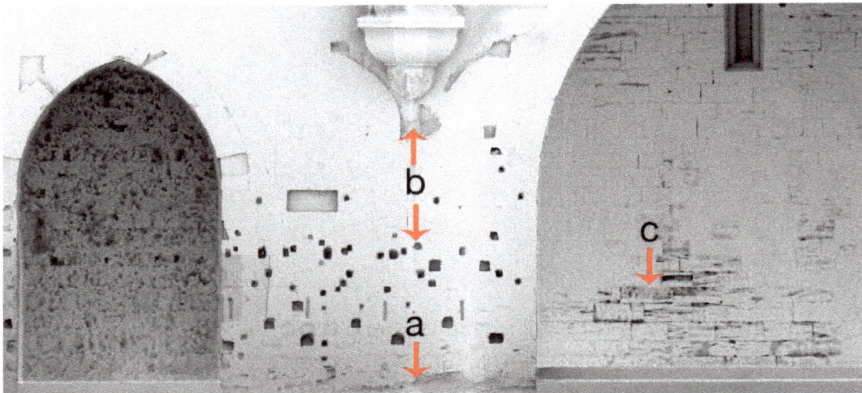

Figure 4. 2D raster from an environmental occlusion generated in 3D: (**a**) loss of material; (**b**) holes and alterations of the stone; and (**c**) disintegration of stone.

4. 3D to 2D and 2.5D Conversions

In order to efficiently communicate the desired information in the results of the analyses, a conversion of the 3D database to 2D must be completed without losing any information in the process, while maintaining the possibility of reverting to a 3D model. For the front façade, the façades perpendicular to this had to be developed in 2D (2.5D).

To do this, the proposed method starts in LiDAR format, and the LAS dataset, which allows the organization of the topology of the points, including variables such as color, elevation, class, and return as a second classification and intensity. The LAS dataset is able to include extensive point clouds, and is structured to allow adequate information management, with different levels of detail, according to the distance. An example is an LAS dataset with depth information and a roughness raster with a roughness range of red to black.

These files allow the generation of the LAS dataset to raster with a GIS. In this case study, the conversion was completed with the ArcGIS program (version 10.1, Environmental Systems Research Institute, Redlands, CA, United States). With a 2-mm subsampled point cloud, this LAS was converted to raster images with a cell value of 2 mm.

The zones that do not have a measurement every 2 mm were interpolated, applying the maximum distance range between cells function, in this case with a maximum of 3 pixels, 6 mm in terms of measurements, so that interpolating surfaces that are not related were avoided.

Differences and Relationships between 3D and 2D (2.5D) Information

Transforming 3D information to 2D as a metric map facilitates on-site work, allows a direct verification of the information developed, and provides true dimensions of all the elements. As a graphic representation, these are all advantages because an architectural plane provides a better reading and references all aspects of the surface of the building, thus giving a better understanding of complex heritage buildings. However, 3D information is also fundamental, since it allows a volumetric understanding of the building; 3D and 2D (2.5) are complementary [43,44]. Working only in 3D limits the inspection of the building and the work of different disciplines involved. Working only in 2D requires over-segmentation of the façade, avoids self-occlusion of the surface, each one with different deformations, and would thus prevent reading the façade as a whole.

The 3D representation of a point cloud depends on the position of the camera from which the 3D point cloud is rendered and the point size (Figure 5a), while in 2D it depends on the chosen metric image resolution (Figure 5b). Both 3D and 2D information allows operations to be performed. For example, the information of the point cloud allows the performance of operations between the data, generating new databases without altering the base information. As scalar values, the 3D point cloud can store as many scalar value fields as needed.

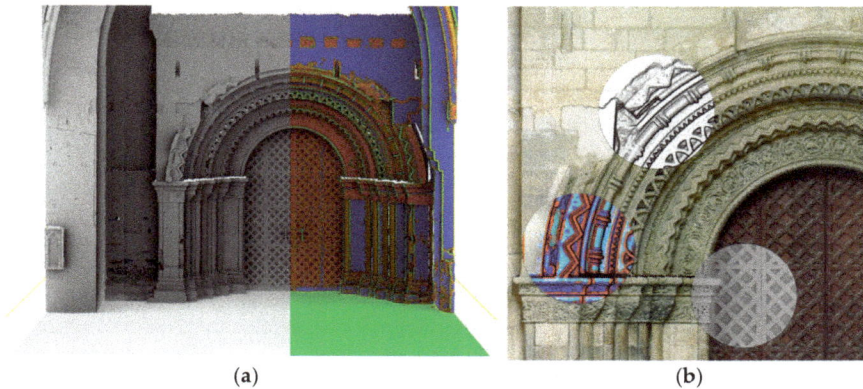

(a)   (b)

Figure 5. Layers of information in: (**a**) 3D point cloud information with factor scales: ambient occlusion and classification; (**b**) superimposed raster images.

In 2D, the images with scale generate specific documentation; each image is a layer of information with a determined purpose, and these images can be superimposed or interpolated to analyze specific aspects.

Since raster information is controlled, it is possible to return that information to a 3D model. In this process, it is essential to maintain the metadata of the data conversions to be able to reverse the information, or to track the alterations of the obtained results. In addition, maintaining the metadata supports other disciplines that work in specific formats by maintaining the metric precision and the position of the data.

5. Raster Information

TLS technology makes it possible to generate images with a high degree of precision, providing geometric information on the façades. This information allows the generation of derivative products, such as a Digital Elevation Model (DEM). The DEM complements the 2D data with the information in 2.5D [45], merging this information with the raster. This allows performance analysis, emphasizing the geometric variables of the building, using topography or aerial LiDAR, while retaining the possibility of returning to a 3D model.

5.1. Geometric Details with Illumination Process

A DEM not only contains explicit information about elevation in a surface, but also regarding distance and neighborhood relationships. Operations can then be performed on the superficies, which is common in topographic analyses, such as hillshade or slope analysis, Figure 6.

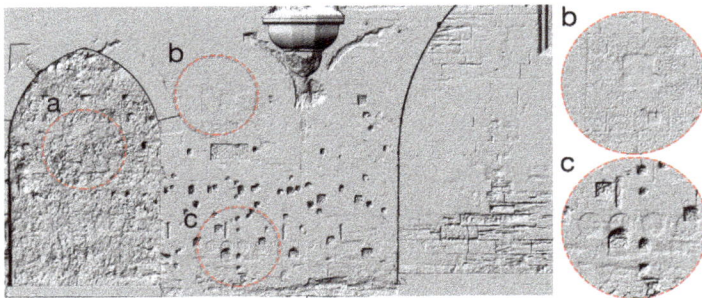

Figure 6. Raster hillshade 315° azimuth 45°: (**a**) surface roughness not visible from photographs, (**b**) surface roughness not visible to the naked eye, and (**c**) carved symbols, barely visible on site.

This process can be avoided by applying a mean filter on the points, eliminating the points outside the average, so in this case, the ones corresponding to the points of the lower plane are removed because they are random data due to lack of data in the upper plane. This mean filter was applied to all the images, ensuring information relevant to the objectives of the case study was not lost.

5.2. Representation of Façade Details from a Slope Analysis

A slope analysis in GIS identifies the boundaries between surfaces; making it possible to obtain details at different scales and with different magnitudes. Slope analysis is understood as the maximum height change ratio between one cell and the eight neighboring cells. Slope analysis can also be defined as the angle between the vector normal to the surface at that point and the vertical.

As metric planes, the images allow a clear interpretation of the building. The images identify, for example, the patterns in the stones, which is useful for different disciplines, such as for the identification of stratigraphic units of a façade in archaeology [46,47]. Details with little relief have a lower line value than jumps between surfaces or larger planes.

In this case study, the slope analysis was performed with the ArcGIS program, with the technique developed by Burrough et al. [48]. A slope analysis, in which an accurate representation of the shapes and geometries of the elements are distinguished, is seen in Figure 7.

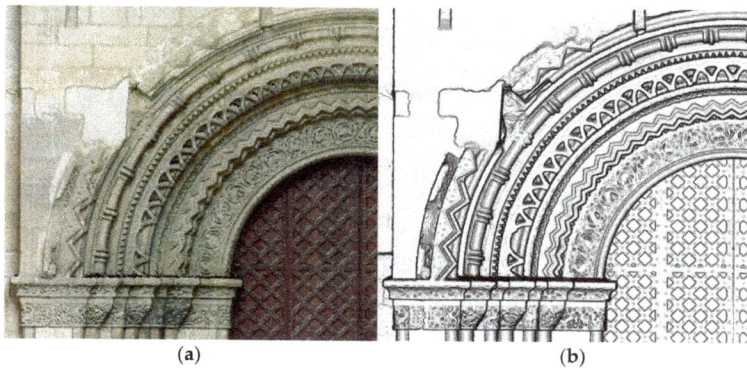

Figure 7. Identification of previous interventions with different materials: (**a**) identification of alterations in geometric patterns and (**b**) slope raster obtained through DEM spatial analysis.

6. Identification of Alterations in the Stone by Combining Raster Images

Stone decay can be either additive or subtractive [1]. The DEM identifies the depth differences between planes of a surface and from this, can identify the boundaries between these surfaces (Figure 8a). Even in this image, whether the material is loose or will gain material as part of the decay mechanism cannot be identified. For this reason, the metric image of Figure 8b is generated. The ambient occlusion is projected onto the 3D model and generates an image that identifies the concavities of the surfaces, and with it, the loss of the material. In this zone, the sandstone and disintegration of the stone is identified. This occurs because of the presence of clays in the Lleida stone; these clays degrade with a certain speed, and the water in the stone causes the clay to expand, creating internal tensions that lead to the disintegration of the material near the surface of the element.

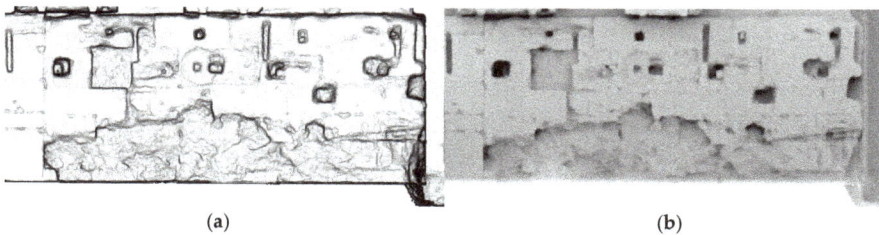

Figure 8. Deterioration of the stone: (**a**) the boundaries between surfaces, the slope, is identified; and (**b**) the sandstone is identified as well as the disintegration of the stone (ambient occlusion).

Identification of different materials, using RGB images combined with images of lighting techniques and observation on site, shows complete blocks of restored stone, different from those around them. This information helps to delimit the area that has been restored and identify blocks that suffer from disintegration (Figure 9). The stone substitutions were made with Vinaixa stone, as it does not suffer from the same degradation as that from Lleida.

The stone degradation is also visible in other areas, like in the lower area on the left side of the Nave del Aepístola (Figure 9). In this area, a different type of material loss is present, seen in the loss of its clay base caused by water [49]. The generated images identify the limits of the alterations and quantify the loss of material. A study of this type would enable management actions of the patrimonial property, would provide more information to help better manage the conservation processes, and verify that the pathology does not worsen.

Figure 9. Identification of previous interventions with different stone: (**a**) RGB left; and (**b**) disintegration of stone (hillshade).

Using only one color image with RGB information is not sufficient to perform an analysis for directing restoration efforts on a heritage building. The singular image lacks important information. For example, the roughness of the surface, shown in Figure 10, is shown as a continuous surface in a RGB image, but the ambient occlusion shows the actual stone deterioration.

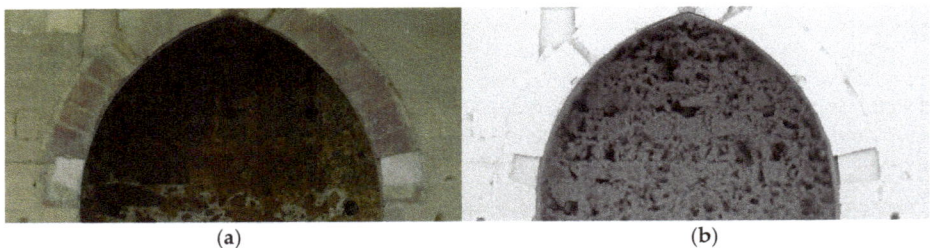

Figure 10. Efflorescence in the stone: (**a**) the RBG image showing a smooth surface; and (**b**) deterioration of the stone by alveolization of the materials shown by ambient occlusion.

This deterioration, the loss of material by alveolization, is a phenomenon characterized by the formation of voids or cavities of different sizes and morphologies, called alveoli. Its appearance depends on the composition of the rock and environmental factors [50].

A partial loss of the volume or an erosion of the stone can occur due to the consolidation of the stone, for example, with the application of replacement mortar. This erosion or loss of material must be documented for its control.

Semiautomatic Process to Convert Vector Planes

To obtain images that can be edited and are ready for specialists, the 2D images must be vector planes. Converting this information to a 2D vector plane is performed through a semiautomatic process, in which a raster image is created and converted to a vector plane.

The first step is performing a curvature analysis in relation to the scale of the desired vector plane, accounting for the detail that would be seen on the determined scale. In this process, it is important to consider the resolution of the data acquisition and the noise generated in the data capture [51].

Next, the raster lines are converted to vector lines, with a program like AutoCAD Raster Design, which allows image analysis and subsequent vectorization (Figure 11). In this process, manual editing is needed for areas requiring interpretation, or areas with occlusions of the data collection, and to simplify or clean areas of difficult interpretation. In the same way, manual editing is necessary to separate the layers [52], taking into account the value of the line and the representation of the plane.

The complete process from point cloud conversion to DEM generation, slope analysis, and vectorization required two hours to complete, plus 15 min of zone correction per occlusion, and 30 min to reorder layers, depending on the value of the line. Drawing only one of the doors in CAD, manually, took a day of work [3]; the entire façade would be a week due to the complexity of the decorations.

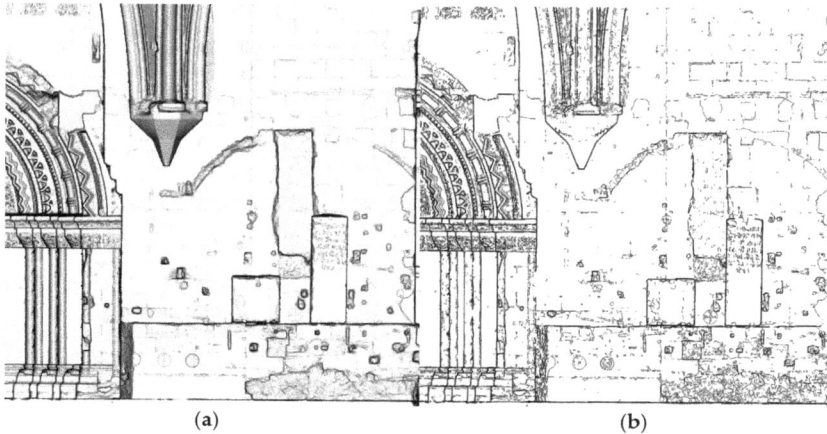

(a)   (b)

Figure 11. Raster to vector conversion: (**a**) curvature analysis; and (**b**) vector planimetry from the curvature analysis.

7. Conclusions

As shown in the literature, there is a growing use of GIS and image processing with TLS surveys, which relate the 3D models to 2D maps; low-resolution models are created and mapped, documenting the starting information in 3D and expanding the information as images [18–24]. A new methodology is developed in this article—a 3D controlled conversion to 2D and 2.5D—that operates without interpolations and reduces the loss of base information, enabling the conversion of new information generated in 2D and 2.5D to a 3D model (point cloud), recording the rasters as a database. This allows for comparison of the differences in the storage of information between raster and point cloud.

According to the methodology, a case study is developed, establishing parameters of graphical representations and highlighting features associated with stone pathologies, such as analysis of the patrimonial building surface, which complement the field work of the restoration.

For the Seu Vella, the raster images showed the delimited areas with deterioration in the stone, and the behavior of the Vinaixa and Lleida stone in the cathedral were identified. We determined that the Lleida stone, the predominant rock in this cathedral, suffers significant erosion, both in loss of material and in alveoli formation in all orientations, produced by the presence of water and by the mineralogy of the stones rich in clays.

This comparison between layers of raster images and their relationship to the 3D model with an on-site review, allows us to detect the change in stone type through surface roughness, stone relief, stone erosion or alteration, the boundaries between surface planes, and the color of the stones.

This study helps work toward achieving preservation of buildings, with a focus on prevention and maintenance work, which contributes to their protection. This raster information allowed us to generate vectorial plans, and the adopted technique leads to an improvement of the heritage register compared to traditional techniques that are currently used to produce stone-by-stone elevations.

Author Contributions: Juan Corso and Felipe Buill designed the methodology, conceived and designed the experiments; Juan Corso performed the experiments; Juan Corso and Josep Roca analyzed the data; Josep Roca contributed the material of the case study; Juan Corso wrote the paper.

Conflicts of Interest: The authors declare no conflict of interest.

References

1. Kottke, J. An Investigation of Quantifying and Monitoring Stone Surface Deterioration Using Three Dimensional Laser Scanning. Master's Thesis, University of Pennsylvania, Philadelphia, PA, USA, 2009.
2. Almagro, A. Simple methods of photogrammetry—Easy and fast. *Int. Arch. Photogramm. Remote Sens. Spat. Inf. Sci.* **2002**, *34*, 32–38.
3. Barber, D.M.; Dallas, R.W.; Mills, J.P. Laser scanning for architectural conservation. *J. Archit. Conserv.* **2006**, *12*, 35–52. [CrossRef]
4. Seif, A.; Santana, M.; KreidI, J. *Protecting Baalbek's Integrity, "A Proposal for an Integrated Risk Preparedness Strategy"*; UNESCO: Beirut, Lebanon, 2011; ISBN 9789081695305.
5. ICOMOS; ISCS. Illustrated Glossary on Stone Deterioration Patterns. Available online: http://www.icomos.org/publications/monuments_and_sites/15/pdf/Monuments_and_Sites_15_ISCS_Glossary_Stone.pdf (accessed on 1 October 2017).
6. Fregonese, L.; Biolzi, L.; Bocciarelli, M. Surveying and monitoring for vulnerability assessment of an ancient building. *Sensors* **2013**, *13*, 9747–9773. [CrossRef] [PubMed]
7. Pescia, A.; Bonalib, E.; Gallib, C. Laser scanning and digital imaging for the investigation of an ancient building: Palazzo d'Accursio study case (Bologna, Italy). *J. Cult. Herit.* **2012**, *13*, 215–220. [CrossRef]
8. Pozo, S.; Herreo, J.; Felipe, B.; Hernández, D.; Rodríguez, P.; González, D. Multispectral radiometric analysis of façades to detect pathologies from active and passive remote sensing. *Remote Sens.* **2016**, *8*, 80. [CrossRef]
9. Drap, P.; Durand, A.; Nidir, M. Photogrammetry and archaeological knowledge: toward a 3D information system dedicated to medieval archaeology: A case study of Shawbak Castle in Jordan. In Proceedings of the 3D ARCH 2007, Zürich, Switzerland, 12–13 July 2007.
10. Willis, A.; Sui, Y. Estimating gothic facade architecture from imagery. In Proceedings of the IEEE Computer Society Conference on Computer Vision and Pattern Recognition Workshops (CVPRW), San Francisco, CA, USA, 13–18 June 2010.
11. Burgerb, A.; Grimm-Pitzinger, A.; Thaler, E. A combination of modern and classic methods of surveying historical buildings—The Church St. Valentin in the South Tyrol. In Proceedings of the XXI International CIPA Symposium, Athens, Greece, 1–6 October 2007.
12. Van Genechten, B. *Theory and Practice on Terrestrial Laser Scanning: Training Material Based on Practical Applications*; Universidad Politecnica de Valencia Editorial: Valencia, Spain, 2008; ISBN 9788483633120.
13. Charter, V. International Charter for the Conservation and Restoration of Monuments and Sites. Second International Congress of Architects and Technicians of Historic Monuments. Available online: http://www.icomos.org/charters/venice_e.pdf (accessed on 1 October 2017).
14. Allen, K.M.S.; Zubrow, E.B.W. *Interpreting Space: GIS and Archaeology*; Taylor & Francis: London, UK, 1990; ISBN 9780850668247.
15. Conolly, J.; Lake, M. Geographical information systems in archaeology. In *Cambridge Manuals in Archaeology*; Cambridge University Press: Cambridge, UK, 2006; ISBN 9780521793308.
16. Pieraccini, M.; Guidi, G.; Atzeni, C. 3D digitizing of cultural heritage. *J. Cult. Herit.* **2001**, *2*, 63–70. [CrossRef]
17. Berndt, E.; Carlos, J. Cultural heritage in the mature era of computer graphics. *IEEE Comput. Graph. Appl.* **2000**, *20*, 36–37. [CrossRef]
18. Salonia, P.; Negri, A. Cultural Heritage emergency: GIS-based tools for assessing and deciding preservation and management. In Proceedings of the Twenty-Third Annual ESRI International User Conference, San Diego, CA, USA, 7–11 July 2003.
19. Bartolomucci, C. Una proposta di cartella clinica per la conservazione programmata. *Arkos Sci. Restaur. Archit.* **2004**, *5*, 59–65.
20. Aldenderfer, M.; Maschner, H.D. (Eds.) *Anthropology, Space, and Geographic Information Systems*; Oxford University Press: Oxford, UK, 1996; ISBN 9780195085754.

21. Campanaro, D.M.; Landeschi, G.; Dell'Unto, N.; Touati, A.M.L. 3D GIS for cultural heritage restoration: A 'white box' workflow. *J. Cult. Herit.* **2016**, *18*, 321–332. [CrossRef]
22. Canciani, M.; Ceniccola, V.; Messi, M.; Saccone, M.; Zampilli, M. A 3D GIS method applied to cataloging and restoring: The case of Aurelian Walls at Rome. *ISPRS Int. Arch. Photogramm. Remote Sens. Spat. Inf. Sci.* **2013**, *1*, 143–148. [CrossRef]
23. Dell'Unto, N.; Landeschi, G.; Apel, J.; Poggi, G. 4D recording at the trowel's edge: Using three-dimensional simulation platforms to support field interpretation. *J. Archaeol. Sci. Rep.* **2017**, *12*, 632–645. [CrossRef]
24. Landeschi, G. Assessing the damage of an archaeological site: New contributions from the combination of image-based 3D modelling techniques and GIS. *J. Archaeol. Sci. Rep.* **2016**, *10*, 431–440. [CrossRef]
25. Pedelì, C. An interdisciplinary conservation module for condition survey on cultural heritages with a 3D information system. *ISPRS Int. Arch. Photogramm. Remote Sens. Spat. Inf. Sci.* **2013**, *2*, 483–487. [CrossRef]
26. Janvier-Badosa, S.; Stefani, C.; Brunetaud, X.; Beck, K.; De Luca, L.; Al-Mukhtar, M. Documentation and analysis of 3D mappings for monument diagnosys. In *Built Heritage: Monitoring Conservation Management*; Springer: Cham, Switzerland, 2014; pp. 347–357. ISBN 9783319085333.
27. Brunetaud, X.; Luca, L.D.; Janvier-Badosa, S.; Beck, K.; Al-Mukhtar, M. Application of digital techniques in monument preservation. *Eur. J. Environ. Civ. Eng.* **2012**, *16*, 543–556. [CrossRef]
28. Stefani, C.; Brunetaud, X.; Janvier-Badosa, S.; Beck, K.; De Luca, L.; Al-Mukhtar, M. Developing a toolkit for mapping and displaying stone alteration on a web-based documentation platform. *J. Cult. Herit.* **2013**, *15*, 1–9. [CrossRef]
29. Salonia, P.; Messina, T.L.; Marcolongo, A.; Appolonia, L. Photo scanner 3D survey for monitoring historical monuments. The case history of Porta Praetoria in Aosta. *Geoinform. FCE CTU* **2011**, *6*, 314–322. [CrossRef]
30. Bartolomucci, C.; Trizio, I.; Bonzagni, D. The reintegration of urban lacunas at castelvecchio calvisio (AQ) after the 2009 earthquake: The use of GIS 3D as a project monitoring tool. In Proceedings of the 4th International Conference on Progress in Cultural Heritage Preservation, Limassol, Cyprus, 29 October–3 November 2012; pp. 586–593.
31. Lock, G. Theorising the practice or practicing the theory: Archaeology and GIS. *Archaeol. Polona* **2001**, *39*, 153–164.
32. Lock, G.R. *Using Computers in Archaeology: Towards Virtual Pasts*; Routledge: New York, NY, USA, 2003; ISBN 9780415167703.
33. Haklay, M.E. Virtual reality and GIS: Applications, trends and directions. In *Virtual Reality in Geography*; Taylor & Francis: New York, NY, USA, 2002; pp. 47–57.
34. Dell'Unto, N. The use of 3D models for intra-site investigation in archaeology. In *3D Recording and Modeling in Archaeology and Cultural Heritage*; Archaeopress: Oxford, UK, 2014; ISBN 9781407312309.
35. Lymer, K. Image processing and visualisation of rock art laser scans from loups's hill, county durham. *Digit. Appl. Archaeol. Cult. Herit.* **2015**, *2*, 155–165. [CrossRef]
36. Pesci, A.; Casula, G.; Boschi, E. Laser scanning the Garisenda and Asinelli towers in Bologna (Italy): Detailed deformation patterns of two ancient leaning buildings. *J. Cult. Herit.* **2011**, *12*, 117–127. [CrossRef]
37. Pesci, A.; Teza, G.; Bonali, E.; Casula, G.; Boschi, E. A laser scanning-based method for fast estimation of seismic-induced building deformations. *ISPRS J. Photogram. Remote Sens.* **2013**, *79*, 185–198. [CrossRef]
38. Arias, P.; Herraez, J.; Lorenzo, H.; Ordonez, C. Control of structural problems in cultural heritage monuments using close-range photogrammetry and computer methods. *Comput. Struct.* **2005**, *83*, 1754–1766. [CrossRef]
39. Gainza, M. El deterioro de la Piedra en el Patrimonio Construido y Aplicación de Nanoformulaciones Para la Conservación de los Mismos. Master's Thesis, Universitat Politècnica de Catalunya, Barcelona, Spain, 2015.
40. Macià, M. *La Seu Vella de Lleida: Guia Històrica i Arquitectónica*; Generalitat de Catalunya, Departament de Cultura: Barcelona, Spain, 1997; ISBN 8439344120, 9788439344124.
41. Brodu, N.; Lague, D. 3D terrestrial lidar data classification of complex natural scenes using a multi-scale dimensionality criterion: Applications in geomorphology. *ISPRS J. Photogramm. Remote Sens.* **2012**, *68*, 121–134. [CrossRef]
42. Aalund, F. A Comparative Study of Screen-Space Ambient Occlusion Methods. Bachelor Thesis, Technical University of Denmark Informatics and Mathematical Modelling, Kgs. Lyngby, Denmark, 2013.
43. Barber, D.; Mills, J.; Andrews, D. *3D Laser Scanning for Cultural Heritage: Advice and Guidance to Users on Laser Scanning in Archaeology and Architecture*; Technical Report; English Heritage: Swindon, UK, 2011.

44. Andrés, M.A.N.; Pozuelo, F.B. Evolution of the architectural and heritage representation. *Landsc. Urban Plan.* **2009**, *91*, 105–112. [CrossRef]

45. De Reu, J.; Plets, G.; Verhoeven, G.; De Smedt, P.; Bats, M.; Cherretté, B.; De Maeyer, W.; Deconynck, J.; Herremans, D.; Laloo, P.; et al. Towards a three-dimensional cost-effective registration of the archaeological heritage. *J. Archaeol. Sci.* **2013**, *40*, 1108–1121. [CrossRef]

46. Mileto, C.; Vegas, F. El análisis estratigráfico: Una herramienta de conocimiento y conservación de la arquitectura. In *España: Arqueología Aplicada al Estudio e Interpretación de Edificios Históricos, Últimas Tendencias Metodológicas*; Ministerio de Cultura: Madrid, España, 2011; pp. 145–157.

47. Fiorini, A.; Urcia, A.; Archetti, V. The digital 3D survey as standard documentation of the archaeological stratigraphy. In Proceedings of the 12th International Symposium on Virtual Reality, Archaeology and Cultural Heritage VAST, Prato, Italy, 18–21 October 2011.

48. Burrough, P.; McDonell, R. *Principles of Geographical Information Systems*; Oxford University Press: New York, NY, USA, 1998.

49. Pancorbo, F. *Corrosión, Degradación y Envejecimiento de los Materiales Empleados en la Edificación*; Marcombo, S.A.: Barcelona, Spain, 2010; ISBN 8426715761, 9788426715760.

50. Valdeón, L.; Esbert, R.; Marcos, R. La alveolización y otras formas de alteración desarrolladas sobre las areniscas del palacio de Revillagigedo de Gijón (Asturias). *Mater. Constr.* **1985**, *35*, 41–48. [CrossRef]

51. Jacobs, G. Understanding spot size for laser scanning. *Prof. Surv. Mag.* **2006**, *26*, 48–50.

52. Almagro, A. *Levantamiento Arquitectónico, volumen 8 de Monográfica (Universidad de Granada): Biblioteca de Arquitectura, Urbanismo y Restauración*; Universidad de Granada: Granada, Spain, 2004; ISBN 8433831909, 9788433831903.

geosciences

MDPI

Article

Accurate Reconstruction of the Roman Circus in Milan by Georeferencing Heterogeneous Data Sources with GIS

Gabriele Guidi * , Sara Gonizzi Barsanti , Laura Loredana Micoli and Umair Shafqat Malik

Department of Mechanical Engineering, Politecnico di Milano, 20156 Milan, Italy;
sara.gonizzi@polimi.it (S.G.B.); laura.micoli@polimi.it (L.L.M.); umairshafqat.malik@polimi.it (U.S.M.)
* Correspondence: gabriele.guidi@polimi.it; Tel.: +39-02-2399-7183

Received: 11 August 2017; Accepted: 14 September 2017; Published: 20 September 2017

Abstract: This paper presents the methodological approach and the actual workflow for creating the 3D digital reconstruction in time of the ancient Roman Circus of Milan, which is presently covered completely by the urban fabric of the modern city. The diachronic reconstruction is based on a proper mix of quantitative data originated by current 3D surveys and historical sources, such as ancient maps, drawings, archaeological reports, restrictions decrees, and old photographs. When possible, such heterogeneous sources have been georeferenced and stored in a GIS system. In this way the sources have been analyzed in depth, allowing the deduction of geometrical information not explicitly revealed by the material available. A reliable reconstruction of the area in different historical periods has been therefore hypothesized. This research has been carried on in the framework of the project Cultural Heritage Through Time—CHT2, funded by the Joint Programming Initiative on Cultural Heritage (JPI-CH), supported by the Italian Ministry for Cultural Heritage (MiBACT), the Italian Ministry for University and Research (MIUR), and the European Commission.

Keywords: digital heritage; knowledge representation; 3D survey; data integration; GIS; 4D reconstruction

1. Introduction

The 3D digital representations through time of cultural heritage sites can be useful in order to study and preserve their memory, to properly communicate them to the public even when bad conservation conditions occur, and to plan their maintenance and promotion.

Usually, 3D modeling involved in the study of Cultural Heritage includes two definitively distinct categories: reality-based and reconstructive [1].

Reality-based 3D models are generally intended, as those obtained with high accuracy thanks to both active and passive 3D acquisition technologies, are nowadays largely available, such as laser scanners or Structure from Motion (SfM)/Image Matching (IM) photogrammetry, that allow one to sample the exterior surfaces of the object, building, site, or landscape of interest at high resolution. The dense 3D cloud provided is then meshed and possibly texturized, originating what is usually indicated as a reality-based 3D model [2,3].

On the other hand, reconstructive 3D models are those originated by a much less dense set of geometrical information, due to the fact that they are referred to objects, buildings, sites, or landscapes not existing anymore in their original form [4,5].

Paul Reilly introduced the concept of Virtual Archaeology in the early 1990s [6], pointing out how reconstructive models play a crucial role in archaeological studies for the possibility to visualize interactively non-existing architecture [7]. It is a tool for the shared synthesis of knowledge, for interactive interpretation of archaeological ruins, and for cultural dissemination [8]. In addition, virtual reconstruction in archaeology has been introduced as a novel method for finding archeological discoveries [9].

However, differently from reality-based models, a virtual reconstruction is necessarily based on a set of assumptions, and the acceptance of a certain level of uncertainty [10] that can be limited by integrating historical sources, providing, even in an indirect way, additional geometrical constraints [11]. In order to define shared methodologies for approaching the problem of 3D reconstruction of partially or totally non-existing archaeological structures, in the past two decades, different relevant initiatives have been taken including the Virtual Archaeology Special Interest Group (VASIG) of the international organization "Computer Applications and Quantitative Methods in Archaeology" (CAA), the Cultural Virtual Reality Organization (CVRO), and the European project EPOCH (www.epoch-net.org). These and other contributions have laid the foundation of the "London Charter", whose main aim is to establish internationally-recognized principles for the use of computer-based visualization of cultural heritage [12].

Different projects have faced the issues of virtual reconstruction in archaeology e.g., aiming at the digital reconstruction of large cities [13], the digital reunification of an ancient structure on site with decoration conserved in a different location [14], the construction of knowledge models behind the studied site [15], the different interaction typology between user and digital contents [16], and the public engagement and the dissemination of archaeological data [17].

Technically, the reconstruction activities might be oriented more to the study of a structure using engineering tools [18,19], to the accurate visualization of the reconstructed scene [20,21], or to the exploration of a particular site through time [22].

Even if a reconstruction can simply be based on drawings originated by previous studies, several works appeared in the literature [23–26] which have shown that a serious reconstruction work should start from the accurate digitization of the existing remains, enriched with additional pieces of information provided by various non-uniform sources.

However, there are cases where the three-dimensional diachronic reconstruction is particularly complex. There are several reasons for this: (i) the nearly total absence of remains to be surveyed; (ii) one or more time periods of an artefact's life with little historical documentation; (iii) uncertainty of sources; (iv) difficulty to correlate documents and data to a three-dimensional representation.

This paper presents an experimental activity defined by the project Cultural Heritage Through Time—CHT2 (http://cht2-project.eu), funded in the framework of the Joint Programming Initiative on Cultural Heritage (JPI-CH) supported by the Italian Ministry for Cultural Heritage (MiBACT), the Italian Ministry for University and Research (MIUR), and the European Commission. Its goal is to develop time-varying 3D products, from landscape to architectural scale, to envisage and analyze lost scenarios, or visualize changes due to anthropic activities or intervention, pollution, wars, earthquakes, or other natural hazards. The main aim of the CHT2 project is to merge heterogeneous information and expertise to deliver enhanced four-dimensional (4D) digital products of heritage sites. CHT2 is working on the full integration of the temporal dimension, its management and visualization, for studying and analyzing Cultural Heritage structures and landscapes through time. The proposed methodology for the whole project described in [27], suggests different ways to reconstruct the diachronic life of an historical object, monument, or landscape.

Regarding the case presented in this paper, given the limited detectable findings, the work began with an in-depth philological analysis from the collection of historical and archival data, integrating the various contributions in the area, beginning in early 1939 [28–31].

The case study covers the south-west area of the city center of Milan that corresponds to the Roman Circus area, where it is possible to see several traces of different historical periods from ancient times, to the densely urbanized structure of the present days. According to some archaeological findings, Milan was inhabited since the 5th century BC in the area corresponding to the current Via Meravigli, Via Valpetrosa, and the Piazza del Duomo. In this area, some protohistoric tracks converge, with traces of the later Roman roads, are still recognizable in the city plan. Such roads connected Milan with some towns and villages to the northwest neighborhoods belonging initially to the Golasecca culture, a prehistoric civilization located in the Ticino River area from the Bronze Age until the 1st century BC.

In the 2nd and 1st centuries BC, excavation and leveling took place, in order to adapt the ground to the Roman urban model. At that time, the first urban planning was put in place, probably maintaining the Golasecca road network. The entire settlement occupied an area of 80 hectares, with a spatial organization influenced by differences in height, watercourses and by the existing routes. In the second half of the 1st century BC, the Romanization of the territory was completed [31,32]. In the first Roman city plan, the investigated area, now occupied by the Archaeological Museum, was first dedicated to housing and production activities. These activities were probably related to metal extraction from the sediments of the Nirone River and the related processing. This hypothesis has been confirmed by recent excavations, which revealed the presence of a pre-imperial domus. At the end of the 1st century BC, good quality residences and production activities would have been present. A subsequent domus, built in the 1st and 2nd centuries AD, was later transformed into a prestigious domus in the 3rd century AD. For none of these buildings, unfortunately, is a complete plan known [33].

In the year 286 AD, Milan became the capital of the Western Roman Empire, under the emperor Maximian. During the imperial period, up to 402 AD, the area was modified by the construction of major buildings, such as the imperial palace, the circus, and the defensive walls [29,34].

The Circus was the open-air venue for chariot and horseraces; or rather, the place dedicated to the celebration of the Emperor's greatness. For this reason, circuses were generally located near the Imperial Palace [35].

Milan's circus was also adjacent to the defensive walls, with which it shared the western part. This particular location probably resulted in a number of peculiarities, such as the absence of a Triumphal Arch on the apex of the curve. Although the circus of Milan was one of the most important in the late Roman empire, only a few traces are still available: a tower of the city walls, a tower of the carceres reused as a bell tower (formerly belonging to the Monastero Maggiore), and some sections of the walls or foundations in private properties nearby [30,35,36]. Historical sources report the existence of the circus until the Lombard era. From that period, as happened to other monuments in Milan, the materials of the Roman structures were reused in new construction projects.

Archaeological studies were conducted mostly at the beginning of the 1900s and after World War II, during the reconstruction of some private and public buildings, when it was possible to see the archaeological remains. Unfortunately, nowadays, only small portions of the monument remain visible; moreover, much of the historical documentation was lost in a fire during World War II. Starting from the 1960s, the area was used as the headquarters of the Archaeological Museum of the Municipality of Milan (Civico Museo Archeologico di Milano), that still takes care to preserve the memory of the various eras represented in this part of the town. The rest of the area occupied by the ancient Roman circus is instead almost entirely occupied by residential buildings [37]. This area also includes the Church of San Maurizio, whose nucleus is of early Christian origin, and which was built in the form we know today at the beginning of the 1500s. Small remains of the circus are still visible in the basements of modern buildings in that area. Many questions are still open about the structure of the building and its relation to the surrounding area, including the imperial palace and the town fortification walls.

2. Materials and Methods

In some cases [26,38], diachronic reconstruction of a monumental complex started with the three-dimensional survey of the actual state of the monument, and an investigation of the traces of different phases of the building, along with suitable philological research. In the case of the Roman circus of Milan, given the limited detectable findings, the work started with an in-depth philological research campaign, beginning with historical and archival data. All kinds of sources (texts, maps, drawings, archaeological reports and restriction decrees, photographs) have been integrated to hypothesize a reconstruction of the monument, by referencing such documents to that specific location of the city.

The research began with the bibliography related to the study of the monument. This was useful for gathering all the information available about the previous research and the excavations done during the past centuries.

Especially important for the study of the monument was the book of De Capitani D'Arzago [28], an archaeologist who thoroughly studied the Roman circus of Milan in the late 1930s. His studies confirmed the existence, location, and essential size of the circus, thanks to the discovery of the parallel walls, some portions of the foundations, and a large part of the curve.

Another important book is that of Humphrey [35], in which many Roman circuses, among them also the circus of Milan, were investigated. Thanks to this information, it was possible to interpret the missing parts of the monument by comparing its shape to others from the same period. This research is fundamental when analyzing and reconstructing an ancient building that is no longer visible.

Another step of the work was the collection of maps, drawings and images concerning the various topics covered in the research (Figure 1). Drawings and historical paintings are fundamental for gathering information no longer available today. This kind of approach was useful for gathering typological indications and validating the reconstructive hypotheses proposed by scholars. In our case, we have found sources regarding: (i) the domus of the pre-imperial era; (ii) Roman circuses built in the Empire in the same period; and (iii) monasteries of the Benedictine order.

(a)

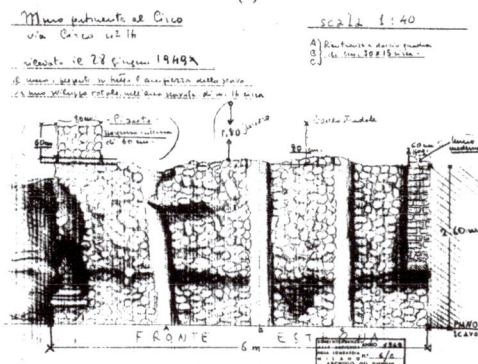

(b)

Figure 1. Different types of archival data: (**a**) reconstructive drawing of the Circus and the Imperial Palace (F. Corni, courtesy of Civico Museo Archeologico di Milano); (**b**) drawing of an archaeological excavation carried out at Via Circo 14 in 1949 (courtesy of Ministero dei beni e delle attività culturali e del turismo).

Unfortunately, in the case of Milan, only modest graphic representations of the involved monuments were available, with reference to their active period. As far as more recent times are concerned, all the drawings of survey campaigns carried out in the area have been collected. Since many buildings destroyed by bombing during World War II (WW2) were rebuilt, in the post war period, the excavations for the foundations of the modern buildings, in some cases, revealed archaeological finds that sometimes were used as the basement for the new building. In other instances, for example, in correspondence to new roads or other unbuilt areas, such findings were simply covered. Sometimes, notes and sketches made during previous excavations revealed crucial pieces of information for defining the structure of the monument.

Next, a thorough iconographic research campaign was carried out, collecting different maps from various periods that could highlight the urban structure of the area. About 60 city maps representing different historical periods from the Renaissance to the present day have been identified at the "Civica Raccolta delle Stampe Achille Bertarelli" in Milan and analyzed to study the evolution of the urban area. Another type of data taken into account were the photographs taken mainly during the post-WW2 excavations, as for example, those shown in Figure 2. Images of artefacts and structures inside the urban area, taken from different points of view and sometimes referred to two or more different periods of their existence, are a valuable support for the three-dimensional reconstruction process.

(a) (b)

Figure 2. Photographs of the remains of the external wall of the circus visible in a private garden of a building in via Vigna 1: (**a**) the entire portion of the remains; (**b**) a detail of the arches sustaining the podium visible on the wall (courtesy of Ministero dei beni e delle attività culturali e del turismo).

In addition, a research campaign in the photographic archive at the Superintendence office was made regarding to the area of interest. About one thousand images were found, and about 100 of them were selected for use in this project. The selection regards artefacts visible during construction projects (e.g., the metro, new skyscrapers) or inspections of the Superintendence office. These images are a valuable documentary heritage because many artefacts, embedded in the foundations of modern buildings, are no longer visible.

The next step of the research, propaedeutic to the 4D reconstruction, involved the identification and survey of all the visible portions of the monument.

Starting from a detailed map of the circus made in the late 1930s, the accurate positioning of all the remains of the circus and its connected structures was checked.

This part of the work dealt also with a capillary search, in connection with the inspectors of the Superintendence of Milan, of all the street numbers of the actual buildings where the remains are still visible in the basements. During this search, it was observed that all the private houses of interest, due to archaeological findings, are under restrictions, but given the period in which these restrictions were defined, most of them are brief and unclear. Hence, it was difficult to identify the individual structures, with regards to their location and their extent, as this required huge archival work.

All areas subjected to archaeological restrictions have been screened along with the Superintendence office, in order to assess the actual presence of remains, their state of conservation and the opportunity to do a three-dimensional survey.

The 3D survey of all the selected remains is a crucial stage for generating a reliable starting point for the reconstruction. Depending on the conditions of operation, such 3D digitization is being made with both SFM photogrammetry and laser scanning, depending on the available conditions of lighting, working space, etc. The monument portions that are detected and suitably georeferenced are used to validate the archaeological excavations of the past, and to give the main constraints over which the three-dimensional reconstruction have been progressively generated.

In addition to the validation of historical plans, the three-dimensional portions are fundamental as elements of proportion, in relation to the examples of the same type of monument highlighted by other sources, to define the elevation of the building, typically, the most critical parameter in the reconstruction of any ancient building not existing anymore. The survey work started first on an unrestricted area belonging to the Archaeological Museum of Milan, which was a supporting partner in the CHT2 project. The first two components of the circus that have been digitized are: (i) the so-called "square tower", originally belonging to the carceres of the circus, and nowadays used as the bell tower of the church dedicated to San Maurizio; and (ii) the so-called "polygonal tower", part of the defensive walls of the city. The two buildings were surveyed both with laser scanning, using a Faro Focus 3D scanner, and with photogrammetry, using a Panasonic DMC GH4 with a 12 mm lens and a Canon 5D MkII with a 20 mm F 2.8 lens. The resulting digital models are shown in Figure 3.

(a)   (b)

Figure 3. 3D models of structures still standing in the area: (**a**) square tower, belonging to the carceres of the circus. The model was created by the integration of photogrammetry for the interiors and laser scanning for the exteriors; (**b**) polygonal tower, which was part of the late roman city walls. In this case the model was originated exclusively by photogrammetric data.

Another useful contribution to our study comes from a significant part of the outer circus wall, nowadays belonging to the garden of a private condominium in Milan, via Vigna 1 (Figures 2, 4 and 5), and a smaller remnant of the inner wall, also called the podium wall. It corresponds on one side to the limit of the circus racetrack, and on the other side to the lower limit of the cavea. This latter portion of the foundations and inner wall are inside the cellar of the above-mentioned condominium.

The external wall was surveyed using both photogrammetry, with a NEX 6 camera coupled with a 24 mm F 1.8 Zeiss lens and laser scanning, using a Faro Focus 3D 120 device. The small portion of the podium wall was surveyed with the Faro laser scanner only, connecting the inside to the outside through a set of redundant scans taken from several positions along the path from the cellars to the exterior garden. The last survey performed relates to a portion of a wall inside the basement of a building in via Luini. It refers to a western portion of the external wall of the circus. What is visible today is a 2–3 m long foundation, cut by a modern wall. The survey was performed using a Faro Focus 3D 120 device.

Figure 4. Documentation of the main portion of the circus outer wall still standing through an orthoimage, providing the measurement of the arches supporting the audience seats.

Figure 5. Documentation of the main portion of the circus outer wall still standing: laser scan including a small remnant of the inner wall found in a private basement, providing the height of the wall in this point of the structure, and its distance from the inner one.

3. Results

These surveys in via Vigna 1 were important for measuring several crucial geometric elements, needed for the 3D reconstruction: (i) the height of the remains of the outer wall of the circus in this area, corresponding to 7 m. Several sources, however, stated that this is not the whole height of the circus.

This was confirmed by other measurements, taken at the curved end of the circus on the side opposite to the carceres, that reached 8.40 m; (ii) the height of the traces of the arches, visible on the inner side of the exterior wall, that identify the starting point of the vaults, located at 3.98 + 0.44 = 4.42 m (Figure 4); (iii) the periodicity of the arches representing the traces of the vaults in the exterior wall, equal to 3.02 + 1.22 = 4.24 m (Figure 4); and (iv) the precise distance between the inner and the outer wall (4.75 m) shown in Figure 5.

This latter information, together with the incline of the seating tiers, which according the bibliographic sources in other circuses of the same period is around 45° [35], allowed us to calculate the positioning of the tiered seats with maximum likelihood, as shown in Figure 6.

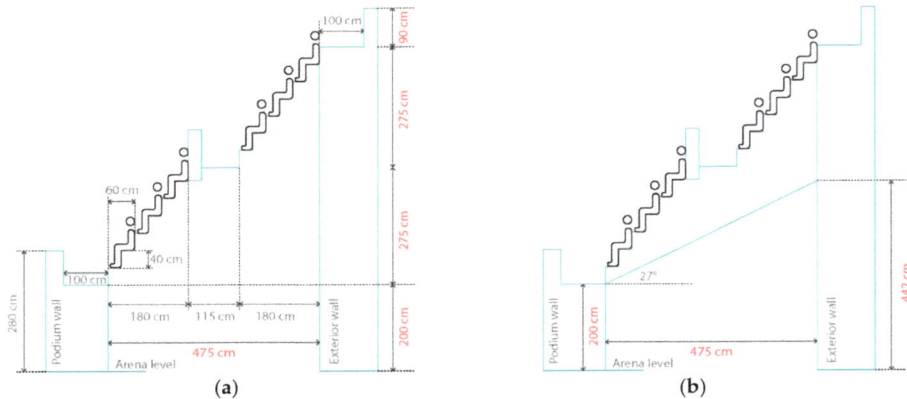

Figure 6. Hypothetical cross-section of the circus showing the distribution of seats, according to the measurements coming out from the laser scanning campaign (in red are the measurements taken on the field): (**a**) subdivision between different social classes (**b**) Hypothetical inclination of the vault.

As already suggested in the literature [29], the typical subdivision of the crowd into social classes would lead to the hypothesis of two orders of three seats, with 30 cm for seating and 30 for the feet, as shown in Figure 6a. This would give six rows of seats divided in two orders, separated by a space, with the double function of a walking lane for reaching the seats, and for separation between the nobles, typically placed closer to the racetrack, and those of the plebeians, in the upper zone.

An element that might seem peculiar in this reconstruction is the positioning of each seat on a plane different from that of the feet, which would be more economical. However, the arrangement shown in Figure 6a was typical of late Roman circuses, like the Circus of Maxentius in Rome, belonging to the same period. In addition, the latter seat organization would give a smaller inclination of the seats zone. Actually, a single large area of 60 cm for the seat of the lower person and for the feet of the upper person would lead to an increase in height of 40 cm, corresponding to the minimum height for a seat with respect to the floor level. This would give an angle of 37°, not compliant with the various considerations in the literature and to the global height of the outer wall. Contrarily, the seat organization shown in Figure 6a would give a global height increase of 5.50 m for 4.75 m of distance, corresponding to an angle of 49°. This reconstruction also fits better with the global height of around 9 m for the exterior structure. The sources give a 2 m height of the planking level of the first (the lower) row of seats, corresponding to the height of the podium wall shown in Figure 6a. This height, plus the 5.50 m needed for accommodating the six rows of seats, gives a height of 7.50 m for the last (the upper) seating. Behind the shoulder of the upper row of seats, a small wall is expected, with a height not lower than 0.9 m for preventing accidental falls from the upper level of seats. This gives a total height of 8.40 m from the arena level that is coherent with the measurement found in another area of the circus.

A second element that could be measured refers to the structure of the arches connecting the outer wall to the podium wall. They are segmental arches 3.02 m wide, reaching a height of 4.42 m from the foundations, as shown in Figure 4 and represented schematically in Figure 6b. This structure refers to the support of the podium, thus also having a corridor function for the passage of the public.

Considering the construction techniques in use during the Roman period, and comparing the remains of the circus of Milan with other performance buildings, the most probable structure was composed of a vault forming the ceiling of the corridor, intersected by architraves sustaining the podium. This structure is still visible, for example, in the Circus Maximus and the Colosseum in Rome (Figure 7). Even if these two buildings are earlier than the Circus in Milan, they represent the best conserved examples to analyze and understand the Roman construction technique of performance buildings.

(a)

(b)

(c)

Figure 7. Examples of the structure built as a support of the podium with a double purpose, upholding the structure of the seats and corridor for the public to reach their seats: (**a**,**b**) Circus Maximus in Rome; (**c**) Flavius Amphitheatre (Colosseum) in Rome.

Although there is evidence of the sustaining arches and pillars on the external wall, it is not possible to identify the same structure on the ruins of the internal one, due to the insufficient height of the remains. For the following 4D reconstruction, the only way is to analyze comparable structures and to utilize the model.

The connection should be close to the top of it on the interior side of the podium wall for limiting the slope of the vault. This defines a lower end of the vault at 2 m from the foundation level, rising to 4.42 m, corresponding to a vertical difference of 4.42 − 2 = 2.42 m. Such displacement, on a horizontal distance of 4.75 m, defines an angle of 27° with respect to the horizontal line, as is indicated in Figure 6b.

3.1. Georeferencing and Harmonization of the Different Sources

All the data collected are crucial for hypothesizing and reconstructing the circus. In order to have a shared basis from which to start, a commercial GIS platform was used (ArcMAP by ESRI). The first step was for georeferencing the most important survey of the past, represented by the hypothetical map of the circus made in the late 1930s [31] using the 3D technologies available at that time. Even if no precise measurement of the underground remains was made with respect to the modern city, this map, nevertheless, represents a reasonable starting point for the following reconstruction phases. The first step was, therefore, the orientation of the De Capitani map with respect to the cadastral map provided by the municipality of Milan. The drawing was georeferenced using the Helmert transformation, using as references, the profiles of some recognizable modern buildings visible also on the De Capitani map. The remains of the circus highlighted on the map were then transformed from raster to vector drawings with AutoCAD, in order to have them as curves to be used for extruding a 3D/4D reconstruction of the circus (Figure 8a).

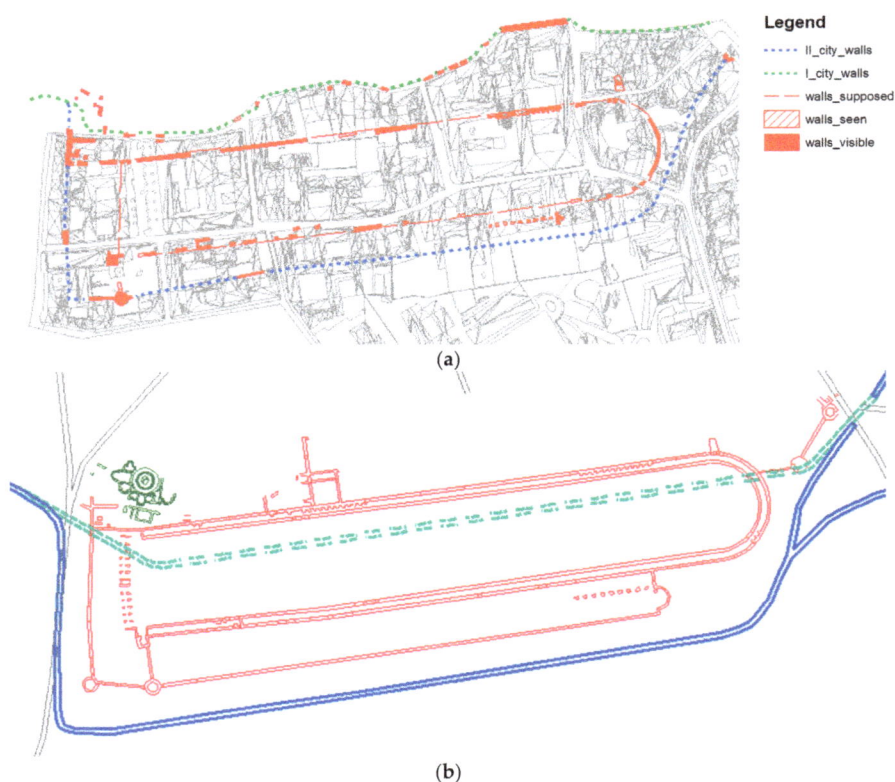

(a)

(b)

Figure 8. Georeferenced representation of the circus maps corresponding to different periods in time: (**a**) De Capitani's map indicating various information about the walls of the circus excavated or guessed; (**b**) Mirabella Roberti's update highlighting the different paths of the river, before (blue dashed line) and after the intervention (blue continuous line) for building the circus.

The map of Mirabella Roberti [29], who updated the one from De Capitani in the 1980s, was then georeferenced and drawn. Added to this was other information about the hypothesized ancient city in the reigns of the emperor Augustus and Maximian (i.e., before and after the circus construction), in order to underline the changes in the topography of the city after some huge interventions occurred in the historical phase before the building of the circus. The most important of these regarded the change of the course of the river Seveso that used to flow in the area where, during Maximian's rule, the circus and the new city walls were built (blue dashed line in Figure 8b).

The route of the river was diverted to make space for the construction of the circus just beside the imperial palace, and follows the itinerary of the new city walls. This is clearly highlighted in Figure 8b by the different paths of the river, before (blue dashed line) and after the intervention (blue continuous line). Here, the only excavated structure is the imperial palace (colored in green in the top left corner of the figure).

The two maps of the circus were then overlapped; as is clearly evident in Figure 9, this makes some differences evident. The first one is that Mirabella Roberti's map seems more complete, with the indication of both the external and internal walls, and the indication of few structures along the Circus length, for example, the tower in correspondence with the southern part of the buildings, or the one with an apse along the western part of the Circus. The main difference, however, is the divergence in length of 10 m between the two maps. The De Capitani map gives a total internal length of 460 m while the Mirabella Roberti one gives 470 m. This information will be confirmed or updated at the end of the surveys of all the visible and existing portions of the circus walls, but, as of now, it seems that Mirabella Roberti's map is more reliable, being drawn around 50 years after De Capitani's, and so, more complete because it includes all the most recent excavations.

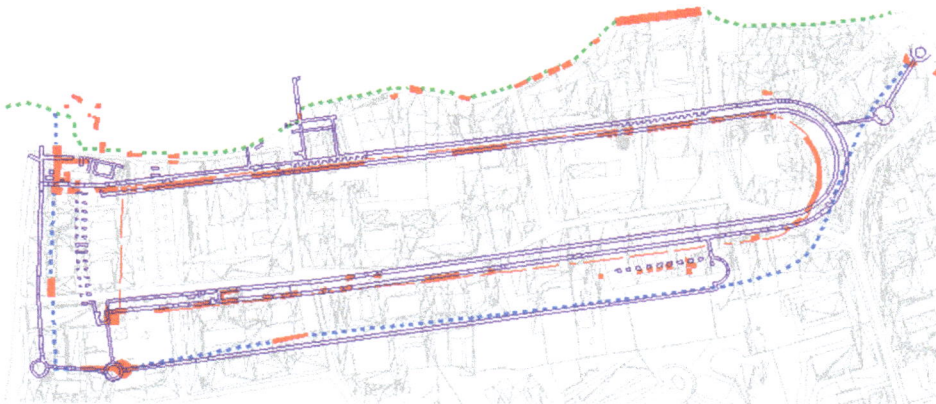

Figure 9. The overlapping of the two maps of the Roman Circus of Milan: De Capitani's map in red, and Mirabella Roberti's map in blue.

This georeferenced map was then connected to a geodatabase that includes all the information collected during the archival review (Figure 10). The fields of the tables were chosen coherently with the previously existing metadata associated with the archival documents. To each portion of wall identified, the information about excavations, documentation, restriction decrees, and images, was connected.

3.2. 4D Reconstruction

The so-called 4D reconstruction is usually intended as the process for obtaining the shape of real objects and their changes along a temporal dimension, based on a methodology for integrating the various sources of non-uniform data [27]. In addition, the surveys described above, georeferenced to the topographic network of the city, helped to identify the accurate positioning of the buildings, and to obtain precise measurements of shape and size of the structures. By merging such data related to the current state of the monument with the vast archival material collected, a rearrangement of the historical representations was made, including the normalization of the historical plans in a uniform scale. From such an integrated base of information, a 4D reconstruction is ongoing, together with the archaeologists' input, in order to better identify the proper reconstruction of the ancient building and all the changes that affected the area from the late Roman period until the present time.

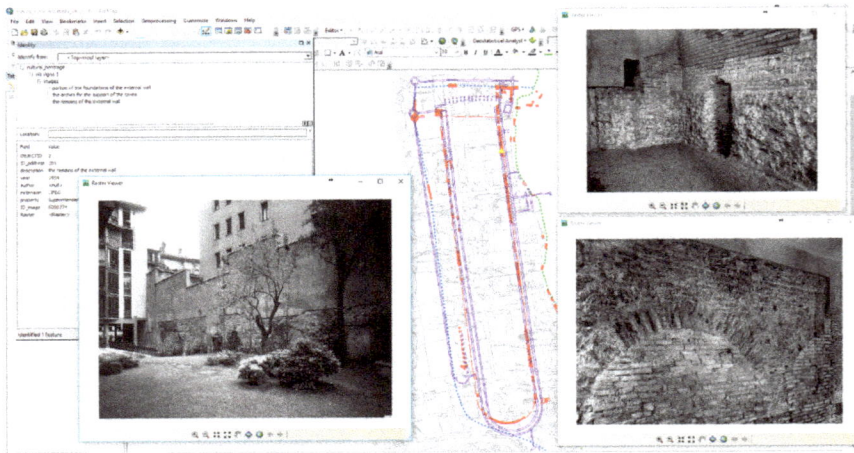

Figure 10. The creation and implementation of the geodatabase connected to all the different portions of the circus' walls identified and surveyed with the metadata related to the ruins and to the archival images collected.

The reconstruction began with the georeferenced maps of De Capitani and Mirabella, which served as a starting point for the 3D model (Figure 11), giving the total length of the monument, which will be confirmed at the end of the surveys, and the connection of all the parts that are not visible in the present time.

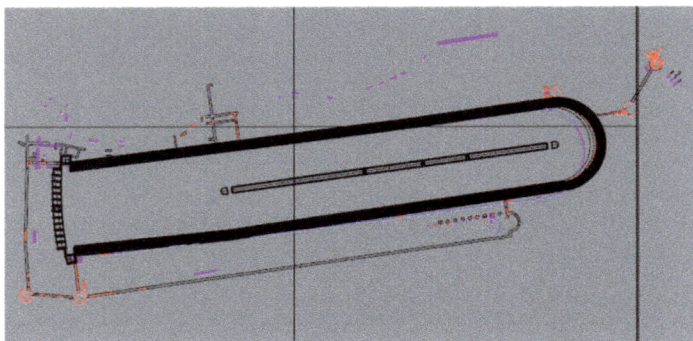

Figure 11. The reconstruction of the circus starting from the documentation provided from the past studies reported on different layers of the same GIS.

Starting with the plan, the 3D reconstruction (Figure 12a) added the information collected during the careful work in the archive of the Superintendence that related to the height of the external wall in different parts of the structure. From the documentation of the excavations, it was possible to know that the external wall was higher in the southern part, where it served also as city walls, while the height in the eastern part was ascertained from the traces of the wall of the circus on the square tower now in the garden of the Archaeological Museum.

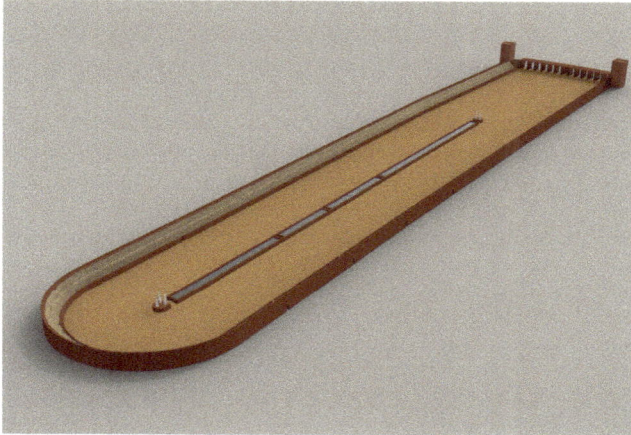

(a)

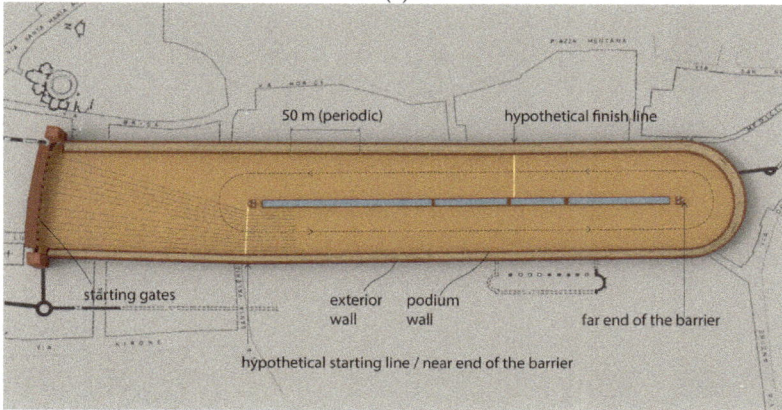

(b)

Figure 12. The reconstruction of the circus: (**a**) prospective view; (**b**) a plan view overlapped to the city map from which it is possible to see the change of direction of the exterior walls in correspondence with the starting line, a technical expedient to permit all the chariots to run the same length before arriving at the starting line.

From the surveys, on the other hand, it was possible to exactly place the two towers, one of the *carceres*, and one related to the city walls, and it is expected to confirm or update the map with the future surveys. Other important information was gathered during the survey in Via Vigna shown in Figures 4 and 5, where the discovery of a few remnants of the internal wall of the Circus allowed us to set the exact distance between the two walls, and consequently, as noted previously, the slope of the seats shown in Figure 6. One part of the main data that has yet to be verified, is the divergence of the walls of the circus in the western part, made to provide the same starting line for all the chariots. As shown in the lower left part of Figure 12b, as of now, the reconstruction has been carried out with the Mirabella Roberti hypothesis as its basis, based in turn on comparisons with other circuses. This hypothesis is under verification, with specific 3D surveys in the basements of private houses containing a few remains of these structures.

Most of the reconstruction, unfortunately, was made by comparing similar structures still visible, such as the circus of Maxentius, the Circus Maximus and the Flavian Amphitheatre in Rome. No traces were found of the *spina*, the central line whose function was to divide the track, and on which, there was the lap-counter. In the Circus of Milan, most probably, it was marked by marble eggs, one of which was lowered for each lap finished by the first chariot. This hypothesis is based on the discovery of one of these eggs during the excavations in the area of the circus.

For this reconstruction, in addition to the discussion and hypothesis with the archaeologists, and the comparison with similar structures built in the Roman period, the archival and bibliographic documentation is used to better identify the proper structure of the building, especially the drawings by Corni (Figure 13). Although these drawings are very hypothetical, and sometimes contain reconstructive errors, they can be used as a starting point to figure out how the elevated structures might have looked like.

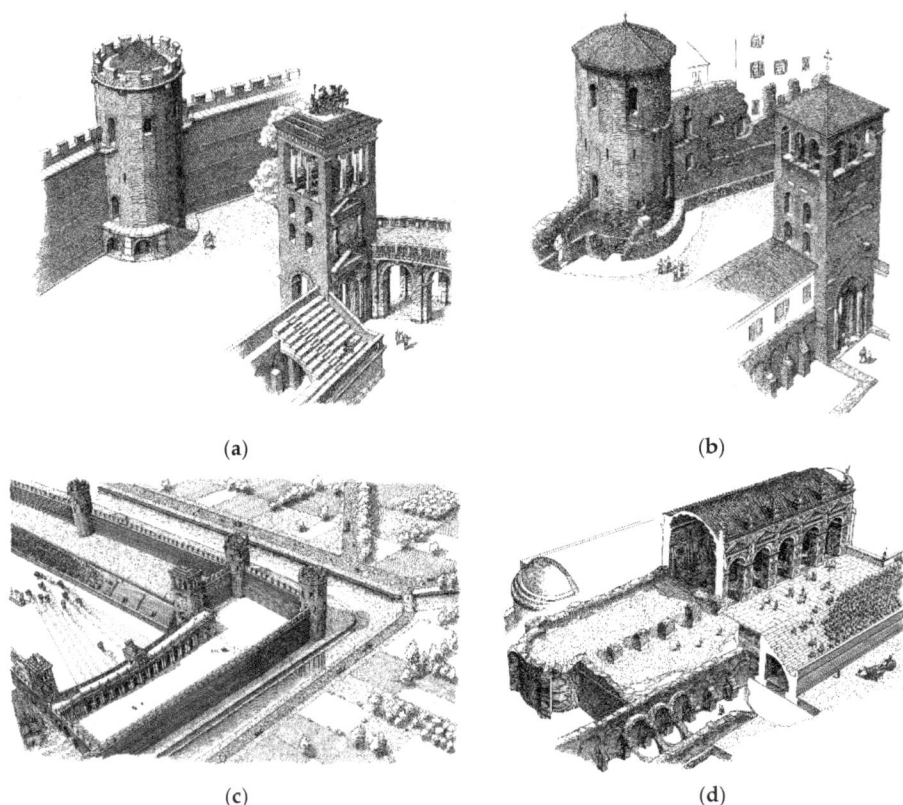

(a)

(b)

(c)

(d)

Figure 13. Drawing by Corni (courtesy of the Directorate of the Archaeological Museum) about the reconstruction of some parts of the Circus: (**a**) the tower of the city walls facing the tower of the *carceres* as they were during the late roman time; (**b**) as they look now inside the garden of the archaeological museum; (**c**) the view from the starting point of the races; (**d**) the hypothesis of the dais where the emperor and his entourage watched the races.

Finally, the 3D reconstruction of the circus was imported in a GIS platform and overlapped on the 3D model of the corresponding portion of the city (Figure 14), georeferenced with markers on the same coordinate system of the GIS map. On the continuation of the project, the georeferencing of the structures of the Circus and the city plan will be improved with other topographical and 3D surveys. Such work involves the laser scanning of each remain in the basements of private apartments. The laser survey covers both the underground structure (where the GPS cannot work) and the external part of the building, in order to catch the GPS signal. On some natural point a differential GPS is used for gathering precise georeferenced coordinates of at least three points of the scanned structure. The same points are then extracted by the laser scanned model, and through the two sets of coordinates, the roto-translation matrix from the laser scanner coordinate system to the georeferenced coordinate system is calculated, thanks to a procedure based on the quaternion method [39].

This alignment of the ancient structure with the existing urban fabric is important for several reasons: (i) the city can have a better perception of a lost monument, probably the second most important after the imperial palace, during the last years of the western Roman Empire; (ii) it will help to define, in an accurate way, the position of the remains, and to identify a buffer zone in which it might be possible to find them; (iii) it will help to better organize the urban intervention in the area; (iv) some elements of the structure, still missing, can be placed in a precise way, helping to identify the probable areas in which to intervene with excavations.

Figure 14. The overlapping of the georeferenced reconstruction of the circus on 3D map of the related area.

4. Discussion and Conclusions

The purpose of the digital reconstruction of the ancient Roman circus of Milan was to give a visual representation to a structure whose formal details, although studied by several authors in the past 70 years, was not very clear.

This large structure of the past had a huge impact on the city at the time of its construction, involving the change of the course of a river, and the complete redefinition of the southern part of the city of Milan.

The digital reconstruction of such a monument was particularly complex, due to the presence of the few remains, in the underground of an active and lively city, mostly in the basements of private houses. This raised the typical problems of archaeological studies in urban areas, such as: (i) the need to agree to a proper survey action with the national authority for the protection of archaeological heritage (Archaeological Superintendence of Milan), adding a heavy load of bureaucratic activity to the typical research work; (ii) the technical issue of having most of the 3D surveys referred to the basements of private houses, imposed to generate laser scanned 3D models of larger areas in order to include in the same model both the underground and parts of the exteriors. In this way, it has been possible to relocate in space each small portion of the circus remains, with a GPS, thus obtaining important clues related to the orientation of the structure in space. Such information allowed us to reconstruct the position of some elements of the circus, never studied before; and (iii) the digitally reconstructed structure of the circus will allow us to communicate to the citizens of Milan, and to visitors in general, the impact on the city and on its future development of such a gigantic historical artefact.

Thanks to geoscience, it was possible to identify and place in space the different portions of the monument, most of them not visible because they are hidden in basements or gardens of private houses. This will be a big step in the study of the Roman city of Milan, helping archaeologists to better define the structure. The work done with the Superintendence regarding the different hypothesis of the reconstruction is truly useful even for scholars, because it forces them to analyze all the documentation. On the other hand, the possibility of seeing the structure in 3D helps them to have a clear idea of the monument, and so to identify the best way to reconstruct the building.

Finally, this work will make it possible for anyone who is interested to understand and appreciate a lost, important monument of their city, because most of the people do not even know of the existence of the circus. Having the perception of the structure among the modern streets and buildings is the best way to help non-experts to understand the dimensions and the importance of the monument, while the tridimensional reconstruction will help people to gain a better comprehension of the building itself.

Acknowledgments: The authors wish to acknowledge the support of the Joint Programming Initiative on Cultural Heritage (JPICH) for funding the project Cultural Heritage through Time (CHT2), with the contribution of MiBACT, MIUR and the European Commission. The author would like to thank Anna Maria Fedeli from the Archaeological Superintendence of Milan, Donatella Caporusso and Anna Provenzali, from the Archaeological Museum of Milan for their precious cooperation in providing documentation on the area, and also for their help with the archival research. Finally, a special thank you goes to the archaeologist Simona Morandi for some useful discussions and for letting us access unpublished surveys related to the square tower that were useful for supporting the reconstruction hypotheses reported here.

Author Contributions: Gabriele Guidi is the coordinator of the project and of the surveys; Laura Loredana Micoli collected and organized the bibliographic and archival data, followed the relation with the Superintendence and performed the surveys; Sara Gonizzi Barsanti designed and created the GIS and the geodatabase, performed the surveys and participated to the collection of bibliographic data; Umair Malik performed the surveys and the 3D reconstruction of the circus. Gabriele Guidi and Sara Gonizzi Barsanti wrote the paper.

Conflicts of Interest: The authors declare no conflict of interest.

References

1. Russo, M.; Guidi, G.; Russo, M. Reality-Based and Reconstructive models: Digital Media for Cultural Heritage Valorization. *SCIRES-IT* **2011**, *1*, 71–86. [CrossRef]
2. Gruen, A. Reality-based generation of virtual environments for digital earth. *Int. J. Digit. Earth* **2008**, *1*, 88–106. [CrossRef]
3. Remondino, F.; Rizzi, A. Reality-based 3D documentation of natural and cultural heritage sites-techniques, problems, and examples. *Appl. Geomat.* **2010**, *2*, 85–100. [CrossRef]
4. De Luca, L.; Veron, P.; Florenzano, M. Reverse engineering of architectural buildings based on a hybrid modeling approach. *Comput. Graph.* **2006**, *30*, 160–176. [CrossRef]
5. Sullivan, E. Potential pasts: Taking a humanistic approach to computer visualization of ancient landscapes. *Bull. Inst. Class. Stud.* **2016**, *59*. [CrossRef]
6. Reilly, P. Towards a virtual archaeology. In *British Archaeological Reports International Series*; Archeopress: Oxford, UK, 1991; Volume 565, pp. 133–139.
7. Dam, A.V.; Forsberg, A.S.; Laidlaw, D.H.; LaViola, J.J.J.; Simpson, R.M. Immersive VR for scientific visualization: A progress report. *IEEE Comput. Graph. Appl.* **2000**, *20*, 26–52. [CrossRef]

8. Barcelò, J.A.; Forte, M.; Sanders, D.H. *Virtual Reality in Archaeology*; British Archaeological Reports; International Series; Archeopress: Oxford, UK, 2000; ISBN 978-1-84171-047-1.

9. *Beyond Illustration: 2D AND 3D Digital Technologies as Tools for Discovery in Archaeology*; Frischer, B.; Dakouri-Hild, A. (Eds.); Archaeopress: Oxford, UK, 2008; ISBN 978-1-4073-0292-8.

10. Smith, J.S.; Rusinkiewicz, S.M.; Alberti, S.; Chen, J.; Coronado, M.; Disco, G.; Georgiou, A.; Pico, T.; Touloumes, G.; Triolo, G. *Modeling the Past Online: Interactive Visualisation of Uncertainty and Phasing*; Springer: Cham, Switzerland, 2014; Volume 8740.

11. Happa, J.; Mudge, M.; Debattista, K.; Artusi, A.; Gonçalves, A.; Chalmers, A. Illuminating the past: State of the art. *Virtual Real.* **2010**, *14*, 155–182. [CrossRef]

12. *The London Charter for the Computer-Based Visualisation of Cultural Heritage (Version 2.1, February 2009)*; Ashgate Publishing Ltd.: Surrey, UK, 2012; ISBN 978-0-7546-7583-9.

13. Frischer, B.; Abernathy, D.; Guidi, G.; Myers, J.; Thibodeau, C.; Salvemini, A.; Müller, P.; Hofstee, P.; Minor, B. Rome reborn. In *ACM SIGGRAPH 2008*; ACM Press: Los Angeles, CA, USA, 2008.

14. Stumpfel, J.; Tchou, C.; Yun, N.; Martinez, P.; Hawkins, T.; Jones, A.; Emerson, B.; Debevec, P. Digital Reunification of the Parthenon and its Sculptures. In Proceedings of the 4th International Conference on Virtual Reality, Archaeology and Intelligent Cultural Heritage, Brighton, UK, 5–7 November 2003; pp. 41–50.

15. Gabellone, F. Ancient contexts and virtual reality: From reconstructive study to the construction of knowledge models. *J. Cult. Herit.* **2009**, *10*, e112–e117. [CrossRef]

16. Allen, P.; Feiner, S.; Troccoli, A.; Benko, H.; Ishak, E.; Smith, B. Seeing into the past: Creating a 3D modeling pipeline for archaeological visualization. In Proceedings of the 2nd International Symposium on 3D Data Processing, Visualization and Transmission (3DPVT 2004), Thessaloniki, Greece, 9 September 2004; pp. 751–758.

17. Welham, K.; Shaw, L.; Dover, M.; Manley, H.; Pearson, M.P.; Pollard, J.; Richards, C.; Thomas, J.; Tilley, C. Google under-the-earth: Seeing Beneath Stonehenge using Google Earth—A tool for public engagement and the dissemination of archaeological data. *Internet Archaeol.* **2015**. [CrossRef]

18. Buna, Z.; Popescu, D.; Comes, R.; Badiu, I.; Mateescu, R. Engineering CAD tools in digital archaeology. *Mediterr. Archaeol. Archaeom.* **2014**, *14*, 83–91.

19. Gonizzi Barsanti, S.; Guidi, G. A Geometric Processing Workflow for Transforming Reality-Based 3D Models in Volumetric Meshes Suitable for Fea. In *ISPRS—International Archives of the Photogrammetry, Remote Sensing and Spatial Information Sciences*; COPERNICUS: Nafplio, Greece, 2017; Volume XLII-2/W3, pp. 331–338.

20. Gutierrez, D.; Seron, F.J.; Magallon, J.A.; Sobreviela, E.J.; Latorre, P. Archaeological and cultural heritage: Bringing life to an unearthed Muslim suburb in an immersive environment. *J. Cult. Herit.* **2004**, *5*, 63–74. [CrossRef]

21. Callieri, M.; Debevec, P.; Pair, J.; Scopigno, R. A realtime immersive application with realistic lighting: The Parthenon. *Comput. Graph.* **2006**, *30*, 368–376. [CrossRef]

22. Laycock, R.G.; Drinkwater, D.; Day, A.M. Exploring cultural heritage sites through space and time. *J. Comput. Cult. Herit.* **2008**, *1*, 1–15. [CrossRef]

23. Guidi, G.; Russo, M. Diachronic representation of ancient buildings: Studies on the "San Giovanni in Conca" Basilica in Milan. *Disegnare Con* **2009**, *2*, 69–80.

24. Micoli, L.L.; Guidi, G.; Angheleddu, D.; Russo, M. A multidisciplinary approach to 3D survey and reconstruction of historical buildings. In *2013 Digital Heritage International Congress (DigitalHeritage)*; IEEE: New York, NY, USA, 2013; pp. 241–248. Available online: http://ieeexplore.ieee.org/abstract/document/6744760/ (accessed on 20 September 2017).

25. Garcia Puchol, O.; McClure, S.B.; Blasco Senabre, J.; Cotino Villa, F.; Porcelli, V. Increasing contextual information by merging existing archaeological data with state of the art laser scanning in the prehistoric funerary deposit of Pastora Cave, Eastern Spain. *J. Archaeol. Sci.* **2013**, *40*, 1593–1601. [CrossRef]

26. Guidi, G.; Russo, M.; Angheleddu, D. 3D Survey and Virtual Reconstruction of archeological sites. *Digit. Appl. Archaeol. Cult. Herit.* **2014**, *1*, 55–69. [CrossRef]

27. Rodríguez-Gonzálvez, P.; Muñoz-Nieto, A.L.; Del Pozo, S.; Sanchez-Aparicio, L.J.; Gonzalez-Aguilera, D.; Micoli, L.L.; Gonizzi Barsanti, S.; Guidi, G.; Mills, J.; Fieber, K.; et al. 4D Reconstruction and visualization of Cultural Heritage: Analyzing our legacy through time. *Int. Arch. Photogramm. Remote Sens. Spat. Inf. Sci.* **2017**, *XLII-2/W3*, 609–616. [CrossRef]

28. De Capitani d'Arzago, A. *Il Circo romAno; Ricerche della Commissione per la forma urbis Mediolani/Istituto di studi romani, Sezione lombarda*; De Capitani d'Arzago, A., Calderini, A., Eds.; Ceschina: Milano, Italy, 1939.

29. Mirabella Roberti, M. *Milano Romana*; Rusconi: Sant'Arcangelo di Romagna, Italy, 1984; ISBN 88-18-33964-8.

30. Blockley, P.; Caporusso, D. *Area del Monastero Maggiore in epoca romana/[testi di Paul Blockley . . . et al.]*; Civico Museo Archeologico: Milano, Italy, 2013; ISBN 978-88-97568-08-7.

31. Caporusso, D.; Donati, M.T.; Masseroli, S.; Tibietti, T. *Immagini di Mediolanum—Archeologia e storia di Milano dal V secolo a.C. al V secolo d.C.*; Civico Museo Archeologico di Milano: Milano, Italy, 2014; ISBN 978-88-97568-00-1.

32. Colombo, A. *Milano preromana, romana e longobarda*; Meravigli: Milano, Italy, 1994; ISBN 88-7955-069-1.

33. *L'area archeologica del Monastero Maggiore di Milano: una nuova lettura alla luce delle recenti indagini*; Pagani, C.; Blockley, P.; Cecchini, N. (Eds.); Quaderni del Civico museo archeologico e del Civico gabinetto numismatico di Milano: Milano, Italy, 2012.

34. Sena Chiesa, G. *Milano capitale dell'impero romano: 286-402 d.C.*; Silvana Editoriale: Milano, Italy, 1990; ISBN 978-88-366-0276-6.

35. Humphrey, J.H. *Roman Circuses: Arenas for Chariot Racing*; University of California Press: Berkeley/Los Angeles, CA, USA, 1986; ISBN 978-0-520-04921-5.

36. Fedeli, A. *Milano Archeologia. I luoghi della Milano antica*; Fondazione Cariplo: Milano, Italy, 2015; ISBN 978-88-86752-69-5.

37. Capponi, C.; Ambrosini, A. *San Maurizio al Monastero Maggiore in Milano: Guida Storico Artistica*; Silvana Editoriale: Cinisello Balsamo, Milano, Italy, 1998; ISBN 88-8215-095-X.

38. Guidi, G.; Russo, M. Diachronic 3D reconstruction for lost Cultural Heritage. *Int. Arch. Photogramm. Remote Sens. Spat. Inf. Sci.* **2011**, *XXXVIII-5/W16*, 371–376. [CrossRef]

39. Horn, B.K.P. Closed-form solution of absolute orientation using unit quaternions. *J. Opt. Soc. Am. A* **1987**, *4*, 629. [CrossRef]

geosciences

MDPI

Article

3D Point Clouds in Archaeology: Advances in Acquisition, Processing and Knowledge Integration Applied to Quasi-Planar Objects

Florent Poux [1,*], Romain Neuville [1], Line Van Wersch [2], Gilles-Antoine Nys [1] and Roland Billen [1]

1. Geomatics Unit, University of Liège (ULiege), Quartier Agora, Allée du six Août, 19, 4000 Liège, Belgium; romain.neuville@ulg.ac.be (R.N.); ganys@ulg.ac.be (G.-A.N.); rbillen@ulg.ac.be (R.B.)
2. Institute of civilizations (Arts and Letters), University of Louvain (UCL), Rue du Marathon 3, 1348 Louvain-la-Neuve, Belgium; line.vanwersch@uclouvain.be
* Correspondence: fpoux@ulg.ac.be; Tel.: +32-4-366-5751

Received: 20 July 2017; Accepted: 25 September 2017; Published: 30 September 2017

Abstract: Digital investigations of the real world through point clouds and derivatives are changing how curators, cultural heritage researchers and archaeologists work and collaborate. To progressively aggregate expertise and enhance the working proficiency of all professionals, virtual reconstructions demand adapted tools to facilitate knowledge dissemination. However, to achieve this perceptive level, a point cloud must be semantically rich, retaining relevant information for the end user. In this paper, we review the state of the art of point cloud integration within archaeological applications, giving an overview of 3D technologies for heritage, digital exploitation and case studies showing the assimilation status within 3D GIS. Identified issues and new perspectives are addressed through a knowledge-based point cloud processing framework for multi-sensory data, and illustrated on mosaics and quasi-planar objects. A new acquisition, pre-processing, segmentation and ontology-based classification method on hybrid point clouds from both terrestrial laser scanning and dense image matching is proposed to enable reasoning for information extraction. Experiments in detection and semantic enrichment show promising results of 94% correct semantization. Then, we integrate the metadata in an archaeological smart point cloud data structure allowing spatio-semantic queries related to CIDOC-CRM. Finally, a WebGL prototype is presented that leads to efficient communication between actors by proposing optimal 3D data visualizations as a basis on which interaction can grow.

Keywords: point cloud; data fusion; laser scanning; dense image-matching; feature extraction; classification; knowledge integration; cultural heritage; ontology

1. Introduction

Gathering information for documentation purposes is fundamental in archaeology. It constitutes the groundwork for analysis and interpretation. The process of recording physical evidence about the past is a first step in archaeological study for a better understanding of human cultures. In general, the goal is to derive spatial and semantic information from the gathered and available data. This is verified in various sub-disciplines of archaeology that rely on archaeometry [1]. In this setting, remote sensing is particularly interesting as a means to not only safely preserve artefacts and their context for virtual heritage [2], but also to complement or replace techniques presenting several limitations [3].

An archaeological breakthrough given by this technique is the moving of interpretation from the field to a post-processing step. The possibility to gather massive and accurate information without transcripts interpretation or in situ long presence is a revolution in archaeological workflows.

It started with stereo-vision and photogrammetry to derive 3D information, but recent development deepened the representativity of digital 3D data through higher resolution, better accuracy and possible contextualization [4]. The study of materials is often linked with on-site related information, forever lost if not correctly transmitted. Digital preservation is therefore necessary to document a state of the findings, and this at different accessible temporal intervals. Visions shared by [5,6] for the digital documentation and 3D modelling of cultural heritage states that any project should include (1) the recording and processing of a large amount of 3D multi-source, multi-resolution, and multi-content information; (2) the management and maintenance of the 3D models for different applications; (3) the visualization of the results to share the information with other users allowing data retrieval "through the Internet or advanced online databases"; (4) digital inventories and sharing "for education, research, conservation, entertainment, walkthrough, or tourism purposes". In this paper, we propose such a solution.

The information as we see it is mostly 3D: "when we open our eyes on a familiar scene, we form an immediate impression of recognizable objects, organized coherently in a spatial framework" [7]. Therefore, tools and methods to capture the 3D environment are a great way to document a 3D state of the archaeological context, at a given time. Analogous to our visual and cognitive system, 3 steps will condition the completeness of the surveyed object. First, the perception, i.e., how the visual system processes the visual information to construct a structured description of the shape of the object/scene. Second, the shape recognition or how the product of the perceptual treatment will contact stored representations in the form of known objects (it will construct a perceptual depiction that will be a representation of the same nature stored in memory). Finally, the identification (labelling), i.e., when a stored structural representation is activated, it will in turn activate the unit of meaning (concept) that corresponds to it, located in the semantic system. Sensors are the analogue to our perception, and aim at extracting the visual stimuli it is sensitive to (spatial information, colour, luminance, movement, etc.). At this stage, neither the information on the shape of the object nor the label is extracted. In the case of 3D remote sensing, the quality of observation is therefore critical to enable high quality and relevant information extraction about the application. As such, the sensory perceptive processing capable of extracting visual primitives must be as objective and complete as possible, making sensors for point cloud generation favourable. Constituted of a multitude of points, they are a great way to reconstruct environments tangibly, and enabling further primitive's extractions (discontinuity, corners, edges, contour,, etc.) as in our perceptive visual system (described in [7]). However, their lack of semantics makes them a bona-fide [8] spatial representation, thus of limited value if not enhanced.

Deriving semantic information is fundamental for further analysis and interpretation. This step is what gives a meaning to the collected data, and allows to reason on sites or artefacts. All this information must be retained and structured for a maximum interoperability. In an archaeological context, many experts must share a common language and be able to exchange and interpret data through ages, which necessitate the creation of formalized structure to exchange such data. Multiple attempts were made, and the CIDOC Conceptual Reference Model (CRM) is a formalization that goes in this direction. "It is intended to promote a shared understanding of cultural heritage information by providing a common and extensible semantic framework that any cultural heritage information can be mapped to. It is intended to be a common language for domain experts and implementers to formulate requirements for information systems and to serve as a guide for good practice of conceptual modelling. In this way, it can provide the semantic glue needed to mediate between different sources of cultural heritage information, such as that published by museums, libraries and archives". It is used in archaeology such as in [9] and provides semantic interoperability. Ontologies offer considerable potential to conceptualize and formalize the a-priori knowledge about gauged domain categories [10] that relies primarily on expert's knowledge about real world objects. If correctly aggregated and linked to spatial and temporal data, digital replicas of the real world can become reliable matters of study that can survive through times and interpretations, which reduces the loss or degradation of information related to any site study in archaeology.

However while promising structures and workflow provide partial solutions for knowledge injection into point clouds [11,12], the integration, the maturation state as well as the link between semantic and spatial information is rudimentary in archaeology. Concepts and tools that simplify this process are rare, which complicates the merging of different experts' perceptions around cultural heritage applications. Being able to share and exchange contextual knowledge to create a synergy among different actors is needed for planning and analysis of conservation projects. In this context, we explore ways to (1) better record physical states of objects of interest; (2) extract knowledge from field observations; (3) link semantic knowledge with 3D spatial information; (4) share, collaborate and exchange information.

This paper is structured in a dual way to provide both a background of 3D used techniques in archaeology, and technical details of the proposed point cloud workflow for quasi-planar heritage objects.

In the first part, we carefully review the state of the art in digital reconstruction for archaeology. This serves as a basis to identify research perspectives and to develop a new methodology to better integrate point clouds within our computerized environment.

Secondly, we propose a framework to pre-process, segment and classify quasi-planar entities within the point cloud based on ontologies, and structure them for fast information extraction. The methodology is illustrated on the case of the mosaics of Germigny-des-Prés (France) and then applied to other datasets (façade, hieroglyphs). Finally, the results are presented, and we discuss the perspectives as well as data visualisation techniques and WebGL integration.

2. Digital Reconstruction in Archaeology: A Review

3D digital exploration and investigations are a proven way to extract knowledge from field observations [13,14]. The completeness and representativity of the 3D data gathered by sensors are critical for such digitalization. Equally, methodologies, materials and methods to "clone" a scene are important for the extensiveness of any reconstruction. The 3D-capturing tools and software drastically evolved the last decade; thus, we review the current state of the art in digital reconstruction for archaeology.

2.1. Archaeological Field Work

Even if an increasing number of archaeological contributions deal with 3D and related management of information, archaeologists are still sceptical about 3D technologies and often use manual drawings for cautious observations and first analyses on the field [15]. The literature gravitates around a controversial or diverging hypothesis which illustrates this reticence to adopt new technologies in remote sensing [14,16]. During an empirical recording of monuments or sites, measurements are taken (by hand), taking distances between characteristic points on the surface of the monument. The definition of the coordinates is done on an arbitrary coordinate system on a planar surface of the structures. The method is simple, reproducible and low-cost but limiting factors such as limited accuracy, time demand and necessary direct contact makes it unfavourable in many scenarios including for inaccessible areas. However, archaeologists will often use such an approach over remote sensing to gather insights that are otherwise considered incomplete. The 3D methods are frequently regarded as intricate, expensive and not adapted to archaeological issues [17]. On most sites, for buildings studies or in excavations, the data gathering and acquisition are made with drawings and pictures in 2D. In some cases, 3D can come after the analyses process and is used as a "fancy" means to present results and rebuilt a virtual past. As noticed by Forte [18]: "there was a relevant discrepancy between bottom-up and top-down processes. The phase of data collecting, data-entry (bottom-up) was mostly 2D and analogue, while the data interpretation/reconstruction (top-down) was 3D and digital". However, the new possibilities given by 3D remote sensing extend the scope of possible conservation and analysis for digital archaeology, and can progressively move to post-processing a part of the interpretation process, making the underlying data (if complete) the source on which different reading and conclusions can be mined. Yet, such techniques cannot replace a field presence when complementary semantics (from other senses such as hearing, taste, smell, touch) are necessary.

The different data types from these remote sensing platforms played a vast role in complexifying the diversification in methodologies to derive the necessary information from the data (data-driven). However, the 3D spatial data extracted from the bottom-up layer for most of these techniques are surveyed points, in mass, creating point clouds. They are driven by the rapid development of reality capture technologies, which become easier, faster and incur lower costs. Use cases in archaeology show the exploration and acceptation of new techniques, which are assessed not only in regard to their accuracy, bust mostly in accordance to their fit to a specific context, and the associated costs. Following the categorization defined in [16], we distinguish "(1) the regional scale, to record the topography of archaeological landscapes and to detect and map archaeological features, (2) the local scale, to record smaller sites and their architecture and excavated features, and (3) the object scale, to record artefacts and excavated finds". In their article, the authors reviewed some passive and active sensors for 3D digitization in archaeology at these different scales. They conclude that the principal limiting factor for the use of the different remote sensing technologies reviewed (Synthetic Aperture Radar (SAR) interferometry, Light Detection and Ranging (LiDAR), Satellite/Aerial/Ground imagery, Terrestrial Laser Scanning (TLS), Stripe-projection systems) is the ratio added value of a digital 3D documentation over the time and training that inexperienced users must invest before achieving good results. In their paper, [19] state that 3D recording is the first step to the digitization of objects and monuments (local and object scales). They state that a 3D recording method will be chosen depending on the complexity of the size and shape, the morphological complexity (Level of Detail—LoD), and the diversity in materials. While this is accurate looking purely at a technical replication, other factors such as user experience, available time or budget envelope will constrain the instrument or technique of choice. They propose a 9-criteria choice selection as follows: cost; material of digitization subject; size of digitization subject; portability of equipment; accuracy of the system; texture acquisition; productivity of the technique; skill requirements; compliance of produced data with standards. While this extends the global understanding and 3D capture planning, it lacks a notion of time management (implied in productivity) or constraints in line with contextual laws and regulations (no contact survey only, etc.). Although they separate "accuracy" from "texture acquisition", both can be related, as well as additional features provided by the sensors (e.g., intensity) that can extend the criterion table.

We note a large discrepancy between scales of the remote sensing and related costs/methods tested for point cloud generation.

At a regional scale, airborne LiDAR is sparsely used in archaeology, mostly as a 2.5 D spatial information source for raster data analysis. It is a powerful tool to analyse past settlement and landscape modification at a large scale. Use cases such as in [20–23] helped remove preconceptions about settlements size, scale, and complexity by providing a complete view of the topography and alterations to the environment, but while it provided new research and analysis directions, the LiDAR data did not leverage 3D point clouds considered too heavy and too raw to provide a source of information.

At both the local scale and the object scale, several use cases exploit active sensing, specifically terrestrial laser scanners (TLS) using phase-based and time-of-flight technologies [24]. Archaeological applications vary such as in [25] to reconstruct a high-resolution 3D models from the point cloud of a cave with engravings dating back to the Upper Palaeolithic era, in [26] to study the damage that affected the granitic rock of the ruins of the Santo Domingo (Spain), or in [27] for the 3D visualization of an abandoned settlement site located in the Central Highlands (Scotland). More recent procedures make use of TLS to reconstruct the Haut-Andlau Castle (France) [28], or in [29] to map the Pindal Cave (Spain). These showed that to capture fine geometric details, laser-scanning techniques provide geometric capabilities that have not yet been exceeded by close-range photogrammetry, especially when concave or convex forms need to be modelled. Rising from the static concept, Mobile laser scanning (MLS) [30] has scaled up the data rate generation of TLS by allowing dynamic capture using other sensors including GNSS position and inertial measurements for rapid street point cloud generation and public domain mapping. New concepts and technology including Solid State LiDAR and simultaneous localization and mapping (SLAM) have pushed dynamic acquisition for quickly mapping with a lower accuracy the surroundings, extending cases using HMLS (Hand-held mobile laser scanning) [31], MMS (Mobile Mapping System) [32], or more recently MMBS (Mobile Mapping Backpack System) [33]. At the object scale, active sensors namely for active triangulation, structured light and computer tomography for 3D modelling is widely used due to its high precision, and adaptation to small isolated objects [34]. Moving to ground technologies, surveys are precise in detecting sub-surface remains. Different geophysical processing techniques and equipment (such as ground penetrating radar (GPR), magnetometry and resistivity) are usually integrated together, to increase the success rate of uncovering archaeological artefacts, for example in [35] to delineate the extent of the remains of a small town that has been submerged (Lake Tequesquitengo, Mexico).

Passive sensing gained a lot of attention in the heritage community following terrestrial use cases and image crowdsourcing, allowing a wide range of professionals and non-expert to recreate 3D content from 2D poses (exhaustive software list from [36–46]). The rapidly growing interest for light aerial platforms such as UAV (Unmanned Aerial Vehicle) based solutions and software based on multi-view dense image matching [47,48] and structure from motion [49] swiftly provided with an alternative to active sensing. Use cases for 3D archaeological and heritage reconstruction are found at the object scale through terrestrial surveys [4,14,50,51] and the local scale through light aerial platforms, making this technique a favourable way to obtain quick and colour balanced point clouds. Moreover, the cost and accessibility (hardware and software) of dense-image matching reconstruction workflows have allowed its spread in archaeological studies. For example, in the Can Sadurní Cave (Spain), Nunez et al. [52] successfully reconstructed an object via dense-image matching and georeferenced the obtained 3D model using TLS point cloud data of the Cave. They state that capture from different positions is fundamental to generate a complete model that does not lack important information.

While reconstruction accuracy is increasing [53], remote sensing via active sensors is favoured in the industry for local scales. There are discussions in which computer vision would replace LiDAR [28,54]; however, practical cases tend to a merging of both (Reconstruction of the Amra and Khar-anah Palaces (Jordan) [55], the castle of Jehay (Belgium) [56]), and predilection applications for each techniques, combining strength of natural light independence with low-cost and highly visual image-based reconstruction [56]. Particularly in the case of mosaics, decorations and ornaments, the combination of features from sensors generating accurate and complementary attributes permits the overcoming of limits arising from a small set of features. Indeed, use case such as in [57] results in a high richness of detail and accuracy when combining TLS and close range photogrammetry which was not achievable otherwise. Thus, multisensory acquisition provides an interesting method that will be investigated.

The high speed and rate generation of 3D point clouds has become a convenient way to obtain instant data, constituting datasets of up to Terabytes, so redundant and rich that control operation can take place in a remote location. However, they often go through a process of filtering, decimation and interpretation to extract analysis reports, simulations, maps, 3D models considered as deliverables. A common workflow in archaeology concerns the extraction of 2D profiles and sections, 2D raster to conduct further analysis or to create CAD deliverables, particularly looking at ornaments, mosaics or façades. This induces several back and forth movements within the pipeline, and the general cohesion, storage system often lack extensibility. This challenge is particularly contradictory, and a solution to automate recognition such as [58] in the context of mosaics would therefore provide very solid ground for tesserae detection, extended to 3D by combining many more sets of features. This will be specifically studied in Section 3.

While all the reviewed literature specifically points out the problems linked with data acquisition and summarize the strength and weakness of each regarding the recorded spatial information, few specifically link additional semantic information. Gathered in situ or indirectly extracted from the observation, the measurements often rely on specific interest points. While this is handy looking at one specific application for one archaeologist, this practice is dangerous regarding the problematic of curators and conservation. Indeed, preserving at a later stage the interpretation through sketches, drawings, painting or text description based on interest points makes any possible data analysis impossible from the raw source. Therefore, 3D point selectivity should not arise at the acquisition step, but in a post-processing manner, to benefit of the flexibility given by 3D data archives, which was impossible before the emergence of automatic objective 3D capturing devices.

We postulate that when designing data processing workflow, specific care must be given to the objectivity linked with the spatial data, which multisensory systems and point clouds specifically answer. As such, they can constitute the backbone of any powerful spatial information system, where the primitive is a 0-simplex [59]. Their handling in archaeology, however, is a considerable challenge (often replace by 3D generalization such as meshes, parametric models, etc.) and thus presents many technical as well as interpretation difficulties.

2.2. Integration of 3D Data

As demonstrated in [60], the evolution of remote sensing for archaeological research and the acceptance in archaeology has grown linearly since 1999 looking at the number of publications (Sources SCOPUS, ScienceDirect & Web of Science search engines) related to remote sensing per year. While this provides new possibilities, the reliability and heterogeneity of the spatial information are issues in heritage for the conservation, interoperability and storage of data. 3D GIS linked to archaeological databases have been thought and proposed for the management of this information at different scales and on different type of sites such as large excavated sites [61–64]. In their paper [62], the authors discuss the possibility and the ultimate goal of having a complete digital workflow from 3D spatial data, to efficiently incorporate the information into GIS systems while relying on formal data model. After stating the limits and difficulties of integrating efficiently 3D data (as well as time variations), they interestingly express the domain specifications and formalization through ontologies. Within a knowledge system, standards and procedures are key to warrant the consistent meaning of collective contents and to trace the "history" of the processed data [63]. In their use case, they create a 3D model segmented regarding semantic information to allow the independent manipulation as well as GIS query between elements. They claim to provide new standards in 3D data capture to be usable by all archaeologist, but their method is empirically defined and the integration of knowledge sources is blurry regarding segmentation and semantic injection. Building archaeology is also a field where 3D applications are used mainly for conservations purpose [65–67]. In these contributions, the authors highlight two characteristics of archaeological cases: the heterogeneity of data and the difficulty of processing 3D spatial entities from irregular archaeological objects (artefacts, buildings, layers, etc.). Several solutions have been offered in the mentioned papers and specific software have been designed (see [68] for a relevant 3D GIS use case and [69] for the most recent summary). In these studies, the definition of archaeological facts rests on their representation as raster data, specific point of interest, polygons and 3D shapes, but never the direct source of spatial information: point clouds.

At this step, several criterions should be considered to choose the most suitable spatial data model. Many researchers proposed 3D grid representation (voxels) as the most appropriate data format for handling volumetric entities and visualizing continuous events [70,71]. Generalizing point cloud entities by volume units such as a voxels allows 3D GIS functionalities such as object manipulation, geometry operations and topology handling regarding [72]. Although Constructive Solid Geometry (CSG) and 3D Boundary representations (B-rep) can roughly depict a spatial entity, the level of generalization of the underlying data has an impact on the accuracy and representativity of GIS functionalities' results. Thus, point cloud brings an additional flexibility by giving the possibility to recover the source spatial data information. The limits with available commercial and open source database GIS systems (which are mostly used by archaeologist) made point clouds a secondary support information for primarily deriving 3D model generalizations. This of course limits the conservation potential of archaeological findings, as the interpretation behind data modelling workflows is unique and irreversible (one-way). As such, to our knowledge, no 3D archaeological GIS system is directly based on 3D point cloud. They are rather considered heavy and uninterpretable datasets. Furthermore, the constitution and leveraging of knowledge sources is still limited, with some experiences by manual injection reviewed in [61–64]. This of course constitutes a major issue that needs to be addressed for scaling up and generalizing workflows. The different literature involved in the constitution of 3D GIS delineates the need of standardization, especially regarding the variety of data types. In this direction, one specific use case in [62] demonstrated that the main advantage of the 3D GIS methodology is the link between attribute information to discrete objects defined by the archaeologist. Their implementation is done regarding the CIDOC-CRM ISO 21127 standard and the design patterns from the ontological model of the workflow of the Centre for Archaeology to achieve semantic compatibility. As opposed, the approach presented in [73] allows linking of 3D models of buildings and graph-based representation of terms. It describes its domain-linked morphology to provide new visual browsing possibilities. In this approach, one expert creates a graph for one specific application. This allows the comparison of semantic descriptions manually established by experts with divergent perspectives but lacks extensibility to match general rules. Indeed, while the description flexibility within one field can benefit from this, it can lead to interoperability problems when a formalization needs to be established, especially regarding geometrical properties or for structuring the semantics according to a pattern. Even though there are several works dedicated to ontology-based classifications of the real-world entities, the ontologies developed so far are rarely integrated with the measurements data (physical data). As such, [74] proposes an observation-driven ontology that plays on ontological primitives automatically identified in the analysed data through geo-statistics, machine-learning, or data mining techniques. These provide a great standpoint to semantic injection and will be further studied. In particular, the possibility to specialize the ontology through extensions such as CIDOC CRMba (an extension of CIDOC CRM to support buildings archaeology documentation) or CIDOC CRMgeo (an extension of CIDOC CRM to support spatiotemporal properties of temporal entities and persistent items) provides new solutions for higher interoperability.

The literature review showed a shift at an acquisition phase toward better means to record physical states of an environment, an object. TLS and dense-image matching showed an increase in popularity, and their combination provide new and promising ways to record archaeological artefacts and will thus be investigated. However, both methods generate heavy point clouds that are not joined directly with knowledge sources or structured analogously to GIS systems. Rather, their use is limited to providing a reference for other information and deliverables (2D or 3D). While this is a step forward toward higher quality documentation regarding other reviewed field methods, this is not a long-term solution when we look at the evolution of the discipline, the quantity of generated data and the ensuing ethics. The identified problem concerns the link between domain knowledge and spatial information: it evolves in parallel, partially intersects or is hardcoded and manually injected. Moreover, the flexibility regarding possible analysis is often null due to interpreted documents that force a vision over elements that no longer physically exist, or which were poorly recorded. Therefore, a strong need for ways to integrate knowledge to point clouds is essential. This "intelligence spring" is categorized regarding 3 sources as identified in [11], being device knowledge (i.e., about tools and sensors), analytic knowledge (i.e., about algorithms, analysis and their results) and domain knowledge (i.e., about a specific field of application). Their rapprochement to point clouds is, however, a bottleneck that arises early in the processing workflow. If we want to better integrate point clouds as intelligent environments [75], we must correctly assemble knowledge sources with the corresponding "neutral" spatial information. This relies on different procedures to (1) pre-process the point cloud; (2) detect the entities of interest within the initial point clouds and (3) attach the knowledge to classify and allow reasoning based on the classification. As such, our work proposes to leverage the use of ontologies as knowledge sources, as well as defining a workflow to directly process and integrate point clouds within 3D GIS systems, creating virtual heritage [2]. In the next part, we describe our technical method for integrating knowledge within reality based point-clouds from TLS and close-range photogrammetry. While the following methodology can be extended to different applications with examples such as in Section 5, it is illustrated and applied to quasi-planar objects of interest.

3. Materials and Methods

The applied workflow of object detection and classification is organized as follows: in the data pre-processing step, the different point clouds are treated using the procedure described in Section 3.1 (Step 1, Figure 1). Subsequently, point cloud descriptors as well as object descriptors such as the extent, shape, colour and normal of the extracted components are computed (Step 2, Figure 1) and imported into the next classification procedure using a converter developed in this study (Step 3, Figure 1). In the last step, the objects are classified based on the features formalized in the ontology (Step 4, Figure 1).

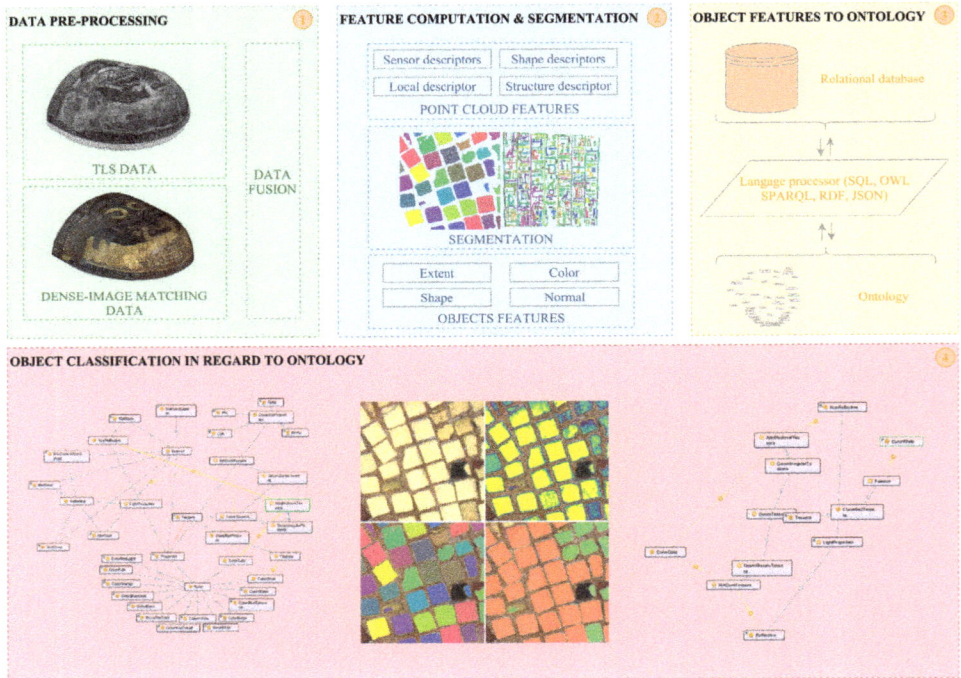

Figure 1. Overview of the methodology developed in this paper. (1) Data pre-processing; (2) Feature computation and segmentation; (3) Object features to ontology; (4) Object classification in regard to ontology.

The process to identify features of interest within the signal is the foundation for the creation of multi-scale ensembles from different datasets. The work described in [76] extensively reviews data fusion algorithms defined by the U.S department of Defense Joint Directors of Laboratories Data Fusion Subpanel as "a multilevel, multifaceted process dealing with automatic detection, association, correlation, estimation and combination of data and information from single and multiple sources to achieve refined position and identify estimates, and complete and timely assessments of situations and threats and their significance". The combination of different sensors generating complementary signatures provides pertinent information without the limitations of a single use and creates a multisensory system [77]. Thus, following the postulate of the state of the art, we decide to adopt a multi-sensory workflow for maximizing information (Section 2.1).

3.1. Point Cloud Data Acquisition and Pre-Processing

A pre-processing step is necessary to obtain a highly representative signal of the value measured as defined in [78]. Indeed, to avoid external influential sources that degrade the information, this step demands adapted techniques to minimize errors including noise, outliers and misalignment. Filtering the data strongly depends on device knowledge [11].

Several sets of data from various contexts were acquired to perform different tests. The Carolingian church located in Germigny-des-Prés (Loiret, France) houses ancient mosaics dating from the 9th century, composed of about one hundred thousand tesserae (the average surface of a tessera is 1 cm^2, square of 1 cm by 1 cm). The preserved works offer a unique opportunity for the study of mosaics and glass. Indeed, the tesserae that composes it are mainly made in this material, which is rare in the archaeological context of the early Middle Ages [79]. However, part of the mosaic was restored in the 19th century; therefore, tesserae are from two periods; thus, we must first distinguish the different tesserae types (based on their age) for accessing alto-medieval glass information. The study could reveal important predicates, considering each tessera taken independently or by analysing different properties, while conjecturing with expert's domain knowledge. The mosaic of the vault culminates at 5402 m above the ground, presenting many challenges for 3D capture from active and passive sensors. The dome is protected, and the limited accessibility tolerates only a light scaffolding, too narrow for the positioning of tripods, illustrating the need to adapt means to the context Figure 2.

Figure 2. The vault of Germigny-des-prés being captured for dense-image matching processing.

The first sample was acquired using a phase-based calibrated terrestrial laser scanner: the Leica P30. The different scans were registered using 1338 reflective targets, of which 127 were shot by a total station (Leica TCRP1205, accuracy of 3 mm + 2 ppm) and used for indirect georeferencing afterwards. The mean registration error is 2 mm, and the mean georeferencing deviation is 2 mm (based on available georeferenced points measured from the total station). Two point cloud segments of the same zone (mosaic) were extracted: one unified point cloud that includes measurements from

8 different positions with varying range and resolutions, and one high resolution point cloud (HPC) from one optimized position by using an extended mounted tribrach. A comparison emphasized the influence of the angle of incidence and the range over the final resolution, precision and intensity of the point cloud. Thus, we chose the HPC for its higher representativity (Figure 3).

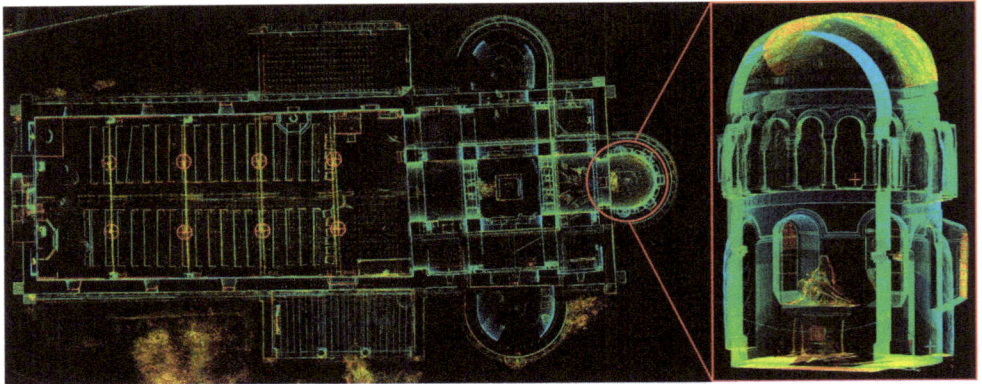

Figure 3. Point cloud of the church of Germigny-des-Prés. Top View (left) and zone of interest (right).

The TLS was operated at 1550 nm for a maximum pulse energy of 135 NJ. Initial filtering was conducted such as deletion of intensity overloaded pixels (from highly retro-reflective surfaces) and mixed pixels to flag problematic multi-peak scan lines and keep the right return via full-waveform analysis. The final accuracy of a single point at 78% albedo is 3 mm. The final HPC is composed of 30,336,547 points with intensity ranging from 0.0023 to 0.9916, and covers solely the mosaic. Several pictures were taken at different positions to obtain a 3D point cloud of the mosaic. These pictures were shot using a Canon EOS 5D mark III camera equipped with a 24–105 mm lens. In total, 286 pictures of 5760 × 3840 in RAW, radiometrically equalized and normalized, were used to reconstruct the photogrammetric point cloud (Figure 4).

Figure 4. Point cloud from a set of 2D poses reconstructed via dense-image matching using the software ContextCapture v 4.4.6, Bentley Systems, Exton, United States [37].

The knowledge around the acquisition methodology provides important information as missing/erroneous data, misadjusted density, clutter and occlusion are problems that can arise from an improper or impossible capture configuration on the scene [80], resulting in a loss of transmitted information or data quality. Combining different sensors with diverse acquisition methodologies allows the overcoming of this challenge and provides a better description of the captured subject through (1) higher quality features (i.e., better colour transcription, better precision, etc.); (2) specific and unique attribute transfer; (3) resolution and scale adaptation, sampling or homogenizing [81]. The knowledge extracted from a device, analytical knowledge or a domain formalisation constitutes the fundamental information repository on which a multi-level data structure is constructed (Section 3.2).

The first step is therefore to correctly reference point clouds, known as data registration. The method is derived from previous work to perform accurate attribute transfer [56]. The main idea is that a priority list processing is established and influences data fusion regarding knowledge. When combining different point clouds, their geometry and attributes in overlapping areas are then properly addressed. The complementary information needs to be combined from the different available sources if relevant, keeping the most precise geometry as a structure. Avoiding heterogeneous precisions is essential, leading to point deletion rather than point caching and fusing. Once correctly registered, every point cloud data source goes through a pixel and attribute level fusion (if not previously fused at the sensory level).

3.2. Knowledge-Based Detection and Classification

Our approach for object extraction relies on domain knowledge that relays through point cloud features. Segmentation [82] and feature extraction are well studied areas within point cloud processes. However, the integration of knowledge is still rare, with few example of hybrid pipelines [83,84]. Our proposed approach constitute a hybrid method inspired by previous work in shape recognition [85–88], region growing pipelines [80,89,90] and abstraction-based segmentation [91–95] relying on 3D connected component labelling and voxel-based segmentation. As such, different features presented in Table 1 constitute the base for segmentation.

Table 1. Point features computed from the point cloud data after data fusion, before segmentation.

Type	Point Features	Range	Explanation
Sensor desc.	X, Y, Z	Bounding-box	Limits the study of points to the zone of interest
	R, G, B [1]	Material Colour	Limited to the colour range that domain knowledge specifies
	I		Clear noise and weight low intensity values for signal representativity
Shape desc.	RANSAC [2]	-	Used to provide estimator of planarity
Local desc.	Nx, Ny, Nz [3]	$[-1, 1]$	Normalized normal to provide insight on point and object orientation
	Density [4]	-	Used to provide insights on noise level and point grouping into one object
	Curvature	$[0, 1]$	Used to provide insight for edge extraction and break lines
	KB [5] Distance map		Amplitude of the spatial error between the raw measurements and the final dataset
Structure desc. [6]	Voxels	-	Used to infer initial spatial connectivity

[1] [96]; [2] [97]; [3] [98]; [4] [98]; [5] Knowledge-based; [6] [11].

The point cloud data processing was implemented using the programming interfaces and languages MATLAB, Python, SQL, SPARQL, OWL, Java as well as the C++ Library CCLib from CloudCompare [93] and the software Protégé [99].

First, the point cloud is segmented regarding available colour information by referring to the database table containing float RGB colour ranges for each material composing the dataset. Then, the gap is enhanced by superimposing intensity values over colour information (this allows us to refine and better access point filtering capabilities) as in Equation (1).

$$R_e = R \times I, \, G_e = G \times I, \, B_e = B \times I \tag{1}$$

A statistical outlier filter based on the computation of the distribution of point to neighbour distances in the input dataset similarly to [100] is applied to obtain a clean point cloud. This step can be avoided if the colour range and the datasets are perfectly in line.

The segmentation developed is a multi-scale abstraction-based routine that decomposes the 3D space in voxels at different abstraction levels and by constructing an octree structure to speed up computations. The three-dimensional discrete topology (3DDT) proposed by [101] generates a voxel coverage by intersection with another representation model (parametric or boundary) of an object. This is possible by playing on the different configurations of voxel adjacencies. A voxel has 6 neighbour voxels by one face, 18 neighbour voxels by a face or an edge and 26 neighbours by a face, an edge or a vertex. Our approach is based on a 26-connectivity study that groups adjacent voxels if not empty (i.e., voxels containing points). It is conditioned by the analytical knowledge where the density information constrains the initial bounding-box containing the point cloud. An initial low-level voxel structure is then computed, retaining the number of points as attribute. Let $v_i \in \mathbb{R}^3$ be a voxel. Let v_i be its neighbour voxel. We define V_{CEL} as the connected element (segment) as in Equation (2):

$$\forall \, v_i \in \mathbb{R}^3, \, \exists \, v_j = n(v_i) \mid V_{CEL} = [v_i, \, v_j] \leftrightarrow v_j \neq \varnothing \tag{2}$$

where $n(v_i)$ is the neighbour voxel of and v_i, V_{CEL} is the group segment from a 26-connectivity adjacency study.

The topological grouping also permits us to clean the remaining noise N from difficult colour extraction regarding the equations Equations (3) and (4). Let $p_n \in \mathbb{R}^3$ be the *n*-th point of V_{CEL}. There exists P_{CEL} as follows:

$$\forall \, p_1, \, \dots, \, p_n \in \mathbb{R}^3, \, \exists \, P_{CEL} \mid P_{CEL} = \{p_1, \, \dots, \, p_n\}^1 \tag{3}$$

$$P_{CEL} = N \leftrightarrow S_{Number\_CEL} < d(P_{CEL}) \times \min(S_m) \,\&\, S_{Size\_CEL} < \min(Vm)^2 \tag{4}$$

[1] where p is a point (x, y, z) in space, [2] where N is the remaining noise, S_{Number_CEL} is the number of points in P_{CEL}, $d(P_{CEL})$ is the point density of P_{CEL}, $\min(S_m)$ is the minimum of the surface of the material S_m, S_{Size_CEL} is the voxel volume occupancy of the CEL, $\min(V_m)$ is the minimum of the volume of the material V_m; therefore, N is the group composed of fewer points than the knowledge-based assumption from the density achievable from the sensor, the minimum surface of the object and the minimum volume of the object.

Then, our multi-scale iterative 3D adjacency algorithm at different octree levels recursively segments under-segmented groups (detected by injecting analytical knowledge regarding minimum bounding-box size of processed material as in Equation (4)), refining the voxel-based subdivision until the number of generated voxels is inferior to the density-based calculation of estimated voxel number. When subgroups dimensions correspond to material's available knowledge, segments are added to the "Independent Tesserae" segments. Otherwise, a convolution bank filter is applied regarding the longest side of the calculated best fit P.C.A Bounding Box. For absorbent materials or objects sensitive to the sensor emitting frequency (implies low intensity, thus high noise), the 3D distance map as in Table 1 is used to detect points that belong to each object of interest. The accuracy of the extracted segments is assessed by ground truth manual counting of different samples.

Then, on each detected segment, every point is projected on the RANSAC best fit plane, and we extract the 2D outline through the convex hull of the projected points p_i to constrain the plane. Let P_{pCEL} be the projected points of P_{CEL} onto the best fit plane. Then, we obtain $\mathrm{Conv}\,(P_{pCEL})$ as in Equation (5):

$$\text{Conv}\left(P_{pCEL}\right) = \left\{ \sum_{i=1}^{|P_{pCEL}|} \alpha_i \times p_i \mid \forall \, p_i \in P_{pCEL}, \, \forall \, \alpha_i \geq 0 \right) : \sum_{i=1}^{|P_{pCEL}|} \alpha_i = 1 \right\} \qquad (5)$$

where $\text{Conv}(P_{pCEL})$ is the convex polygon of P_{CEL} as a finite point set (x, y) in R^2 (x, y, z). It can be extended to 3D, nD if necessary.

We calculate the compactness (CS) and complexity (CP) of the generated polygon in regard to the work of [102], as well as its area, and its statistically generalized (gaussian mixture) colour (Table 2).

Table 2. Segment features computed from the extracted segments.

Type	Point Features	Range	Explanation
Sensor generalization	Xb, Yb, Zb	barycentre	Coordinates of the barycentre
	Rg, Gg, Bg [1]	-	Material unique colour from statistical generalization
	I	-	Intensity unique value from statistical generalization
Shape desc.	CV [2]	-	Convex Hull, used to provide a 2D shape generalization of the underlying points
	Area	-	Area of the 2D shape, used as a reference for knowledge-based comparison
	CS, CP	[0, 1]	Used to provide insight on the regularity of the shape envelope
Local generalization	Nx, Ny, Nz [3]	[−1, 1]	Normalized normal of the 2D shape to provide insight on the object orientation

[1] [96]; [2] Convex-hull; [3] [98].

The final classification of the delineated objects is based on the available and constituted domain ontology of point cloud features for archaeology. The idea behind the ontology is that the integrated cultural information from a variety of sources is brought together into an integrated environment where we can ask broader questions than we can ask from individual pieces.

Ontologies can be expressed using different knowledge representation languages, such as the Simple Knowledge Organization System (SKOS), the Resource Description Framework (RDF), or the Web Ontology Language 2 (OWL2) specification. These languages contrast in terms of the supported articulateness. The SKOS specification, for instance, is widely used to develop thesauri, the CIDOC CRM is mainly used for describing heritage sites, the Basic Formal Ontology [103] at a higher conceptualization level to incorporates both 3D and 4D perspectives on reality within a single framework. The OWL2 ontology language is based on the Description Logics (DL) for the species of the language called OWL-DL. DL thus provides the formal theory on which statements in OWL are based and then statements can be tested by a reasoner. The OWL semantics comprise three main constructs: classes, properties and individuals. Individuals are extensions of classes, whereas properties define relationships between two classes (Object Properties), an individual and a data type (Data Properties).

We used the OWL and the RDF languages to define ontologies for their high flexibility and interoperability within our software environment (Protégé & Java). As for the study of mosaics, the ontology is set upon the point cloud data and its attributes, thus indirectly leveraging domain ontologies. Indeed, sensor related knowledge is needed to understand the link between features and their representativity. The following meta-model is formalised in UML and provides a conceptual definition for implementations. We therefore used the model to provide a clear vision and comprehension of the underlying system, but the ontology creation slightly differs from privilege performances; therefore, adaptations are made at the relation scheme modelling level.

The characterization (knowledge representation and data modelling) in Figure 5 is a Level-2 domain meta-model, that can plug to a Smart Point Cloud structure [104]. The general idea is that different hierarchical levels of abstraction are constituted to avoid overlapping with existing models and to enhance the flexibility and opening to all possible formalized structure. The core instruction is that the lower levels are closer to a domain representation than higher levels (level-0 being the higher level), but they impose their constraints. The overall structure is a pyramidal assembly, allowing the resolution of thematic problems at lower levels with reference to constraints formally imposed by the higher levels.

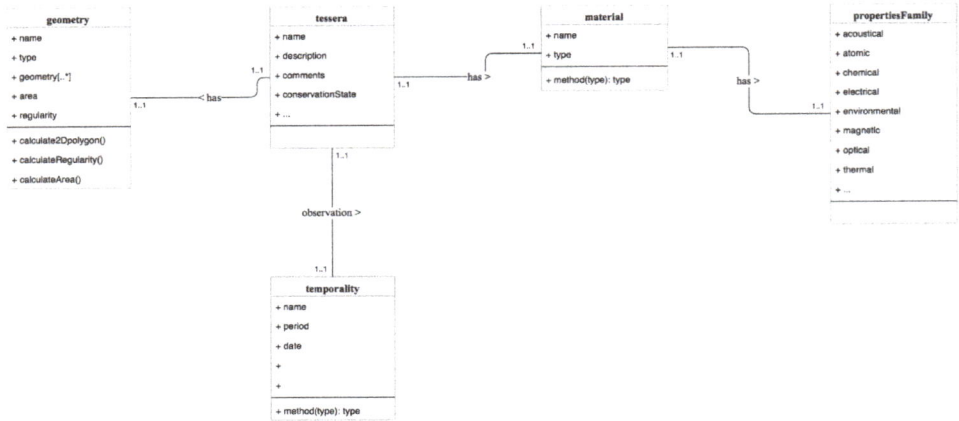

Figure 5. UML meta-model of the ontology. Tesserae have one or multiple geometries, which are characterized by their regularity (determined by the ontological reasoning framework), and an area. Tesserae also have a temporality (characterized as a time interval, being placed at early Middle Ages or during a restoration at the 19th century) and different materials. These materials retain various properties including light sensitivity.

The ontology implementation was structured as triplets. Each triplet corresponds to a relation (subject, predicate and object), which expresses a concept. The end goal is to reason based on the constituted ontology to extract information about the tessera geometric regularity, its material and temporal classification (ClassifiedTessera) as in Figure 6.

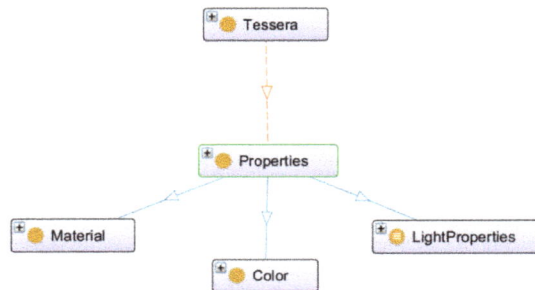

Figure 6. Sub-ontology for the classification of point cloud tessera objects. Blue arrows represent links regarding the tree structure (these are "subClassOf"). The oranges links represent the "hasProperty" relationship that we created to describe the relationship between a Tessera and its properties. It is a simple relationship from domain (Tessera) to Range (Properties).

The ontology is then populated with the domain knowledge as detailed in Figure 7, and the different predicates are established to obtain a final classification of the point cloud. Note that a tolerance of 20% regarding the definition of geometries was used to allow relative variations within one tesserae family. Analogously, any quasi-planar object may be substituted and described by the afore-mentioned properties, thus extending the provided ontology.

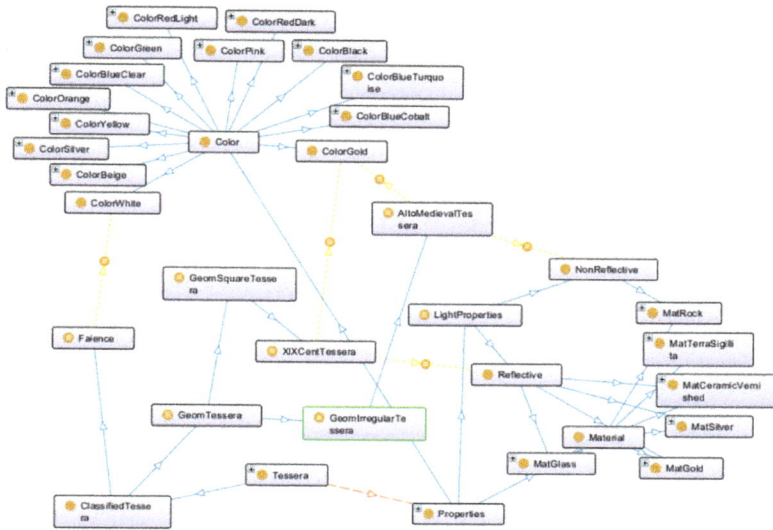

Figure 7. Detailed ontology for the classification of the mosaic's point cloud. The yellow lines are the links of sub-assumptions, of reasoning. They are in fact links of equivalence between a class and its definition.

The different results allow to classify the point cloud, after determining the regularity of the 2D outline regarding different constraints (examples in Table 3).

Table 3. Example of tessera classification using RDF constraints.

RDF Triple Store	Effect
((CS some xsd:double[> "1.1"^^xsd:double]) or (CS some xsd:double[< "1.05"^^xsd:double])) and (CP some xsd:double[> "4.0E-4"^^xsd:double])*	Tessera is irregular *(1)*
(CP some xsd:double[<= "4.0E-4"^^xsd:double]) and (CS some xsd:double[>= "1.05"^^xsd:double, <= "1.1"^^xsd:double])	Tessera is square
(1) and (hasProperty some ColorGold) and (hasProperty some NonReflective) and (Area some xsd:double[<= "1.2"^^xsd:double])	Tessera is alto-medieval
(hasProperty some ColorWhite) and (Area some xsd:double[>= "16.0"^^xsd:double, <= "24.0"^^xsd:double])	Material is Faience

The domain knowledge including size, geometry and spatial distribution leads to object classification. For enhancing its interoperability, the developed DSAE (Digital Survey-based Architectural Element) ontology can directly be extended using the well-established CIDOC-CRM formal ontology. Indeed, the CIDOC-CRM is purely descriptive, and does provide only "factual" tests (a node is linked to an arc, which is linked to another node). The provided DSAE ontology can reason based on complex declaration of conditions (such as AND, OR, ONLY, etc.), thus is much more structured than the CIDOC-CRM, and allows to reason. As such, it permits automatic classification that can be plugged to the CIDOC-CRM enabling archaeologists to better understand the underlying point cloud data. In the case of tesserae, each tessera material is then considered as a E57 Material specialization which comprises the concept of materials (Specialization of E55 Type), LightProperties and Color can be seen as S9 Property (it describes in a parametric way what kind of properties the values are) and Area, CP and CS as SP15 Geometry attributes (which comprises the union of geometric definitions and the linked declarative places) from the extension CIDOC-CRMgeo (based on the ontology GeoSPARQL), as in Figure 8.

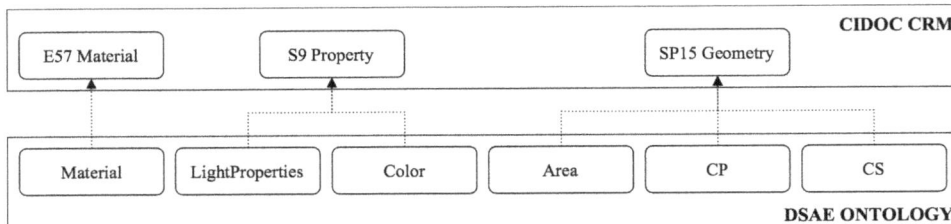

Figure 8. Connectivity relationship between the CIDOC-CRM and the DSAE ontology to extend interoperability and allow descriptive knowledge for archaeologist to be included.

Finally, semantic information is transferred to the point cloud that can be used for information extraction. Once extracted semantics have been successfully linked to the spatial information, we address structuring for interaction purposes. The data structuration is made in regard to [104]. The main idea is that the structure is decomposed in three meta-models acting at three different conceptual levels to efficiently manage massive point cloud data (and by extension any complex 3D data) while integrating semantics coherently. The Level-0 describes a meta-model to efficiently manage and organise pure point cloud spatial data information. The Level-1 is an interface between the level-0 and the level 2 (specific domain-based knowledge). As such, the data integration methodology relies on incorporating the point cloud data in the Smart Point Cloud data structure [104] in regard to the workflows described in Section 3. The structuration therefore follows the object decomposition, where points of each object are grouped together to form world objects (i.e., Independent Tesserae) once concepts and meaning have been linked. This constitutes the entry point of the ontology which acts as a Level-2 specialization to inject relevant knowledge. To facilitate the dissemination of information, query results from specific queries need to be visualized properly. For users to access and share a common viewpoint result of a semantic query, we enhanced the approach in [105] by applying over each object (i.e., tessera) one unique colour per instance for each class (e.g., faience pieces); all non-requested tesserae are coloured in black as in Figure 9.

For each class of object, we compute a bounding box and we locate its centre. The bounding box centre becomes the centre of the sphere on which the camera will move to determine the optimal camera position. The coordinates of the camera on the sphere are computed according to the following formulae [105]:

$$X = x_{center} + r \times \cos(\phi) \times \cos(\theta), \; Y = y_{center} + r \times \sin(\phi), \; Z = z_{center} + r \times \cos(\phi) \times \sin(\theta)^1 \quad (6)$$

[1] where X, Y and Z are the camera coordinates, x_{center}, y_{center} and z_{center} are the coordinates of the centre of the sphere, r is the radius of the sphere, ϕ is the vertical angle, θ is the horizontal angle.

From each camera position, we compute the number of visible tesserae from the user request observed in the produced image. Since each instance of one sort of tesserae is coloured uniquely, the algorithm performs by counting the number of different pixels colours. Hence, the number of distinct colours in the image corresponds to the number of tesserae seen from this camera position. The camera position that maximises the view of requested tesserae corresponds to the optimal viewpoint. If two camera locations present the same number of observed tesserae, we apply a maximisation criterion regarding the pixels to determine the optimal camera position.

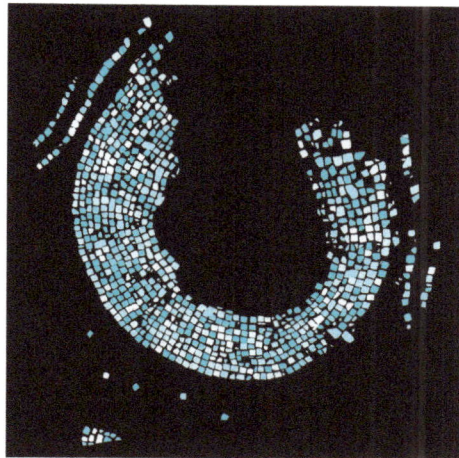

Figure 9. Unique colourisation of a group of golden tesserae (bottom-up view).

4. Results

We tested the method on different samples from different zones of the mosaic to identify the influence of the segmentation and the classification in different scenarios, as well as another point cloud from terrestrial laser scanner captured in Jehay (Belgium). To assess the quality of the segmentation, knowledge-based tessera ground truth was extracted from the point cloud and compared to the segmentation method extracts. Results (Table 4) show an average 95% segmentation accuracy for point cloud gold tesserae, 97% for faience tesserae, 94% for silver tesserae and 91% for coloured glass.

Table 4. Segmentation accuracy of tesserae samples.

Tesserae	Segmentation Number of Points		Accuracy
	Ground truth	Tesserae C.	
Gold			
Sample NO. 1	10,891	10801	99%
Sample NO. 2	10,123	11,048	91%
Sample NO. 3	10,272	10,648	96%
Sample NO. 4	11,778	12,440	94%
Faience			
Sample NO. 1	27,204	28,570	95%
Sample NO. 2	23,264	22,978	99%
Sample NO. 3	23,851	24,440	98%
Sample NO. 4	22,238	22,985	97%
Silver			
Sample NO. 1	1364	1373	99%
Sample NO. 2	876	931	94%
Sample NO. 3	3783	3312	88%
Sample NO. 4	1137	1098	97%
C. Glass			
Sample NO. 1	1139	1283	87%
Sample NO. 2	936	1029	90%
Sample NO. 3	821	736	90%
Sample NO. 4	598	625	95%

The tesserae recognition pipeline including segmentation, classification and information extraction was conducted over 3 different representative zones of the point cloud to be exhaustive and to be able to count manually each tessera for assessing the results. In the first zone containing 12,184,307 points, three types of tesserae were studied: 138 Gold tesserae from the 19th century renovation (NG), 239 ancient gold (AG) and 11 faience pieces (FT) (Figure 10). The automatic segmentation correctly recognized all FT (100% accuracy) and 331 golden tesserae (GT) (88% accuracy), remaining ones being 5% of under-segmentation (in groups of 2/3 tesserae), 7% of tesserae not detected. The classification correctly labelled respectively 100% FT, 98% NG, and 99% AG.

Figure 10. Zone 1: Classification workflow of tesserae in Zone 1. From left to right: Colour point cloud; abstraction-based segmented point cloud; classified point cloud; 2D geometry over point cloud.

In the second zone containing 12,821,752 points, 313 gold tesserae (195 NG and 118 AG) and 269 silver tesserae (ST) were processed. In total, 284 (91%) golden tesserae were correctly segmented, of which 93% were correctly labelled NG and 95% AG, and 93% of ST were correctly segmented, of which 87% were correctly labelled. The third larger sample composed of 34,022,617 points includes 945 gold tesserae and 695 CG (coloured glass) tainted in black. The other tesserae in the sample had an insufficient resolution for ground truth generation. In total, 839 (89%) golden tesserae were correctly segmented, of which 86% were correctly labelled NG and 95% AG. Concerning CG, (494) 71% were correctly segmented, and 98% were correctly labelled. While classification results are very high, segmentation is heavily influenced by the quality of the data; hence, CG shows lower results because of its harsh sensor representation (tesserae are not easily discernible).

Globally, 59,028,676 points and 2610 tesserae were processed; 2208 (85%) were correctly detected and segmented, of which 2075 (94%) were correctly labelled (Table 5).

Table 5. Recapitulation of tesserae detection results.

ID	Tesserae			Segmentation		Classification		Res.
		Type	Nb	Nb	%	Nb	%	Nb
1		NG	138	331	88%	131	98%	7
		AG	239			196	99%	43
		FT	11	11	100%	11	100%	0
2		NG	155	284	91%	128	93%	27
		AG	158			139	95%	19
		ST	269	249	93%	216	87%	53
3		NG	396	839	89%	297	86%	99
		AG	549			471	95%	78
		CG	695	494	71%	486	98%	209
	Total		2610	2208	85%	2075	94%	535

The full workflow was also conducted over a point cloud acquired using a TLS at a local scale to detect specific stones and openings of the façade of the castle of Jehay (Belgium). The point cloud comprises around 95 million points and has an uneven density due to the acquisition set-ups. The segmentation allowed us to correctly detect calcareous stones as in Table 6, as well as openings regarding the surface of reference (best fit plane through convolutional bank filter) and the full limestone bay frames.

Table 6. Segmentation accuracy of the façade of the castle of Jehay over calcareous stones.

Elements	Segmentation In Number of Points		Accuracy
	Ground truth	Method	
Calcareous Stones			
Sample NO. 1	37,057	35,668	96%
Sample NO. 2	30,610	27,100	88%
Sample NO. 3	34,087	32,200	99%
Sample NO. 4	35,197	30,459	86%

The same reasoning engine was used based on the DSAE ontology. The DSAE-based classification first studied the material Limestone (related to the property colour, same as S9 from CIDOC-CRM) and the geometry regularity (related to SP15 attribute from geometry) in regard to CS, CP and Area (in the case of 3D objects, the area was extended to a volume feature by taking into account every spatial dimension.), then differentiated openings through dimension-based predicates (SP15 Geometry) as presented in Figure 11. The CIDOC-CRM and its extension CIDOC-CRMba [106] provide an added descriptive value for archaeologists that can be directly plugged as in Figure 11.

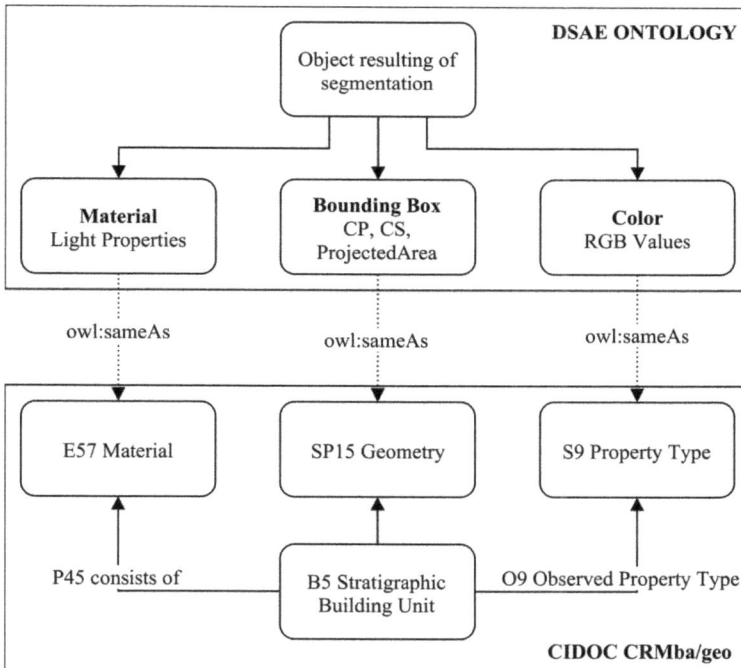

Figure 11. DSAE ontology and plugged CIDOC-CRM + CIDOC-CRMba for the detection of objects of interest: calcareous stones, openings and limestone bay frames.

The detected segments are classified, with 85% accuracy for independent calcareous stones, and 100% for woodworking openings (differentiated by size and geometric regularities) and limestone bay frames. The results over the Renaissance façade recognition pipeline are illustrated in Figure 12. We notice the fine detection for each element and the irregularity for some stones due to the uneven quality of the point cloud colorization. Calcareous stones classification was largely impacted by the segmentation inaccuracy within certain zones that led to over-segmentation and thus incorrect labels due to shape irregularity. These influential factors are discussed in Section 5.

Figure 12. Ontology-based classification of the South-South-West façade of the castle of Jehay. From left to right: the façade studied; the result of the segmentation (stones only); the result of the full segmentation; the result of the classification for quasi-planar objects of interest.

It is interesting to note that the DSAE ontology can be further used for distinguishing wide woodworking openings from smaller ones based on Area (or Volume) properties, and their geometric regularity. However, their label is considered weak for archaeological purposes, and extending the ontology as in Section 5 would provide a better automatic characterization for archaeological analysis.

The established data infrastructure gravitate around a client-server protocol that allows maximum flexibility and extensibility in regard to the 4 prerequisites of digital archaeology as defined in [5,6]. The platform can scale up to multiple simultaneous connexions and handles multi-source datasets. Every client that connects to the server as in Figure 13 benefits of functionalities from both the ontology reasoner and SQL statements (e.g., in Section 4). The implementation was made using PostgreSQL DBMS enhanced with plugins (PostGIS and pgPointCloud). The software Protégé alongside the programming toolkit JENA (Java) was also used to create and link ontologies.

As for the client-side, it was constructed to be as open and accessible as possible. As such, the World Wide Web is a democratized way to share and exchange information. It constitutes a long-term means to collaborate, and is independent of the location which is very important considering the need to be able on site to work with digital copies. Indeed, an application accessible anywhere and by multiple users at the same time is key for an archaeological 3D platform. Thus, we implemented the application in WebGL, a JavaScript API for rendering 3D graphics within any compatible web browser. We used Three.js, a cross-browser JavaScript library which uses the WebGL framework and enhances it. By a simple interaction with the GUI, the users can access and share a common viewpoint result of a semantic query. Figure 14 presents the optimal viewpoint for two classes of tesserae similarly to [12].

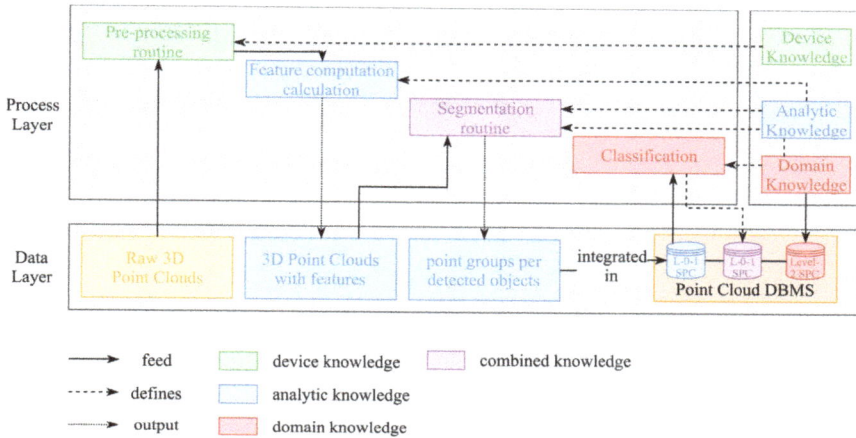

Figure 13. Server-side data management system. Point clouds go through different processing steps regarding Section 3, and point groups based on the definition of objects regarding domain knowledge are constituted and populate the SPC database.

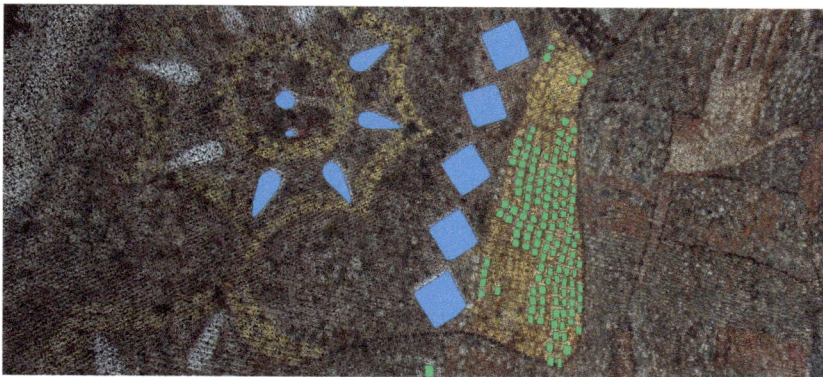

Figure 14. Query result in the WebGL prototype of the optimal viewpoint for faience pieces and gold tesserae in an extracted zone through SQL query.

The complete workflow therefore us allows to (1) pre-process multi-sensory point cloud data, (2) compute features of interest, segment and classify the point cloud according to domain knowledge formalized in ontologies; (3) structure the data in a server-side SPC point cloud 3D GIS; (4) disseminate the information through a client-side app built upon WebGL with a specific visual processing engine to provide optimal viewpoints from queries.

5. Discussion

The democratization of TLS and dense-image matching in archaeological workflows makes them a preferred way to record spatial information. Point clouds are very interesting for their objectivity and flexibility in interpretation processes. If the acquisition is complete, they transcript every visual element that was observed on the field. However, other components that can arise to our other senses such as mechano-reception (touching, hearing) or chemo-reception (taste, smell) are not captured by these remote sensors. However, their integration and link to the point clouds can be important as they constitute another source for better comprehension of the observed subject. Sensors that can capture such information as objectively as possible would be another step toward a possible better acquisition automatization. Today, archaeologists rely mostly on field-work to extract necessary information from human senses, eventually with the use of other sensors to detect additional patterns (e.g., x-fluorescent characterization in Germigny-Des-Prés). Exploring combination of multisensory surveys with sensor-level data fusion provides a great opportunity for further research and to keep a record of a more complete context. Indeed, archaeological studies deal with more and more information including archaeological observations but also data coming from other sciences (e.g., geology, chemistry, physics, etc.) and all these must be organized and considered together for an optimal understanding of the site. To avoid loss of information, recording of the fact and interpretation must be integrated in the same process [18] and specific tools should be investigated.

Regarding spatial information, 3D point clouds constitute a very exhaustive source for further archaeological investigations. However, their lack of integration in workflows narrows the possibilities and interpretation work. We identified their main weakness to propose better handling and combinatory potential between different information sources: how to coherently aggregate semantics, spatial (and temporal information in a later stage). With respect to the number of observations (points), autonomous processing is very important. When dealing with thousands of archaeological objects of interest (composed of millions of points) in a scene (composed of billions of points), manually segmenting and classifying would be a very time consuming and an error prone process. In this paper, we presented an effective approach to automate tesserae recognition from terrestrial laser scanning data and dense image-matching. Knowledge-based feature constraints are defined to extract gold, silver, coloured glass and ceramic tesserae from a hybrid point cloud. Then convex hull polygons are fitted to different segment separately. Knowledge is introduced again to generate assumptions for problematic parts. Finally, all polygons, both directly fitted and assumed, are combined to classify and inject semantics into the point cloud. Tests on three datasets showed automated classification procedures and promising results (Figure 15).

Figure 15. Classification results over the different zones of the mosaics.

The developed method tackles data quality challenges including heterogeneous density, surface roughness, curvature irregularities, and missing, erroneous data (due to reflective surfaces for example). We see that in zones where the colour quality is good and blur is low, classification results exceeds 95% accuracy. However, the method is very sensitive to 3D capture conditions and representativity such as colour, intensity, resolution and sharpness. Therefore, segmentation will fail when the input data does not allow correct feature extraction and abstraction-based connectivity estimation. More complete tesserae knowledge will help to better understand and detect complex shapes and patterns. While the classification results using domain knowledge are promising, the full point cloud labelling scheme could be enhanced by improving specifically the segmentation step. The data quality influences the final results. As illustrated, a challenge is brought about by varying densities and poor point-feature quality that can lead to over-segmentation when predominant features rely on point-proximity/density criterions. While this is not an issue for dense point clouds that describe continuous surfaces, it can constitute a hindrance for heterogeneous density or uneven datasets. Equally, colour/intensity that create imprecise colorization/featuring leads to rough classification. A solution would be to move the colour-based segmentation to the DSAE ontology to provide new discriminative possibilities. Also, the combination of dense image matching with laser data and 3D distance map improves the outline generation in a later stage, and allows a better shape estimation (Figure 16). Yet, an efficient registration is mandatory for accurate results. To improve the classification results, the segmentation can be enhanced using a watershed algorithm as well as obtaining higher representativity colour attribute for example. These are research directions that will be investigated. Also, to improve the robustness of the segmentation, a region-growing from a seed point located at every centroid of each detected connected element potentially provides a solution to under-segmentation, and investigations are necessary in this direction.

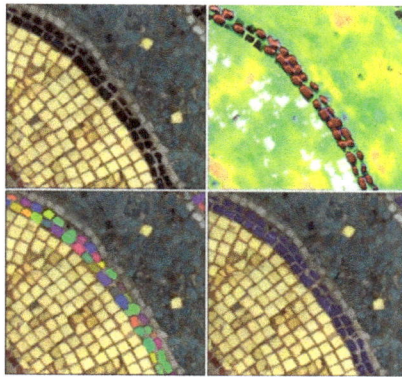

Figure 16. Classification and semantization of dark coloured glass.

The constituted ontology provides a reasoning engine based on available information that can be further enhanced to integrate new triple stores. As such, an acquisition campaign using a portable X-ray Fluorescent device was carried out to quantify the relative quantity of chemical component within some tesserae. Integrating this semantic information could provide new reasoning capabilities such as detecting every gold tesserae that contain a quantity X of Plumb. The method will also be refined and extended to the full point cloud by implementing a machine learning framework using obtained labelled data as training data. First results are encouraging using supervised classification [107], and other approaches such as reinforcement learning will be investigated for they high reasoning potential and complementarity to ontologies. However, the computer memory-demand of point clouds may impose a link to 2D projective raster's and to leverage existing training datasets (e.g., DeepNet).

The data structure relies on PostgreSQL RDBMS while indirectly integrating ontology reasoning results. It allows specific queries over the classified point cloud to extract spatial, semantic or a combination of both information. The blend of SPARQL and SQL allows us to combine efficiently the strength of both the relational database structuration and block-wise storage capabilities with the powerful reasoning proficiencies provided by ontologies. Different queries are therefore available, which are big leap forward regarding point cloud processing for archaeology (e.g., in Table 7).

However, while the temporal integration was inferred, only static intervals and fixed point in time were treated. Better integration such as continuous data or the storage and reasoning over datasets covering one location at different time intervals has yet to be further investigated. Indeed, new descriptors emerging from change detection could provide new insights and possibilities for cultural heritage conservation.

Table 7. Example of queries over the point cloud.

Language	RDF Triple Store	Effect
SPARQL	PREFIX rdf:<http://www.w3.org/1999/02/22-rdf-syntax-ns#> PREFIX npt: <http://www.geo.ulg.ac.be/nyspoux/> SELECT ?ind WHERE { ?ind rdf:type npt:AltoMedievalTessera } ORDER BY ?ind	Return all alto-medieval tesserae (regarding initial data input)
SQL	SELECT name, area FROM worldObject WHERE ST_3DIntersects(geomWo::geometry, polygonZ::geometry);	Return all tesserae which are comprised in the region defined by a selection polygon and gives their area
SPARQL & SQL	SELECT geomWo FROM worldObject WHERE ST_3DIntersects(geomWo::geometry, polygon2Z::geometry) AND area > 0,0001; PREFIX rdf: <http://www.w3.org/1999/02/22-rdf-syntax-ns#> PREFIX npt: <http://www.geo.ulg.ac.be/nyspoux/> SELECT ?ind WHERE { ?ind rdf:type npt: XIXCentTessera } ORDER BY ?ind	Return all renovated tesserae in the region 2 where the area is superior to 1 cm^2

The proposed methodology (described in Section 3) was as general as possible to be extended to other use cases, at the object and local scales. It provides a potential solution for bringing intelligence to spatial data, specifically point clouds as seen in [104]. For example, we tested a point cloud from dense-image matching captured in Denmark (Ny-Calsberg Museum) and processed using Bentley ContextCapture (Figure 17). It constitutes an interesting object scale dataset where the interest lies in deciphering the hieroglyphs.

Figure 17. 3D point cloud of the statue of the Egyptian priest Ahmose and his mother, Baket-re. Diorite. C.1490–1400 BC. 18th Dynasty. New Empire. Ny Carlsberg Glyptotek Museum. Copenhagen. Denmark. From left to right: 3D point cloud; feature extraction and segmentation; 3D visualization.

The methodology was applied, and each hieroglyph was successfully detected independently. As the spatial context is conserved, we can locate the relative position of each hieroglyph regarding the others, and using a lexicon or a structured ancient hieroglyph ontology, each sign could be detected by shape matching (e.g., RANSAC), and a reading order extracted as in [108]. Thus, the methodology is suitable to reason from information extraction, and possibilities are very encouraging. Deepening the classification through well-established ontologies such as CIDOC-CRMba as illustrated in Section 4 is possible, and the extension to other use cases requires us to identify specific specializations and the level of detail within the tree depth. If we consider the Renaissance façade of the castle of Jehay, the CIDOC-CRM ontology as well as the CIDOC-CRMba and the CIDOC-CRMgeo add flexibility for moving deduction capabilities from the analytic part to the ontology. This is very interesting as it maximizes the DSAE reasoning capabilities instead of determining analytically discriminative features (such as bounding-box "is contained in" relationship from coordinates). As an example, the classification of the façade can be related to the specialization levels from B1 to B5 of the CIDOC-CRMba, and directly plugged as in Figure 12. Then, specifically looking at full limestone bay frames (same as B5 Stratigraphic Building Unit), an element that is contained within a limestone bay frame is classified as an empty section regarding Figure 18. The topological relations are introduced with the use of the well-known GeoSPARQL ontology to allow the detection of openings based on a AP12 "contains" relationship.

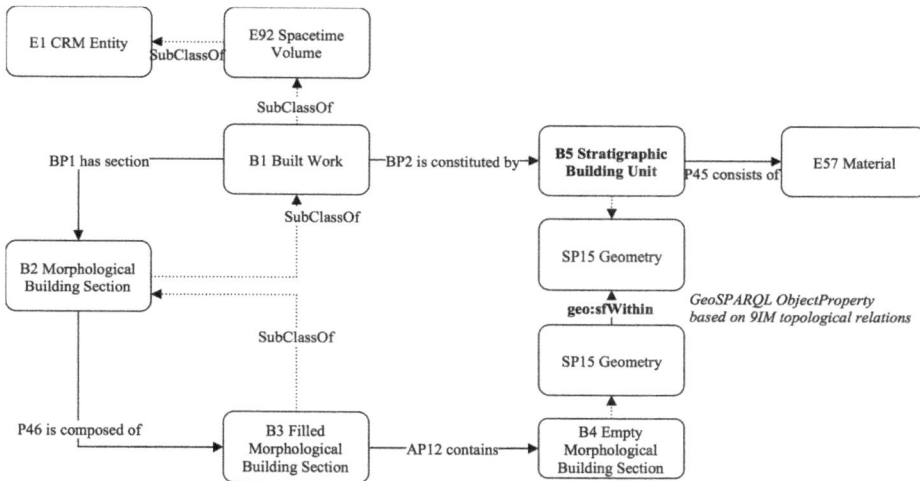

Figure 18. CIDOC-CRM ontology for the detection of objects of interest: calcareous stones and openings. Considering the castle of Jehay (B1), it has a building section (BP1) being the studied façade (B2), composed of different elements such as calcareous stones (B3), embrasures (B3) and openings (B4).

Therefore, by integrating attributes such as Color and ProjectedArea of the different elements (as well as topological "is Within" test), the ontology can be used for reasoning. Based on general axioms, it semantically recognises building parts of a façade as in Table 8.

Table 8. Classification of elements based on numerical attributes and topological relations.

Language	Equivalent To Definition	Effect
OWL (Protégé)	(hasProperty some ColorLimestone) and (hasProperty some NonReflective) and (CP some xsd:double[<= "4.0E-4"^^xsd:double]) and (CS some xsd:double[>= "1.05"^^xsd:double, <= "9.0"^^xsd:double]) and (ProjectedArea some xsd:double[>= "0.05"^^xsd:double, <= "0.4"^^xsd:double])	Defines an element as a BayFrame
OWL (Protégé)	(not (hasProperty some ColorLimestone)) and (sfWithin some BayFrame) and (BoundingBox some xsd:double[>= "2.9"^^xsd:double, <= "3.5"^^xsd:double])	Defines an element as a DoorSection

It is interesting to note that further reasoning is made possible due to extended knowledge over Renaissance-style mullioned windows. Indeed, double mullioned openings are a complex architectural element present over this façade. They are 2 mullioned openings where the separation by stones is inexistent. Each can be described regarding CIDOC CRMba as: 1 frame (B5) and 6 openings (B4 Empty morphological Building Section). Thus, an extended ontology can recognize double mullioned windows and a reason such as the one presented in [109] would provide extended automatization.

The final step for the visualization and presentation of the results is to share and distribute the information to other users and relies on virtual environments with specific interaction. The perception in 3D spaces is a dynamic phenomenon and concerns firstly behaviours and effects [110]. Data visualization is important to explore the data, to obtain some idea of what they contain, and therefore, to develop some intuitions about how to go about solving a problem from that data, determining what features are important and what kinds of data are involved. Visualization is also important when looking at the output of data science systems: data summarization for creating useful exploratory statistics, essential to understanding what was collected and observed. Although used before for tackling models and algorithms to avoid missing crucial information, data visualization is important for translating what might be interpretable only to a specialist for a general audience. In the context of point cloud, semantics and domain can highly influence the type of rendering used to directly transmit the correct information in a correct way to the end user. New ways of interacting with the data—Virtual reality, augmented reality, real time exploration and collaboration, holograms—are redefining possible interactions and exploration. Remondino, 2003 [111] list different surface representations that can be used to represent and use a point cloud, including parametric modelling, implicit and simplicial representation, approximated and interpolated surfaces. The time-consuming task of accurate 3D surface reconstruction from point cloud requires many steps of pre-processing, topology determination, triangular mesh generation, post-processing and assessing. For example, Hussain, 2009 [112] propose two simplification algorithms for LoD generation by decimating and simplifying meshes, thus reducing accuracy and quality. The development of an Internet browser-based solution allows maximum flexibility regarding theses identified problems, including data indexation vis-à-vis [113] to provide streaming capabilities independently of the size of the dataset.

Based on the algorithm developed by [105], we manage the 3D viewpoint so as to determine an optimal position and orientation of the camera for the visualisation of three kinds of tesserae distinguished by their material: faience, gold and silver. Through the previous steps of recognition and semantization described, we are now able to exploit the semantically rich point cloud data structure [11] to visualise efficiently the different sorts of tesserae. To achieve this, we performed a pre-processing step, totally transparent to users, in which we compute the optimal camera positions on a 3D COLLADA model of the mosaic which is constituted of the minimum convex hull of each tesserae information stored in the database. This technical implementation will be enhanced to enable more direct integration of the geometry generalizations from the database. The algorithm looks at the pixels of the computational display which avoids the under-object recognition phenomenon. It also allows us to directly work on the final rendering of the 3D model which already integrates the use of an algorithm to process hidden faces. Finally, it can be used on any kind of 3D data structure (vector, raster or point cloud). It is worth mentioning that additional viewpoints could be computed which depends on the initial query. For instance, we can calculate multiple optimal camera positions for one specific sort of tesserae, depending on a needed surface, distance to rotation center, density estimate, etc. The latest could be particularly interesting for the golden tesserae since they are quite scattered in space. Furthermore, we can also investigate the impact of the statistical parameter used when two viewpoints present the same number of objects (maximum, average, etc.).

To integrate the semantically rich point cloud and the viewpoint management of queried tesserae, we developed web software using jQuery, Three.js, Potree (an Open Source JavaScript library for point cloud rendering) and tween.js. The platform includes a tool to directly allow semantic extraction and visualisation of pertinent information for the end users. It enables efficient information relay between actors. The web application is accessible on any HTML5-compatible browser. It enables real time point cloud exploration of the mosaics in the Oratory of Germigny-des-Prés, and emphasises the ease of use as well as performances. However, the integration of a natural language processor would allow us to extend the possibilities for users to formulate queries that are translated into SQL and SPARQL analogues.

6. Conclusions

In this paper, we first reviewed the state of the art in digital archaeology. We pointed out gaps in the integration of spatial information with semantic components and the limited management of 3D point clouds within 3D GIS. The recording and processing of 3D multi-source complex data were addressed, as well as their management, conservation, visualization and presentation for different users. In this paper, we propose a new solution to integrate archaeological knowledge within point cloud processing workflows. Specifically, we decompose point clouds regarding available features and estimated geometric properties that generate ontologies to classify and reason based on information extraction. We developed a data-driven ontology for point cloud analysis to facilitate interoperability to other formal ontologies such as the CIDOC-CRM, and applied the workflow over different point clouds. Quasi-planar objects (doors, windows, tesserae, calcareous stones, hieroglyphs) were successfully detected, and an HTML-5 cross-platform web application was created to facilitate the knowledge dissemination such as ancient mosaic located in the oratory of Germigny-des-Prés. Then, we extracted the necessary requested information from the semantically rich point cloud data to efficiently visualise user's request based on computed optimal camera positions and orientations that maximise the visibility of requested objects (e.g. tesserae). Then, the optimal viewpoints are dynamically rendered to users through the platform on which interactions can grow.

Supplementary Materials: The ontology is available online at: http://www.geo.ulg.ac.be/nyspoux/.

Acknowledgments: This project was financed by "prix de la fondation Comhaire"—lauréate L. Van Wersch, FRS-FRNS. The authors would like to thank the anonymous reviewers for their in-depth suggestions.

Author Contributions: Florent Poux conceived and designed the experiments, validated the algorithms and documented their formulas, acquired and processed the images and laserscan data, performed various analyses of the data and wrote most of the paper. Line Van Wersch introduced ceramic and glass knowledge for this study as well as archaeological insights. Gilles-Antoine Nys designed the specific ontologies and participated in their integration within the workflows. Romain Neuville designed the visual query engine. Roland Billen participated in ontological reasoning and interpretation of the results. All the authors participated in proof-reading and reviewing of the paper.

Conflicts of Interest: The authors declare no conflict of interest. The founding sponsors had no role in the design of the study; in the collection, analyses, or interpretation of data; in the writing of the manuscript, and in the decision to publish the results.

References

1. Leute, U. *Archaeometry: An Introduction to Physical Methods in Archaeology and the History of Art*; VCH Verlagsgesellschaft mbH: Weinheim, Germany, 1987; ISBN 3527266313.
2. Koller, D.; Frischer, B.; Humphreys, G. Research challenges for digital archives of 3D cultural heritage models. *J. Comput. Cult. Herit.* **2009**, *2*, 1–17. [CrossRef]
3. Joukowsky, M. Field archaeology, tools and techniques of field work for archaeologists. In *A complete Manual of Field Archaeology: Tools and Techniques of Field Work for Archaeologists*; Prentice Hall: New York, NY, USA, 1980; ISBN 0-13-162164-5.
4. Remondino, F. Heritage Recording and 3D Modeling with Photogrammetry and 3D Scanning. *Remote Sens.* **2011**, *3*, 1104–1138. [CrossRef]
5. Tucci, G.; Bonora, V. GEOMATICS & RESTORATION—Conservation of Cultural Heritage in the Digital Era. In Proceedings of the ISPRS—International Archives of the Photogrammetry, Remote Sensing and Spatial Information Sciences, Florence, Italy, 22–24 May 2017; Volume XLII-5/W1, p. 1.

6. Patias, P. Cultural heritage documentation. In Proceedings of the International Summer School "Digital recording and 3D Modeling", Crete, Greece, 24–29 April 2006; pp. 24–29.

7. Treisman, A.M.; Gelade, G. A Feature-Integration of Attention. *Cogn. Psychol.* **1980**, *136*, 97–136. [CrossRef]

8. Smith, B.; Varzi, A. Fiat and bona fide boundaries. *Philos. Phenomenol. Res.* **2000**, *60*, 401–420. [CrossRef]

9. Binding, C.; May, K.; Tudhope, D. Semantic interoperability in archaeological datasets: Data mapping and extraction via the CIDOC CRM. In *Lecture Notes in Computer Science (Including Subseries Lecture Notes in Artificial Intelligence and Lecture Notes in Bioinformatics)*; Springer: Berlin/Heidelberg, Germany, 2008.

10. Lüscher, P.; Weibel, R.; Burghardt, D. Integrating ontological modelling and Bayesian inference for pattern classification in topographic vector data. *Comput. Environ. Urban Syst.* **2009**, *33*, 363–374. [CrossRef]

11. Poux, F.; Hallot, P.; Neuville, R.; Billen, R. Smart point cloud: Definition and remaining challenges. *ISPRS Ann. Photogramm. Remote Sens. Spat. Inf. Sci.* **2016**, *IV-2/W1*, 119–127. [CrossRef]

12. Poux, F.; Neuville, R.; Hallot, P.; Van Wersch, L.; Luczfalvy Jancsó, A.; Billen, R. Digital Investigations of an Archaeological Smart Point Cloud: A Real Time Web-Based Platform To Manage the Visualisation of Semantical Queries. *Int. Arch. Photogramm. Remote Sens. Spat. Inf. Sci.* **2017**, *XLII-5/W1*, 581–588. [CrossRef]

13. Pieraccini, M.; Guidi, G.; Atzeni, C. 3D digitizing of cultural heritage. *J. Cult. Herit.* **2001**, *2*, 63–70. [CrossRef]

14. Remondino, F.; Rizzi, A. Reality-based 3D documentation of natural and cultural heritage sites—Techniques, problems, and examples. *Appl. Geomat.* **2010**, *2*, 85–100. [CrossRef]

15. Hartmann-Virnich, A. Transcrire l'analyse fine du bâti: Un plaidoyer pour le relevé manuel dans l'archéologie monumentale. In *La Pierre et sa Mise en Oeuvre Dans L'art Médiéval: Mélanges d'Histoire de L'art Offerts à Eliane Vergnolle*; Gallet, Y., Ed.; Brepols Publishers: Turnhout, Belgium, 2011; pp. 191–202.

16. Lambers, K.; Remondino, F. Optical 3D Measurement Techniques in Archaeology: Recent Developments and Applications. In *Layers of Perception, Proceedings of the 35th International Conference on Computer Applications and Quantitative Methods in Archaeology (CAA), Berlin, Germany, 2–6 April 2007*; Posluschny, A., Ed.; Habelt: Bonn, Germany, 2008; pp. 27–35.

17. Boto-varela, G.; Hartmann-virnich, A.; Nussbaum, N.; Reveyron, N.; Boto-varela, P.D.D.G.; Hartmann- virnich, A.; Nussbaum, N. Archéologie du bâti: Du mètre au laser. *Perspectives* **2012**, *2*, 329–346.

18. Forte, M.; Dell'Unto, N.; Di Giuseppantonio Di Franco, P.; Galeazzi, F.; Liuzza, C.; Pescarin, S. The virtual museum of the Western Han Dynasty: 3D documentation and interpretation. In *Space, Time, Place, Third International Conference on Remote Sensing in Archaeology*; Archaeopress: Oxford, UK, 2010; Volume 2118, pp. 195–199.

19. Pavlidis, G.; Koutsoudis, A.; Arnaoutoglou, F.; Tsioukas, V.; Chamzas, C. Methods for 3D digitization of Cultural Heritage. *J. Cult. Herit.* **2007**, *8*, 93–98. [CrossRef]

20. Chase, A.F.; Chase, D.Z.; Weishampel, J.F.; Drake, J.B.; Shrestha, R.L.; Slatton, K.C.; Awe, J.J.; Carter, W.E. Airborne LiDAR, archaeology, and the ancient Maya landscape at Caracol, Belize. *J. Archaeol. Sci.* **2011**, *38*, 387–398. [CrossRef]

21. Chase, A.F.; Chase, D.Z.; Fisher, C.T.; Leisz, S.J.; Weishampel, J.F. Geospatial revolution and remote sensing LiDAR in Mesoamerican archaeology. *Proc. Natl. Acad. Sci. USA* **2012**, *109*, 12916–12921. [CrossRef] [PubMed]

22. Evans, D.H.; Fletcher, R.J.; Pottier, C.; Chevance, J.-B.; Soutif, D.; Tan, B.S.; Im, S.; Ea, D.; Tin, T.; Kim, S.; et al. Uncovering archaeological landscapes at Angkor using lidar. *Proc. Natl. Acad. Sci. USA* **2013**, *110*, 12595–12600.

23. Coluzzi, R.; Lanorte, A.; Lasaponara, R. On the LiDAR contribution for landscape archaeology and palaeoenvironmental studies: The case study of Bosco dell'Incoronata (Southern Italy). *Adv. Geosci.* **2010**, *24*, 125–132. [CrossRef]

24. Reshetyuk, Y. Self-Calibration and Direct Georeferencing in Terrestrial Laser Scanning. Ph.D. Thesis, Royal Institute of Technology (KTH), Stockholm, Sweden, 2009.

25. Lerma, J.L.; Navarro, S.; Cabrelles, M.; Villaverde, V. Terrestrial laser scanning and close range photogrammetry for 3D archaeological documentation: The Upper Palaeolithic Cave of Parpalló as a case study. *J. Archaeol. Sci.* **2010**, *37*, 499–507. [CrossRef]

26. Armesto-González, J.; Riveiro-Rodríguez, B.; González-Aguilera, D.; Rivas-Brea, M.T. Terrestrial laser scanning intensity data applied to damage detection for historical buildings. *J. Archaeol. Sci.* **2010**, *37*, 3037–3047. [CrossRef]

27. Entwistle, J.A.; McCaffrey, K.J.W.; Abrahams, P.W. Three-dimensional (3D) visualisation: The application of terrestrial laser scanning in the investigation of historical Scottish farming townships. *J. Archaeol. Sci.* **2009**, *36*, 860–866. [CrossRef]

28. Grussenmeyer, P.; Landes, T.; Voegtle, T.; Ringle, K. Comparison methods of terrestrial laser scanning, photogrammetry and tacheometry data for recording of cultural heritage buildings. In Proceedings of the International Archives of the Photogrammetry, Remote Sensing and Spatial Information Sciences, Beijing, China, 15 October 2008; Volume XXXVI, pp. 213–218.

29. Gonzalez-Aguilera, D.; Muñoz-Nieto, A.; Rodriguez-Gonzalvez, P.; Menéndez, M. New tools for rock art modelling: Automated sensor integration in Pindal Cave. *J. Archaeol. Sci.* **2011**, *38*, 120–128. [CrossRef]

30. Barber, D.; Mills, J.; Smith-Voysey, S. Geometric validation of a ground-based mobile laser scanning system. *J. Photogramm. Remote Sens.* **2008**, *63*, 128–141. [CrossRef]

31. James, M.R.; Quinton, J.N. Ultra-rapid topographic surveying for complex environments: The hand-held mobile laser scanner (HMLS). *Earth Surf. Process. Landf.* **2014**, *39*, 138–142. [CrossRef]

32. Thomson, C.; Apostolopoulos, G.; Backes, D.; Boehm, J. Mobile Laser Scanning for Indoor Modelling. *ISPRS Ann. Photogramm. Remote Sens. Spat. Inf. Sci.* **2013**, *II-5/W2*, 289–293.

33. Lauterbach, H.; Borrmann, D.; Heß, R.; Eck, D.; Schilling, K.; Nüchter, A. Evaluation of a Backpack-Mounted 3D Mobile Scanning System. *Remote Sens.* **2015**, *7*, 13753–13781. [CrossRef]

34. Sansoni, G.; Carocci, M.; Rodella, R. Calibration and performance evaluation of a 3-D imaging sensor based on the projection of structured light. *IEEE Trans. Instrum. Meas.* **2000**, *49*, 628–636. [CrossRef]

35. Galindo Domínguez, R.E.; Bandy, W.L.; Mortera Gutiérrez, C.A.; Ortega Ramírez, J. Geophysical-Archaeological Survey in Lake Tequesquitengo, Morelos, Mexico. *Geofísica Int.* **2013**, *52*, 261–275. [CrossRef]

36. Pix4D Terrestrial 3D Mapping Using Fisheye and Perspective Sensors—Support. Available online: https://support.pix4d.com/hc/en-us/articles/204220385-Scientific-White-Paper-Terrestrial-3D-Mapping-Using-Fisheye-and-Perspective-Sensors#gsc.tab=0 (accessed on 18 April 2016).

37. Bentley 3D City Geographic Information System—A Major Step Towards Sustainable Infrastructure. Available online: https://www.bentley.com/en/products/product-line/reality-modeling-software/contextcapture (accessed on 26 September 2017).

38. Wenzel, K.; Rothermel, M.; Haala, N.; Fritsch, D. SURE—The ifp Software for Dense Image Matching. In *Photogrammetric Week 2013*; Institute for Photogrammetry: Stuttgart, Germany, 2013; pp. 59–70.

39. Agisoft Agisoft Photoscan. Available online: http://www.agisoft.com/pdf/photoscan_presentation.pdf (accessed on 26 September 2017).

40. nFrames GmbH, nFrames. Available online: http://nframes.com/ (accessed on 26 September 2017).

41. Capturing Reality, RealityCapture. Available online: https://www.capturingreality.com (accessed on 26 September 2017).

42. Wu, C. VisualSFM. Available online: http://ccwu.me/vsfm/ (accessed on 26 September 2017).

43. Moulon, P. openMVG. Available online: http://imagine.enpc.fr/~moulonp/openMVG/ (accessed on 26 September 2017).

44. Autodesk, Autodesk: 123D Catch. Available online: http://www.123dapp.com/catch (accessed on 26 September 2017).

45. EOS Systems Inc. Photomodeler. Available online: http://www.photomodeler.com/index.html (accessed on 26 September 2017).

46. Profactor GmbH, ReconstructMe. Available online: http://reconstructme.net/ (accessed on 26 September 2017).

47. Remondino, F.; Spera, M.G.; Nocerino, E.; Menna, F.; Nex, F. State of the art in high density image matching. *Photogramm. Rec.* **2014**, *29*, 144–166. [CrossRef]

48. Haala, N. The Landscape of Dense Image Matching Algorithms. Available online: http://www.ifp.uni-stuttgart.de/publications/phowo13/240Haala-new.pdf (accessed on 26 September 2017).

49. Koenderink, J.J.; van Doorn, A.J. Affine structure from motion. *J. Opt. Soc. Am. A* **1991**, *8*, 377. [CrossRef] [PubMed]

50. Nex, F.; Remondino, F. UAV for 3D mapping applications: A review. *Appl. Geomat.* **2014**, *6*, 1–15. [CrossRef]

51. Guarnieri, A.; Remondino, F.; Vettore, A. Digital photogrammetry and TLS data fusion applied to cultural heritage 3D modeling. In *International Archives of Photogrammetry and Remote Sensing*; Maas, H.-G., Schneider, D., Eds.; ISPRS: Dresden, Germany, 2006; Volume XXXVI.

52. Nunez, M.A.; Buill, F.; Edo, M. 3D model of the Can Sadurni cave. *J. Archaeol. Sci.* **2013**, *40*, 4420–4428. [CrossRef]

53. Novel, C.; Keriven, R.; Poux, F.; Graindorge, P. Comparing Aerial Photogrammetry and 3D Laser Scanning Methods for Creating 3D Models of Complex Objects. In *Capturing Reality Forum*; Bentley Systems: Salzburg, Austria, 2015; p. 15.

54. Leberl, F.; Irschara, A.; Pock, T.; Meixner, P.; Gruber, M.; Scholz, S.; Wiechert, A. Point Clouds: Lidar versus 3D Vision. *Photogramm. Eng. Remote Sens.* **2010**, *76*, 1123–1134. [CrossRef]

55. Al-kheder, S.; Al-shawabkeh, Y.; Haala, N. Developing a documentation system for desert palaces in Jordan using 3D laser scanning and digital photogrammetry. *J. Archaeol. Sci.* **2009**, *36*, 537–546. [CrossRef]

56. Poux, F.; Neuville, R.; Hallot, P.; Billen, R. Point clouds as an efficient multiscale layered spatial representation. In Proceedings of the Eurographics Workshop on Urban Data Modelling and Visualisation, Liège, Belgium, 8 December 2016.

57. Koch, M.; Kaehler, M. Combining 3D laser-Scanning and close-range Photogrammetry—An approach to Exploit the Strength of Both methods. In Proceedings of the Computer Applications to Archaeology, Williamsburg, VA, USA, 22–26 March 2009; pp. 1–7.

58. Stanco, F.; Battiato, S.; Gallo, G. *Digital Imaging for Cultural Heritage Preservation: Analysis, Restoration, and Reconstruction of Ancient Artworks*; CRC Press: Boca Raton, FL, USA, 2011; ISBN 1439821739.

59. Billen, R. Nouvelle Perception de la Spatialité des Objets et de Leurs Relations. Ph.D. Thesis, University of Liège, Liege, Belgium, 2002.

60. Agapiou, A.; Lysandrou, V. Remote sensing archaeology: Tracking and mapping evolution in European scientific literature from 1999 to 2015. *J. Archaeol. Sci. Rep.* **2015**, *4*, 192–200. [CrossRef]

61. Rollier-Hanselmann, J.; Petty, Z.; Mazuir, A.; Faucher, S.; Coulais, J.-F. Développement d'un SIG pour la ville médiévale de Cluny. *Archeol. Calc.* **2014**, *5*, 164–179.

62. Katsianis, M.; Tsipidis, S.; Kotsakis, K.; Kousoulakou, A. A 3D digital workflow for archaeological intra-site research using GIS. *J. Archaeol. Sci.* **2008**, *35*, 655–667. [CrossRef]

63. Apollonio, F.I.; Gaiani, M.; Benedetti, B. 3D reality-based artefact models for the management of archaeological sites using 3D Gis: A framework starting from the case study of the Pompeii Archaeological area. *J. Archaeol. Sci.* **2012**, *39*, 1271–1287. [CrossRef]

64. Galeazzi, F.; Callieri, M.; Dellepiane, M.; Charno, M.; Richards, J.; Scopigno, R. Web-based visualization for 3D data in archaeology: The ADS 3D viewer. *J. Archaeol. Sci. Rep.* **2016**, *9*, 1–11. [CrossRef]

65. Stefani, C.; Brunetaud, X.; Janvier-Badosa, S.; Beck, K.; De Luca, L.; Al-Mukhtar, M. Developing a toolkit for mapping and displaying stone alteration on a web-based documentation platform. *J. Cult. Herit.* **2014**, *15*, 1–9. [CrossRef]

66. De Reu, J.; Plets, G.; Verhoeven, G.; De Smedt, P.; Bats, M.; Cherretté, B.; De Maeyer, W.; Deconynck, J.; Herremans, D.; Laloo, P.; et al. Towards a three-dimensional cost-effective registration of the archaeological heritage. *J. Archaeol. Sci.* **2013**, *40*, 1108–1121. [CrossRef]

67. Scianna, A.; Gristina, S.; Paliaga, S. Experimental BIM applications in archaeology: A work-flow. In *Lecture Notes in Computer Science (Including Subseries Lecture Notes in Artificial Intelligence and Lecture Notes in Bioinformatics)*; University of Illinois at Urbana-Champaign: Champaign, IL, USA, 2014; Volume 8740, pp. 490–498. ISBN 978-3-319-13695-0; 978-3-319-13694-3.

68. Campanaro, D.M.; Landeschi, G.; Dell'Unto, N.; Leander Touati, A.M. 3D GIS for cultural heritage restoration: A "white box" workflow. *J. Cult. Herit.* **2016**, *18*, 321–332. [CrossRef]

69. Soler, F.; Melero, F.J.; Luzon, M.V. A complete 3D information system for cultural heritage documentation. *J. Cult. Herit.* **2017**, *23*, 49–57. [CrossRef]

70. Lieberwirth, U. 3D GIS Voxel-Based Model Building in Archaeology. In Proceedings of the 35th International Conference on Computer Applications and Quantitative Methods in Archaeology (CAA), Berlin, Germany, 2–6 April 2007; Volume 2, pp. 1–8.

71. Orengo, H.A. Combining terrestrial stereophotogrammetry, DGPS and GIS-based 3D voxel modelling in the volumetric recording of archaeological features. *J. Photogramm. Remote Sens.* **2013**, *76*, 49–55. [CrossRef]

72. Clementini, E.; Di Felice, P. Approximate topological relations. *Int. J. Approx. Reason.* **1997**, *16*, 173–204. [CrossRef]

73. Moscati, A.; Lombardo, J.; Losciale, L.V.; De Luca, L. Visual browsing of semantic descriptions of heritage buildings morphology. In Proceedings of the Digital Media and its Applications in Cultural Heritage (DMACH) 2011, Amman, Jordan, 16 March 2011; pp. 1–16.

74. Janowicz, K. Observation-Driven Geo-Ontology Engineering. *Trans. GIS* **2012**, *16*, 351–374. [CrossRef]

75. Novak, M. *Intelligent Environments: Spatial Aspects of the Information Revolution*; Droege, P., Ed.; Elsevier: Amsterdam, The Nederland, 1997; ISBN 0080534848.

76. Klein, L.A. *Sensor and Data Fusion: A Tool for Information Assessment and Decision Making*; SPIE Press: Bellingham, WA, USA, 2004; ISBN 0819454354.

77. Petrie, G. Systematic oblique ae using multiple digitarial photography i frame cameras. *Photogramm. Eng. Remote Sens.* **2009**, *75*, 102–107.

78. Otepka, J.; Ghuffar, S.; Waldhauser, C.; Hochreiter, R.; Pfeifer, N. Georeferenced Point Clouds: A Survey of Features and Point Cloud Management. *Int. J. Geoinf.* **2013**, *2*, 1038–1065. [CrossRef]

79. Van Wersch, L.; Kronz, A.; Simon, K.; Hocquet, F.-P.; Strivay, D. Matériaux des Mosaïques de Germigny-Des-Prés. Germigny-des-prés. Available online: http://pointcloudproject.com/wp-content/uploads/2017/01/Unknown-Van-Wersch-et-al.-Mat%C3%A9riaux-des-mosa%C3%AFques-de-Germigny-des-Pr%C3%A9s.pdf (accessed on 26 September 2017).

80. Dimitrov, A.; Golparvar-Fard, M. Segmentation of building point cloud models including detailed architectural/structural features and MEP systems. *Autom. Constr.* **2015**, *51*, 32–45. [CrossRef]

81. Poux, F.; Neuville, R.; Billen, R. Point cloud classification of tesserae from terrestrial laser data combined with dense image matching for archaeological information extraction. *ISPRS Ann. Photogramm. Remote Sens. Spat. Inf. Sci.* **2017**, *IV-2/W2*, 203–211. [CrossRef]

82. Weinmann, M.; Jutzi, B.; Hinz, S.; Mallet, C. Semantic point cloud interpretation based on optimal neighborhoods, relevant features and efficient classifiers. *J. Photogramm. Remote Sens.* **2015**, *105*, 286–304. [CrossRef]

83. Pu, S.; Vosselman, G. Knowledge based reconstruction of building models from terrestrial laser scanning data. *J. Photogramm. Remote Sens.* **2009**, *64*, 575–584. [CrossRef]

84. Ben Hmida, H.; Boochs, F.; Cruz, C.; Nicolle, C. Knowledge Base Approach for 3D Objects Detection in Point Clouds Using 3D Processing and Specialists Knowledge. *Int. J. Adv. Intell. Syst.* **2012**, *5*, 1–14.

85. Schnabel, R.; Wahl, R.; Klein, R. Efficient RANSAC for Point Cloud Shape Detection. *Comput. Graph. Forum* **2007**, *26*, 214–226. [CrossRef]

86. Lin, H.; Gao, J.; Zhou, Y.; Lu, G.; Ye, M.; Zhang, C.; Liu, L.; Yang, R. Semantic decomposition and reconstruction of residential scenes from LiDAR data. *ACM Trans. Graph.* **2013**, *32*, 1. [CrossRef]

87. Ochmann, S.; Vock, R.; Wessel, R.; Klein, R. Automatic reconstruction of parametric building models from indoor point clouds. *Comput. Graph.* **2016**, *54*, 94–103. [CrossRef]

88. Chaperon, T.; Goulette, F. Extracting cylinders in full 3D data using a random sampling method and the Gaussian image. In *Computer Vision and Image Understanding*; Academic Press: San Diego, CA, USA, 2001; pp. 35–42.

89. Nurunnabi, A.; Belton, D.; West, G. Robust Segmentation in Laser Scanning 3D Point Cloud Data. In Proceedings of the 2012 International Conference on Digital Image Computing Techniques and Applications, Fremantle, Australia, 3–5 December 2012; IEEE: Fremantle, WA, USA, 2012; pp. 1–8.

90. Rusu, R.B.; Blodow, N. Close-range scene segmentation and reconstruction of 3D point cloud maps for mobile manipulation in domestic environments. In Proceedings of the IEEE/RSJ International Conference on Intelligent Robots and Systems (IROS 2009), St. Louis, MO, USA, 10–15 October 2009.

91. Samet, H.; Tamminen, M. Efficient component labeling of images of arbitrary dimension represented by linear bintrees. *IEEE Trans. Pattern Anal. Mach. Intell.* **1988**, *10*, 579–586. [CrossRef]

92. Girardeau-Montaut, D.; Roux, M.; Thibault, G. Change Detection on points cloud data acquired with a ground laser scanner. In *ISPRS Workshop Laser Scanning*; ISPRS: Enschede, The Netherlands, 2005.

93. Girardeau-Montaut, D. *Détection de Changement sur des Données Géométriques Tridimensionnelles*; Télécom ParisTech: Paris, France, 2006.

94. Douillard, B.; Underwood, J. On the segmentation of 3D LIDAR point clouds. In Proceedings of the 2011 IEEE International Conference on Robotics and Automation (ICRA), Shanghai, China, 9–13 May 2011; pp. 2798–2805.

95. Aijazi, A.; Checchin, P.; Trassoudaine, L. Segmentation Based Classification of 3D Urban Point Clouds: A Super-Voxel Based Approach with Evaluation. *Remote Sens.* **2013**, *5*, 1624–1650. [CrossRef]

96. Sapkota, P. *Segmentation of Coloured Point Cloud Data*; International Institute for Geo-Information Science and Earth Observation: Enschede, The Netherlands, 2008.

97. Fischler, M.A.; Bolles, R.C. Random sample consensus: A paradigm for model fitting with applications to image analysis and automated cartography. *Commun. ACM* **1981**, *24*, 381–395. [CrossRef]

98. Poux, F.; Hallot, P.; Jonlet, B.; Carre, C.; Billen, R. Segmentation semi-automatique pour le traitement de données 3D denses: Application au patrimoine architectural. *XYZ Rev. L'Assoc. Fr. Topogr.* **2014**, *141*, 69–75.

99. Knublauch, H.; Fergerson, R.W.; Noy, N.F.; Musen, M.A. The Protégé OWL Plugin: An Open Development Environment for Semantic Web Applications. Proceedings of the 3rd International Semantic Web Conference (ISWC2004), Hiroshima, Japan, 7–11 November 2004; pp. 229–243.

100. Point Cloud Library (PCL). Statistical Outlier Fliter. Available online: http://pointclouds.org/documentation/tutorials/statistical_outlier.php (accessed on 26 September 2017).

101. Kaufman, A.; Yagel, R.; Cohen, D. Modeling in Volume Graphics. In *Modeling in Computer Graphics*; Springer: Berlin/Heidelberg, Germany, 1993; pp. 441–454.

102. Brinkhoff, T.; Kriegel, H.; Schneider, R.; Braun, A. Measuring the Complexity of Polygonal Objects. In Proceedings of the 3rd ACM International Workshop on Advances in Geographic Information Systems, Baltimore, MD, USA, 1–2 December 1995; p. 109.

103. Arp, R.; Smith, B.; Spear, A.D. *Building Ontologies with Basic Formal Ontology*; The MIT Press: Cambridge, MA, USA, 2015; ISBN 9788578110796.

104. Poux, F.; Neuville, R.; Hallot, P.; Billen, R. Model for reasoning from semantically rich point cloud data. *ISPRS Ann. Photogramm. Remote Sens. Spat. Inf. Sci.* **2017**, in press.

105. Neuville, R.; Poux, F.; Hallot, P.; Billen, R. Towards a normalized 3D geovizualisation: The viewpoint management. *ISPRS Ann. Photogramm. Remote Sens. Spat. Inf. Sci.* **2016**, *IV-2/W1*, 179–186. [CrossRef]

106. Ronzino, P.; Niccolucci, F.; Felicetti, A.; Doerr, M. CRMba a CRM extension for the documentation of standing buildings. *Int. J. Digit. Libr.* **2016**, *17*, 71–78. [CrossRef]

107. Garstka, J.; Peters, G. Learning Strategies to Select Point Cloud Descriptors for 3D Object Classification: A Proposal. In *Eurographics 2014*; Paulin, M., Dachsbacher, C., Eds.; The Eurographics Association: Strasbourg, France, 2014.

108. Franken, M.; van Gemert, J.C. Automatic Egyptian hieroglyph recognition by retrieving images as texts. In Proceedings of the 21st ACM International Conference on Multimedia—MM '13, Barcelona, Spain, 21–25 October 2013; ACM Press: New York, NY, USA, 2013; pp. 765–768.

109. Hmida, H.B.; Cruz, C.; Boochs, F.; Nicolle, C. From 9-IM topological operators to qualitative spatial relations using 3D selective Nef complexes and logic rules for bodies. In Proceedings of the International Conference on Knowledge Engineering and Ontology Development (KEOD 2012), Barcelona, Spain, 4–7 October 2012; SciTePress: Setubal, Portugal, 2012; pp. 208–213.

110. Forte, M. Cyber-archaeology: An eco-approach to the virtual reconstruction of the past. In *International Symposium on "Information and Communication Technologies in Cultural Heritage"*; Earthlab: Ioannina, Greece, 2008; pp. 91–106.

111. Remondino, F. From point cloud to surface. In Proceedings of the International Workshop on Visualization and Animation of Reality-based 3D Models, Tarasp-Vulpera, Switzerland, 24–28 February 2003; Volume XXXIV.

112. Hussain, M. Efficient simplification methods for generating high quality LODs of 3D meshes. *J. Comput. Sci. Technol.* **2009**, *24*, 604–608. [CrossRef]

113. Scheiblauer, C.; Wimmer, M. Out-of-core selection and editing of huge point clouds. *Comput. Graph.* **2011**, *35*, 342–351. [CrossRef]

geosciences

MDPI

Article

Ontology-Based Photogrammetry Survey for Medieval Archaeology: Toward a 3D Geographic Information System (GIS)

Pierre Drap [1,*,†], Odile Papini [1,†], Elisa Pruno [2,†], Micchele Nucciotti [2,†] and Guido Vannini [2,†]

[1]   Aix-Marseille Université, CNRS, ENSAM, Université De Toulon, LSIS UMR 7296,
     Domaine Universitaire de Saint-Jérôme, Bâtiment Polytech, Avenue Escadrille Normandie-Niemen,
     13397 Marseille, France; Odile.Papini@univ-amu.fr
[2]   SAGAS Department, University of Florence, 50121 Florence, Italy; elisapruno53@gmail.com (E.P.);
     michele.nucciotti@unifi.it (M.N.); guido.vannini@unifi.it (G.V.)
*   Correspondence: Pierre.Drap@univ-amu.fr; Tel.: +33-4-91-82-85-20
†   These authors contributed equally to this work.

Received: 3 September 2017; Accepted: 17 September 2017; Published: 26 September 2017

Abstract: This paper presents certain reflections concerning an interdisciplinary project between medieval archaeologists from the University of Florence (Italy) and computer science researchers from CNRS, National Center for Scientific Research, (France), aiming towards a connection between 3D spatial representation and archaeological knowledge. We try to develop an integrated system for archaeological 3D survey and all other types of archaeological data and knowledge by incorporating observable (material) and non-graphic (interpretive) data. Survey plays a central role, since it is both a metric representation of the archaeological site and, to a wider extent, an interpretation of it (being also a common basis for communication between the two teams). More specifically, 3D survey is crucial, allowing archaeologists to connect actual spatial assets to the stratigraphic formation processes (i.e., to the archaeological time) and to translate spatial observations into historical interpretation of the site. It is well known that laser scanner, photogrammetry and computer vision are very useful tools for archaeologists, although the integration of the representation of space, as well as archaeological time has not yet found a methodological standard of reference. We propose a common formalism for describing photogrammetric survey and archaeological knowledge stemming from ontologies: indeed, ontologies are fully used to model and store 3D data and archaeological knowledge. We equip this formalism with a qualitative representation of time, starting from archaeological stratigraphy. Stratigraphic analyses (both of excavated deposits and of upstanding structures) are closely related to Edward Cecil Harris's theory of the "Unit of Stratigraphication" (referred to as "US", while a stratigraphic unit of an upstanding structure *Unita Stratigrafica Murale*, in Italian, will be referred to as "USM"). Every US is connected to the others by geometric, topological and, eventually, temporal links, and these are recorded by the 3D photogrammetric survey. However, the limitations of the Harris matrix approach led us to use another formalism for representing stratigraphic relationships, namely Qualitative Constraints Networks (QCN), which was successfully used in the domain of knowledge representation and reasoning in artificial intelligence for representing temporal relations.

Keywords: medieval archeology; photogrammetry; 3D survey; ontology; stratigraphic analysis; Harris matrix; temporal relations

1. Introduction

This paper presents a set of reflections based on sixteen years of interdisciplinary cooperation between medieval archaeologists from the University of Florence (Italy) and computer science researchers from CNRS of Marseille (France), aiming towards a connection between 3D spatial representation and archaeological knowledge for interpreting the Mediterranean Middle Ages [1]. In the last decade, we witnessed significant improvements in photogrammetric techniques, starting with the SIFT descriptor in 2004 [2], used to automatically match thousands of homologous points. This opened the way to increased automatization of the photogrammetric process, auto-calibration and automatic 3D dense point cloud generation. These new photogrammetric tools now can easily replace the terrestrial laser scanner still used at the turn of this century in archeology, but offer no improvements and remain unable to link semantic data with this plethora of accurate geometric data. At the same time, the stratigraphic analysis applied to upstanding buildings in the framework of medieval archeology is, on the one hand, absolutely mandatory and, on the other hand seems, to have reached a crucial point in its evolution. Italian Medieval Archeology, in parallel with a close relationship with historical research, contributed to the development of archaeological methodologies at European and world levels. A specific and particularly outstanding contribution was provided in the field of non-destructive urban and territorial analysis. Such was indeed the (successful) attempt to embed stratigraphic theory in the study of historical buildings, extending the principles of the site-formation process to the architectural-formation process [3]. The translation of the Harris paradigm [4] in upstanding structures [3,5], i.e., in full 3D context, brings some inconsistencies, which become more and more visible with the development of new 3D surveying tools. Many scholars argue the possibility to increase the Harris matrix with specific fields suitable for better defining and analyzing upstanding building stratigraphy. In Italy, where the use of upstanding building stratigraphy is commonly used, especially among medieval archaeologists, see the status quaestionis in Gallina [6]. At the same time, recent advances in knowledge representation, as the development of ontologies also in cultural heritage, change the way to manage knowledge and extend the possibilities to connect with other research fields. The presented work aims at merging photogrammetric survey and temporal relations in order to propose a new representation of temporal relations stemming from temporal qualitative networks thanks to the Allen approach. This is done by developing a common framework describing knowledge used in photogrammetry, as well as in stratigraphy fully based on ontologies. An ontology describing the photogrammetric process and the measured artifact (ashlar block, observed relevant surface and then the Unit of Stratigraphication (US) and connected concepts) is aligned with the well-known ontology used now since a long time in cultural heritage: International Committee for Documentation, Conceptual Reference Model (CIDOC CRM) [7]. The use of ontologies to manage cultural heritage is increasing every year, opening the door for many interesting perspectives [8–11]. Significant advances developed in CIDOC CRM [12–14] (and now with CIDOC-CRMba, an ontology to encode metadata about the documentation of archaeological buildings [15]) are also very useful for exploring certain theoretical concepts underlying the construction of the Harris matrix. CIDOC CRM is now well adopted by cultural heritage actors from a theoretical point of view [16–19], as well as applicative works [16] and a very interesting direction toward Geographic Information Systems (GIS) applications [20].

We have developed an ontology to represent both the photogrammetric process and the measured objects, ashlar blocks and US. The objects are modeled according to the point of view of the measurement process but indeed, these artifacts or concepts as US can also be seen from a cultural heritage or conservation point of view. This is the reason why we have aligned our ontology with CIDOC CRM. This modeling work is based on a previous study that started from the premise that collections of measured items are marred by a lack of precision concerning their measurement, assumptions about their reconstruction, their age and origin. It was therefore important to ensure the coherence of the measured items and potentially propose a possible revision. This previous work was done in the context of underwater archeology with similar problems. For more information, see [21–24]. The case study of this work is the Montreal Castle in Showbak, Jordan. Where the University of Florence has been working since the 1980s and LSIS since 2001 [25]. We are thus working on a huge quantity of data acquired over time according to the evolution of the technology and the team evaluation. We present in this paper both photogrammetric surveys and their link with archaeological data through a common formalism based both on pure XML and on ontology.

2. Managing Time in Archeology

One of the first interesting computing applications in archeology concerning the temporal domain was probably due to David George Kendall [26,27], a statistician at the University of Cambridge. He worked on seriation ("The purpose of seriation is to order the individuals of a population in order that exhibits in the most coherent way the evolutionary structure of their descriptive traits. The principles according to which order—which reflects temporal evolution—is obtained may be diverse, even if they explicitly express the notion of evolution" [27], cited by Guenoche and Tchernia in [28]) in archeology by treating the data of Flinders Petrie. Petrie, a pioneer of modern and scientific archeology, proposed a method for determining a relative chronology of artifacts according to their characteristics and to the context in which they were discovered. The analysis of the relationships between all the significant details of the objects found during the excavation also makes it possible to group the objects into families and to highlight evolutionary trends in a group of objects. Seriation is still relevant [29,30], although the management of time in archeology has been recast with the work of Harris and the stratigraphic approach [31] where the relations between stratigraphic units are both temporal and topological.

Edward Cecil Harris, a British archaeologist, developed in 1979 a matrix method for the management and verification of spatial-temporal relations between stratigraphic units. The three temporal relations (anteriority, posteriority and synchronism) associated with topological relations describing the physical relations (resting on, superimposing, overlapping, etc.) serve to organize the stratigraphic units in relative time. The stratigraphic unit, referred to as the US, is defined as the smallest unit of land that can be located before and after neighboring units by examining its physical boundaries.

The essential difference between Harris's approach and seriation lies in the fact that the establishment of the relative chronology of the stratigraphy (Harris method) does not take into account the artifacts present in the stratigraphic units [32], but only the physical position of each US and their relationship. The main objective in the field is to recognize the elementary stratigraphic units and to understand their topological relationships (and, of course, how they were formed). From these relations, the archaeologist deduces the stratigraphic relations that make it possible to propose a chronological order according to their physical position. For example, with regard to the stratigraphic analysis of a building, the different stages of its construction can be observed and decomposed into stratigraphic units.

During archaeological excavations, hundreds or even thousands of stratigraphic units can be identified and listed. The development of a graph and its translation into a Harris diagram or matrix has enabled the detection of inconsistencies in the data, mainly by detecting cycles in the graphs.

Harris's approach therefore proposes a relative scheduling as a function of time. However, the use of the typology of artifacts (the typology that is the basis of all seriation) present in the stratigraphic units may be useful to determine the time that can be quantified using absolute benchmarks. The two approaches, the relatively ordered time by Harris and the absolute time given by the typology, but also by the correlation with texts of the time or by scientific methods such as carbon 14 (Radiocarbon dating is a method developed by Willard Frank Libby used to calculate the age of objects containing organic material. The measurement of the ^{14}C decay is a standard way to roughly calculate the date of death or fixation of an object containing organic material) dating or dendrochronology (The dendrochronology dating method, first developed by Andrew Ellicott Douglas, is based on the growth of tree rings. This method can only be used if one knows the growth curve of the region where the object that we want to date was found), are both linear approaches.

Djindjian and Desachy resituated Harris's work from the perspective of graphs resulting from operational research [33–36]. The PERT (Program Evaluation and Review Technique) method and the

MPM (Potential Metrics and Methods) method consist of breaking down the production process into a series of precise operations, or tasks, separated by stages, each of which constitutes the end of one task and the beginning of another [33].

2.1. Harris's Methods for the Temporal Representation of Stratigraphic Units

Figure 1 shows the PERT diagram (a) and the MPM diagram (b) in the corresponding Harris matrix (c) and Harris diagram (d). It illustrates the relationship between the Harris method and the pre-existing PERT and MPM methods. In both cases, the graph accounts for the scheduling of the tasks required to execute a program. In the PERT method, the steps for executing the program are represented by the vertices, while the arcs represent the task. In the case of the MPM method, it is the opposite.

The analogy with the archaeological process is strong: the stratigraphic units are represented as temporally-related tasks, and the very close relationship between the Harris diagram (d) and the MPM diagram (b) is shown in Figure 1.

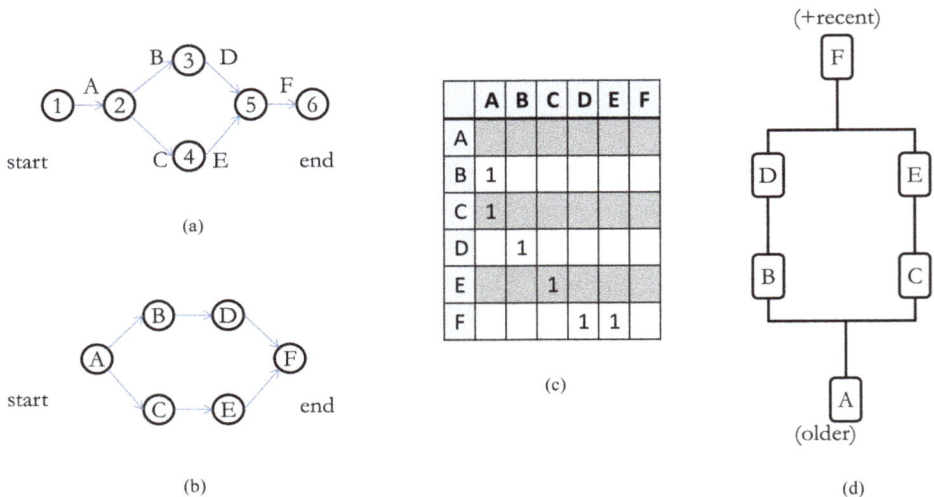

Figure 1. Relative temporal representations: (**a**) PERT diagram; (**b**) MPM diagram; (**c**) Harris matrix; (**d**) Harris graph.

The relative position of the five stratigraphic units can be seen in the example in Figure 2. On the left, the perspective view allows us to propose a chronology; a cross-section of the same US is in the center; and on the right, the Harris diagram is shown.

It is important to note US 12, which is a cut in the layer labeled as US 15. US 12 is a negative US because it is only a trace of an action of cut. US 12 now separates US 13 and 14, but the archaeological investigation must be able to understand if these two US before the cut of US 12 are the same US or not. If this is the case, we will add in the diagram the synchronous relation between US 13 and 14. This is followed by US 11, which fills US 12 visible at that time.

The Harris formalism which allows one to manage and visualize these US, is well distributed throughout the community by software, mostly open source, which helps to visualize and store these graphs.

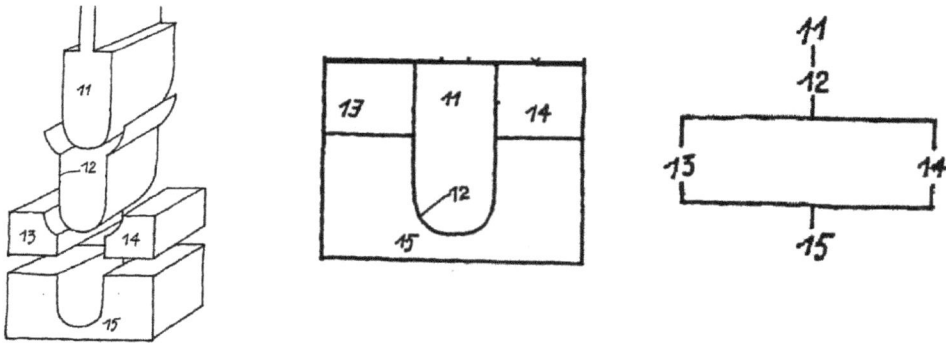

Figure 2. Different representations of Unit of Stratigraphication (US) relations, cited from ([37], p. 83).

Earlier software such as ABC FlowCharter 4.0 (originally from Micrografx, in 1995 and then by iGraphx) was mainly presented as flowchart software with an adaptation for the Harris formalism. Also in 1985, the old Strata software program from the Computing Laboratory, University of Kent at Canterbury, used a set of recorded observations of stratigraphic relationships as input and was the foundation stone of jnet software [38] based on semi-structured formalism (XML). In the same way, we can mention the Bonn Archaeological Software Package (BASP) [39] developed since 1973 included in many software packages dedicated to archeology (orthophoto, level curve, etc.). None of these software packages are available any longer.

Still used by the community, we can cite ArchEd from the University of Vienna, Austria [40], which allowed one to build stratigraphic graphs, but they were completely disconnected from the survey. More recently, Intrasis, a Swedish GIS software package dedicated to archeology, is able to manage 3D files coming from terrestrial laser scanners and linked using ArchEd [41]. In the same direction, based on the GIS platform, the Harris matrix Composer from Christoph Traxler and Wolfgang Neubauer [42] tries to fill the gap between survey and the Harris matrix. The most recent one, developed by Jerzy Sikora and his team [43], plans to use mobile devices in the field that are able to merge visualization, data and analysis.

The main works on US formalization and visualization have been done as part of archaeological digs. The latest works include ADS (Archeology Data Service, U.K.) and ISTI-CNR, Italy, where archaeological dig reports, 3D surveying, 3D visualization on the web and the Harris formalism can be found in the same tool [44]. Indeed, the Harris formalism has been developed for archaeological excavation and a transposition on an upstanding structure, in a real 3D context, with stratigraphic units still functional at the time they were studied, which implies the development of many new issues to be considered, as well as an accurate formalism.

However, Harris's representation of time is also limited. Indeed, taking into account only points on the temporal scale, it cannot express the notion of duration. This restriction due to the linear representation of time was already perceived in the Nineteenth Century by Flinders Petrie: "The main value of dates is to show the sequence of events; and it would matter very little if the time from Augustus to Constantine had occupied six centuries instead of three, or if Alexander had lived only two centuries before Augustus. The order of events and the relation of one country to another is the main essential in history. Indeed, the tacit common-sense of historians agrees in treating the periods of great activity and production more fully than the arid ages of barbarism, and so substituting practically a scale of activity as the standard rather than a scale of years.", Petrie in 1899, as quoted by Lucas ([45], p. 8).

Thus, the formalism proposed by Harris responds well to the need for formatting and control of the data, but does not allow representing the complexity of the temporal data used in archeology.

Since the 1980s, the concept of palimpsest seems to be necessary to describe the complexity of temporal superposition already directly observable. "The house where I am writing this paper was built towards the beginning of this century, in the courtyard of an ancient farm whose structure is still visible. From my open window, I see an interweaving of houses and constructions, most of them dating back to the 19th century, sometimes including parts of earlier constructions from the 18th or 17th century. The 20th century here looks so localized, so secondary: it is reduced to details, such as windows, doors or, within houses and flats, furniture.... Right now, the present here is made up of a series of past durations that makes the present multi-temporal." Laurent Olivier cited by Lucas ([45], p. 37).

Since the 1990s, a new trend has been developing and proposes to use other temporal models in archeology to take this complexity into account. Ann Ramenofsky, an archaeologist at the University of New Mexico, USA, suggests using intervals to take into account measurement uncertainties in absolute chronology assessments ([46], p. 79). Other teams propose to extend the temporal relations proposed by Harris to the time management formalism proposed by Allen [46]. This results in the number of temporal relations increasing from 3–13 (see Figure 3), as well as the ability to reason in terms of duration over intervals and no longer on a succession of ordered instants [47,48]. However, even if this formalism is much richer (13 relations are proposed to manage time intervals), this approach has not yet spread in the world of archeology [49].

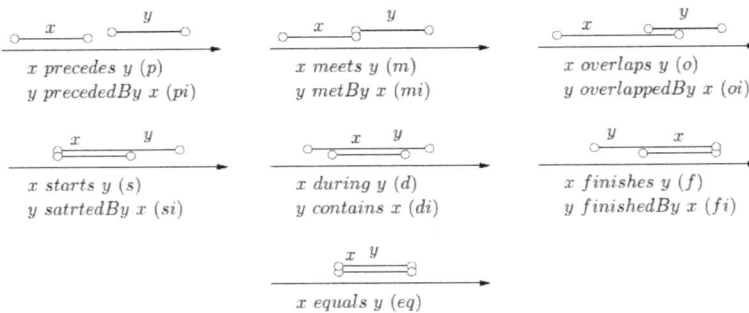

x precedes y (p)
y precededBy x (pi)

x meets y (m)
y metBy x (mi)

x overlaps y (o)
y overlappedBy x (oi)

x starts y (s)
y satrtedBy x (si)

x during y (d)
y contains x (di)

x finishes y (f)
y finishedBy x (fi)

x equals y (eq)

Figure 3. Allen's intervals algebra atomic relations.

As we have seen, the topic of time in archeology is most important and very complex, though it has been too often neglected. The work conducted by G. Lucas [45,50] has been, until now, the most exhaustive presentation and reflection about it. Following Lucas, time is a theoretical concept used in different ways in contemporary archeology, but its meanings are not always the topic of discussion.

2.2. Chronology

Therefore, by analyzing and breaking up an archaeological deposit into basic, discrete stratigraphic units, archaeologists are able to create (relative) chronologies of all activities performed (by man or by nature) at a site over time, in order to describe and interpret it. The physical relationships between deposits and interfaces help archaeologists deduce stratigraphic relationships, i.e., the relative chronology of each US. To allow archaeologists to organize all the stratigraphic relationships between the different US, Harris also designed a diagram, the matrix. "The Matrix changed the paradigm of the archeology from the one-dimensional concept embodied in the Wheelerian section drawing to a four-dimensional model that combines the three physical dimensions with that of time, the forth dimension. In this sense, it is like the clock face of twelve hours and the Gregorian calendar of a twelve-month year, which are diagrammatic ways in which time, which does not exist in any material form, can be seen. More than any other science, archeology is a time-related discipline and the Harris

Matrix has given that emphasis a revolutionary force in its ability to translate the physical evidence of stratification into calendar of relative time, unique to each site, but universally comparable through the Matrix diagrams." ([51], p. 10).

As previously stated in this paper, the Harris matrix is an important and not yet out-dated tool, but which suffers from its inability to express the concept of duration, as well as upstanding building stratigraphy.

In agreement with certain scholars [52], especially in Italian Medieval Archeology, who wonder how it applies to the study of upstanding building archeology, we try to work on:

1. Maximum readability and immediacy of the drawings and schedules, avoiding, as much as possible, obscure symbolism and hypertrophy of the documentation
2. Simultaneous presence in the same media of all kinds of information (text, drawing, pictures), in order to really simplify the information sharing. It is clear that the Harris matrix itself is not very easy to understand and lacks meaning, and without photographs and drawings, the US are equally unclear. This rigid separation of the information must be properly addressed
3. The formalism used to store and represent this complex and heterogeneous knowledge cannot be rigid and too structured because archaeological data and the knowledge itself change over time [53].

Despite its lack of the concept of duration, the Harris matrix has been a useful tool for analyzing archaeological stratigraphy up to now.

Since in archeology the representation of space is the same as the representation of time, our team started to work on the modeling of archaeological time after 2009 [25,54] in order to embed this aspect within the system and to eliminate the separation of duration from the temporal sequence of a site. We were also trying to develop a tool able to automatically generate Harris matrices, as well as integrate nonlinear time models [45,52,55,56] into the chronological structure of the Harris matrix itself.

3. Photogrammetric Analysis of the Shawbak Castle

Archaeological excavations are always irreversibly destructive, so it is important to accompany them with detailed documentation reflecting the accumulated knowledge of the excavation site. However, not only the excavation needs an accurate documentation, also the upstanding structures research [3], especially before any restoration work, which often deletes the previous historical information. This documentation is usually iconographic and textual. Graphical representations of archaeological upstanding structures such as drawings, sketches, watercolors, photographs, topography and photogrammetry are indispensable for such documentation and are an intrinsic part of an archaeological survey. However, as pointed out by Olivier Buchsenschutz in the introduction to the symposium Images and archaeological surveys, in Arles, France, in 2007 [57], even a very precise drawing only retains certain observations that support a demonstration, just as a speech retains only some arguments, but this choice is not usually explicit. This somewhat lays the foundation of this work: a survey is both a metrics document and an archaeologist's interpretation.

The Castle of Shawbak (Ma'an region, Jordan), also known as the "Krac de Montréal" (Figure 4), is one of the best preserved rural settlements in the entire Middle East [58]. As the core site of the Florence archaeologists' research in Jordan, it has been chosen to develop a new documentation system. This extensive site (230 m × 80 m) is stratigraphically complex both in depth and in upstanding structures. According to Brogliolo's rules of upstanding structures stratigraphy [3], it is subdivided into 23 CF (Corpo di Fabbrica, which means the same parts of structure, which form a single part of the building). In order to determine the site's chrono-stratigraphic sequences, the Florentine archaeologists started by analyzing the stratigraphy of all CF, which of course, is the first step of the study.

All the examples presented here are only from CF5, a very important archaeological feature of the castle, which summarizes its different temporal phases (starting with the Crusader and Ayyubid periods). It has also been used to produce a full-scale 3D model showcased during the exhibition From Petra to Shawbak. Archeology of a frontier (see Figure 5) [59].

Figure 4. The castle of Shawbak (Ma'an region, Jordan).

Figure 5. The full-scale 3D model of Corpo di Fabbrica 5 (CF5) built for the exhibition From Petra to Shawbak. Archeology of a frontier.

3.1. How Photogrammetry Helps Archaeological Research

The photogrammetric approach is used to solve two kinds of problems: the first one is to build a 3D or 2D representation, or part of the site, in order to have a representation of the site that can be used as an interface for the large amount of textual and iconographic data collected and computed by archaeologists. The second one is the necessity to collect metric data on identified artifacts in order to be able to perform a dimensional analysis, clustering and other statistical computations.

Indeed, two families of objects must be surveyed:

- The area that we need to study, which can be represented by an orthophoto, a digital terrain model (DTM) or, more generally when the studied site is complex and a full 3D approach is required, by a 3D surface, mainly meshes.
- The artifact that we seek to position in space and for which we have good a priori knowledge (in our case, the atomic element is the ashlar block, the smallest measurable element of each USM)

Throughout this work, we deal with these two aspects, artifacts and unstructured surface, by addressing two different approaches; one using a priori knowledge through measurements and the second based solely on geometry. The first approach, based on the a priori knowledge that we have about the measured artifact, uses our knowledge of the object to compute its size and position in space. This method can also reduce the time required for measurements. The second approach, used to survey land for example, but also the main structure of the castle, uses automatic tools coming from photogrammetry to compute a dense cloud of 3D points.

Finally, a very important point is the link between geometry and knowledge; a model, 3D or 2D, representing a site is a relevant interface to access the data known about the site. 3D site representations provide important added value to archaeologists who are then able to study a three-dimensional overall picture. Moreover, it should be noted that, by the nature of archaeological research, archaeological data are incomplete, heterogeneous, discontinuous and subject to possible updates and revisions to each field season campaign. The documentation system, linked to archaeological data, must be able to manage these constraints. The archaeological knowledge in this work is represented by ontologies and we have developed some 3D viewer software able to manage links with ontologies, as well as display graphical requests on these ontologies. This will be discussed in the next section.

3.2. Dense Cloud of 3D Points and Meshes

Since 2002, with one three-month campaign per year, more than 90,000 photographs have been taken of the Shawbak site, following the continuous progress of digital camera technology, as well as photogrammetry software.

These campaigns have produced photographic sets acquired with calibrated digital cameras both in convergent and parallel coverage with the survey of control points using a total station and DGPS. These control points are used to reference the photogrammetric models in a common geodetic system. We are still processing these photographs, using the set of control points, in order to obtain a complete 3D model with dense point clouds of the entire site.

This first 3D elaboration produces a huge quantity of 3D points and meshes with the help of several photogrammetric software programs mainly Photomodeler from EOS and Photoscan from Agisoft. The full collection of images is split into several sets of 3000–9000 photographs, which are oriented independently and then computed in the global reference system with the help of control points.

In 2012, a helium balloon was used to survey the site, which resulted in a small-scale model. Oriented with DGPS control points, this model of course is not suitable for measuring artifacts, but allows us to position all the studied buildings in a common 3D model. Figure 6 shows the whole model and some large-scale images of the buildings surveyed.

Figure 6. 3D models of the entire castle.

3.3. Measuring Artifacts and Querying the Model: Spatial Considerations

Specific photogrammetry tools dedicated to measuring stones individually have been under development since 2002 to help archaeologists to easily produce photogrammetric surveys. These tools are now integrated in a more complex system, which allows automatic production of 2D or 3D representations from archaeological database queries. The graphic 2D documents produced through this process look like the handmade drawings done by archaeologists using ortho-rectified photos.

Once all the photographs are oriented, the I-MAGE process (Image processing and Measure Assisted by GEometrical primitive), developed in 2001, is used to support the user during the measuring process in photogrammetric surveys. Users can make a 3D measurement using one single photograph, without altering the precision of the result. Previously published at a CIPA Heritage Documentation symposium [60], (CIPA was founded in 1968 as the *Comité International de la Photogrammétrie Architecturale*, in English: "International Committee of Architectural Photogrammetry"), this method allows the user to concentrate on the archaeological aspects of the survey and pay less attention to the photogrammetric ones.

We use this approach also to produce 3D models of building blocks (i.e., ashlars) based only on the visible sides. The morphology of each ashlar block is expressed as a polyhedron with two parallel sides, or faces. In most of the cases, only one side is visible, sometimes two, rarely three. The survey process can inform about the dimensions of one face, then the entire polyhedron is computed according to the architectural entity's morphology (extrude vector) and the data (depth, shape, etc.) provided by the archaeologist.

Computing an extrusion vector can be easy when the architectural entity's morphology is obvious. During a wall survey, for example, an extrusion vector can be computed by a least squares adjustment. This is the plane used by I-MAGE. In this case where the entity's geometrical properties are simple, the extrusion vector is calculated before the survey phase, and the block is extruded directly from the measured points (see Figures 7 and 8).

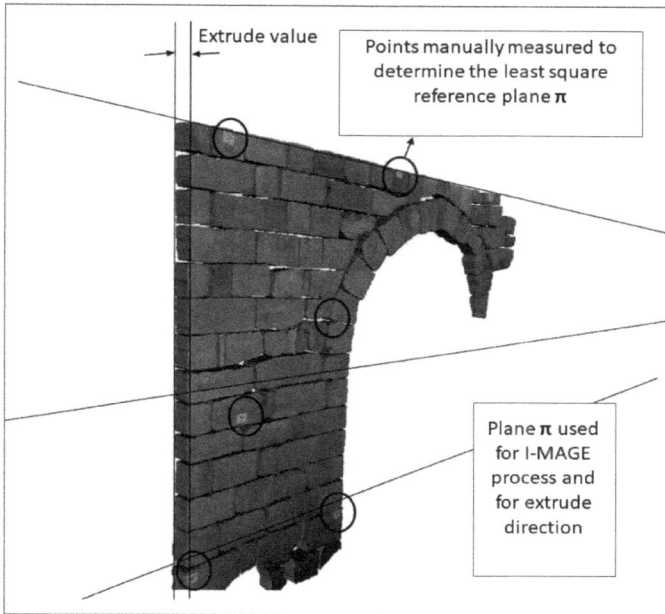

Extrude value

Points manually measured to determine the least square reference plane π

Plane π used for I-MAGE process and for extrude direction

Figure 7. Ashlar blocks using a plane as an approximation of the exterior face of the wall [53].

In the case of the survey of an arch, the extrusion should be radial, needs the geometrical features of the entity (intrados, radius, axis) and is therefore processed afterwards.

This approach for measuring blocks was already published in an ISPRS (International Society for Photogrammetry and Remote Sensing) congress [61] and has been combined with the I-MAGE process in order to obtain an integrated tool.

The main problem is determining whether a block is an edge block when observing photogrammetric data. To do this, archaeologists need a document of the highest quality.

We have tried developing an automatic edge detection system for ashlar blocks, but indeed, the problem is more related to expert knowledge. Since 2016, we have developed a new version of the I-MAGE tool using both an oriented image and the dense point clouds. The user can describe the block geometry observing a good quality photograph, and the 3D block geometry is computed using the dense cloud of 3D points. The extrude value is still given as an hypothesis by the archaeologist.

Since 2014, we stopped using a database to store ashlar block measurements and are currently only using an ontology describing both the photogrammetric process and the archaeological knowledge related to the blocks and stratigraphy. These aspects are detailed in the next section.

Figure 8. An example of the stone-by-stone survey using an extrusion vector from CF34 (the Ayyubid Palace in Shawbak).

The survey process produces a set of measured ashlar blocks linked with both the 3D model and the archaeological knowledge (stratigraphy, lithotype, stone tool analysis, etc.).

3.4. The Use of Stratigraphic Units in Archeology

Starting from the most important concept of the stratigraphic archeology, the US, we try to document directly in the survey the main characteristics of each one. Following the publication of the first edition of E. Harris's book Principles of archaeological stratigraphy (1979) [4], many archaeologists follow the idea that all archaeological sites, to a greater or lesser degree, are stratified, and for this reason, it is necessary to know the main principles of the archaeological stratigraphy to obtain all the possible information, as we could see before.

As we can see, the forms of US are the result of either natural (deposition or erosion) or human (construction or destruction) actions. Their position on the entire wall, as well as the physical and stratigraphical relationships with the other US are necessary to detect the relative chronological sequence of the entire building (and, when comparing all structures, the relative chronological sequence of the entire site). The characteristics that distinguish each US of the upstanding building are mainly the stones' lithology, the ashlar blocks' dimension and shape and the kind and the quantity (or also the absence) of the mortar. Of course, it is extremely relevant to note the physical relationships between the different US (see Figure 9). Therefore, we need to survey all the dimensional and technological data and also, of course, to survey the position of all the US.

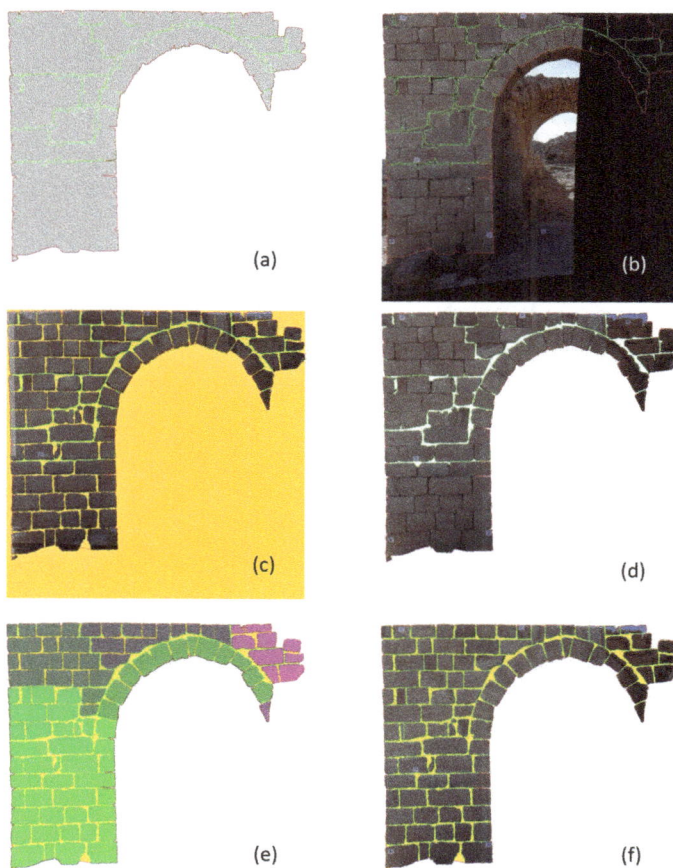

Figure 9. Several orthophoto/map generation images based on several queries: (**a**) the design of USM perimeters; (**b**) same as (a) with a reprojection of one of the oriented images; (**c**) Image (a) with a reprojection of the texture inside the block perimeter; (**d**) texture reprojection inside of the US perimeter (computed with blocks); (**e**) Image (a) query with a uniform color inside each block according to their US and an automatic extraction of the cement (computed as the Boolean difference between the global perimeter of all US and all the individual blocks); (**f**) the same as (e), but with texture mapping for each block.

4. Photogrammetry, Knowledge and Time

The 3D GIS, merging photogrammetry and ontologies, is the last step of this chain and aims at the automatic production of 3D (or 2D) models through ontological queries: these 3D models are in fact at the same time a graphic image of the archaeological knowledge and the current interface through which the user can edit the dataset.

This approach enables automatic 3D thematic representation and new archaeological analysis, through bidirectional links between 3D representation and archaeological data. In this section, we will see in Section 4.1 the ontology development, merging photogrammetry and archeology knowledge, then in Section 4.2, we present briefly certain tools used to easily manage these ontologies. Section 4.3 will show the link with stratigraphy and the use of time relations, and Section 4.4 proposes a research direction to represent duration in stratigraphy and to solve time constraints in an archaeological context.

4.1. Ontology for Photogrammetry Process

The ontology developed in the framework of our research takes into account the manufactured items surveyed, as well as the method used to measure them, in this case, photogrammetry. The surveyed item is therefore represented from the measurement point of view and has access to all the photogrammetric data that contributed to its measurement in space. Two ontologies are aligned in this context; one dedicated to photogrammetric measurement and the geo-localization of the measured items, whereas the other is dedicated to the measured items, principally the archaeological artifacts, describing their dimensional properties, ratios between the main dimensions and default values. These ontologies are developed closely linked to the Java class data structure that manages the photogrammetric process, as well as the measured items. Each concept or relationship in the ontology has a counterpart in Java (the opposite is not necessarily true). Moreover, surveyed items are also archaeological items studied and possibly managed by archaeologists or conservators in a museum. It is therefore important to be able to connect the knowledge acquired when measuring the item with the ontology designed to manage the associated archaeological knowledge. CIDOC CRM is a generic ontology that does not support the items that it represents from a photogrammetric point of view; a simple mapping would not be sufficient, and an extension with new concepts and new relationships would be necessary. This modeling work is based on a previous study that started from the premise that collections of measured items are marred by a lack of precision concerning their measurement, assumptions about their reconstruction, their age and origin. It was therefore important to ensure the coherence of the measured items and potentially propose a possible revision. This previous work was done in the context of underwater archeology with similar problems. For more information, see [21–24].

The extension of the CIDOC-CRM ontology is structured around the concept of *Man-Made Object*. The root of *ItemMesurable* developed in this project extends this concept (see Figure 10). The mapping operation is done in Java by interpreting a set of data held by the Java classes as a current identification of the object: 3D bounding box, specific dimension. These attributes are then computed in order to express the correct CRM properties.

Several methodologies can be chosen regarding mapping two ontologies. For example, Nicola Amico and his team [13] chose to model the survey location with the concept of *Activity* in CRM. They also developed a formalism for the digital survey tool mapping the digital camera definition with *Digital Machine Event*. We see here that the mapping problem is close to an alignment problem, which is really problematic in this case. Aligning two ontologies dealing with digital camera definition is not obvious; a simple observation of the lack of interoperability between photogrammetric software shows the wildness of the problem. We are currently working on an alignment/extension process with Sensor ML which is an ontology dedicated to sensors. Although some work has already been achieved [62], it is not enough to clearly hold the close range photogrammetry process, from image measurement to artifact representation.

In addition, we use the concepts of time representation present in CIDOC CRM in order to represent the Harris relations and then Allen's formalism about the time intervals.

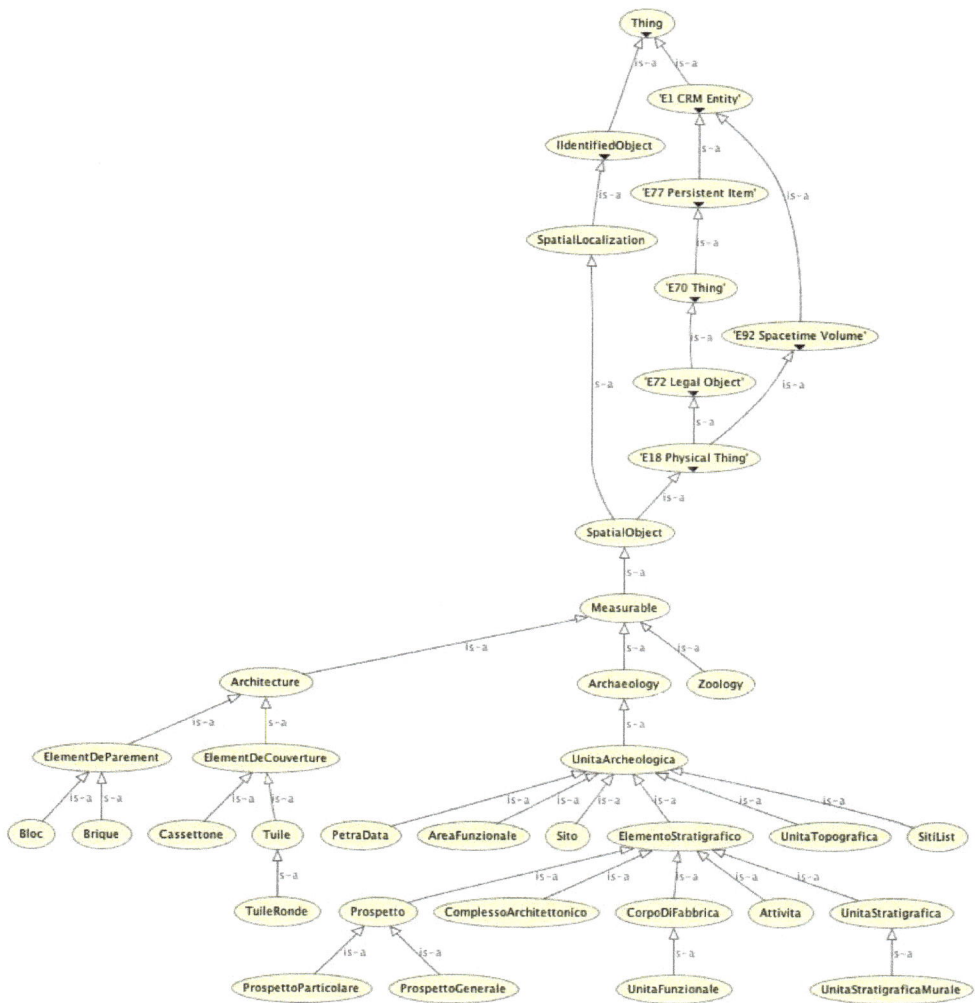

Figure 10. Partial ontology visualization with CIDOC CRM alignment.

4.2. Tools for Managing Ontologies

The use of ontologies in archeology and cultural heritage, as we have already stated, is becoming increasingly widespread. Indeed, this formalism is particularly suitable for heterogeneous data and offers concepts and tools to manage incompleteness, updates and revisions of the involved knowledge. Therefore, the way ontologies are structured is far from a traditional relational database, and managing them can be difficult for a person without a solid background in computer science. In addition, even if many research papers are published in the field of ontology for cultural heritage, the use of ontologies is not yet a common or widespread tool for managing archaeological excavations. Much research shows that the ontologies are very suitable for archeology, but only a very few archaeological missions use ontologies to manage data coming from excavations.

This is normal: even if we can demonstrate that a new approach is very interesting, changing technology is always a high cost decision, and ensuring the continuity of the data, analysis and student training is a real challenge.

We decided to develop some tools in order to display data stored in ontologies as a virtual relational database in order to have a simplified view of the stored data. These tools can provide a static point of view on these ontologies and allow simple manipulations similar to those possible on a relational database.

The first one is an editor that displays classes, instances and properties present in an ABox (a set of assertional axioms in an ontology) as if the data were structured by a table in a relational database (see Figure 11). By reading the ABox, the editor is able to display the classes present in the ontology, all their instances and offers a simple and dynamic way to display data properties as normal fields of these instances and the possibility to modify them (also, as we are in the ontology context, to add or remove fields). This of course gives us many other possibilities as the archaeological work is always in progress, and a modification of the data structure is always possible and easy to implement through this interface.

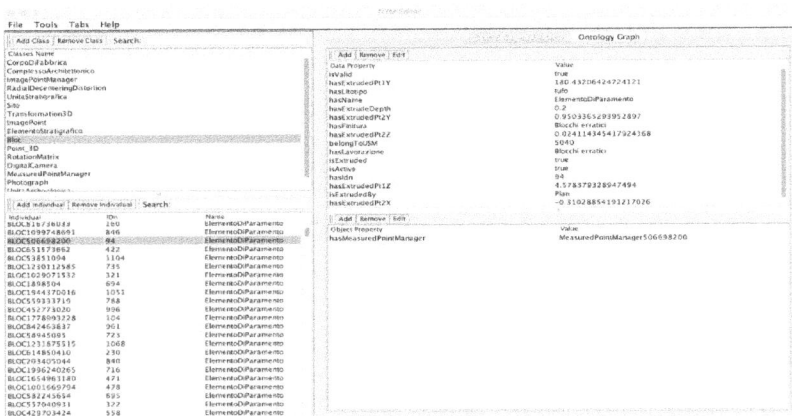

Figure 11. The ontology editor: classes, instances and properties edition from the OWL file.

In addition, the editor is able to display graphs showing the relations between instances classes and properties. A full bidirectional access is provided and allows one to modify the value of the properties accessed by clicking on the graph.

The same approach allows us to build an editor dedicated to physical and stratigraphic relations between US.

This tool is dedicated to stratigraphic relations between US, reading an ABox and displaying an exhaustive list of the US present in the ABox with all the physical and stratigraphic relations between them. This tool allows one to add, cancel, modify the relations between US and visualize them through a graph. We are still working on a consistency check tool to control on the fly the digit (see Figure 12).

Finally, we developed a tool for 3D visualization of the artifacts stored in the ABox. Indeed, currently, ontologies are used also as a serialization tool to store all the data relative to the survey: oriented photographs, 2D and 3D points, relevant points for stone-by-stone surveys, computed ashlar blocks and archaeological concepts often non-measured directly as stratigraphic units, *Corpo di Fabbrica*, etc. The tool is able to read the ABox and, thanks to the procedural attachment (as detailed in the next section), can instantiate the corresponding Java instance and produce a 3D representation of the instances stored in the ABox. This tool was used to produce the two images shown in Figure 13 according to the specific request.

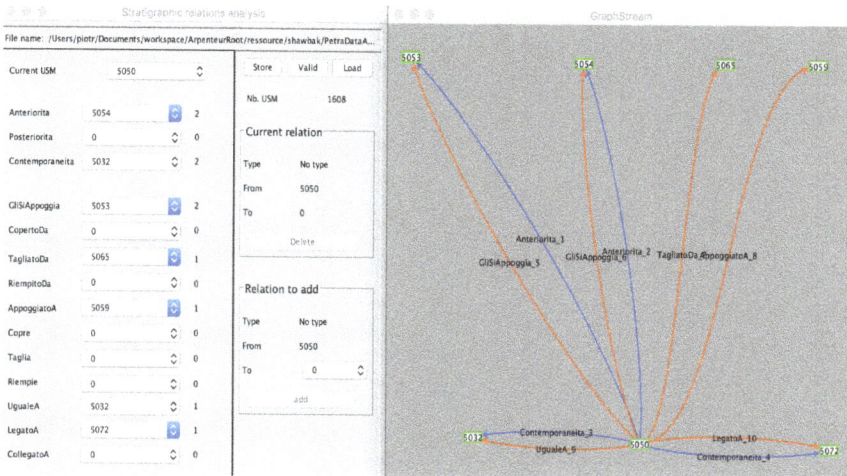

Figure 12. Physical and stratigraphic relations editor with graph visualization of all the relations in which the current USM is involved. A color code is used to separate physical relations from stratigraphic relations.

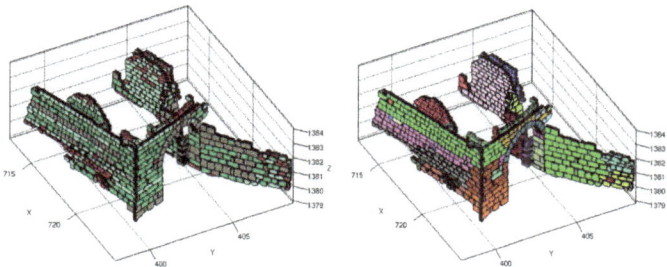

Figure 13. Ashlar block visualization, on the left; the colors show the clusterization result based on the block's height. Right: the colors show the clusterization result based on USM.

4.3. ARPENTEUR Ontology: A Link between Photogrammetry and Stratigraphy

We propose now to work on the link between 3D measurement and temporal relations. Indeed, archaeologists use a set of rules to determine temporal relationships by observing physical relationships. Table 1 describes the physical conditions necessary for establishing the temporal relationships. The temporal relations are deduced from the physical relations, and often, these physical relations can be deduced from the relative position of the US between them using the photogrammetric measurement of their components. As the US are built on measured objects, it will be possible to compute the veracity of the physical relations in order to deduce the temporal relations if they have not been specified by the archaeologist or to ensure a coherence of time relations.

Table 1. Link between physical and stratigraphical relations.

Physical Relations	Stratigraphical Relations
Gli si appoggia (ad A si appoggia B)	A Anteriore B
Coperto da (A coperto da B)	A Anteriore B
Tagliato da (A tagliato da B)	A Anteriore B
Riempito da (A riempito da B)	A Anteriore B
Appoggiato a (A appoggiato a B)	A Posteriore B
Copre (A copre B)	A Posteriore B
Taglia (A taglia B)	A Posteriore B
Riempie (A riempie B)	A Posteriore B
Uguale a (A uguale a B)	A Contemporaneo B
Legato a (A legato a B)	A Contemporaneo B
Collegato a (A collegato a B)	A Contemporaneo B

4.3.1. Ontology and Graphical Representation

This work is based on a close link between, on the one hand, the software engineering aspects and the operative modeling of the photogrammetry process, the needed computation and artifacts measurable by photogrammetry in the context of this project and, on the other hand, the ontological representation of the same photogrammetry process and surveyed artifacts. The present implementation is based on a double formalism: the Java programming language, used for computation, photogrammetric algorithms, 3D visualization of photogrammetric models and cultural heritage objects, then OWL (Web Ontology Language) for the definition of ontologies describing the concepts involved in this photogrammetric process, as well as the surveyed artifacts.

For several years, OWL has been used as a standard for the implementation of ontologies (W3C, 2004). In its simplest form, it allows for representing concepts (class), instances (individual), attributes (data properties) and relations (object properties). The ontology construction in OWL, symmetric to the Java taxonomy, cannot be produced automatically. Each concept of the ontology has been constructed with a concern for the representation of accurate knowledge from a particular point of view: measurement. The same point of view presides over the elaboration of the Java taxonomy, but software engineering constraints involve differences in the two hierarchies of concepts. For each ontology concept, a procedural attachment method has been developed with OWLAPI. As each concept present in the ontology has a homologous class in the Java tree, each individual of the ontology can produce a Java instance and can benefit from its computational capabilities. In the same way, each Java class has a counterpart in the ontology and can produce an individual of that ontology.

We have abandoned automatic mapping using Java annotation and Java beans for a manual extraction even if this is commonly done in the literature [63–65].

Thus, reading an XML file used to serialize a Java instance set representing a statement can immediately (upon reading) populate the ontology; similarly, reading an OWL file can generate a set of Java instance counterparts of the individuals present in the ontology. The link between individuals and instances persists and it can be used dynamically. The huge advantage of this approach is that it is possible to perform logical queries on both the ontology and the Java representation. We can thus read the ontology, visualize in 3D the artifacts present in the ontology and graphically visualize the result of SQWRL queries in the Java viewer.

The approach we have chosen so far, using OWLAPI and the Pellet reasoner, allows for handling SQWRL queries using an extension of SWRL built-in [66] packages. SWRL provides a very powerful extension mechanism that allows for implementing user-defined methods in the rules [67]. The photogrammetric survey is expressed as an ABox. An ontology describing the photogrammetric process, as well as the measured objects was populated by the measurements of each block and a set of corresponding data (USM owner, etc.). It is therefore this ontology that contains the metric information such as the geo-positioning of each block and all the physical and stratigraphic relationships provided by archaeologists.

The reading and the geometrical interpretation of this ontology can lead us to validate or invalidate the physical relations by a topological analysis of the relations between the various components of the USM, but also to check the consistency of temporal relations by extracting a graph of constraints and then using a logical solver.

For example, the 3D visualization shown in Figure 13 is done using the ABox and on-the-fly instantiation of corresponding Java instances. The colorization is the result of an SQWRL request done on the ontology.

Moreover, a first possible interpretation of these data should be the generation of a graph of temporal relations where the nodes, instead of having a coded position as in the Harris graphs, should have as the coordinate of the USM center of mass projected in the USM main plane.

As illustrated in Figures 14 and 15, the graph is superimposed on an unrealistic representation of the stone-to-stone reading of the building.

4.3.2. Positive and Negative US

As we said before, the US can be positive (i.e., layers), negative (i.e., cuts) or neutral (i.e., interfaces). In this regards, until now, we have been working mostly on the survey and representation of positive USM, although since 2009–2010, the research for modeling negative and interface units was also started (see Figure 16).

The latter ones are in fact 3D surfaces (not volumes) possessing stratigraphic relationships with other units. They are not formed by ashlars and mortar, since they are only signs of, say, a destruction, like with the traces left by an earthquake or by human (partial to total) destruction of a material feature. Figure 16 shows a negative USM represented by a cloud of 3D points extracted using the perimeters drawn on the photographs by an archaeologist.

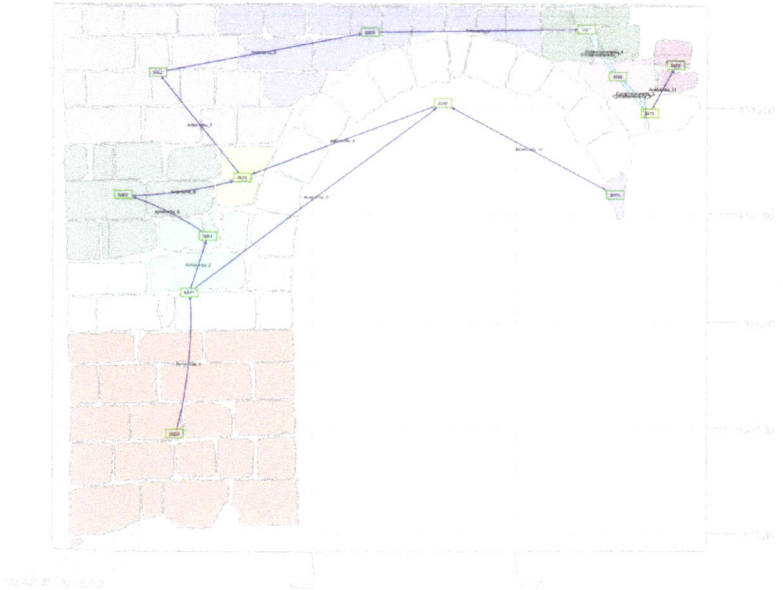

Figure 14. Positive stratigraphic relations with the USM node coordinate coming from photogrammetric measurement.

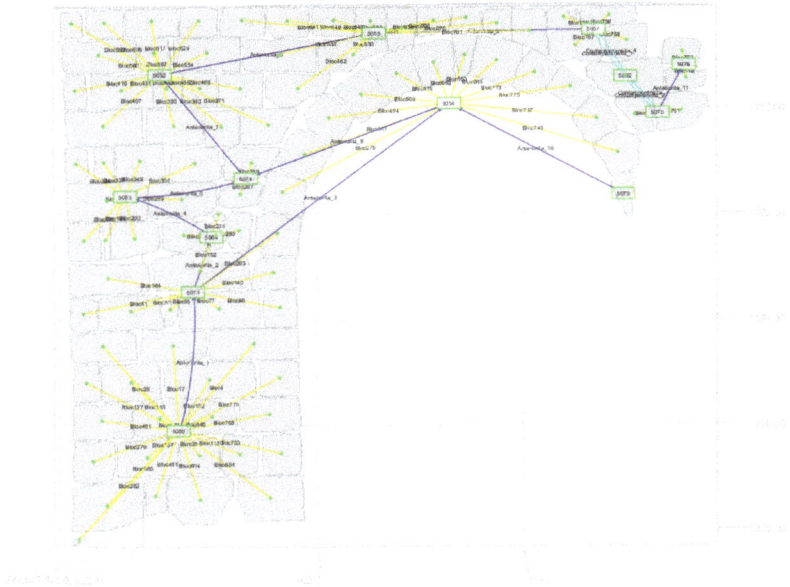

Figure 15. 2D bloc representation with the stratigraphic graph and bloc position. The nodes of the graphs are computed as centroids of US or block.

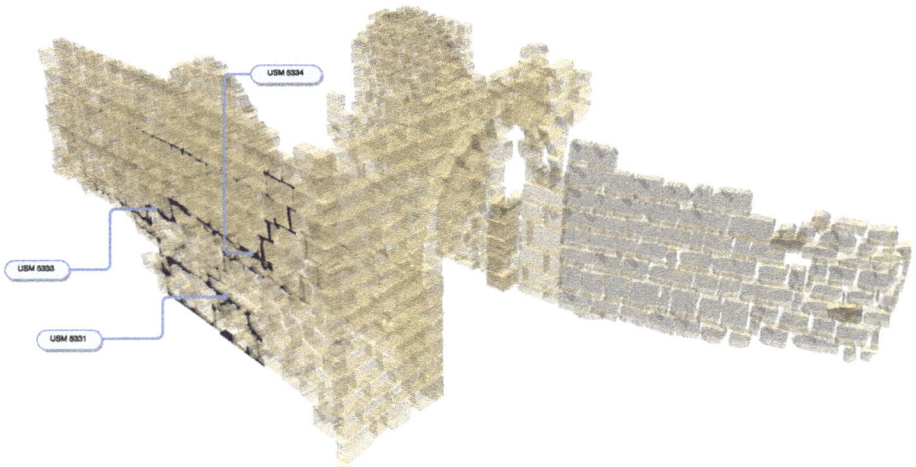

Figure 16. Negative USM represented by the set of 3D points extracted using the perimeters designed on the photographs by an archaeologist.

4.4. Toward Duration for Representing the Stratigraphic Unit

Once the problem of surveying and representing stratigraphic units is solved through ontology and then graphically, we have to face some limits of the Harris paradigm. As we saw before, one of the main improvements we are working on is considering US as intervals of time and no longer as instants.

Indeed, inferring all the Allen's relations directly from the physical relations between US is not trivial, and currently, the only way to do that is by using the expertise of archaeologists.

Nevertheless, we start with a few deduced relationships from Harris's formalism to that of Allen. Considering Stratigraphic Unit A, it should have a start point named A_{min} and an end point named A_{max}.

For example, the Harris relation A is Contemporaneo to B can be translated by A starts B in Allen's relations and the Harris relation A Anteriorita B can be translated by A precedes B or A meets B using the bound of intervals as expressed in Table 2 (For a typological simplification, here, A and B designate USM and not instants or intervals as should be done to respect the various formalisms of Harris and Allen. The determination of the time interval associated with a USM is a difficult choice, and this aspect is discussed later in this paper.).

In order to represent the stratigraphic unit duration using time intervals, we use Qualitative Constraints Networks (QCN), which have been successfully used in the domain of knowledge representation and reasoning in artificial intelligence for representing temporal relations.

Table 2. Harris relations seen from an interval point of view.

Stratigraphical Relations	Start and End of US
A Contemporaneo B	$A_{min} = B_{min}$
A Anteriore B	A_{max} anterior B_{min}
A Anteriore B	$A_{max} = B_{min}$
A Posteriore B	B_{max} anterior A_{min}
A Posteriore B	$B_{max} = A_{min}$

Allen's interval algebra has been used for representing and reasoning with archaeological information in the context of archaeological documentation [68] and the dating process [69]; however, as far as we know, this formalism has not been used yet for stratigraphy in archeology.

Definition 1. *Let B be a set of basic relations; a QCN N is a tuple (V, C) where $V = \{v_1, \cdots, v_n\}$ is a set of n variables representing temporal entities; C is a mapping that assigns to each tuple $(v_{i_1}, \cdots, v_{i_j})$ of variables of V a relation $C(v_{i_1}, \cdots, v_{i_j}) \in 2^B$.*

The stratigraphic relations between US are temporal binary relations, and several QCN could be proposed according the choice of the temporal entities and the set of basic relations.

4.4.1. Time Points Algebra

Considering temporal entities as time points, archaeologists can provide temporal relations between the centers of mass of the US, deduced from stratigraphic relations between US. More formally, the domain D is defined by the set of rational numbers (line points) $D = Q$ equipped with the linear order $<$, and the time points algebra stems from three atomic relations $B = \{precedes, follows, same\}$ (see Figure 17). These relations are defined as follows:
$precedes = \{(x, y) \in Q \times Q : x < y\}$,
$follows = \{(x, y) \in Q \times Q : y < x\}$,
$same = \{(x, y) \in Q \times Q : x = y\}$.

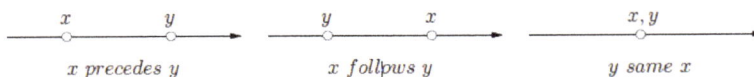

Figure 17. Time points algebra atomic relations

The set of time points algebra relations is denoted by 2^B, where each relation is a disjunction of atomic relations. The set 2^B is equipped with the operations: union ($\cup$), intersection ($\cap$), composition ($\circ$) and inverse ($^-$). The following example illustrates the notion of QCN when the temporal entities are time points.

Example 1. *Let A, B, C, D, E, F be US; the stratigraphic relations between these six US according to the Harris matrix approach are illustrated in Figure 18.*

Let a, b, c, d, e, f be the centers of mass of these six US; considering temporal entities as time points, the QCN denoted by (V, C) is such that the variables are time points, which are the centers of mass of the US, and the constraints are the temporal relations between US. More formally, the set of variables is $V = \{v_1, v_2, v_3, v_4, v_5, v_6\}$, where $v_1 = a$, $v_2 = b$, $v_3 = c$, $v_4 = d$, $v_5 = e$, $v_6 = f$, and the set of constraints is $C = \{C_{12}, C_{23}, C_{26}, C_{34}, C_{45}\}$ (for the sake of readability, the inverse relations and the transitive relations are omitted), where:

The constraint $C_{12} = \{precedes\}$ states that v_1 precedes v_2;
The constraint $C_{23} = \{same\}$ states that v_2 is the same as v_3;
The constraint $C_{26} = \{precedes\}$ states that v_2 precedes v_6;
The constraint $C_{34} = \{precedes\}$ states that v_3 precedes v_4;
The constraint $C_{45} = \{precedes\}$ states that v_4 precedes v_5.

The QCN is represented in Figure 19.

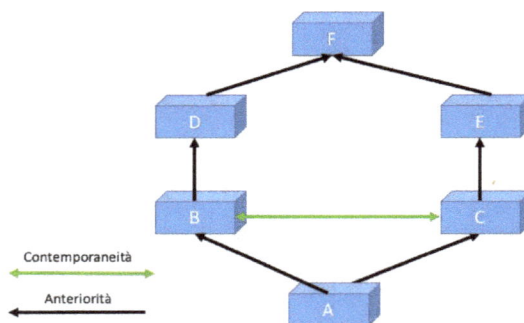

Figure 18. Example of relations between US.

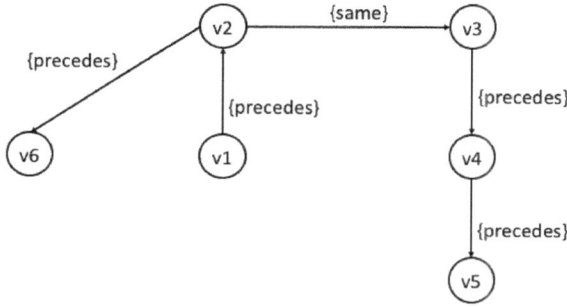

Figure 19. Qualitative Constraints Networks (QCN) when temporal entities are time points.

4.4.2. Intervals Algebra

However, the stratigraphic relations between *US* often involve the notion of duration, which cannot be represented using the Harris matrix approach.

In this case, intervals are more suitable temporal entities. From stratigraphic relations between US, the archaeologists can assign intervals to US, and we can consider relations between intervals using Allen's interval algebra [56]. This algebra stems from 13 atomic temporal relations, $B = \{p, m, o, s, d, f, pi, mi, oi, si, di, fi, eq\}$. More formally, the domain D is defined by the set of intervals of the rationals line $D = \{x = (x^-, x^+) \in Q \times Q : x^- < x^+\}$. Each atomic relation is defined by constraints on the bounds of intervals. For example, the atomic relation s (for starts) is defined by $s = \{(x, y) \in D \times D : x^- = y^-$ and $y^+ > x^+\}$; the atomic relation m (for meets) is defined by $m = \{(x, y) \in D \times D : x^+ = y^-\}$.

The set of Allen's intervals algebra relations is denoted by 2^B, where each relation is a disjunction of basic relations. The set 2^B is equipped with the operations: union ($\cup$), intersection ($\cap$), composition ($\circ$) and inverse ($-$). The following example illustrates the notion of QCN when the temporal entities are intervals.

Example 2. *We come back to Example 1; however, we now consider temporal entities as intervals. The new QCN denoted by (V, C) is such that the variables are intervals consisting of the lower and upper bounds of duration of the US, and the constraints are the temporal relations between these intervals. More formally, the set of variables is $V = \{v_1, v_2, v_3, v_4, v_5, v_6\}$, where $v_1 = (a^-, a^+)$, $v_2 = (b^-, b^+)$, $v_3 = (c^-, c^+)$, $v_4 = (d^-, d^+)$, $v_5 = (e^-, e^+)$, $v_6 = (f^-, f^+)$. The set of constraints is $C = \{C_{12}, C_{23}, C_{26}, C_{34}, C_{45}\}$ (for the sake of readability, the inverse relations and the transitive relations are omitted) where:*
The constraint $C_{12} = \{b, m\}$ states that v_1 precedes v_2 or v_1 meets v_2 since from Figure 18 and Table 2, one can deduce that $a^+ < b^-$ or $a^+ = b^-$;
The constraint $C_{23} = \{f\}$ states that v_2 finishes v_3 since from Figure 18 and Table 2, one can deduce that $b^- = c^-$;
The constraint $C_{26} = \{b, m\}$ states that v_2 precedes v_6 or v_2 meets v_6 since from Figure 18 and Table 2, one can deduce that $b^+ < f^-$ or $b^+ = f^-$;
The constraint $C_{34} = \{b, m\}$ states that v_3 precedes v_4 or v_3 meets v_4 since from Figure 18 and Table 2, one can deduce that $c^+ < d^-$ or $c^+ = d^-$;
The constraint $C_{45} = \{b, m\}$ states that v_4 precedes v_5 or v_4 meets v_5 since from Figure 18 and Table 2, one can deduce that $d^+ < e^-$ or $d^+ = e^-$.
The QCN is represented in Figure 20.

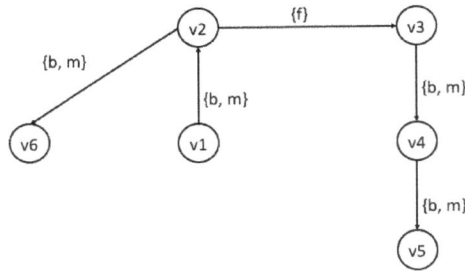

Figure 20. QCN when temporal entities are intervals.

Note that information on the bounds of the intervals may be incomplete, as illustrated in Example 2, the constraint C_{23} only follows from the lower bounds of the intervals v_2 and v_3 assigned to US. Table 2 does not provide any relations between upper bounds, and we need more information than what is provided by Figure 18. Archaeologists should be equipped with suitable tools allowing them to complete information on the bounds in the best possible way.

Moreover, an important issue is the QCN consistency checking problem. Indeed, a QCN may be inconsistent for several reasons. Errors can be made on the interpretation of stratigraphic relations; discontinuities involved with the 3D complexity of studied structures may exist; the survey may be incomplete (for instance, lack of photographs, etc.). Efficient tools have to be developed in order to check the consistency and to pinpoint the inconsistencies in each step of the process from the formal temporal representation to the photogrammetric 3D survey.

Concerning the QCN, three main approaches have been proposed for consistency checking in the artificial intelligence community, stemming from discrete CSP (Constraint Satisfaction Problem) [70], from SAT (Boolean Satisfiability Problem) [71] and, more recently, from Answer Set Programming (ASP) [72].

4.5. From Harris to Allen

The choice we made is to express duration in a US model trying to use an Allen interval to model the current knowledge expressed with Harris relations. Passing from three time-relations to thirteen interval relations is not trivial and of course not bijective. More than one solution is possible.

According to the current stratigraphic problem in upstanding building structures, we can propose the two sets of solutions in Table 3 for passing from Harris to Allen and in Table 4 for the reciprocal relations. Of course, these proposals reflect the archaeological problem studied, and as we can see, there are some temporal relationship that we were not able to better refine than using subsequent and anteriority (without any other specification proposed by Allen temporal relationships) (see Table 4, Rows 2, 3 and 4).

To study temporal relations between US belonging to a building that we can observe, measure and survey, we need to define a temporal range in which this analysis will be relevant. The first point of view we can have, the most trivial, is to consider that all the observed US are still 'alive' because we are able to observe and measure them. In this case, most of the Harris relations jump into a unique Allen interval: "X finishes y" (see Figure 3); and the end of the temporal interval is the current time of observation. This point of view was developed by Lucas as stressed in the Introduction of this paper ([45], p. 48).

This solution, even if it sounds correct, does not fit in with the current state of our work (we need more time to study and discuss these theoretical models); at this point of our research, we prefer to limit the duration life of each US to the end of its formation.

This point of view is the one used by Harris considering the US relations; we simply added the concept of duration preserving the main goal of Harris's model: to study the constructive relations between US in order to understand the evolution of the building over time.

We propose two tables of correspondence (Tables 3 and 4), which is the starting point of a possible translation of the temporal relations by Harris in intervals expressing the duration proposed by Allen.

The "translation" of the temporal relations proposed by Harris in the thirteen interval relations proposed by Allen is too ambiguous to be formalized by a function passing from three states to 13 states. In fact, to obtain these three temporal relationships proposed by Harris, archaeologists rely on the physical relationship between the US (defined in Table 5).

We therefore propose a conversion using the physical relations to Allen relations rather than a direct conversion of Harris relations to Allen relations.

Such a conversion table is shown in Table 3. Thus, according to the physical relations that were conditioned the choice of the Harris relation, different Allen relations can be proposed.

Of course, the elaboration of this translation between Harris's proposition and Allen's formalism is based on a clear and univocal definition of these relations (see Figure 21).

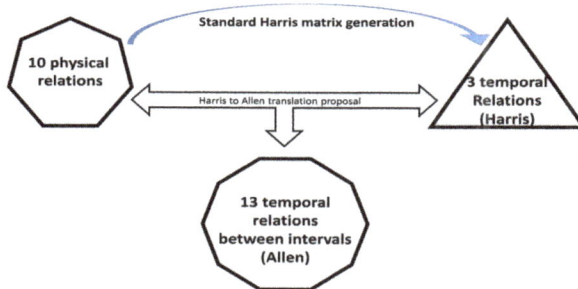

Figure 21. Harris to Allen proposal schema.

These concepts have been used for years by the archaeologists of the Florence team, and their application to the particular case of standing buildings has made it possible to define them in detail. In order not to deviate from the sense in which they are used in this work and in future work, we have left their Italian name in the text (in Table 3, Column 2 and 3; in Table 4, Column 2).

Table 3. From Harris to Allen.

n	(Harris) Physical Relations	(Harris) Stratigraphical Relation	Allen (More Than One Relation Is Possible)
1	It leans on (y leans on x) Appoggiato a (y appoggiato a x)	y is later than x y Posteriorità x	y preceded by x y met by x
2	leaned on by (x is leaned on by y) Gli si appoggia (ad x si appoggia y)	x is earlier than y x Anteriorita y	x precedes y x meets y
3	It connects (x is connected with y) CollegatoA (ad x si CollegatoA y)	x is contemporary to y x Contemporaneo y	x equals y x starts y x during y x finishes y
4	Linked to (x is linked to y) Legato a (x è legato a y)	x is contemporary to y x Contemporaneo y	x equals y x starts y x during y x finishes y
5	Covered by (y is covered by x) Coperto da (y è coperto da x)	x is earlier than y x Anteriorita y	y preceded by x y meetby x y overlaped by x

Table 3. *Cont.*

n	(Harris) Physical Relations	(Harris) Stratigraphical Relation	Allen (More Than One Relation Is Possible)
6	It covers (x covers y) Copre (x copre y)	x is later than y x Posteriorità y	x precedes y x meets y x overlaps y
7	It fills (x fills y) Riempie (x riempie y)	x is later than y x Posteriorità y	x preceded by y
8	Filled by (y is filled by x) Riempito da (y è riempito da x)	y is earlier than x y Anteriorita x	y preceded x
9	It cuts (x cuts y) Taglia (x taglia y)	x is later than y x Posteriorità y	x preceded by y
10	Cut by (y is cut by x) Tagliato da (y è tagliato da x)	y is earlier than x y Anteriorita x	y precedes x
11	Equal to (x is equal to y) Uguale a (x è uguale a y)	x is contemporary to y x Contemporaneo y	x equals y

Table 4. From Allen to Harris.

n	Allen	Harris (Physical)	Harris (Temporal)	Example
1	x precedes y y preceded by x x precedes y (p) y precededBy x (pi)	(a) x covers y Copre y is covered by da x Coperto da (b) x fills up y Riempie y is filled up by x Riempito da	(a) x is later than y y is earlier than x (b) x is later than y y is earlier than x	(a) Construction of upper floor, preparation layer of a brick floor, etc. (b) Filling of a wall post-hole; filling of a door, window, arch, etc.
2	x meets y y met by x x meets y (m) y metBy x (mi)	x leans on y Gli si appoggia y is leaned on by x Appoggiato a	x is earlier than y y is later than x	Consecutive building phases. Until now, we are not able to distinguish between meets and overlaps. We use for both subsequent and anteriority.
3	x overlaps y y overlapped by x x overlaps y (o) y overlappedBy x (oi)	x leans on y Gli si appoggia y is leaned on by x Appoggiato a	x is earlier than y y is later than x	Building of wall and angle wall in the same phase; generally speaking, every time, we are in front of USM built in the same phase Until now, we were not able to distinguish between MEETS and OVERLAPS. We use for both subsequent and anteriority.
4	x starts y y started by x x starts y (s) y satrtedBy x (si)	x is linked to y Legato a y is linked to x Legato a	x is contemporary to y y is contemporary to x	Two linked walls (their building is started at the same time, but we cannot know if they are also finished at the same time); wall and arch (both are started at the same moment, but one of them is finished before the other): we cannot recognize this stratigraphically, so we use the contemporary relationship.
5	x during y y contains x x during y (d) y contains x (di)	x cuts y Taglia y is cut by x Tagliato da	x is later than y y is earlier than x	Cut in a wall to realize a post-hole or a window or a door in the same building phase of a wall.
6	x finishes y y finished by y x finishes y (f) y finishedBy x (fi)	x is linked with y Legato a y is linked with x Legato a	x is contemporary to y y is contemporary to x	Two walls linked with "wait joint" ashlars.
7	X equals y x equals y (eq)	X equals y UgualeA	x is contemporary to y	Post-hole equal for typology and for function.

The definition of these concepts given in Table 5 is based on a previous work of the Florentine team when defining the database named PetraData [73].

This is the point of view chosen for the future work that we plan to develop on the use of duration for US.

Table 5. Harris physical and temporal relation definition.

n	Harris Relation	Definition
1	It is leaned against (y is leaned against x) Appoggiato a (y appoggiato a x)	One US A is 'AppoggiatoA' on a US B if: A is in contact with B; A is 'Posteriorità' to B; There is a horizontal plane that touches both A and B. This connection is mainly used to define the physical relationship between walls or structures (between US and USM).
2	It leans against (x leans against y) Gli si appoggia (ad x si appoggia y)	A US A 'GliSiAppoggia' a US B if: A is in contact with B; A is 'anteriorità' B; there is a horizontal plane that touches both A and B. This connection is mainly used to define the physical relationship between walls or between layers and structures (between US and USM)
3	It connects (x is connected with y) CollegatoA (ad x si CollegatoA y)	A US A is 'CollegatoA' to a US B if: A can be in contact with B or there is a third US in contact with both A and B; A is 'contemporaneita' with B; this connection describes a link, for example, in the case of two wall edges crossed by a passing bridge pit. Used only for USM. This is a ternary relationship where the US responsible for the link does not exist, or no longer exists, or is not observable.
4	Linked to (x is linked to y) Legato a (x è legato a y)	One US A is 'LegatoA' to US B if: A is in contact with B; A is 'contemporaneita' with B; the convex volumes of A and B intersect. Often designed to merge A and B. Only used for USM.
5	Covered by (y is covered by x) Coperto da (y è coperto da x)	A US A is 'CopertoDa' by a US B if: A is in contact with B; A is 'anteriorita' to B; there is a vertical plane passing through both A and B. This connection is mainly used to define the physical relationship between layers and/or structures (between US and USM).
6	It covers (x covers y) Copre (x copre y)	A US A 'Copre' a US B if: A is in contact with B; A is 'posteriorita' to B; there is a veritable plane passing through both A and B. This connection is mainly used to define the physical relationship between layers and/or structures (between US and USM).
7	It fills (x fills y) Riempie (x riempie y)	A US A 'Riempie' US B if: A is in contact with B; B has a solid concavity, even partially from A. B is generally, but not always, a negative US.
8	Filled by (y is filled by x) Riempito da (y è riempito da x)	A US A is 'RiempitoDa' a US B if: A is in contact with B; A is partially or in totality concave and even partially filled by B. A is generally, but not always, a negative US.
9	It cuts (x cuts y) Taglia (x taglia y)	A US 'Taglia' a US B if: A and in contact with B; it physically subtracts material from B, and A is a negative US.
10	Cut by (y is cut by x) Tagliato da (y è tagliato da x)	A US A is 'TagliatoDa' US B if: A is in contact with B; A was physically subtracted from material B, and B is a negative US.
11	Equal to (x is equal to y) Uguale a (x è uguale a y)	A US A is 'UgualeA' to US B if: A is not necessarily in contact with B; A is consecutive to B; A and B are physically the same US, but there may be a physical discontinuity between them.
12	x is earlier than y x Anteriorita y	An USM A is 'Anteriorita' to USM B if A is formed before USM B
13	x is contemporary to y x Contemporaneo y	A USM A is in 'Contemporaneita' with B if A and B have been formed at the same time.
14	x is later than y x Posteriorità y	An USM A is in 'Posteriorita' compared to B if A was formed when B was already formed. Posteriority

Even if only considered an 'intellectual challenge', we are studying the possibility to control from the beginning the periodization of the archaeological stratigraphy. In fact, after creating the matrix, the archaeologists' second step is to build a coherent periodization that considers all the known elements. We are currently testing this system to ensure better control of the entire pipeline.

5. Conclusions and Future Work

This paper focuses on the main goals obtained by the collaboration between medieval archaeologists from the University of Florence (Italy) and ICT researchers from CNRS LIS laboratory of Marseille (France). From the beginning, this fifteenth collaboration is built on a mutual exchange of knowledge. Our first common field of interest is survey, and our main purpose is to add knowledge to it. The future of this collaboration can be expressed in this way: first of all, the changing paradigm and producing new tools mean deep changes in the way we work. We change both how photogrammetry is used and how the observed data are managed; as new photogrammetric tools are developed and traditional relational databases are replaced by ontologies, new tools become available to manage and infer data.

Beyond the scientific problems, an important obstacle addressed by this work is that a major paradigm shift will significantly impact the on-going work. The changing paradigm means also changing vocabulary, managing new concepts, changing well-known tools for unknown and not completely debugged tools.

The main problem in fact of such an interdisciplinary project is to both develop new concepts and new tools while continuing to produce clear, relevant and verified data. On the one hand, the previous and already existing management system is still used to complete existing data coming from previous campaigns; and on the other hand, we need to be able to feed the new system with enough data in order to test it. Of course when, as is the case here, the two systems are very different, the data from the actual system currently in use will have to be 'ported' to the new one. Finally, the decision has to be made to change systems, which always has a heavy cost at the beginning.

In the framework of this collaboration, archaeologists have learned to create 3D models using photogrammetry since 2001 when this method was not yet very widespread [2,74].

We have developed a tool for digitizing ashlar blocks in 3D using only one single image so that archaeologists can remain concentrated on their main task: adding semantics to the 3D model.

Due to the fact that today, photogrammetry can easily produce a dense cloud of 3D points, we are currently adapting this tool to use both a set of oriented photographs and a dense cloud of 3D points to digitize ashlar blocks and USM. The underlying ontological model is used here to immediately produce ABox instances and feed the ontological survey with all relevant data (stone tool used, lithotype, material and other archaeological and non-graphic observations).

The first step was to obtain a 3D GIS, and we worked very hard to make it possible [25]. Now, our group is starting to work on the field of ontologies, especially to manage the US and to add knowledge to the survey [74]. Even if there are some theoretical works on the ontologies, until now, not many archaeologists have studied and tested ontologies to manage heterogeneous data. To do this, it will be possible to replace the database, which is a very important tool, but often too rigid. The intensive use of ontologies to manage both photogrammetric geometry and archaeological data will allow us to first check the consistency of digitized data and then infer new relationships between US using an inference engine, such as Pellet, directly from the ontology. For example, we are currently working on producing Harris relations using known physical relationships between US, and then, we will be able to check and produce new temporal relationships between US according to Allen's formalism. We are now working on developing a complete pipeline based on ontologies to manage both the photogrammetric process and archaeological knowledge. This pipeline will allow a common formalism for all the heterogeneous data involved in the archaeological analysis and will allow specific queries on these data with several kinds of output (textual, 3D, 2D, Harris-like matrix). In doing this, we continue to follow our method of collaboration, which is truly interdisciplinary, building together our new knowledge base, useful both for archaeologists and ICT researchers.

The second approach, developed in parallel, is to build a Qualitative Constraints Networks (QCN) based on the physical relationships between US. This will allow us to check the consistency between relations. The QCN will be developed in ASP using the solver ASPERIX [75,76]. We plan to investigate the ASP approach for several reasons. ASP is a unified formalism for both representing and implementing the consistency checking problem expressed by a logic program that is very close to natural language. We already used this formalism when dealing with the revision of geographical information within the framework of the European project REVIGIS (Uncertain knowledge, maintenance and revision in geographic information systems, ID: 27781, 1998–1999) [77]. More precisely, we compared the CSP, SAT and ASP approaches. CSP focuses on the direct resolution of inconsistencies by means of propagation mechanisms, while SAT and ASP concentrate on the identification of inconsistencies. However, an experimental study on a real-world application benchmark showed that ASP gave better results than SAT [78]. Moreover, ASP can easily interface with ontologies and Java.

As all the data collected using photogrammetry are in 3D, we will propose to express all the relations between US and other concepts by 3D graphs to be aware of and overcome several well-known ambiguities due to the 2D expression.

Finally, this work tries to analyze the temporal problems connected to the Harris matrix, which, as Lucas and Gallina pointed out, can be divided into two different fields. The first one concerns the different kinds of time relationships between USM, starting from the physical-temporal relationships used in the Harris matrix.

The proposed charts (Tables 3 and 4) aim to create an interconnection table between temporal intervals (Allen) and temporal relationships (Harris). As we can see, there are some temporal relationships that we are unable to better refine than using subsequent and anteriority (without any other specification proposed by Allen's interval relations). On the other hand, the most important criticality concerning the Harris matrix's temporality is the lack of expression of duration.

When we analyze a matrix, we are convinced that we will see many instants linked to each other. When we analyze a matrix, we are convinced to see many instants linked to each other. We propose in this paper a first step to fill this gap and a way to use Allen's interval formalism that will enable archaeologists, as well as researchers to further advance science in regards to knowledge representation and reasoning in artificial intelligence for temporal relationships.

Acknowledgments: The University of Florence (UniFI) and the French National Center for Scientific Research (CNRS) have worked together for more than 15 years in the framework of the ARPENTEUR project (an Architectural PhotogrammEtry Network Tool for Education and Research http://www.arpenteur.org), which has been developed since 1997 in Marseille, France, by the CNRS. The ontology aspect of this work is developed in the context of the GROPLAN project (http://www.groplan.eu), partially funded by the French National Agency for Research (ANR). The authors wish to thank Yaaqoub Semlali for the development of the ontological editor during his internship at the CNRS.

Author Contributions: Guido Vannini is the Director of the Medieval Petra archaeological mission at the Crusader Ayyubid settlement in Transjordan, which is the core of the archaeological data of this paper. Michele Nucciotti, Vice-Director and Elisa Pruno, responsible for the field research, processed all the archaeological data used in this paper. Odile Papini, Computer Science professor at Aix-Marseille University, initiated the idea of using the Allen formalism, as well as the Qualitative Constraints Networks (QCN) for the representation of temporal relations in this context. Pierre Drap, researcher at the CNRS, Marseille, has worked for more than 15 years with the Italian team and has developed and supervised the photogrammetric and ontological aspects. All the authors contributed equally in writing the manuscript.

Conflicts of Interest: The authors declare no conflict of interest.

References

1. Drap, P.; Merad, D.; Gaoua, L.; Pruno, E.; Marcotulli, C.; Vannini, G. Underwater Photogrammetry in a Terrestrial Excavation: San Domenico (Prato-Italy). *Int. Arch. Photogramm. Remote Sens. Spat. Inf. Sci.* **2015**, *40*, 171–176.
2. Drap, P.; Durand, A.; Seinturier, J.; Vannini, G.; Nucciotti, M. Full XML Documentation from Photogrammetric Survey to 3D Visualization. The Case Study of Shawbak Castle in Jodan. In Proceedings of the CIPA 2005 XX International Symposium, Torino, Italy, 26 September–1 October 2005; Volume 2, pp. 771–776.
3. Brogiolo, G.P. *Archeologia dell'Edilizia Storica*; Aedes Muratoriana: Modena, Italy, 1988; p. 120.
4. Harris, E.C. *Principles of Archaeological Stratigraphy*, 2nd ed.; Academic Press: Waltham, MA, USA, 1979; p. 170.
5. Francovich, R.; Parenti, R. *Archeologia e Restauro dei Monumenti. I Ciclo di Lezioni Sulla Ricerca Applicata in Archeologia*; All'Insegna del Giglio: Siena, Italy, 1987.
6. Gallina, D. Sillogismo deduttivo o abduzione? Alcune proposte per l'abbandono/superamento del matrix di Harris nell'analisi dell'architettura. In Proceedings of the VI Cngresso Nazionale di Archeologia Medievale, L'Aquila, Italy, 12–15 September 2012; Giglio, A.D., Ed.; All'Insegna del Giglio: Firenze, Italy, 2012.

7. Doerr, M. The CIDOC CRM An ontological approach to semantic interoperability of metadata. *AI Mag.* **2001**, *24*, 75–92.

8. Lodi, G.; Asprino, L.; Nuzzolese, A.G.; Presutti, V.; Gangemi, A.; Recupero, D.R.; Veninata, C.; Orsini, A. Semantic Web for cultural heritage Valorisation. In *Data Analytics in Digital Humanities*; Hai-Jew, S., Ed.; Springer International Publishing: Cham, Switzerland, 2017; pp. 3–37.

9. Noardo, F. A Spatial Ontology for Architectural Heritage Information. Revised Selected Papers. In Proceedings of the Geographical Information Systems Theory, Applications and Management: Second International Conference (GISTAM 2016), Rome, Italy, 26–27 April 2016; Grueau, C., Laurini, R., Rocha, J.G., Eds.; Springer International Publishing: Cham, Switzerland, 2017; pp. 143–163.

10. Niang, C.; Marinica, C.; Markhoff, B.; Leboucher, E.; Malavergne, O.; Bouiller, L.; Darrieumerlou, C.; Laissus, F. Supporting Semantic Interoperability in Conservation-Restoration Domain: The PARCOURS Project. *J. Comput. Cult. Herit.* **2017**, *10*, 1–20.

11. Bing, L.; Chan, K.C.C.; Carr, L. Using Aligned Ontology Model to Convert cultural heritage Resources into Semantic Web. In Proceedings of the 2014 IEEE International Conference on Semantic Computing, Newport Beach, CA, USA, 16–18 June 2014; pp. 120–123.

12. Niccolucci, F.; D'Andrea, A. An Ontology for 3D Cultural Objects. In Proceedings of the 7th International Symposium on Virtual Reality, Archaeology and cultural heritage VAST, Nicosia, Cyprus, 30 October–4 November 2006.

13. Amico, N.; Ronzino, P.; Felicetti, A.; Niccolucci, F. Quality Management of 3D Cultural Heritage Replicas with CIDOC-CRM. Available online: http://ceur-ws.org/Vol-1117/paper6_slides.pdf (accessed on 3 September 2017).

14. Niccolucci, F.; Hermon, S.; Doerr, M. The Formal Logical Foundations of Archaeological Ontologies. In *Mathematics and Archaeology*; Barcelo, J.A., Bogdanovic, I., Eds.; CRC Press: Abingdon, UK, 2015; pp. 86–99.

15. Ronzino, P.; Niccolucci, F.; Felicetti, A.; Doerr, M. CRMba a CRM extension for the documentation of standing buildings. *Int. J. Digit. Libr.* **2016**, *17*, 71–78.

16. Araújo, C.; Martini, R.G.; Henriques, P.R.; Almeida, J.J. Annotated Documents and Expanded CIDOC-CRM Ontology in the Automatic Construction of a Virtual Museum. In *Developments and Advances in Intelligent Systems and Applications*; Rocha, A., Reis, L.P., Eds.; Springer International Publishing: Cham, Switzerland, 2018; pp. 91–110.

17. Niccolucci, F. Documenting archaeological science with CIDOC CRM. *Int. J. Digit. Libr.* **2016**, *18*, 223–231.

18. Gaitanou, P.; Gergatsoulis, M.; Spanoudakis, D.; Bountouri, L.; Papatheodorou, C. Mapping the Hierarchy of CIDOC CRM. In Proceedings of the Metadata and Semantics Research: 10th International Conference (MTSR 2016), Gottingen, Germany, 22–25 November 2016; Garoufallou, E., Subirats Coll, I., Stellato, A., Greenberg, J., Eds.; Springer International Publishing: Cham, Switzerland, 2016; pp. 193–204.

19. Niccolucci, F.; Hermon, S. Expressing reliability with CIDOC CRM. *Int. J. Digit. Libr.* **2016**, doi:10.1007/s00799-016-0195-1.

20. Hiebel, G.; Doerr, M.; Eide, O. CRMgeo: A spatiotemporal extension of CIDOC-CRM. *Int. J. Digit. Libr.* **2016**, doi:10.1007/s00799-016-0192-4.

21. Curé, O.; Sérayet, M.; Papini, O.; Drap, P. Toward a novel application of CIDOC CRM to underwater archaeological surveys. In Proceedings of the 4th IEEE International Conference on Semantic Computing (ICSC 2010), Pittsburgh, PA, USA, 22–24 September 2010; pp. 519–524.

22. Papini, O.; Drap, P. The Revision of Partially Preordered Information in Answer Set Programming. Proceedings of ECSQARU. In *Lecture Notes in Computer Science*; Springer: Berlin, Germany, 2011; pp. 421–433.

23. Seinturier, J. Fusion de Connaissances: Applications aux Relevés PhotogramméTriques de Fouilles ArchéOlogiques Sous-Marines. Ph.D. Thesis, University of Toulon, Toulon, France, 2007.

24. Serayet, M. Raisonnement à Partir d'Information Structurées et Hiérarchisées: Application à l'Information Archéologique. P.h.D. Thesis, Aix Marseille, Marseille, France, 2010.

25. Drap, P.; Seinturier, J.; Chambelland, J.C.; Gaillard, G.; Pires, H.; Vannini, G.; Nucciotti, M.; Pruno, E. Going To Shawbak (Jordan) And Getting The Data Back: toward a 3D GIS dedicated to medieval archeology. *INCT Bull. Sci. Géogr.* **2009**, *24*, 40–50.

26. Kendall, D.G. A statistical approach to Flinders Petries' sequence-dating. *Bull. Int. Stat. Inst.* **1963**, *34*, 657–680.

27. Kendall, D.G. Abundance matrices and seriation in archeology. *Probab. Theory Relat. Fields* **1971**, *17*, 104–112.

28. Guénoche, A.; Tchernia, A. L'analyse descriptive dans la construction d'un modèle typologique des amphores Dressel 20. In *Archéologie et Calcul*; book Section 2; Borillo, M., Ed.; Union Generale D'editions: Paris, France, 1978; pp. 167–181.

29. Regnier, S. Sériation des Niveaux de Plusieurs Tranches de Fouille Dans Une Zone ArchéOlogique HomogèNe. In *Raisonement et MéThodes MathéMatiques en Archéologie*; American Anthropological Association: Arlington, VA, USA, 1977.

30. Halekoh, U.; Vach, W. A Bayesian approach to seriation problems in archeology. *Comput. Stat. Data Anal.* **2004**, *45*, 651–673.

31. Harris, E.C. The laws of archaeological stratigraphy. *World Archaeol.* **1979**, *11*, 111–117.

32. Djindjian, F. *Méthodes Pour l'Archéologie*; Armand Colin: Paris, France, 1991; p. 401.

33. Djindjian, F.; Desachy, B. Sur l'aide au traitement des données stratigraphiques des sites archéologiques. *Hist. Mes.* **1990**, *11*, 51–88.

34. Desachy, B. Traitement multidimensionnel (analyse factorielle des correspondances) des poids groupes de céramique du site des Hallettes à Compiègne (Oise). *Rev. ArchéOl. Picardie* **1997**, *13*, 169–170.

35. Desachy, B. Du terrain au temps archeologique, vers un systeme d'information stratigraphique. In *Temps et Espaces de l'Homme en Société: Analyses et Modèles Spatiaux en Archéologie*; Association pour la Promotion et la Diffusion des Connaissances archéologiques: Valbonne, France, 2012; pp. 269–272.

36. Desachy, B. De la Formalisation du Traitement des DonnéEs Stratigraphiques en Archéologie de Terrain. Ph.D. Thesis, Pantheon-Sorbonne University, Paris, France, 2008.

37. Carandini, A. *Storie Dalla Terra. Manuale di Scavo Archeologico*; Einaudi, Ed.; Piccola biblioteca Einaudi: Torino, Italy, 1991.

38. Ryan, N. *Jnet, a Successor to Gnet.* *Archaologie und Computer*; Workshop 6; Borner, W., Ed.; Forschungsgesellschaft Wiener Stadtarchaologie: Wien, Austria, 2001; pp. 182–196.

39. Scollar, I. Bonn Archaeological Software Package (BASP). 2017. Available online: http://www.uni-koeln. de/~al001/ (accessed on 22 August 2017).

40. Hundack, C.; Mutzel, P.; Pouchkarev, I.; Reitgruber, B.; Schuhmacher, B.; Thome, S. Arched | Technical University of Vienna, Austria. 2017. Available online: http://www.ads.tuwien.ac.at/arched (accessed on 22 August 2017).

41. Museums, F.N.H. Intrasis Is a GIS Designed to Handle and Structure Geographical Data. 2017. Available online: http://www.intrasis.com/index.htm (accessed on 22 August 2017).

42. Traxler, C.; Neubauer, W. The Harris matrix Composer—A New Tool to Manage Archaeological Stratigraphy. In *Kulturelles Erbe und Neue Technologien Workshop, Archaeologie und Computer*; Wolfgang Borner, S.U., Ed.; Museen der Stadt Wien, Phoibos Verlag: Vienna, Austria, 2008. Available online: http://www. stadtarchaeologie.at/ (accessed on 3 September 2017).

43. Sikora, J.; Tyszkiewicz, J. Strati5—Open Mobile Software for Harris matrix. In Proceedings of the 43rd Annual Conference on Computer Applications and Quantitative Methods in Archaeology (CAA 2015), Siena, Italy, 30 March–3 April 2015; Campana, S., Scopigno, R., Eds.; Archeopress: Oxford, UK, 2015; Volume 2, pp. 182–196.

44. Galeazzi, F.; Callieri, M.; Dellepiane, M.; Charno, M.; Richards, J.D.; Scopigno, R. Web-based visualization for 3D data in archeology: The ADS 3D viewer. *J. Archaeol. Sci.* **2016**, *9*, 1–24.

45. Lucas, G. *The Archaeology of Time*; Routledge: Abingdon, UK, 2005.

46. Ramenofsky, A. The Illusion of Time. In *Unit Issues in Archaeology: Measuring Time, Space and Materia*; Ramensofsky, A.F., Steffen, A., Eds.; University of Utah Press: Salt Lake City, UT, USA, 1998; pp. 74–86.

47. Accary, T.; Bénel, A.; Calabretto, S. Modélisation de connaissances temporelles en archéologie. *Rev. Sci. Technol. l'Inf. RSTI Journ. Francoph. d'Extr. Gest. Connaiss. ECG* **2003**, *17*, 503–508.

48. Desachy, B. Du temps ordonne au temps quantifié : Application d'outils mathématiques au modèle d'analyse stratigraphique d'Edward Harris. *Bull. Soc. Préhist. Française* **2005**, *102*, 729–740.

49. Rodier, X.; Saligny, L. Modélisation des objets urbains pour l'étude des dynamiques urbaines dans la longue durée. In Proceedings of the SAGEO 2007 Colloque International de Géomatique et d'Analyse Spatiale, Clermont-Ferrand, France, 18–22 June 2007.

50. Lucas, G. Time and the archaeological archive. *Rethink. Hist.* **2010**, *14*, 343–359.

51. Harris, E.C. The stratigraphy of standing structures con alcune considerazioni in nota di R. Parenti. In *Archeologia dellArchitettura*; All'Insegna del Giglio: Sesto, Italy, 2003; pp. 9–16.

52. Althusser, L. *For Marx*; Verso: Stockholm, Sweden, 1969.
53. Drap, P.; Durand, A.; Nedir, M.; Seinturier, J.; Papini, O.; Gabrielli, R.; Peloso, D.; Kadobayashi, R.; Gaillard, G.; Chapman, P.; et al. Photogrammetry and archaeological knowledge: Toward a 3dinformation system dedicated to medieval archeology: A case Study of Shawbak castle in Jordan. In Proceedings of the 3D Virtual Reconstruction and Visualization of Complex Architectures (3D-ARCH 2007), Zurich, Switzerland, 12–13 July 2007.
54. Pruno, E.; Drap, P. Dalla stratigrafia all'archeologia teorica: il matrix di Harris e l'archeologia del tempo. In Proceedings of the VI Congresso Nazionale di Archeologia Medievale, L'Aquila, Italy, 12–15 September 2012; Giglio, A.D., Ed.; All'Insegna del Giglio: Firenze, Italy, 2012.
55. Nicolis, G.; Prigogine, I. *Self-Organization in Nonequilibrium Systems. From Dissipative Structures to Order through Fluctuations*; Wiley: Hoboken, NJ, USA, 1977; p. 491.
56. Allen, J.F. Maintaining knowledge about temporal intervals. *Commun. ACM* **1983**, *26*, 832–843.
57. Buchsenschutz, O. Images et RelevéS ArchéOlogiques, de la Preuve à la Démonstration. Available online: http://cths.fr/ed/edition.php?id=4703 (accessed on 3 September 2017).
58. Vannini, G. *Archeologia dell'Insediamento Crociato-Ayyubide in Transgiordania: Il Progetto Shawbak*, All'insegna del Giglio: Firenze, Italy, 2007.
59. Vannini, G.; Nucciotti, M. Da Petra a Shawbak, Archeologia di una frontiera. *Archeologia Viva* **2009**, *135*, 18–31.
60. Drap, P.; Gaillard, G.; Grussenmeyer, P. Simple photogrammetric methods with arpenteur. 3-d plotting and orthoimage generation : The I-MAGE process. In Proceedings of the CIPA 2001 International Symposium, Potsdam, Germany, 18–21 September 2001.
61. Drap, P.; Gaillard, G.; Grussenmeyer, P.; Hartmann-Virnich, A. A stone-by-stone photogrammetric survey using architectural knowledge formalised on the ARPENTEUR Photogrammetric workstation. In Proceedings of the XIXth Congress of the International Society for Photogrammetry and Remote Sensing (ISPRS), Amsterdam, The Netherlands, 16–23 July 2001.
62. Xueming, P.; Beckman, P.; Havemann, S.; Tzompanaki, K.; Doerr, M.; Fellner, D.W. A Distributed Object Repository for cultural heritage. In Proceedings of the 11th International Conference on Virtual Reality, Archaeology and Cultural Heritage (VAST 2010), Paris, France, 21–24 September 2010.
63. Kalyanpur, A.; Pastor, D.J.; Battle, S.; Padget, J.A. Automatic Mapping of OWL Ontologies into Java. *SEKE* **2004**, *4*, 98–103.
64. Stevenson, G.; Dobson, S. Sapphire: Generating Java Runtime Artefacts from OWL Ontologies. In *Advanced Information Systems Engineering Workshops, Proceedings of the CAiSE 2011 International Workshops, London, UK, 20–24 June 2011*; Salinesi, C., Pastor, O., Eds.; Springer: Berlin/Heidelberg, Germany, 2011; pp. 425–436.
65. Horridge, M.; Knublauch, H.; Rector, A.; Stevens, R.; Wroe, C. A Practical Guide To Building OWL Ontologies Using The Protege-OWL Plugin and CO-ODE Tools Edition 1.0. Available online: http://mowl-power.cs.man.ac.uk/protegeowltutorial/resources/ProtegeOWLTutorialP4_v1_3.pdf (accessed on 3 September 2017).
66. O'Connor, M.J.; Das, A. A Mechanism to Define and Execute SWRL Built Ins in Protege OWL. Available online: https://protege.stanford.edu/conference/2006/submissions/abstracts/7.3_Martin_oConnor_BuiltInBridge.pdf (accessed on 3 September 2017).
67. Keßler, C.; Raubal, M.; Wosniok, C. Semantic Rules for Context-Aware Geographical Information Retrieval. In *Smart Sensing and Context, Proceedings of the 4th European Conference (EuroSSC 2009), Guildford, UK, 16–18 September 2009*; Barnaghi, P., Moessner, K., Presser, M., Meissner, S., Eds.; Springer: Berlin/Heidelberg, Germany, 2009; pp. 77–92.
68. Accary-Barbier, T.; Calabretto, S. Building and using temporal knowledge in archaeological documentation. *J. Intell. Inf. Syst.* **2008**, *31*, 147–159.
69. Belussi, A.; Migliorini, S. *Modeling Time in Archaeological Data: The Verona Case Study*; University of Verona: Verona, Italy, 2014.
70. Dechter, R.; Meiri, I.; Pearl, J. Temporal Constraint Networks. *Artif. Intell.* **1991**, *49*, 61–95.
71. Condotta, J.; Nouaouri, I.; Sioutis, M. A SAT Approach for Maximizing Satisfiability in Qualitative Spatial and Temporal Constraint Networks. In *Principles of Knowledge Representation and Reasoning, Proceedings of the Fifteenth International Conference (KR 2016), Cape Town, South Africa, 25–29 April 2016*; AAAI Press: San Jose, CA, USA, 2016; pp. 432–442.

72. Brenton, C.; Faber, W.; Batsakis, S. Answer Set Programming for Qualitative Spatio-Temporal Reasoning: Methods and Experiments. In Proceedings of the Technical Communications of the 32nd International Conference on Logic Programming (ICLP 2016), New York, NY, USA, 16–21 October 2016; USAVos, M.C., King, A., Saeedloei, N., De, M., Eds.; Schloss Dagstuhl–Leibniz-Zentrum fuer Informatik: Dagstuhl, Germany, 2016; Volume 52, pp. 1–15.

73. Niccolucci, F.; Tonghini, C.; Vannini, G.; Crescioli, M. PETRA: Un sistema integrato per la gestione dei dati archeologici. *Archeol. Calcolatori* **2000**, *XI*, 49–67.

74. Drap, P.; Papini, O.; Pruno, E.; Nucciotti, M.; Vannini, G. Surveying medieval archeology: A new form for harris paradigm merging photogrammetry and temporal relations. In Proceedings of the 3D Arch, 3D Virtual Reconstruction and Visualization of Complex Architectures and Scenarios, Argolida, Greece, 1–3 March 2017.

75. Lefèvre, C.; Nicolas, P. The First Version of a New ASP Solver: ASPeRiX. In *Logic Programming and Nonmonotonic Reasoning, Proceedings of the 10th International Conference (LPNMR 2009), Potsdam, Germany, 14–18 September 2009*; Erdem, E., Lin, F., Schaub, T., Eds.; Springer: Berlin/Heidelberg, Germany, 2009; pp. 522–527.

76. Lefèvre, C.; Béatrix, C.; Stéphan, I.; Garcia, L. *ASPeRiX, a First Order Forward Chaining Approach for Answer Set Computing*; CoRR: Ithaca, NY, USA, 2015.

77. Benferhat, S.; Ben-Naim, J.; Papini, O.; Wurbel, E. Answer Set Programming encoding of Prioritized Removed Sets Revision: Application to GIS. *Appl. Intell.* **2010**, *32*, 60–87.

78. Benferhat, S.; Ben-Naim, J.; Jeansoulin, R.; Kelfallah, M.; Lagrue, S.; Papini, O.; Wilson, N.; Wurbel, E. Revising geoinformation: The results of REV!GIS. In *Lecture Note in Artificial Intelligence, Proceedings of the Eighth European Conference on Symbolic and Quantitative Approaches to Reasoning with Uncertainty, Barcelona, Spain, 6–8 July 2005*; Springer: Berlin, Germany, 2005.

MDPI AG

St. Alban-Anlage 66

4052 Basel, Switzerland

Tel. +41 61 683 77 34

Fax +41 61 302 89 18

http://www.mdpi.com

Geosciences Editorial Office

E-mail: geosciences@mdpi.com

http://www.mdpi.com/journal/geosciences

www.ingramcontent.com/pod-product-compliance
Lightning Source LLC
Chambersburg PA
CBHW051704210326
41597CB00032B/5364